Hyperspectral Remote Sensing

Hyperspectral Remote Sensing

Theory and Applications

Edited by

Prem Chandra Pandey
Center for Environmental Sciences & Engineering, School of Natural Sciences, Shiv Nadar University, Greater Noida, India

Prashant K. Srivastava
Remote Sensing Laboratory, Institute of Environment and Sustainable Development, Banaras Hindu University, Varanasi, India

Heiko Balzter
University of Leicester, Leicester, United Kingdom

Bimal Bhattacharya
Space Applications Centre, Indian Space Research Organization, Ahmedabad, India

George P. Petropoulos
Department of Geography, Harokopio University of Athens, Athens, Greece

ELSEVIER

Elsevier
Radarweg 29, PO Box 211, 1000 AE Amsterdam, Netherlands
The Boulevard, Langford Lane, Kidlington, Oxford OX5 1GB, United Kingdom
50 Hampshire Street, 5th Floor, Cambridge, MA 02139, United States

British Library Cataloguing-in-Publication Data

A catalogue record for this book is available from the British Library

Library of Congress Cataloging-in-Publication Data

A catalog record for this book is available from the Library of Congress

ISBN: 978-0-08-102894-0

For Information on all Elsevier publications
visit our website at https://www.elsevier.com/books-and-journals

Publisher: Candice Janco
Acquisitions Editor: Amy Shapiro
Editorial Project Manager: Redding Morse
Production Project Manager: Joy Christel Neumarin Honest Thangiah
Cover Designer: Mark Rogers

Idea by Prem Pandey and Design by Mr. Kumar Krishna Mohan - Sr. Graphic Designer, Shiv Nadar University

Typeset by MPS Limited, Chennai, India

Contents

List of contributors

Akash Anand Remote Sensing Laboratory, Institute of Environment and Sustainable Development, Banaras Hindu University, Varanasi, India

V. Andronis Laboratory of Remote Sensing, National Technical University of Athens, Athens, Greece

Paul Arellano School of Earth Sciences, Energy and Environment—Center of Earth Observation, Yachay Tech University, Urcuqui, Ecuador; Department of Postgraduate Studies, Faculty of Biological Sciences, Central University of Ecuador, Quito, Ecuador

Nour-El-Islam Bachari Faculty of Biological Sciences, University of Sciences and Technology Houari Boumediene, Bab Ezzouar, Algeria

Heiko Balzter Centre for Landscape and Climate Research, School of Geography, Geology and the Environment, University of Leicester, Leicester, United Kingdom; National Centre for Earth Observation, University of Leicester, Leicester, United Kingdom

Yangson Bao Collaborative Innovation Center on Forecast and Evaluation of Meteorological Disasters, Nanjing, University of Information Science & Technology, Nanjing, P.R. China

Oswaldo Bayona Escuela Superior Politécnica del Litoral, ESPOL, Guayaquil, Ecuador

Jon Atli Benediktsson Department of Electrical and Computer Engineering, University of Iceland, Reykjavík, Iceland

Juan Carlos Berrío School of Geography, Geology and Environment, University of Leicester, Leicester, United Kingdom

Bimal Bhattacharya Space Applications Centre, Indian Space Research Organization, Ahmedabad, India

Paul A. Brewer Department of Geography and Earth Sciences, University of Aberystwyth, Ceredigion, United Kingdom

Juan Cevallos-Cevallos Escuela Superior Politécnica del Litoral, ESPOL, Guayaquil, Ecuador

Shard Chander Space Applications Centre, Ahmedabad, India

Ronald Criollo Escuela Superior Politécnica del Litoral, ESPOL, Guayaquil, Ecuador

Hilda Deborah Department of Computer Science, Norwegian University of Science and Technology, Gjøvik, Norway

Khidir Abdala Kwal Deng Collaborative Innovation Center on Forecast and Evaluation of Meteorological Disasters, Nanjing, University of Information Science & Technology, Nanjing, P.R. China

P.K. Garg Civil Engineering Department, Indian Institute of Technology, Roorkee, India

Barbara Gerič Stare Agricultural Institute of Slovenia, Plant Protection Department, Ljubljana, Slovenia

Przemysław Głomb Institute of Theoretical and Applied Informatics, Polish Academy of Sciences, Gliwice, Poland

Arindam Guha Geosciences Group, National Remote Sensing Centre, Indian Space Research Organization, Hyderabad, India

Ashwin Gujrati Space Applications Centre, Ahmedabad, India

Ayushi Gupta Remote Sensing Laboratory, Institute of Environment and Sustainable Development, Banaras Hindu University, Varanasi, India

Jon Yngve Hardeberg Department of Computer Science, Norwegian University of Science and Technology, Gjøvik, Norway

Mazlan Hashim Geoscience and Digital Earth Centre (INSTeG), Research Institute of Sustainable Environment, Universiti Teknologi Malaysia, UTM Johor, Johor, Malaysia; Department of Geoinformatics, Faculty of Built Environment and Surveying, Universiti Teknologi Malaysia, UTM Johor, Johor, Malaysia

Noordyana Hassan Geoscience and Digital Earth Centre (INSTeG), Research Institute of Sustainable Environment, Universiti Teknologi Malaysia, UTM Johor, Johor, Malaysia; Department of Geoinformatics, Faculty of Built Environment and Surveying, Universiti Teknologi Malaysia, UTM Johor, Johor, Malaysia

V. Karathanassi Laboratory of Remote Sensing, National Technical University of Athens, Athens, Greece

G. Sandhya Kiran The Maharaja Sayajirao University of Baroda, Vadodara, India

Matej Knapič Agricultural Institute of Slovenia, Plant Protection Department, Ljubljana, Slovenia

K. Koloniatis Laboratory of Remote Sensing, National Technical University of Athens, Athens, Greece

Bo Kong Institute of Mountain Hazards and Environment, Chinese Academy of Sciences, Chengdu, P.R. China

Nikos Koutsias Department of Environmental Engineering, University of Patras, Agrinio, Greece

Aswathy V. Krishna Space Applications Centre, Ahmedabad, India

Salim Lamine Faculty of Natural Sciences and Life and Earth Sciences, University Akli Mohand Oulhadj of Bouira, Bouira, Algeria; Department of Geography and Earth Sciences, University of Aberystwyth, Ceredigion, United Kingdom

Wenzhi Liao Sustainable Materials Management, Flemish Institute for Technological Research (VITO), Antwerp, Belgium; IPI-TELIN, Ghent University, Ghent, Belgium

Xian Liu Department of Plant Biology, Southern Illinois University, Carbondale, IL, United States

Xiangmeng Liu College of Earth Sciences, Chengdu University of Technology, Chengdu, P.R. China

Mark G. Macklin School of Geography, College of Science, University of Lincoln, Lincoln, United Kingdom

Ramandeep Kaur M. Malhi Remote Sensing Laboratory, Institute of Environment and Sustainable Development, Banaras Hindu University, Varanasi, India

Kiril Manevski Department of Agroecology, Aarhus University, Tjele, Denmark; Sino-Danish Center for Education and Research, Eastern Yanqihu Campus, Beijing, P.R. China

Manoj K. Mishra Space Applications Centre, Indian Space Research Organization, Ahmedabad, India

Onisimo Mutanga Department of Geography, School of Agricultural, Earth and Environmental Science, University of KwaZulu-Natal, Pietermaritzburg, South Africa

Rowan Naicker Department of Geography, School of Agricultural, Earth and Environmental Science, University of KwaZulu-Natal, Pietermaritzburg, South Africa

Rajashree Naik Department of Environmental Science, School of Earth Sciences, Central University of Rajasthan, Ajmer, India

Shinya Numata Department of Tourism Science, Graduate School of Urban Environmental Sciences, Tokyo Metropolitan University, Minami-Osawa 1-1, Hachiouji, Tokyo, Japan

Daniel Ochoa Escuela Superior Politécnica del Litoral, ESPOL, Guayaquil, Ecuador

Nkeiruka N. Onyia Centre for Landscape and Climate Research, School of Geography, Geology and the Environment, University of Leicester, Leicester, United Kingdom

Mahesh Pal Department of Civil Engineering, National Institute of Technology, Kurukshetra, India

Manish Kumar Pandey Remote Sensing Laboratory, Institute of Environment and Sustainable Development, Banaras Hindu University, Varanasi, India

Prem Chandra Pandey Center for Environmental Sciences & Engineering, School of Natural Sciences, Shiv Nadar University, Greater, Noida, India

Andrew Pavlides School of Mineral Resources Engineering, Technical University of Crete, Chania, Greece

Kabir Peerbhay Department of Geography, School of Agricultural, Earth and Environmental Science, University of KwaZulu-Natal, Pietermaritzburg, South Africa

George P. Petropoulos Department of Geography, Harokopio University of Athens, Athens, Greece

Noël Richard Laboratory XLIM, UMR CNRS 7252, University of Poitiers, Poitiers, France

Michał Romaszewski Institute of Theoretical and Applied Informatics, Polish Academy of Sciences, Gliwice, Poland

Arvind Sahay Space Applications Centre, Ahmedabad, India

L.K. Sharma Department of Environmental Science, School of Earth Sciences, Central University of Rajasthan, Ajmer, India

Mbulisi Sibanda Department of Geography, School of Agricultural, Earth and Environmental Science, University of KwaZulu-Natal, Pietermaritzburg, South Africa

Prachi Singh Remote Sensing Laboratory, Institute of Environment and Sustainable Development, Banaras Hindu University, Varanasi, India

R.P. Singh Space Applications Centre, Ahmedabad, India

Saša Širca Agricultural Institute of Slovenia, Plant Protection Department, Ljubljana, Slovenia

Prashant K. Srivastava Remote Sensing Laboratory, Institute of Environment and Sustainable Development, Banaras Hindu University, Varanasi, India; DST-Mahamana Centre for Excellence in Climate Change Research, Banaras Hindu University, Varanasi, India

Dimitris Stratoulias Department for Management of Science and Technology Development, Ton Duc Thang University, Ho Chi Minh City, Vietnam; Faculty of Applied Sciences, Ton Duc Thang University, Ho Chi Minh City, Vietnam

Nik Susič Agricultural Institute of Slovenia, Plant Protection Department, Ljubljana, Slovenia

Mohamad Zakri Tarmidi Geoscience and Digital Earth Centre (INSTeG), Research Institute of Sustainable Environment, Universiti Teknologi Malaysia, UTM Johor, Johor, Malaysia; Department of Geoinformatics, Faculty of Built Environment and Surveying, Universiti Teknologi Malaysia, UTM Johor, Johor, Malaysia

Leonidas Toulios Department of Soil Water Resources, Institute of Industrial & Forage Crops, Hellenic Agricultural Organization (HAO) "Demeter" (former NAGREF), Directorate General of Agricultural Research, Larissa, Greece

Gregor Urek Agricultural Institute of Slovenia, Plant Protection Department, Ljubljana, Slovenia

Qing Wang Department of Geography and Environmental Resources, Southern Illinois University, Carbondale, IL, United States

Huan Yu College of Earth Sciences, Chengdu University of Technology, Chengdu, P.R. China

Uroš Žibrat Agricultural Institute of Slovenia, Plant Protection Department, Ljubljana, Slovenia

Biography

Dr. Prem Chandra Pandey received his Ph.D. (2015) from Centre for Landscape and Climate Research (CLCR), Department of Geography, University of Leicester, Leicester, United Kingdom, under the Commonwealth Scholarship and Fellowship Plan (CSFP). He received his B.Sc. in Botany (Hons.) from Institute of Sciences, Banaras Hindu University, M.Sc. in Environmental Sciences from Institute of Sciences, BHU, and M.Tech. in Remote Sensing from Birla Institute of Technology (BIT) Mesra, Ranchi, India. He is currently working as Assistant Professor in Center for Environmental Sciences & Engineering, Shiv Nadar University, Greater Noida, India. He did his postdoctoral with the Department of Geography and Human Environment, Faculty of Exact Sciences, Tel Aviv University Israel. He has worked as a National Postdoctoral Fellow with Remote Sensing Laboratory, Institute of Environment and Sustainable Development (IESD), Banaras Hindu University (BHU), Varanasi. He worked on remote sensing applications as professional research fellow in National Urban Information System (NUIS) funded by National Remote Sensing Centre (NRSC) Government of India at BIT Mesra, Ranchi. He has been a recipient of several prestigious International and National awards including Commonwealth Fellowship, United Kingdom, INSPIRE Fellowship, Ministry of Human Resource and Development (MHRD), Government of India, University Grant Commission (UGC) Fellowships, Science and Engineering Research Board (SERB)-NPDF, and IT Merit awards from the Government of India. He has authored more than 35+ research articles published in international peer-reviewed journal articles, edited three books, contributed chapters to nine books, and presented his work in several conferences. Additionally, he is also a member of Indian Society of Geomatics (ISG), Indian Society of Remote Sensing (ISRS), SPIE, and Association of American Geographers (AAG). His research interests include forestry, environmental pollutant modeling, urban studies, and agricultural studies.

Dr. Prashant K. Srivastava received his Ph.D. from Department of Civil Engineering, University of Bristol, Bristol, United Kingdom, sponsored by British High Commission, United Kingdom under the Commonwealth Scholarship and Fellowship Plan (CSFP) and MHRD, GOI. He received his B.Sc. in Agriculture Sciences from the Institute of Agricultural Sciences, Banaras Hindu University (BHU), and his M.Sc. in Environmental Sciences from School of Environmental Sciences (SES), Jawaharlal Nehru University (JNU), India. Currently he is working as an Assistant Professor at the Institute of Environment and Sustainable Development, Banaras Hindu University, Varanasi, India. He has worked as a Research Scientist with NASA Goddard Space Flight Center, Hydrological Sciences Branch on SMAP satellite soil moisture retrieval algorithm development, instrumentation and its applications. He is the recipient of several awards such as NASA Fellowship, United States of America; University of Maryland Fellowship, United States of America; Commonwealth Fellowship, United Kingdom, and received JRF-NET fellowships from Council of Scientific and Industrial Research and University Grant Commission, Government of India. He is leading many national and international projects funded by reputed agencies. He has published over 140 peer-reviewed journal papers, seven books, many book chapters, and has presented his work in several conferences. Currently, he is also serving as Associate editor of many scientific peer reviewed journals. He is also a member of European Geosciences Union (EGU), Indian Society of Geomatics, Indian Society of Remote Sensing, Indian Association of Hydrologists (IAH), and International Society for Agro-meteorology (INSAM), International Association for Hydro-Environment Engineering and Research (IAHR), and International Association of Hydrological Sciences (IAHS).

Prof. Heiko Balzter received his Dipl. -Ing. Agr. (equivalent to M. Sc.) and Dr. Agr. (Ph.D.) from Justus-Liebig-University, Giessen, Germany, in 1994 and 1998, respectively. He is a research professor and the director of the Centre for Landscape and Climate Research at the University of Leicester, United Kingdom, and Official Development Assistance (ODA) Programme Leader in the NERC National Centre for Earth Observation. Before joining the University of Leicester he was Head of the Section for Earth Observation at the Centre for Ecology and Hydrology, Monks Wood, United Kingdom, where he worked from 1998 to 2006. His research interests include interactions of the water cycle with ecosystems across multiple spatial and temporal scales, pressures from climate change and land use change on ecosystem services, and the effects of spatial patterns and processes on biological populations in evolving 3D landscapes. He holds the Royal Society Wolfson Research Merit Award (2011) and the

Royal Geographical Society's Cuthbert Peek Award "for advancing geographical knowledge of human impact through Earth Observation" (2015). He received the President's Cup for the Best Paper Annual Remote Sensing and Photogrammetry Society Conference (2009). He chairs the BESS-EO working group on biodiversity, ecosystem services and Earth Observation, is a member of the NERC Future Landscapes Scoping Group, and UK representative on the Group on Earth Observations (GEO) Programme Board. He is leading the Copernicus Land Monitoring Service for the United Kingdom, which produces the CORINE land cover map. In the Forests 2020 project he is leading the Forest Cover Change Detection task. In GLOBBIOMASS, he is in charge of the regional case studies work package. He is Principal Investigator of the European Centre of Excellence in Earth Observation Research Training GIONET (€3.5m). He is a fellow of the Higher Education Academy, the Royal Statistical Society, a member of the American Geophysical Union, British Ecological Society, Fellow of the Royal Geographical Society, and member of the Remote Sensing and Photogrammetry Society as well as the Chartered Management Institute. He serves on the International Geosphere/Biosphere Program, UK National Committee, the European Space Sciences Committee of the European Science Foundation, the LULUCF Scientific Steering Committee for the Department for Energy and Climate Change, the AATSR Science Advisory Group to Department for Environment, Food and Rural Affairs, and the Natural Environment Research Council Peer Review College. His research interests focus on Earth Observation applications to advance the quantitative understanding of climate change and land use change impacts on ecosystem services, and the effects of spatial-temporal patterns and processes. He has extensive expertise in Earth observation and remote sensing of forests, land cover, and lakes.

Dr. Bimal Bhattacharya is associated with SAC ISRO Ahmedabad as Scientist SG. He received his Ph.D. (Agriculture Physics) in 1995 from Indian Agricultural Research Institute, New Delhi, India. He received his M.Sc. (Agriculture Physics) from Indian Agricultural Research Institute, New Delhi, India in 1991. He served as a scientist in Indian Council of Agricultural research (ICAR) from 1995 to 2000. He is involved in developing satellite-based agro-met products and their utilization in National Agro-met Advisory Services in India. He led the development of satellite-based surface energy balance modeling over Indian terrestrial ecosystems especially with remote sensing data from a suite (INSAT VHRR, 3A CCD, 3D Imager) of Indian geostationary sensors. He led an ISRO-Geosphere-Biosphere Programme (GBP) National-Scale Science Project on Energy and Mass Exchange in Vegetative Systems. He is also the Principal Investigator on the Land Surface Processes in the ISRO-CNES MEGHA-TROPIQUE Mission and Science Member of MoES-NERC Indo-UK Monsoon Project called INCOMPASS. He is the science leader of the AVIRIS-NG Airborne Hyperspectral Mission in Indian and also the Science Co-Chair of ISRO-CNES (Indo-French) Thermal Infrared

Mission. He is currently ISRO representative in CEOS for land surface imaging. He is the recipient of the P.R. Pisharoty award from Indian Society of Remote Sensing and young scientist award from AsiaFlux, Japan sponsored by Japanese Space Agency, JAXA. He also received the ISRO team member award. He is the teaching faculty in UN-sponsored course on "Satellite Meteorology and Global Climate" under Centre for Space Science Technology and Education for Asia-Pacific (CSSTEAP). He has published more than 50 research papers in peer-reviewed international and national journals.

Dr. George P. Petropoulos is an Assistant Professor in Geoinformation at Harokopio University of Athens, Greece. He received his Ph.D. (2008) from King's College London, his M.Sc. in Remote Sensing (2003) from University College London (UCL), and his B.Sc. in Natural Resources Development and Agricultural Engineering (1999) from the Agricultural University of Athens, Greece. His research focuses on the use of EO alone or synergistically with land surface process models in deriving key state variables of the Earth's energy balance and water budget. He also has strong interests in the development of EO and GIS geospatial analysis techniques in geohazards (mainly floods, wildfires, and frost) and in quantifying land use/cover and its changes using technologically advanced EO sensing systems and synergistic multisensor modeling techniques. He also contributes to the development of open source software tools in EO modeling and on the benchmarking of EO operational algorithms/products and surface process models. Dr. Petropoulos currently serves as a council member of the Remote Sensing and Photogrammetric Society (RSPSoC). He is the editor of SENSED (the RSPSoc Newsletter), associate editor or editorial board member of several international scientific journals in EO and environmental modeling. He has edited six books and has coauthored more than 80 research articles in international peer-reviewed journals.

Foreword

Remote sensing technologies in general and hyperspectral remote sensing technology in particular have emerged as an important new source of data for environmental applications over the past few years. Hyperspectral remote sensing applications have gained significant momentum. From early multispectral sensors flown in the 1960s by NASA which gave birth to many other remote sensing satellites, including hyperspectral satellites such as Hyperion, PRISMA, HySI, and others by United States of America, Italy, and India, among others. This unparalleled rapid progress in sensors and their platforms has provided a major impetus for the use of hyperspectral applications in many fields such as, but not limited to, geology, agriculture, water quality, forestry, urban, biodiversity, and so forth. The results of these applications have been spectacular as many researchers have made interesting observations about environmental phenomena, and decision makers have considered these data for sustainable environmental management.

Over 30 years ago, I stated in one of my early research papers that *"remote sensing has added a new dimension to the analysis and studies of environmental processes, issues and decision-making."* This statement is still true today as evidenced by the content of *Hyperspectral Remote Sensing: Theory and Applications* edited by Dr. Prem Chandra Pandey, Dr. Prashant K. Srivastava, Prof. Heiko Balzter, Dr. Bimal Bhattacharya, and Dr. George P. Petropoulos. This book is a welcome, significant, and timely contribution to the growing field of hyperspectral remote sensing.

This book provides an excellent compilation of several methods, techniques, and applications as well as illustrative examples of advancements of hyperspectral remote sensing. The contributors have covered a wide variety of theoretical background, algorithms, applications, and discussed a set of new approaches for hyperspectral data analysis and algorithm development. In addition, an informed view of the future challenges of hyperspectral remote sensing will offer food for thought.

The editors are expert researchers in environmental applications of hyperspectral sensors data. They have skillfully presented this detailed and technical content in a user-friendly and coherent format. I am confident that the readers of this book will quickly appreciate the rapid advancements being made in this field. I am positive that researchers and students alike will gain insights into the novel dimensions of applying hyperspectral remote sensing to environmental analysis and understanding. I wish the editors and publishers of this important volume great success.

Kamlesh Lulla
Houston, TX, United States
Dr. Kamlesh Lulla served as Chief Scientist for Earth Observations and Chief, Earth Science Branch at NASA Johnson Space Center in Houston, TX, United States.

Preface

This book aims to provide an all-inclusive overview of state-of-the-art hyperspectral remote sensing and its applications in different research areas. The book is designed in such a way that it will be used by the people in their respective research domains who will realize that hyperspectral technology may offer a solution to their application area. Readers will have a better understanding of how to incorporate and evaluate different approaches to hyperspectral analyses, as well as which approaches may or may not work for the applications of interest. *Hyperspectral Remote Sensing: Theory and applications* is the first volume of the "Series of Earth Observation" by Elsevier. The purpose of this book is not to provide a tutorial in hyperspectral remote sensing; rather, its aim is to provide an illustration of the potential applications and analysis techniques that can be used, addressing the unique challenge in different applications across the globe.

The aim of the book is to make researchers aware of and enrich their understanding of the concept of hyperspectral data processing. This book has 21 chapters addressing the principles, techniques, and applications of hyperspectral remote sensing under different research themes with field spectroscopy and airborne and spaceborne imaging spectroscopy.

The first section of the book explains the basic concept and underlying principles of errors and their correction methods. This covers all the major aspects of hyperspectral data, source of errors, types of errors, and their correction methods. Advanced classification techniques such as radial basis function neural network (RBFN) and Support Vector Machine (SVM) for feature selection are also provided.

Further sections of book will take readers through all the major applications of hyperspectral remote sensing in vegetation, water, soil, and minerals, as well as pollution detection. Data fusion with other remote sensor images and utilization of spectral indices for different applications are also presented. This section will also demonstrate narrow band and selected bands of hyperspectral data to detect and interpret the level of hydrocarbon pollution in water resources as well as in forest regions. The book describes case studies that have applied this information to the use of hyperspectral remote sensing in forestry, agriculture, water, soil, and mineral applications. The final chapter deals with the future perspective and challenges in hyperspectral remote sensing community.

The case studies in each chapter illustrate how hyperspectral remote sensing is being used to solve many of the disturbing environmental issues of our society. These include identification of functionally distinct plants, chengal tree abundance, precision agriculture, tropical grassland discrimination under different inorganic fertilizers, snow, inland water and wetland mapping, meteorological studies, soil contamination and hydrocarbon pollution detection, and monitoring, detection of crop parasites, and image fusion for cocoa bean fermentation analysis. For water application, topics include snow mapping and parameter retrieval (such as snow grain size, contamination), inland water quality mapping and a case study on a wetlands ecosystem. For soil and land applications, topics include heavy-metal contamination in soil, soil parameters, soil properties presenting the efficacy of hyperspectral data and multiimage fusion datasets.

We are grateful to the reviewers who made the time to review the chapter manuscripts and Elsevier editorial acquisition members including Morse Redding, Marisa Lafleur, Honest Joy, Robertson Naomi for their constant support and help. Last but not the least, the editors thank the publisher for providing the opportunity to set down the thoughts of several contributors to produce this book.

I hope *Hyperspectral Remote Sensing* will provide insight into the breadth of the hyperspectral application-related topics. Users of this book are encouraged to adapt it and use it the way it best fits their own needs that would help them in understanding the capabilities and potentials of hyperspectral remote sensing and applications.

Editors
Prem Chandra Pandey, Greater Noida, India
Prashant K. Srivastava, Varanasi, India
Heiko Balzter, Leicester, United Kingdom
Bimal Bhattacharya, Ahmedabad, India
George P. Petropoulos, Athens, Greece

SECTION

I

Introduction to Hyperspectral Remote Sensing and Principles of Theory and Data Processing

1

Revisiting hyperspectral remote sensing: origin, processing, applications and way forward

Prashant K. Srivastava[1,2], Ramandeep Kaur M. Malhi[1], Prem Chandra Pandey[3], Akash Anand[1], Prachi Singh[1], Manish Kumar Pandey[1], Ayushi Gupta[1]

[1]REMOTE SENSING LABORATORY, INSTITUTE OF ENVIRONMENT AND SUSTAINABLE DEVELOPMENT, BANARAS HINDU UNIVERSITY, VARANASI, INDIA

[2]DST-MAHAMANA CENTRE FOR EXCELLENCE IN CLIMATE CHANGE RESEARCH, BANARAS HINDU UNIVERSITY, VARANASI, INDIA

[3]CENTER FOR ENVIRONMENTAL SCIENCES & ENGINEERING, SCHOOL OF NATURAL SCIENCES, SHIV NADAR UNIVERSITY, GREATER, NOIDA, INDIA

1.1 Introduction

Multispectral remote sensors with a few broad spectral bands were evolved during the 1970s to monitor natural resources. Looking at the existing limitations of multispectral images in different applications and recognizing the demand for better and more advanced images with higher spectral resolution, there was an urgent need for research and development of hyperspectral imaging systems. Hyperspectral remote sensing is an output of imaging spectroscopy that is facilitated by rapid advancement in technologies and the development of detectors, optical design and components, atmospheric radiative transfer and processing capability. Imaging spectroscopy, in turn, utilizes two sensing techniques, namely spectroscopy and imaging. An imaging system captures the spatial distribution of a scene and measures the relative concentration of the objects, while spectroscopy offers the ability to differentiate the elusive absorption features of divergent materials for a scene. The initial or conventional approaches of using multispectral or broadband sensors has the limitation of dividing a discontinuous spectral coverage into numerous broadbands. Hyperspectral remote sensing gained growth and popularity because it collects data that span over a vast region in umpteen contiguous narrow spectral bands of the electromagnetic spectrum, ranging from visible (VIS)—near-infrared (NIR) to shortwave infrared (SWIR) and can achieve a spectral resolution of $10^{-2}\lambda$. Based on platform type, hyperspectral remotely sensed data can be classified as non-imaging or imaging *in situ* measurements, airborne images, and space-borne images.

Hyperspectral Remote Sensing. DOI: https://doi.org/10.1016/B978-0-08-102894-0.00001-2

1.2 Origin of hyperspectral remote sensing

A landmark step was achieved in 1979 when hybrid array detectors, mercury cadmium telluride on silicon charge-coupled devices was made available for the first time leading to the construction of an imaging spectrometer that operated at wavelengths beyond 1.0 of μm. The airborne imaging spectrometer (AIS) was developed at the National Aeronautics and Space Administration (NASA) Jet Propulsion Laboratory (JPL) in 1983 and operated at wavelengths between 0.8 and 2.5 μm. AIS was replaced by the airborne visible/infrared imaging spectrometer (AVIRIS) in the early 1990s that covered the entire spectrum from 0.4 μm to 2.45 μm at a high spectral rate plus high spatial resolution over an 11 km swath. The primary objectives of these missions were identifying and assessing the characteristics of surface materials. It was with AVIRIS data that the first vegetation analysis was carried out through near-infrared spectroscopy (NIRS) analysis by John Aber and Mary Martin of the University of New Hampshire. Although AVIRIS has offered the bulk of high-quality hyperspectral data, it lacks regularity.

With known multispectral specifications and properties, one wants to gain information from spectral information, and there must be several narrow spectral channels in the same electromagnetic spectrum (400–2500 nm) compared to a few broad spectral channels in multispectral images (Pandey et al., 2019b). Thus the idea behind hyperspectral imaging systems is straightforward, while offering even more information to understand our environment better. This led to further innovations and advancement in hyperspectral imaging systems in the early 1980s. Hyperspectral remote sensing or imaging spectrometry has evolved from ERTS-1 [1] (Earth Resources Technology Satellite)-1 (Landsat-1) 4-band images, to AIS, AVIRIS. AVIRIS was the first hyperspectral imaging system developed at NASA JPL in the United States as a result of innovations in spectrometers technologies, then AVIRIS-NG (Airborne Visible Infrared Imaging Spectrometer-Next Generation) and HyspIRI (Hyperspectral Infrared Imager) were introduced.

The first hyperspectral mission Earth Observation-1 (EO-1) Hyperion sensor was launched by NASA's Earth Observing-1 satellite in November 2000. This space-borne mission brought significant and dramatic changes in the applications on forestry, agriculture, mineral identifications and land-use or land-cover classifications (Pandey et al., 2018, 2019a). Enhanced spectral resolution in the 400–2500 nm range was extensively and widely used by researchers, and provided more reliable and accurate results (Kramer, 2002). In order to achieve high spectral resolution, multispectral sensors were effectively enhanced in their image acquisition in the spectral range, reduction in the spectral sampling rate, and several contiguous spectral channels. Later on, the Compact High-Resolution Imaging Spectrometer (CHRIS) for the European Space Agency (ESA) Project for On-Board Autonomy-1 (PROBA-1) in 2001 was launched by ESA for the global mapping of natural resources. Hyperion has been decommissioned and CHRIS is currently working and operating in the orbit. It has been designed for a life operational time period of a one year only and has 62 contiguous channels with a spatial resolution of 17 m.

These hyperspectral imaging satellites are accompanied by several other missions later on (Puschell, 2000). The information on current and future space-borne hyperspectral missions were compiled with references to several published articles (Nieke et al., 1997; Puschell, 2000; Kramer, 2002; Staenz and Held, 2012). There is a focus to dramatically increase contiguous spectral bands in upcoming space-based hyperspectral imaging systems (Nieke et al., 1997). A medium spectral resolution hyperspectral imaging system called PRecursore IperSpettrale della Missione Applicativa (PRISMA) was launched by Italy's ASI on March 22, 2019 (Longo, 2020). PRISMA's hyperspectral sensors will be capable to acquire images in ~235 contiguous spectral channels in visible (VIS)—near-infrared (NIR)—short-wave infrared (SWIR) ranges. The Environmental Mapping and Analysis Program (EnMAP) for hyperspectral sensors was launched by Deutsche Zentrum für Luft-and Raumfahrt (DLR) (The German Aerospace Center) the German Research Centre for Geosciences [GeoForschungsZentrum (GFZ)] in 2015 to acquire images in over 200 narrow contiguous bands. In the meantime, NASA has scheduled the HyspIRI mission launch for 2022, which will acquire images with 237 spectral bands with 60 m spatial resolution. Detailed information on present and future hyperspectral missions are presented in Table 1—1. This also presents the major space-borne missions launched or planned by several countries or space agencies. Apart from those listed in Table 1—1, there are HISUI (The Japanese Hyperspectral Imager Suite) onboard the Advanced Land Observing Satellite-3 (ALOS-3), and the French HYPXIM hyperspectral missions (Briottet et al., 2011; Matsunaga et al., 2013) were also in orbit. HISUI is a space-borne instrument suite designed to work both as hyperspectral as well as multispectral imagers. Therefore the importance of these planned hyperspectral missions or initiatives should be understood to ensure guaranteed, uninterrupted hyperspectral image acquisition and coverage beyond 2020.

The focus of the HyspIRI mission will be on the world's ecosystem studies offering information on future ecosystems, disasters, and the carbon cycle. This mission will offer VIS—NIR and SWIR hyperspectral data and multispectral thermal data regularly. The combination of spatiotemporal and spectral data will offer a challenge to researchers.

1.3 Atmospheric correction: a primary step in preprocessing hyperspectral images

Passive remote sensing is basically the study of interaction between light source and Earth surface features, in which every feature has an unique spectral response. The spectral resolution and continuous wavelength range of an image play an important role in distinguishing the feature; the higher the spectral resolution the higher will be its feature-detection capability (Okada and Iwashita, 1992; Clark, 1999). Hyperspectral imaging holds the potential to distinguish surface features more precisely because of its high-spectral resolution and narrow bandwidths. Hyperspectral images have numerous narrow bands covering ultra-violet, VIS—NIR, and SWIR regions of the electromagnetic spectrum, which makes it more susceptible to atmospheric distortions. Atmospheric gases and aerosols absorb and transmit the incoming light in regards to its wavelength and this causes distortions in the image. Therefore atmospheric correction is required in all hyperspectral data in which the raw

Table 1–1 Major earth observation hyperspectral missions (launched or planned future missions).

	Sensors	Platform	Spectral range	Spatial resolution/ spectral resolution	Channels	Revisit time	Organization/ Nation	Launch year
01	Hyperion	EO-1	400–2500 nm	30 m/10 nm	220	200 days	NASA	November 2000
02	CHRIS	PROBA	400–1050 nm	18–36 m/(1.25–11 nm) –62 spectral to provide 34 m	150	2 (mid-latitudes)	ESA-UK	October 2001
03	EnMap		~VNIR (420–1000 nm) ~SWIR I (900–1390 nm) ~SWIR II (1480–1760 nm) ~SWIR III (1950–2450 nm)	30 m at Nadir Spectral sampling VNIR: 5–10 nm (6.5 nm average) SWIR: 10 nm (average)	92 ~420–1030 108 ~950–2450	23 days and 4 days (across track ± 30°)	DLR Germany/GFZ	2015
04	HyspIRI[d]		Hyperspectral imaging ~400–2500 nm Multispectral IR at 8–12 μm	60 m at 150 km Swath (after 2013-30 m)	217	VSWIR—19 days (after 2013- 16 days) TIR—5 days[f]	NASA—United States	2022
05	HySIS		400–1200 nm	30 m/~10 nm	55	5/19 days	ISRO India	November 2018
06	SHALOM	MBT Space	400–2500 nm	10 m	241	2 days	ISA—Israel	June 2019[a]
07	PRISMA[e]	AVIO Italian launcher	400–2500 nm	20–30 m	240	<29 days Relook day <7 days	Italy Space Agency and VEGA	March 2019[a]
08	FLEX	Earth explorer		300 m		28 days	ESA	2022[a]
09	HySI	Chandrayaan-1	400–950 nm (15 nm)	80 m	32	~	ISRO India	2008
10	HJ-1A	CAST	[a]WVC—0.43–0.90 μm [b]HSI—0.45–0.95 μm	30 m 100 m/5 nm	4 115	4 days 4–31 days (side look ± 30°)	China	September 2008
	HJ-1B		WVC—approximately 0.43–0.90 μm [c]IRMSS—approximately 0.75–1.10 μm 1.55–1.75 μm 3.50–3.90 μm 10.5–12.5 μm	30 m 150 m 150 m 150 m 300 m	4 4	4 days 4 days		

11	Hero (CASI)		400–2500 nm	30 m/∼10 nm	>200	3	Canada	
12	VENUS		415–910 nm	5.3 m	12	2	CNES/Israel	2016
13	SumbandilaSat/ MSI		440—2350 nm	15 m/5.7 nm	200	—	South Africa	September 2009
14	HICI		350–1081 nm	90 m/∼5.7 nm	128	—	NASA United States	2009
15	MSMI	Sunspace[g]	440–2350 nm	15 m/∼10 nm	200	—	South Africa	March 2010
16	HISUI	ALOS-3	400–2500 nm	30 m/∼10 nm VNIR and 12.5 SWIR	185 (VNIR-57, SWIR128)		Japan	2015
				5 m (for multispectral part)	4 bands			
17	HYPXIM-CA		400–2500 nm	15 m ≤ 14 VIS and ≤ 10 NIR/SWIR	>200		France—CNES	2019

[a]Wide View CCD Cameras (WVC).
[b]Hyperspectral Imager (HSI).
[c]Infrared Multispectral Scanner (IRMSS).
[d]https://hyspiri.jpl.nasa.gov/.
[e]https://www.unoosa.org/documents/pdf/copuos/2019/copuos2019tech11E.pdf.
[f]TIR measures both day and night data with 1 daytime image and 1 night-time image every 5 days.
[g]After launch, mission control was handed over from **SunSpace** to SAC (Satellite Applications Centre) at Hartebeetshoek near Pretoria in **South Africa.**

radiance image is converted to a reflectance image considering all spectra is shifted to a similar albedo.

The techniques for accurate removal of atmospheric absorption and scattering from a hyperspectral image is mainly divided into two categories, namely relative and absolute (also termed as empirical) atmospheric corrections (San and Suzen, 2010). Further the technique of relative atmospheric correction is subdivided into three methods: (1) Internal average reflectance correction (Kruse, 1988; Ben-Dor and Kruse, 1994), (2) flat field correction (Gao et al., 2009), and (3) empirical line correction (Gao et al., 2009). In the relative atmospheric correction technique there is no need of providing a priori information about surface or atmosphere as it uses the data statistics and runs mathematical operations to correct the image. On the other hand absolute atmospheric correction techniques take a priori information such as water vapor, atmospheric gas content, and topography effects into consideration to run radiative transfer codes. On a per-pixel basis, the difference between radiation leaving Earth and received at the sensor along with the a priori inputs are used by the radiative transfer codes to correct the atmospheric distortions. Absolute atmospheric correction has been proven to be the better technique over the relative correction technique because it considers the local geographical and atmospheric a priori inputs for correcting the data (Nikolakopoulos et al., 2002).

Several radiative transfer codes have been introduced by researchers over the past decades for atmospheric corrections. Many of these codes are developed for a satellite specific imaging system and for a specific spectral and spatial range. Some of these radiative transfer codes include, Atmospheric Correction Now (ACORN) that is based on MODerate resolution atmospheric TRANsmission-4 (MODTRAN-4) (Miller, 2002). Atmospheric Correction (ATCOR) is also a MODTRAN-4 based code and is integrated with ERDAS Imagine (Earth Resource Development Assessment System) software (Adler-Golden et al., 1999), Atmospheric Removal (ATREM) (Gao et al., 1993), High Accuracy Atmospheric Correction for Hyperspectral data (HATCH) (Qu et al., 2003) which is an improved version of ATREM, and Fast Line-of-sight Atmospheric Analysis of Spectral Hypercube (FLAASH) (Cooley et al., 2002) which is integrated in ENVI (Environment for Visualizing Images) software. The major drawbacks of these radiative transfer codes are their complex algorithms, precalculated lookup tables, inputs based on interpolation, and the requirement of a priori data related to pixel-wise atmospheric and topographic information.

Radiative transfer codes based on atmospheric correction was introduced by Gao et al. (1992, 1993) in the 1990s and is termed ATREM. Using this algorithm, from AVIRIS data scaled surface reflectance is retrieved assuming a horizontal surface having Lambertian reflectance. The ATREM radiative transfer code is formulated by deriving the pixel-wise water vapor absorption bands at 0.94 and 1.4 μm, the incoming solar radiation, azimuth angle, and narrow bandwidth. Accordingly the transmission spectrum of atmospheric gases namely, carbon dioxide (CO_2), carbon monoxide (CO), methane (CH_4), nitrogen oxide (NO_2) and oxygen (O_2) is simulated. The scattering effect caused by suspended atmospheric molecules and aerosols is modeled using the Simulation of the Satellite Signal in Solar Spectrum

(5S) model (Tanré et al., 1990). In early 2000 a line-by-line atmospheric transmittance model (Gao and Davis, 1997) was introduced into ATREM code which uses 6S module (Vermote et al., 1994) that included the effect of NO_2 in the 0.4–0.8 μm spectral range. Several other radiative transfer codes having advanced features like topographic corrections, spectral smoothing, and feature adjacency correction such as ACRON, FLAASH, and HATCH have recently been introduced.

Taking into consideration the advancements in atmospheric correction techniques used for hyperspectral images, FLAASH is one of the most accurate and easy to use radiative transfer model. FLAASH is a software package introduced by Air Force Research Laboratory, Space Vehicles Directorate (AFRL/VS) and is currently integrated with ENVI software for commercial use. FLAASH is able to more accurately analyze the VIS to SWIR region of the electromagnetic spectrum. FLAASH is a MODTRAN-4 based model with physics-based derivations of different atmospheric interaction properties that include atmospheric pressure, water vapor column, and aerosol and cloud cover, which is further used as a priori information for converting radiance to reflectance (Berk et al., 1998, 2000). Hyperspectral data, its sensor type, and metadata (including sensor altitude, viewing, and solar angle) is used as input for FLAASH to retrieve preliminary water vapor column data at a per-pixel basis. Then a water vapor Look-Up Table (LUT) is generated using MODTRAN simulation to retrieve reflectance from radiance. FLAASH also generates a cloud mask to classify the pixels having cloud cover and flag those pixels to be removed before calculation. Considering the variables are all wavelength dependent, image polishing and renormalization is performed last to generate a new reflectance cube (Boardman, 1998).

Despite these atmospheric correction models, there are still possibilities for further improvement in radiative transfer codes. The built-in smoothing modules need improvement in order to identify any artificial broad absorption features in the retrieved reflectance spectra. A hybrid approach using image-based empirical methods and radiative transfer codes can increase the retrieval accuracy and lower the computational complexities.

1.4 Empirical and radiative transfer models

Inverse modeling plays a vital role for quantitative remote sensing, which mainly uses physical- or empirical-based models to estimate unknown parameters (Wang, 2012). Over the past few years various types of models have been developed related to atmosphere, vegetation, and radiation and a model-based inverse problem was also introduced. The introduction of multispectral and hyperspectral remote sensors with greater spectral and spatial information came into the account to solve model-based inverse problems (Liang, 2005; Wang et al., 2009). Inverse modeling is mainly useful to retrieve various types of vegetation parameters using physical or empirical models. These models also can be used in direct (forward) mode to compute canopy-reflectance values after providing leaf and canopy traits [chlorophyll content, water content, Leaf Area Index (LAI)]. Moreover, such physical-based radiative transfer models can be inverted from reflectance or EO data for retrieval of biophysical and biochemical variables like LAI and fraction of photosynthetically active radiation, which are mainly used to monitor the health status and enumerate the vegetation influence. Narrow-band information in hyperspectral sensors can aid in providing quantitative estimates of canopy biochemical properties (such as chlorophyll and nitrogen concentrations) when compared to multispectral (broadband) sensors (Goodenough et al., 2004, 2006).

Inversion of the canopy Radiative Transfer Model (RTM) is known as one of the best approaches for the retrieval of biophysical and biochemical parameters. During the past few years various types of studies have been done on RTMs. Most useful RTMs are the leaf PROSPECT (leaf optical properties model) and canopy-based SAIL models (canopy bidirectional reflectance model) that combine the PROSAIL model and are useful for inverse modeling (Verrelst et al., 2019). Advance methods such as PROSPECT (COSINE) and PROSAIL have maximum computational power and numbers of inversion algorithms but are still time consuming to provide efficient results. Further optimization and the robustness of LUT inversion application have also been explored against spectroscopic data (Banskota et al., 2013, 2015; Bao et al., 2017). LUT-based inversion approaches provide fast and efficient results because of before-hand completion of the inversion itself which is assumed to be the most computationally expensive part of the inversion procedure. The LUT-based inversion toolbox is a part of the ARTMO (Automated Radiative Transfer Models Operator) software package that can be run on MATLAB and also provides essential tools and information for running as well as inverting a collection of plant RTMs, both at the leaf (PROSPECT, DLM, Liberty, and Fluspect) and canopy level (4SAIL, INFORM, FLIGHT). Leaf and canopy RTMs are used as an input for the generation of class-based LUTs (Fig. 1–1). Moreover ARTMO is capable of evaluating LUT-based inversions on land-cover classification, which allows reasonable retrieval of biophysical parameters over land surface (Rivera et al., 2013; Verrelst et al., 2013).

A study has demonstrated the use of CASI (Compact Airborne Spectrographic Imager) hyperspectral reflectance data for monitoring some forest sites with dense canopies (LAI >4) and has also shown the achievability of RTMs for pigment estimation. The RTM inversion technique exhibited promising results with RMSE values from 3.0 to 5.5 $\mu g/cm^2$ for a leaf chlorophyll range of 19.1–45.8 $\mu g/cm^2$ (Zarco-Tejada et al., 2001).

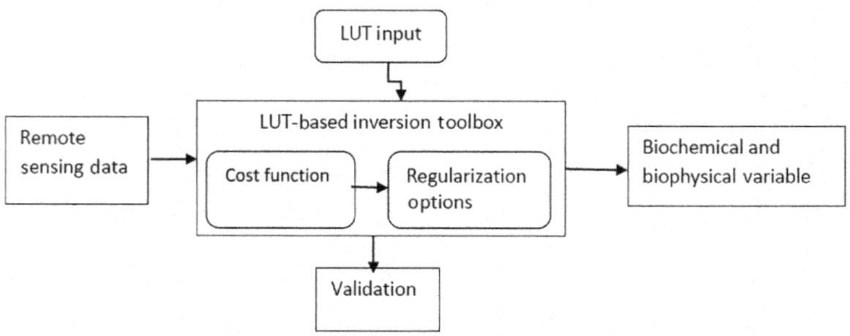

FIGURE 1–1 Basic principle of the Look-Up Table-based inversion toolbox (Rivera et al., 2013).

This study has confirmed that the success of LUT-based inversion strongly depends on the retrieval parameters and applied regularization options. Performance of LUT-based inversion is directly related to the used-cost functions (Verrelst et al., 2013). LUT-based inversion on other types of hyperspectral sensors such as AVIRIS and Hyperion are also useful to retrieve vegetation parameters and pigments for the reorganization of the plant and crop species.

1.5 Applications of hyperspectral remote sensing

There are a plethora of applications possible using hyperspectral remote sensing, but herein only a number of important areas such as vegetation, urban, mineral, water and agriculture are provided with respect to hyperspectral applications.

1.5.1 Vegetation analysis

Terrestrial vegetation is considered as one of the Earth's most significant natural resources and is spread over a large surface. It encompasses different landscapes namely forest, agriculture, rangeland, wetland, and urban vegetation (Jensen, 2009). Monitoring terrestrial vegetation at spatio-temporal scale is crucial for studying different terrestrial resource management applications. Field measurements of foliar and canopy vegetation's biophysical and biochemical parameters are the best employed approaches for its monitoring (Mutanga et al., 2004; Chmura et al., 2007; Tilling et al., 2007). Conventional field estimations of these parameters involve time, money, and labor and it cannot be easily extended to big geographical areas.

Hyperspectral remote sensing is the most appropriate alternative technique for retrieving vegetation biophysical and biochemical parameters since narrow-band measurements of spectral reflectance in hyperspectral remote sensing allow finer determination of the changes in vegetation's biophysical and biochemical parameters. Numerous authors have explored the capability of hyperspectral remote sensing in providing valuable information on these parameters. Both ground-level hyperspectral data acquired using hand-held spectroradiometers as well airborne or spaceborne hyperspectral instruments have been widely used in ample vegetation studies. Available airborne hyperspectral instruments include AVIRIS, AVIRIS-NG, and HyMap; whereas spaceborne hyperspectral instruments are Hyperion, CHRIS, EnMAP, and the HJ-1A hyperspectral imager product.

Additionally other vegetation analyses including diversity parameters such as species diversity (Peng et al., 2018; Malhi et al., 2020) and species richness (Psomas et al., 2011; Peng et al., 2019) are also successfully retrieved using this advanced remote sensing technique. A large number of narrow bands of hyperspectral data also encourages the classification and mapping of different vegetation types at varied taxonomic scales, mostly down to the species level. For example, species-level classification with high accuracy is achieved for grapevine varieties (Fernandes et al., 2015) using hyperspectral data (Clark et al., 2005). Classification using hyperspectral data was successfully carried out for different vegetation types such as annual gramineous weeds (Deng et al., 2016), food crops (Mariotto et al., 2013), shrubs of arid zones (Lewis, 2002), and montane or subalpine trees (Sommer et al., 2016) and forest tree species (Hycza et al., 2018). The hyperspectral technique is also used to determine different plant functional types that are functionally similar plant species in terms of resource utility, ecosystem function, and response to environment conditions. Such classification is carried out using methods like spectral mixture analysis (Schaaf et al., 2011). Studies were also performed where hyperspectral remote sensing was used in the detection and classification of the early onset of plant disease and stress (Lowe et al., 2017).

1.5.2 Urban analysis

The world's population is continuously shifting toward urban centers resulting in the regular modification in the landscape at regional level. The densification of urban areas is making the urban fabric (such as buildings, roads, etc.) more complex and it is also leading to the generation of urban-heat islands (Weber et al., 2018). Hence there is a great need to understand the relation between urban systems with respect to biotic and abiotic components. To understand the dynamics of the urban system, the identification of building materials, impervious surfaces, mineral composition, water quality, vegetation, and fallow lands at very fine spectral resolution is crucial, and can be archived by hyperspectral data. Hyperspectral data provides a very fine spectral resolution that helps in the identification of building and road materials, microclimate model development, vegetation health monitoring, pollution monitoring, water quality assessment, and so forth, because all the elements have a unique absorption feature in their reflectance spectra. There are several studies that have been done using hyperspectral data to enhance the urban management system, including the study done by Karoui et al. (2018) in which they used the hyperspectral data to identify photovoltaic panels within the region using the spectral unmixing technique. Roof-top mapping is a common application of hyperspectral data in urban studies (Chisense et al., 2012) where the endmembers of rooftop materials were generated and classification on HyMap hyperspectral data was performed to identify the rooftop materials. This study is also crucial in urban management as it can help in the identification of buildings that are the major source of urban heat islands. Urban Land Use Land Cover (ULULC) mapping is one of the most important applications of hyperspectral data. Again a large number of spectral bands provides better feature detection capability with respect to multispectral data. There are several classification techniques such as Support Vector Machine (Tratt et al., 2016), Artificial Neural networks (Goel et al., 2003), *K*-means (Filho et al., 2003; Licciardi et al., 2011), and Principle Component Analysis (PCA) (Licciardi et al., 2011) and Segmented PCA (Pandey et al., 2014) that are commonly used for hyperspectral data. The NIR, thermal and SWIR regions of electromagnetic spectrum in hyperspectral data are sensitive to different gases. This aids in identifying and tracking gaseous emission from compact sources present in urban-industrial areas (Tratt et al., 2016). Hyperspectral data also has a major role in water-quality mapping (Olmanson et al., 2013), identifying soil properties (Hively et al., 2011), and pest detection

(Glaser et al., 2009). Apart from the different applications of hyperspectral data for urban studies, there are certain limitations also, including the data complexity and lack of spatio-temporal hyperspectral data as there are limited spaceborne hyperspectral sensors.

1.5.3 Mineral identification

Hyperspectral remote sensing can identify minerals precisely compared to conventional remote sensing techniques. This is again because of the high number of contiguous bands that help to construct a complete reflectance spectrum of every pixel in the scene. Studies by Ting-ting and Fei (2012) take porphyry copper deposit as an example to demonstrate the usefulness of hyperspectral remote sensing in this mineral deposit exploration. The power of hyperspectral remote sensing can also be used for detection of hydrothermal alteration or alteration mineral zones related to different types of mineralization systems (Carrino et al., 2018). In one case study, the mineral assemblies containing high sulfidation epithermal targets was mapped using the HyMap images combined with petrography, magnetic data and X-ray diffraction (XRD) (Carrino et al., 2018). In another study, an integrated approach was performed using hyperspectral data and geochemical properties to identify limestone and siliciclastic rocks, argillization and pyrite oxidation in the callville limestone and is well related to gold mineralization (Sun et al., 2019). Hyperspectral remote sensing has huge potential in the mining industry as a fast and noninvasive analytical approach for mineralogical characterization (Tuşa et al., 2020). The mineral target can be identified using the principal components and also through broadband spectral analysis. Some focused algorithms such as the Spectral Angle Mapper and Mixture Tuned Matched Filtering techniques can be used for discrimination and mapping in detail. Therefore, hyperspectral remote sensing is found to be very feasible and advantageous for mineral exploration in remote areas where primary information is limited (Bishop et al., 2011).

1.5.4 Water quality

Monitoring water quality is very important for maintaining ecosystem health and the livelihood of the population. It reflects the health of surface water bodies as a snapshot in time (weeks, months, and years). Therefore, best practices and efforts are needed to monitor and improve water quality. As hyperspectral remote sensing can capture water quality parameters, it could be viable solution for water quality management. For example, Zhang et al. (2020) in their study, provided a self-adapting selection method in integration of artificial neural networks to quantitatively predict water quality parameters such as phosphorus, nitrogen, biochemical oxygen demand, chemical oxygen demand and chlorophyll a. Similarly for lake water quality, specific heavy metals were detected and predicted using portable FieldSpec_3 ASD spectroradiometer and various spectra were taken in the laboratory (Rostom et al., 2017). Studies by Shafique et al. (2003) used the hand-held spectrometer data and collected water spectrum from rivers directly. They established and used correlations between the ground-truth data and spectrum for developing spectral indices to estimate chlorophyll a, turbidity, and phosphorus. In other study (Sailaja et al., 2017), five water quality parameters (i.e. chlorophyll, turbidity, Secchi depth, total phosphorus, and total suspended solids) were estimated through regression models. They took field spectroradiometer and hyperion reflectance values and related these with *in situ* ground data.

Wang and Yang (2019) provided a quantitative systematic review to identify existing challenges and future directions. Their review identified that the semiempirical method was used by most of the researchers and is the most frequently used inversion method. They concluded that the ground object spectrometer is a highly applied data source and most of the study provided estimates of chlorophyll, suspended solid, and so forth, but rarely considered the human induced factors which is a drawback of the model's robustness.

1.5.5 Agricultural applications

Hyperspectral remote sensing or imaging spectroscopy shows great potential in agriculture applications. High-resolution spectral data has the advantage of discriminating crop types and species (Sahoo et al., 2015) as well as biophysical and biochemical parameters. Various crop parameters that are possible to retrieve are LAI (Gong et al., 2003; Asano et al., 2009; Ke et al., 2016; Din et al., 2017; Li et al., 2017; Zhang et al., 2017), leaf pigments like chlorophyll, carotenoids, xanthophylls, and anthocyanins (Qian et al., 2003; Wang et al., 2014; Sonobe et al., 2018; Zarco-Tejada et al., 2019; Amirruddin et al., 2020), nitrogen content (Blackburn, 2007; Knyazikhin et al., 2013), biomass (Psomas et al., 2011; Bratsch et al., 2017), height (Xavier et al., 2006) and others. These parameters are retrieved from hyperspectral remote sensing by mainly employing three approaches. The first approach is based on statistical models wherein empirical relationships between spectral parameters and biophysical or biochemical parameters can be established (Gupta et al., 2014). Parametric regression uses vegetation indices, shape indices, and spectral transformation, while nonparametric regression uses linear and nonlinear machine learning regression algorithms. Conventional multivariate regression tends to raise the issue of multicollinearity and overfitting because of the high correlation among spectral bands (Kumar, 1975; Hawkins, 2004) that can be overcome using partial least square regression (Li et al., 2014; Yu et al., 2015; Ryan and Ali, 2016) and are thus highly used in hyperspectral remote sensing field. Different machine learning regression algorithms such as the Gaussian process, Artificial Neural Network (Srivastava et al., 2020), and support vector regression have also been extensively used (Halme et al., 2019). The second approach utilizes the physical models or radiative transfer models wherein simulation of foliar or canopy reflectance is executed using biophysical or biochemical parameters as inputs in the forward run. Then, inversion of radiative transfer models is performed using numerical optimization and LUT methods. The model inversion is carried out for retrieving the desired parameters. These physical models are more capable in terms of robustness and universality for varied climatic conditions, vegetation types and regions. They deliver better systematic information on relationships between vegetation parameters and vegetation reflectance compared to statistical models. The third is a hybrid approach that uses a combination of statistical and physical models to estimate biophysical or biochemical parameters.

Table 1–2 Ultraspectral missions.

	Ultraspectral sensors	Platform	Agency/country
01	AIRS	Aqua satellite	NASA
02	MIPAS '8	ENVISAT-1	ESA
03	CrIS Up to 2378 spectral channels	As part of the National Polar-Orbiting Operational Environmental Satellite System	NASA
04	GIFTS"	Earth Observing-3 (EQ-3) mission	NASA

1.6 Way forward

High-dimensional hyperspectral remote sensing data makes the task of information extraction a tough one as traditional techniques used with broadband sensors are of limited use. Small satellites can be the future of the remote sensing community as less powerful vehicles are needed for their launch, thus offering significant cost savings for the future missions. Less-powerful vehicles are less expensive compared to powerful vehicles, and constitute a significant reduction of the total cost of missions, which can be diverted to other purposeful research work. Therefore, the fundamental idea behind small satellites is cost savings, including a lighter payload in launch vehicles (i.e. below 100 kg). The utmost requirement in small hyperspectral satellites is the sensor's size, mass, its output, data downlink, and transmission to ensure a successful mission. It is crucial to have images in the VIS−NIR spectral ranges with the best permissible spectral resolutions; however it is constrained by the microsatellite size and mass parameters (approximately <100 kg and 0.5 min any direction) (Levi and Washabaugh, 2001; Gerber et al., 2005; Hartley et al., 2005).

In future, ultraspectral imaging systems will overtake the hyperspectral missions. Ultraspectral sensors have more than ~300 narrow contiguous spectral channels with very narrow spectral sampling (<1 nm) for each individual pixel (Cutter, 2005). The concept behind the ultraspectral channels will enable scientists for physical analysis of gas absorption characterizing atmospheric temperature and structures. To accomplish the aim, sensors such as the Atmospheric InfraRed Sounder (AIRS), the Michelson Interferometer for Passive Atmospheric Sounding (MIPAS)'8, the Crosstrack Infrared Sounder (CrIS), and Geostationary Imaging Fourier Transform Spectrometer (GIFTS) are made operational as ultraspectral imaging systems. CrIS is a part of the National Polar-orbiting Operational Environmental Satellite System of NASA. AIRS is advanced with 2378 spectral channels and having cutting-edge infrared technology to create 3-D maps of air and surface temperature, water vapor, and cloud properties (https://airs.jpl.nasa.gov/). Details are listed in Table 1−2. GIFTS can be optimized to function in different modes such as multispectral, hyperspectral, or ultraspectral modes, thus is unique to any other satellite sensors working in a single mode.

Acknowledgment

The authors thank the National Mission on Himalayan Studies, G.B. Pant National Institute of Himalayan Environment (NIHE), Almora, Uttarakhand for funding this work.

References

Adler-Golden, S.M., Matthew, M.W., Bernstein, L.S., Levine, R.Y., Berk, A., Richtsmeier, S.C., et al., 1999. Atmospheric correction for shortwave spectral imagery based on MODTRAN4. Imaging Spectrometry V. International Society for Optics and Photonics, pp. 61–69.

Amirruddin, A.D., Muharam, F.M., Ismail, M.H., Ismail, M.F., Tan, N.P., Karam, D.S., 2020. Hyperspectral remote sensing for assessment of chlorophyll sufficiency levels in mature oil palm (*Elaeis guineensis*) based on frond numbers: analysis of decision tree and random forest. Comput. Electron. Agric. 169, 105221.

Asano, T., Kosugi, Y., Uto, K., Kosaka, N., Odagawa, S., Oda, K., 2009. Leaf area index estimation from hyperspectral data using group division method. In: 2009 IEEE International Geoscience and Remote Sensing Symposium. IEEE, pp. IV-817–IV-820.

Banskota, A., Serbin, S.P., Wynne, R.H., Thomas, V.A., Falkowski, M.J., Kayastha, N., et al., 2015. An LUT-based inversion of DART model to estimate forest LAI from hyperspectral data. IEEE J. Sel. Top. Appl. Earth Observ. Remote Sens. 8, 3147–3160.

Banskota, A., Wynne, R.H., Thomas, V.A., Serbin, S.P., Kayastha, N., Gastellu-Etchegorry, J.P., et al., 2013. Investigating the utility of wavelet transforms for inverting a 3-D radiative transfer model using hyperspectral data to retrieve forest LAI. Remote Sens. 5, 2639–2659.

Bao, S., Cao, C., Chen, W., Tian, H., 2017. Spectral features and separability of alpine wetland grass species. Spectrosc. Lett. 50, 245–256.

Ben-Dor, E., Kruse, F., 1994. The relationship between the size of spatial subsets of GER 63 channel scanner data and the quality of the Internal Average Relative Reflectance (IARR) atmospheric correction technique. Remote Sens. 15, 683–690.

Berk, A., Bernstein, L., Anderson, G., Acharya, P., Robertson, D., Chetwynd, J., et al., 1998. MODTRAN cloud and multiple scattering upgrades with application to AVIRIS. Remote Sens. Environ. 65, 367–375.

Berk, A., Acharya, P.K., Bernstein, L.S., Anderson, G.P., Chetwynd Jr, J.H., Hoke, M.L., 2000. Reformulation of the MODTRAN band model for higher spectral resolution. Algorithms Multispectral, Hyperspectral, Ultraspectral Imagery VI. International Society for Optics and Photonics, pp. 190–198.

Bishop, C.A., Liu, J.G., Mason, P.J., 2011. Hyperspectral remote sensing for mineral exploration in Pulang, Yunnan Province, China. Int. J. Remote Sens. 32, 2409–2426.

Blackburn, G.A., 2007. Hyperspectral remote sensing of plant pigments. J. Exp. Bot. 58, 855–867.

Boardman, J.W., 1998. Post-ATREM polishing of AVIRIS apparent reflectance data using EFFORT: a lesson in accuracy versus precision. In: Summaries of the Seventh JPL Airborne Earth Science Workshop. Jet Propul. Lab. Pasadena, CA, p. 53.

Bratsch, S., Epstein, H., Buchhorn, M., Walker, D., Landes, H., 2017. Relationships between hyperspectral data and components of vegetation biomass in Low Arctic tundra communities at Ivotuk, Alaska. Environ. Res. Lett. 12, 025003.

Briottet, X., Marion, R., Carrere, V., Jacquemoud, S., Chevrel, S., Prastault, P., et al., 2011. HYPXIM: a new hyperspectral sensor combining science/defence applications. In: 2011 3rd Workshop on Hyperspectral Image and Signal Processing: Evolution in Remote Sensing (WHISPERS). IEEE, pp. 1–4.

Carrino, T.A., Crósta, A.P., Toledo, C.L.B., Silva, A.M., 2018. Hyperspectral remote sensing applied to mineral exploration in southern Peru: a multiple data integration approach in the Chapi Chiara gold prospect. Int. J. Appl. Earth Observ. Geoinf. 64, 287–300.

Chisense, C., Hahn, M., Engels, J., 2012. Classification of roof materials using hyperspectral data. Int. Arch. Photogramm. Remote Sens. Spat. Inf. Sci. 39, 103.

Chmura, D.J., Rahman, M.S., Tjoelker, M.G., 2007. Crown structure and biomass allocation patterns modulate aboveground productivity in young loblolly pine and slash pine. For. Ecol. Manag. 243, 219–230.

Clark, R.N., 1999. Spectroscopy of rocks and minerals, and principles of spectroscopy. Man. Remote Sens. 3, 2–2.

Clark, M.L., Roberts, D.A., Clark, D.B., 2005. Hyperspectral discrimination of tropical rain forest tree species at leaf to crown scales. Remote Sens. Environ. 96, 375–398.

Cooley, T., Anderson, G.P., Felde, G.W., Hoke, M.L., Ratkowski, A.J., Chetwynd, J.H., et al., 2002. FLAASH, a MODTRAN4-based atmospheric correction algorithm, its application and validation. In: IEEE International Geoscience and Remote Sensing Symposium. IEEE, pp. 1414–1418.

Cutter, M., 2005. Review of a small satellite hyper-spectral mission. Proceedings of the AIAA/USU Conference on Small Satellites, 19th Annual AIAA/USU, The Year in Retrospect, SSC05-IV-2. 1-4.

Deng, W., Huang, Y., Zhao, C., Chen, L., Wang, X., 2016. Bayesian discriminant analysis of plant leaf hyperspectral reflectance for identification of weeds from cabbages. Afr. J. Agric. Res. 11, 551–562.

Din, M., Zheng, W., Rashid, M., Wang, S., Shi, Z., 2017. Evaluating hyperspectral vegetation indices for leaf area index estimation of *Oryza sativa* L. at diverse phenological stages. Front. Plant Sci. 8, 820.

Fernandes, A.M., Melo-Pinto, P., Millan, B., Tardáguila, J., Diago, M.P., 2015. Automatic discrimination of grapevine (*Vitis vinifera* L.) clones using leaf hyperspectral imaging and partial least squares. J. Agric. Sci. 153, 455–465.

Filho, S., Frery, A.G., de Araujo, A.C., Alice, C.C., Cerqueira, H., Loureiro, J., et al., 2003. Hyperspectral images clustering on reconfigurable hardware using the *k*-means algorithm. In: Proceedings of 16th Symposium on Integrated Circuits and Systems Design, 2003. SBCCI 2003. IEEE, pp. 99–104.

Gao, B.-C., Davis, C.O., 1997. Development of a line-by-line-based atmosphere removal algorithm for airborne and spaceborne imaging spectrometers. Imaging Spectrometry III. International Society for Optics and Photonics, pp. 132–141.

Gao, B., Heidebrecht, K., Goetz, A., 1992. Atmospheric Removal Program (ATREM) User's Guide. Center for the Study of Earth from Space, Cooperative Institute for Research in Environmental Sciences, University of Colorado, Boulder.

Gao, B.-C., Montes, M.J., Davis, C.O., Goetz, A.F., 2009. Atmospheric correction algorithms for hyperspectral remote sensing data of land and ocean. Remote Sens. Environ. 113, S17–S24.

Gao, B.-C., Heidebrecht, K.B., Goetz, A.F., 1993. Derivation of scaled surface reflectances from AVIRIS data. Remote Sens. Environ. 44, 165–178.

Gerber, A., Lemmerman, L., Hartley, J., 2005. Advanced Instruments and Their Impact on Earth Science Missions (II). In: 5th IAA Symposium on Small Satellites for Earth Observations.

Glaser, J., Casas, J., Copenhaver, K., Mueller, S., 2009. Development of a broad landscape monitoring system using hyperspectral imagery to detect pest infestation. In: 2009 First Workshop on Hyperspectral Image and Signal Processing: Evolution in Remote Sensing. IEEE, pp. 1–4.

Goel, P.K., Prasher, S.O., Patel, R.M., Landry, J.-A., Bonnell, R., Viau, A.A., 2003. Classification of hyperspectral data by decision trees and artificial neural networks to identify weed stress and nitrogen status of corn. Comput. Electron Agric. 39, 67–93.

Gong, P., Pu, R., Biging, G.S., Larrieu, M.R., 2003. Estimation of forest leaf area index using vegetation indices derived from Hyperion hyperspectral data. IEEE Trans. Geosci. Remote Sens. 41, 1355–1362.

Goodenough, D.G., Pearlman, J., Chen, H., Dyk, A., Han, T., Li, J., et al., 2004. Forest information from hyperspectral sensing. In: IGARSS 2004. 2004 IEEE International Geoscience and Remote Sensing Symposium. IEEE, pp. 2585–2589.

Goodenough, D., Li, J., Asner, G., Schaepman, M., Ustin, S., Dyk, A., 2006. Combining hyperspectral remote sensing and physical modeling for applications in land ecosystems. In: 2006 IEEE International Symposium on Geoscience and Remote Sensing. IEEE, pp. 2000–2004.

Gupta, M., Srivastava, P.K., Mukherjee, S., Sandhya Kiranet, G., 2014. Chlorophyll retrieval using ground based hyperspectral data from a tropical area of India using regression algorithms. Remote Sens. Appl. Environ. Res. 177–194.

Halme, E., Pellikka, P., Mõttus, M., 2019. Utility of hyperspectral compared to multispectral remote sensing data in estimating forest biomass and structure variables in Finnish boreal forest. Int. J. Appl. Earth Observ. Geoinf. 83, 101942.

Hartley, J., Komar, G., Lemmerman, L., Gerber, A., 2005. Advanced Instruments and Their Impact on Earth Science Missions (I). In: Hans-Peter Röser, Rainer San (Eds.), Small Satellites for Earth Observation: Selected Proceedings of the 5th International Symposium of the International Academy of Astronautics Berlin, April 4–8, 2005, Session 7-Instruments.

Hawkins, D.M., 2004. The problem of overfitting. J. Chem. Inf. Comput. Sci. 44, 1–12.

Hively, W.D., McCarty, G.W., Reeves, J.B., Lang, M.W., Oesterling, R.A., Delwiche, S.R., 2011. Use of airborne hyperspectral imagery to map soil properties in tilled agricultural fields. Appl. Environ. Soil. Sci. 2011.

Hycza, T., Stereńczak, K., Bałazy, R., 2018. Potential use of hyperspectral data to classify forest tree species. N. Zealand J. Forest. Sci. 48, 18.

Jensen, J.R., 2009. Remote Sensing of the Environment: An Earth Resource Perspective 2/e. Pearson Education, India.

Karoui, M.S., Zohra Benhalouche, F., Deville, Y., Djerriri, K., Briottet, X., Le Bris, A., 2018. Detection and area estimation for photovoltaic panels in urban hyperspectral remote sensing data by an original NMF-based unmixing method. In: IGARSS 2018—2018 IEEE International Geoscience and Remote Sensing Symposium. IEEE, pp. 1640–1643.

Ke, L., Zhou, Q.-b, Wu, W.-B., Tian, X., Tang, H.-J., 2016. Estimating the crop leaf area index using hyperspectral remote sensing. J. Integr. Agricul. 15, 475–491.

Knyazikhin, Y., Schull, M.A., Stenberg, P., Mõttus, M., Rautiainen, M., Yang, Y., et al., 2013. Hyperspectral remote sensing of foliar nitrogen content. Proc. Natl Acad. Sci. USA 110, E185–E192.

Kramer, H.J., 2002. Observation of the Earth and its Environment: Survey of Missions and Sensors. Springer Science & Business Media.

Kruse, F.A., 1988. Use of airborne imaging spectrometer data to map minerals associated with hydrothermally altered rocks in the northern grapevine mountains, Nevada, and California. Remote Sens. Environ. 24, 31–51.

Kumar, T.K., 1975. Multicollinearity in regression analysis. Rev. Econ. Stat. 57, 365–366.

Levi, J., Washabaugh, P., 2001. A Feasibility Study of Micro-Satellites for Earth Observation. Proceedings of the AIAA/USU Conference on Small Satellites, 15th Annual conference, SSC01-IV-1, Bold New Missions Using Cutting Edge Technologies, pp. 1–13.

Lewis, M., 2002. Spectral characterization of Australian arid zone plants. Can. J. Remote Sens. 28, 219–230.

Li, X., Zhang, Y., Bao, Y., Luo, J., Jin, X., Xu, X., et al., 2014. Exploring the best hyperspectral features for LAI estimation using partial least squares regression. Remote Sens. 6, 6221–6241.

Li, G., Wang, C., Feng, M., Yang, W., Li, F., Feng, R., 2017. Hyperspectral prediction of leaf area index of winter wheat in irrigated and rainfed fields. PLoS ONE 12.

Liang, S., 2005. Quantitative Remote Sensing of Land Surfaces. John Wiley & Sons.

Licciardi, G., Marpu, P.R., Chanussot, J., Benediktsson, J.A., 2011. Linear versus nonlinear PCA for the classification of hyperspectral data based on the extended morphological profiles. IEEE Geosci. Remote Sens. Lett. 9, 447–451.

Longo, F., 2020. PRISMA, The Italian Hyperspectral Mission. Italy Space Agency, Italy.

Lowe, A., Harrison, N., French, A.P., 2017. Hyperspectral image analysis techniques for the detection and classification of the early onset of plant disease and stress. Plant. Methods 13, 80.

Malhi, R.K.M., Anand, A., Mudaliar, A.N., Pandey, P.C., Srivastava, P.K., Sandhya Kiran, G., 2020. Synergetic use of in situ and hyperspectral data for mapping species diversity and above ground biomass in Shoolpaneshwar Wildlife Sanctuary, Gujarat. Trop. Ecol. 61, 106–115. https://doi.org/10.1007/s42965-020-00068-8.

Mariotto, I., Thenkabail, P.S., Huete, A., Slonecker, E.T., Platonov, A., 2013. Hyperspectral versus multispectral crop-productivity modeling and type discrimination for the HyspIRI mission. Remote Sens. Environ. 139, 291–305.

Matsunaga, T., Iwasaki, A., Tsuchida, S., Tanii, J., Kashimura, O., Nakamura, R., et al., 2013. Current status of hyperspectral imager suite (HISUI). In: 2013 IEEE International Geoscience and Remote Sensing Symposium-IGARSS. IEEE, pp. 3510–3513.

Miller, C.J., 2002. Performance assessment of ACORN atmospheric correction algorithm. Algorithms and Technologies for Multispectral, Hyperspectral, and Ultraspectral Imagery VIII. International Society for Optics and Photonics, pp. 438–449.

Mutanga, O., Skidmore, A.K., Prins, H., 2004. Predicting in situ pasture quality in the Kruger National Park, South Africa, using continuum-removed absorption features. Remote Sens. Environ. 89, 393–408.

Nieke, J., Schwarzer, H.H., Neumann, A., Zimmermann, G., 1997. Imaging spaceborne and airborne sensor systems in the beginning of the next century. Sensors, Systems, and Next-Generation Satellites. International Society for Optics and Photonics, pp. 581–592.

Nikolakopoulos, K.G., Vaiopoulos, D.A., Georgios, A., 2002. A comparative study of different atmospheric correction algorithms over an area with complex geomorphology in Western Peloponnese, Greece. In: IEEE International Geoscience and Remote Sensing Symposium. IEEE, pp. 2492–2494.

Okada, K., Iwashita, A., 1992. Hyper-multispectral image analysis based on waveform characteristics of spectral curve. Adv. Space Res. 12, 433–442.

Olmanson, L.G., Brezonik, P.L., Bauer, M.E., 2013. Airborne hyperspectral remote sensing to assess spatial distribution of water quality characteristics in large rivers: the Mississippi River and its tributaries in Minnesota. Remote Sens. Environ. 130, 254–265.

Pandey, P.C., Tate, N.J., Balzter, H., 2014. Mapping tree species in coastal portugal using statistically segmented principal component analysis and other methods. IEEE Sens. J. 14, 4434–4441.

Pandey, P., Manevski, K., Srivastava, P.K., Petropoulos, G.P., 2018. The use of hyperspectral earth observation data for land use/cover classification: present status, challenges and future outlook. In: Thenkabail, P. (Ed.), Hyperspectral Remote Sensing of Vegetation, first ed. pp. 147–173.

Pandey, P.C., Anand, A., Srivastava, P.K., 2019a. Spatial distribution of Mangrove Forest species and biomass assessment using field inventory and Earth Observation hyperspectral data. Biodiversity and Conservation, 28. Springer, pp. 2143–2162.

Pandey, P.C., Koutsias, N., Petropoulos, G.P., Srivastava, P.K., Ben Dor, E., 2019b. Land use/land cover in view of earth observation: data sources, input dimensions, and classifiers—a review of the state of the art. Geocarto Int. 1–32.

Peng, Y., Fan, M., Song, J., Cui, T., Li, R., 2018. Assessment of plant species diversity based on hyperspectral indices at a fine scale. Sci. Rep. 8, 1–11.

Peng, Y., Fan, M., Bai, L., Sang, W., Feng, J., Zhao, Z., et al., 2019. Identification of the best hyperspectral indices in estimating plant species richness in Sandy Grasslands. Remote Sens. 11, 588.

Psomas, A., Kneubühler, M., Huber, S., Itten, K., Zimmermann, N., 2011. Hyperspectral remote sensing for estimating aboveground biomass and for exploring species richness patterns of grassland habitats. Int. J. Remote Sens. 32, 9007–9031.

Puschell, J.J., 2000. Hyperspectral imagers for current and future missions. Visual Information Processing IX. International Society for Optics and Photonics, pp. 121–132.

Qian, C., Jing-feng, H., Xiu-zhen, W., Ren-chao, W., 2003. In situ hyperspectral data analysis for pigment content estimation of rice leaves. J. Zhejiang Univ.-Sci. A 4, 727–733.

Qu, Z., Kindel, B.C., Goetz, A.F., 2003. The high accuracy atmospheric correction for hyperspectral data (HATCH) model. IEEE Trans. Geosci. Remote Sens. 41, 1223–1231.

Rivera, J.P., Verrelst, J., Leonenko, G., Moreno, J., 2013. Multiple cost functions and regularization options for improved retrieval of leaf chlorophyll content and LAI through inversion of the PROSAIL model. Remote Sens. 5, 3280–3304.

Rostom, N.G., Shalaby, A.A., Issa, Y.M., Afifi, A.A., 2017. Evaluation of Mariut Lake water quality using Hyperspectral Remote Sensing and laboratory works. Egypt J. Remote Sens. Space Sci. 20, S39–S48.

Ryan, K., Ali, K., 2016. Application of a partial least-squares regression model to retrieve chlorophyll—a concentrations in coastal waters using hyper-spectral data. Ocean Sci. J. 51, 209–221.

Sahoo, R.N., Ray, S., Manjunath, K., 2015. Hyperspectral remote sensing of agriculture. Curr. Sci. 108, 848–859.

Sailaja, V.H., Babu, P.S., Reddy, M.A., 2017. Hyper-spectral Remote Sensing of Water Quality Parameters in Lakes: A Case Study of Hyderabad City, Telangana State, India. International Journal of Technology & Engineering 8 (2), 5–14.

San, B.T., Suzen, M.L., 2010. Evaluation of different atmospheric correction algorithms for EO-1 Hyperion imagery. *International Archives of the Photogrammetry, Remote Sensing and Spatial Information Science,* **Volume XXXVIII, Part 8**, Kyoto Japan 2010. Commission VIII, WG VIII/3.

Schaaf, A.N., Dennison, P.E., Fryer, G.K., Roth, K.L., Roberts, D.A., 2011. Mapping plant functional types at multiple spatial resolutions using imaging spectrometer data. GIScience Remote Sens. 48, 324–344.

Shafique, N.A., Fulk, F., Autrey, B.C., Flotemersch, J., 2003. Hyperspectral remote sensing of water quality parameters for large rivers in the Ohio River basin. In: First Interagency Conference on Research in the Watershed, Benson, AZ, pp. 216–221.

Sommer, C., Holzwarth, S., Heiden, U., Heurich, M., Müller, J., Mauser, W., 2016. Feature-based tree species classification using hyperspectral and lidar data in the Bavarian Forest National Park. EARSeL eProc 14, 49–70.

Sonobe, R., Miura, Y., Sano, T., Horie, H., 2018. Estimating leaf carotenoid contents of shade-grown tea using hyperspectral indices and PROSPECT–D inversion. Int. J. Remote Sens. 39, 1306–1320.

Staenz, K., Held, A., 2012. Summary of current and future terrestrial civilian hyperspectral spaceborne systems. In: 2012 IEEE International Geoscience and Remote Sensing Symposium. IEEE, pp. 123–126.

Srivastava, P.K., Gupta, M., Ujjwal Singh, Rajendra Prasad, Prem Chandra Pandey, Raghubanshi, A.S., et al., 2020. Sensitivity analysis of artificial neural network for chlorophyll prediction using hyperspectral data, Environ. Dev. Sus. https://doi.org/10.1007/s10668-020-00827-6.

Sun, L., Khan, S., Shabestari, P., 2019. Integrated hyperspectral and geochemical study of sediment-hosted disseminated gold at the Goldstrike District, Utah. Remote Sens. 11, 1987.

Tanré, D., Deroo, C., Duhaut, P., Herman, M., Morcrette, J., Perbos, J., et al., 1990. Technical note Description of a computer code to simulate the satellite signal in the solar spectrum: the 5S code. Int. J. Remote Sens. 11, 659–668.

Tilling, A.K., O'Leary, G.J., Ferwerda, J.G., Jones, S.D., Fitzgerald, G.J., Rodriguez, D., et al., 2007. Remote sensing of nitrogen and water stress in wheat. Field Crop. Res. 104, 77–85.

Ting-ting, Z., Fei, L., 2012. Application of hyperspectral remote sensing in mineral identification and mapping. In: Proceedings of 2012 2nd International Conference on Computer Science and Network Technology. IEEE, pp. 103–106.

Tratt, D.M., Buckland, K.N., Keim, E.R., Johnson, P.D., 2016. Urban-industrial emissions monitoring with airborne longwave-infrared hyperspectral imaging. In: 2016 8th Workshop on Hyperspectral Image and Signal Processing: Evolution in Remote Sensing (WHISPERS). IEEE, pp. 1–5.

Tuşa, L., Khodadadzadeh, M., Contreras, C., Rafiezadeh Shahi, K., Fuchs, M., Gloaguen, R., et al., 2020. Drill-core mineral abundance estimation using hyperspectral and high-resolution mineralogical data. Remote Sens. 12, 1218.

Vermote, E., Tanre, D., Deuze, J., Herman, M., Morcrette, J., 1994. Second Simulation ofthe Satellite Signal in the Solar Spectrum (6S), 65 User Guide Version 6.0. NASA-GSFC, Greenbelt, Maryland 134.

Verrelst, J., Rivera, J.P., Leonenko, G., Alonso, L., Moreno, J., 2013. Optimizing LUT-based RTM inversion for semiautomatic mapping of crop biophysical parameters from Sentinel-2 and -3 data: role of cost functions. IEEE Trans. Geosci. Remote Sens. 52, 257–269.

Verrelst, J., Malenovský, Z., Van der Tol, C., Camps-Valls, G., Gastellu-Etchegorry, J.-P., Lewis, P., et al., 2019. Quantifying vegetation biophysical variables from imaging spectroscopy data: a review on retrieval methods, Surv. Geophys., 40. pp. 589–629.

Wang, Y., 2012. Regularization for inverse models in remote sensing. Prog. Phys. Geogr. 36, 38–59.

Wang, X., Yang, W., 2019. Water quality monitoring and evaluation using remote-sensing techniques in China: a systematic review. Ecosyst. Health Sustain. 5, 47–56.

Wang, Y., Yang, C., Li, X., 2009. Kernel-based quantitative remote sensing inversion. Kernel Methods for Remote Sensing Data Analysis. Wiley & Sons, p. 271.

Wang, M., Niu, X., Chen, S., Guo, P., Yang, Q., Wang, Z., 2014. Inversion of chlorophyll contents by use of hyperspectral CHRIS data based on radiative transfer model. IOP Conference Series: Earth and Environmental Science. IOP Publishing, p. 012073.

Weber, C., Aguejdad, R., Briottet, X., Avala, J., Fabre, S., Demuynck, J., et al., 2018. Hyperspectral imagery for environmental urban planning. IGARSS 2018–2018 IEEE International Geoscience and Remote Sensing Symposium. IEEE, pp. 1628–1631.

Xavier, A.C., Rudorff, B.F.T., Moreira, M.A., Alvarenga, B.S., Freitas, J.Gd, Salomon, M.V., 2006. Hyperspectral field reflectance measurements to estimate wheat grain yield and plant height. Sci. Agricola 63, 130–138.

Yu, K., Gnyp, M.L., Gao, L., Miao, Y., Chen, X., Bareth, G., 2015. Estimate leaf chlorophyll of rice using reflectance indices and partial least squares. Photogramm. Fernerkundung Geoinform. 2015, 45–54.

Zarco-Tejada, P.J., Miller, J.R., Noland, T.L., Mohammed, G.H., Sampson, P.H., 2001. Scaling-up and model inversion methods with narrowband optical indices for chlorophyll content estimation in closed forest canopies with hyperspectral data. IEEE Trans. Geosci. Remote Sens. 39, 1491–1507.

Zarco-Tejada, P.J., Hornero, A., Beck, P., Kattenborn, T., Kempeneers, P., Hernández-Clemente, R., 2019. Chlorophyll content estimation in an open-canopy conifer forest with Sentinel-2A and hyperspectral imagery in the context of forest decline. Remote Sens. Environ. 223, 320–335.

Zhang, Z., He, G., Yang, H., Long, T., Wang, G., Zhang, X., et al., 2017. Hyper-spectral estimation of forest leaf area index from Earth Observing 1 (EO-1) hyperion imagery based on empirical–statistical approach and grey relational analysis. In: Proceedings of the International Conference on Graphics and Signal Processing, pp. 44–47.

Zhang, Y., Wu, L., Ren, H., Liu, Y., Zheng, Y., Liu, Y., et al., 2020. Mapping water quality parameters in Urban Rivers from hyperspectral images using a new self-adapting selection of multiple artificial neural networks. Remote Sens. 12, 336.

2

Spectral smile correction for airborne imaging spectrometers

K. Koloniatis, V. Andronis, V. Karathanassi

LABORATORY OF REMOTE SENSING, NATIONAL TECHNICAL UNIVERSITY OF ATHENS, ATHENS, GREECE

2.1 Introduction

In the past few years, the demand for high spectral and spatial resolution data from hyperspectral imagers is rising in a wide range of scientific and professional fields. Along with that, the need for high quality reliable hyperspectral data (HIS) is rising as well. To ensure the proper quality of the collected data, a series of corrections should be applied during the preprocessing stage in order to eliminate spectral and spatial distortions in the data. These corrections include radiometric, atmospheric, and geometric corrections, and the application of destriping techniques and spectral resampling for bad pixels.

Radiometric corrections are necessary to generate absolute reflectance and emissivity data, which will lead to reliable relationships with the physical properties of the measured surface. They include sensor calibration, sun angle correction methods, which mainly use the sine of the solar elevation angle, as well as methods for the removal of solar illumination effects, which are caused by topographic slope and aspect (empirical method, the Minnaert method, etc.). Atmospheric correction focuses on the estimation of the radiance leaving the ground. It removes the atmospheric components of the radiance received at the sensor such as aerosol and water vapor contributions. Several software packages such as ATREM, ACORN, FLAASH, ATCOR, etc., have been developed for the atmospheric correction of hyperspectral data; some of them include corrections for haze and bidirectional reflectance as well. Stripe noises are caused by detectors that do not have identical transfer functions. The widely used multichip charge-coupled device butting technology in hyperspectral sensors lead to even more serious stripe noises. Calibration and image statistic methods have been developed for destriping (Duan et al., 2013). Partial missing lines are caused by errors either in sampling or scanning equipment, in transmission or recording of image data, or in the reproduction of the media containing the data. Interpolation methods are used for spectral resampling. Any remotely sensed image suffers from some degree of geometric distortion. These distortions are mostly due to changes in sensor position, Earth rotation, and terrain relief displacement. Geometric corrections aim to remove these errors.

Hyperspectral Remote Sensing. DOI: https://doi.org/10.1016/B978-0-08-102894-0.00003-6

Accurate knowledge of the spectral calibration of imaging spectrometers is required for optimum data processing and interpretation. Spectral calibration requires a comparison between the measuring instrument and an "absolute" reference standard of known accuracy. It provides information about the central wavelength of the channels and the corresponding spectral response functions of the instruments. However, the calibration information of the nominal spectral characteristics provided by the manufacturer may be different from true spectral characteristics of the instrument due to the aging of the instrument components and misalignments between the detector array and the instrument slit due to mechanical disturbances and environmental conditions during operation (Guanter et al., 2006). To address this issue, regular spectral recalibration is needed. Several times this may not be possible due to the high cost or the time needed, while for spaceborne hyperspectral imagers this may be physically impossible.

Hence scene-based spectral characterization of imaging spectrometers is frequently necessary to update or replace the preflight laboratory-based spectral characterization supplied by the data provider and to remove spectral and spatial misregistrations such as spectral smile and keystone effect. Spectral smile or the spectral curvature effect is a spectral distortion that commonly affects hyperspectral data collected from pushbroom hyperspectral images due to misalignments at the detector array of the imager. Specifically, it is a shift of the central wavelength in the spectral domain and can be described as a function of the cross-track column number (Yokoya et al., 2010). It appears as a cross-track brightness variation at the first eigenvalue of the minimum noise fraction (MNF) transformation (Green et al., 1988), and in extreme cases, the brightness variation is visible even in the original image. Spectral smile can lead to erroneous results after implementation of preprocessing and processing methods such as negatively affecting the results of the atmospheric correction (Saenz et al., 2002), reducing the accuracy of the classification results (Dadon et al., 2010), and/or compromising the results of vegetation indexes like normalized difference vegetation index or other narrow-band vegetation indexes (Aktaruzzaman, 2008; Meroni et al., 2010). Therefore the development of reliable scene-based correction methods is needed.

Most spectral smile correction methods so far consist of two discrete stages, namely the detection and quantification stage and the implementation of the correction stage (Aktaruzzaman, 2008; Busetto et al., 2011; Dadon et al., 2010; Guanter et al., 2009; Richter et al., 2011; Yokoya et al., 2010). Generally, for the detection and quantification of the spectral smile, most approaches exploit the characteristics of the atmospheric absorption features and mainly that of O_2-A for sensors operated in visible and near infrared (VNIR) because it is expected that oxygen would contribute equally to all pixels of the image in the O_2-A absorption bands if the spectral smile did not affect the image significantly. Comparison of the measured spectral data in the matching bands with simulated at-sensor radiance data derived from a radiative transfer code (e.g., MODTRAN4, 6S) leads to smile quantification. Although the smile is calculated at one single wavelength in which the highest accuracy is expected, it is assumed to describe the complete spectral range covered by the sensor.

Aktaruzzaman (2008) proposed a scene-based smile correction method. In that approach, in order to accurately detect and quantify the spectral smile, the average column spectra of the bands corresponding to the O_2-A absorption feature in a homogeneous area of a Hyperion image were calculated and compared to a set of reference spectra with known shifts that were calculated using a MODTRAN4 radiative transfer code. The minimum spectral angle pointed to the particular reference spectrum, which when compared to image data, will provide information about the amount of shift per column of the image. The correction of spectral smile was implemented by calculating new resampled spectra using the spectral response functions of the instrument, which were shifted in the reverse direction of the corresponding smile shift for each column.

Another interesting method for a scene-based spectral characterization was presented by Guanter et al. (2009). This method exploits the characteristics of the O_2-A absorption feature in order to retrieve the central wavelength shift and the band broadening for every column of the image. The average at-sensor radiance value of each column is calculated and the corresponding surface reflectance values are calculated after atmospheric correction using the nominal spectral calibration of the instrument. It was assumed that after atmospheric correction, the surface reflectance of natural targets is expected to be smooth and any spikes or dips in atmospheric absorption bands are due to spectral artifacts caused by spectral shift and band broadening. In order to properly quantify these artifacts, the surface reflectance spectra for each column is smoothed using a low pass filter and interpolation techniques. Then the smoothed reflectance spectrum is taken as a reference for the calculation of spectral shift and channel broadening. Calculation employs the minimization of the merit function.

Based on this method, Richter et al. (2011) parameterized the spectral smile shift across the detector array taking into consideration the instrument optical design model or measured laboratory data. They proposed an offline approach to derive the fourth-order polynomial smile fit coefficients for appropriate channels in atmospheric absorption regions. These coefficients describe the spectral smile shift of the spectral response functions for the bands of the atmospheric absorption features. The coefficients are then interpolated and optionally extrapolated to all other bands within the same detector or spectrometer unit. As with Guanter et al. (2009), the proposed method uses calibrated image data, column averaging of the image, calculation of a set of spectrally shifted reference radiances, and an optimization procedure for calculating the spectral shift. In this method, the highest Pearson's correlation coefficient was used.

Yokoya et al. (2010) employed subpixel image registration methods to detect smile and keystone properties. Smile is detected by estimating the distortion of the atmospheric absorption line in the spectrum image, and keystone is detected by estimating the band-to-band misregistration of all subscenes. They applied two subpixel estimation methods, namely the normalized cross-correlation (NCC) method and the phase correlation method. When a scene is assumed to be physically homogeneous at each along-track position, the subpixel image registration method based on applying the NCC to this line in the cross-track direction enables the detection of spectral distortion as a function of the cross-track number. After the detection of the smile and keystone properties, cubic spline interpolation is used in order to correct the spectral signatures from the distortions.

Another method, the trend line smile correction (TLSC) method, was proposed by Dadon et al. (2010) for the correction of the smile distortions in Hyperion images. Their approach assumes that, after an MNF rotation, the first eigenimage (MNF-1) embodies the cross-track brightness gradient of all spectral channels that are affected by the smile effect. The eigenvalue of the MNF-smile can be then considered as a measure of the artifact energy due to its relation to the signal to noise ratio. In order to detect and quantify the spectral smile at the images, an indicator derived from the O_2-A absorption feature is derived. Specifically, the average derivative resulting from the corresponding bands of the right shoulder of the O_2-A absorption feature is calculated for every column. The resulting graph of the average derivative values is used as an internal indicator in order to quantify the spectral smile distortions at the image. In addition, using in situ measurements and a MODTRAN4 radiative transfer model, the at-sensor reflectance values at the absorption feature are calculated and compared to image reflectance for retrieving the column, which is least affected by the smile distortions (minimum scale column; MSC). Afterwards, a trend line (TL) is fitted to the graph of the derivative values. This TL is moved in order to set the MSC derivative value equal to zero and then is scaled to the range of the MNF-1 values. Every pixel of the MNF-1 is corrected according to its distance from the MSC. The corrected image is obtained through inverse MNF transform.

The assumption of the TLSC method about the MNF-1 band provides fast and efficient processing. However, in situ measurements are often unavailable, and the conversion of radiance to at-sensor reflectance values requires the MODTRAN code, which calculates atmospheric transmittance and radiance for frequencies from 0 to 50,000 cm^{-1} (wavelength of 200 nm to ∞) at moderate spectral resolution, primarily 2 cm^{-1} (20 cm^{-1} in UV) (Xu et al., 2008). Furthermore, reflectance values are used for MSC retrieval, although the method is applied on radiance data. In this study, in order to overcome the mentioned drawbacks, a modified TLSC method is proposed. The proposed scene-based spectral characterization method uses atmospheric absorption features to retrieve information on the spectral nonuniformity of high spectral resolution imaging spectrometers and applies two criteria on the graph of the average derivative values in order to determine the MSC. Considerable observations are made about the impact of in-flight illumination variations on the spectral smile indicators. Finally, conclusions have been made on the impact of land cover categories on the corrected images.

2.2 Illumination effects on the spectral smile of airborne hyperspectral images

To collect a hyperspectral image (HSI) data cube, pushbroom sensors are commonly used. Pushbroom technology uses a two-dimensional detector; the first direction simultaneously collects intensity values for pixels located in the across-track direction, whereas the second dimension of the detector simultaneously collects for these pixels all the spectral information. Scanning in the spatial dimension along the direction of the motion of the platform is performed through the motion of the platform itself (Fig. 2–1).

The two-dimensional sensor frame has imperfections, which are mainly formed by optical aberrations and sampling inconsistencies in the spectral and the first spatial domain. The

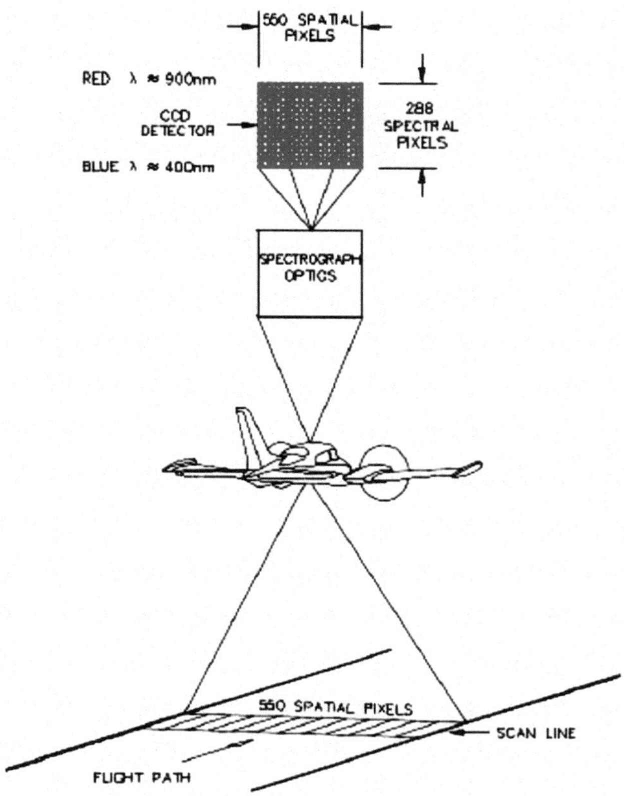

FIGURE 2–1 The pushbroom Casi-550 features. From *Instrument Manual ITRES CASI-550.*

result is a nonuniformity of the point spread function (PSF) through spectral and spatial misregistrations, which correspond to smile and keystone (Nieke et al., 2008). Thus an attempt has been made to minimize PSF variation, or equivalently, to minimize the variation of the spatial response function (SRF) of the detector. Since it is impossible to produce identical PSFs throughout wavelength and field, the ideal spectrum produced by a pushbroom imaging sensor will present PSF variations in the spectral and spatial directions with several characteristics (Mouroulis, 1999); namely all image points will be inside perfect rectangles and aligned with the photodetector elements, and the width of the PSF for any wavelength will be constant, independent of field (Fig. 2–2A). For the shorter and longer wavelengths, all the energy ends up in the corresponding rectangle, whereas for the middle wavelengths, part of the energy ends up in the neighboring rectangle (Høye et al., 2015). The height of the PSF is constant, independent of wavelength and of field (Mouroulis, 1999).

However, in a realistic scenario, spectrum presents certain defects (Fig. 2–2B); namely, along the $\lambda 1$ and $\lambda 5$ columns, the PSF centroids are not aligned, which is smile. The top spectrum is not aligned with a row, although the middle one is, which is keystone. In Fig. 2–2B, columns $\lambda 2$ and $\lambda 4$ show variation in the width of the PSF with field location; this

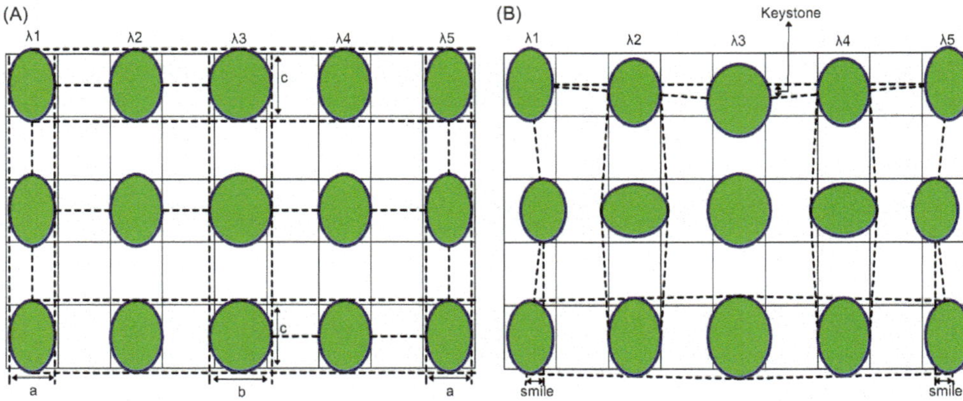

FIGURE 2–2 (A) The ideal spectrum produced by a pushbroom imaging sensor and (B) the effects of smile, keystone, and corresponding point spread function distortions on the spectrum.

causes the width of the SRF of a pixel to vary with field, and produces an effect similar to that of smile. Finally, the bottom row shows a variation of the PSF height with wavelength; this variation has an effect similar to keystone (Mouroulis, 1999). Hence spatial misregistration includes the effects of both keystone and PSF variations and spectral misregistration is caused by smile and corresponding SRF variations.

The spectral smile can be approximated by a quadratic function with one inflection point. The value of the function is the spectral information, which varies according to the location of the pixel in the across-track direction. The inflection point can be placed either within the region defined by the first and last pixel in the across-track direction or out of this region (Hong et al., 2017). Smile can be described as the maximum spectral shift between the inflection point and the pixels in the across-track direction. For Hyperion imagery, for example, with 10 nm spectral resolution, the spectral smile has been reported to vary from 2.6 to 3.6 nm in VNIR and ~1 nm in the short-wave infrared (SWIR) range (Oskouei and Babakan, 2016). It is evident that, the higher the spectral resolution of the HSI, the higher the smile effect on the image. Nevertheless, even for HSI with high spectral resolution, smile is not obvious in a single band. Thus two indicators are used for smile detection and quantification, namely the brightness gradient at the first MNF-1 band indicator, and the brightness gradient of the mean column values of the image, which results from the subtraction of the band affected from an absorption feature of gases (O_2 or CO_2) and a neighbor band around this absorption feature (*D*-values indicator). Typical images of smile indicators are shown in Fig. 2–3. Brightness MNF-1 variations and the across-track profile of the subtraction image vary proportionally to smile effects. These variations are used for smile quantification and postcalibration of the sensor.

However, experiments conducted within this study showed that external factors can also influence the smile indicators. Fig. 2–4A shows four possible orientations of the air-platform during the capture of HSI data, namely (1) South–North, (2) North–South, (3) West–East, and (4) East–West. The position of the hyperspectral sensor is marked in red. Yellow areas are those that are illuminated by the sun, while gray areas are shaded areas. It is evident that the

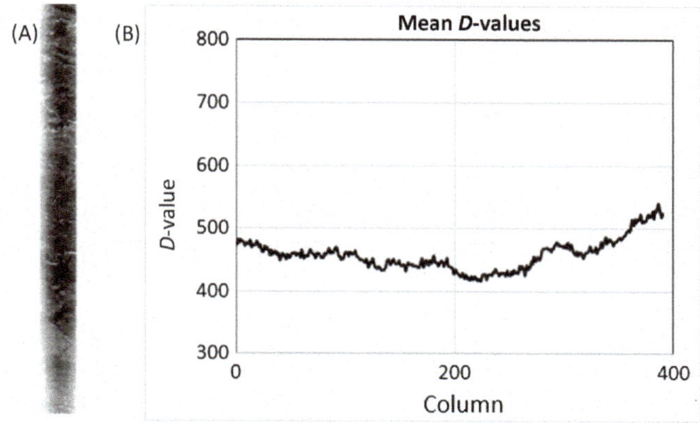

FIGURE 2–3 (A) The MNF-1 indicator and (B) the *D*-value indicator.

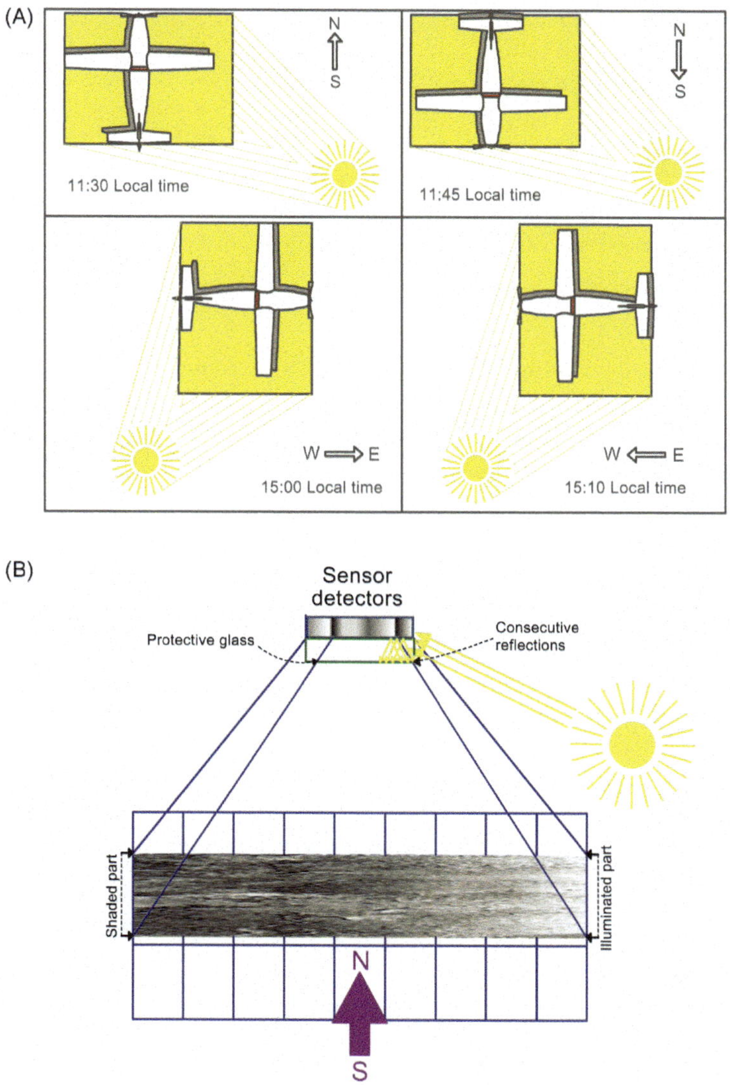

FIGURE 2–4 (A) Examples of sun illumination effects for different air-platform orientations and times. (B) Nonuniform illumination of the protected glass and its impact on image quality.

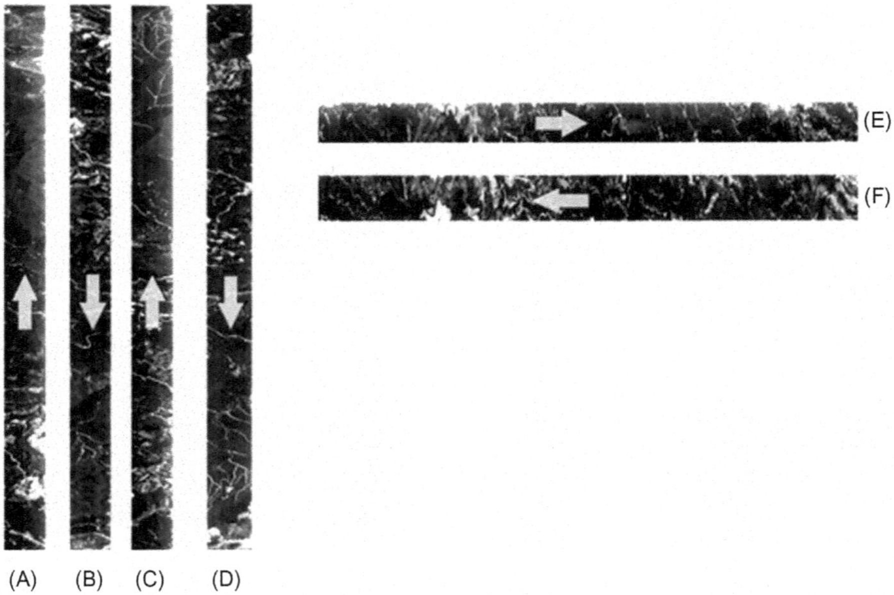

FIGURE 2–5 Airborne hyperspectral images. The arrows indicate the flight direction of the platform.

underside of the fuselage is not uniformly illuminated and depends on the platform orientation. As can be observed in the acquired HSIs, shown in Fig. 2–5, illumination is more uniform in the cases of (2) South–North and (4) East–West orientations, whilst this does not occur in the cases of (1) North–South and (3) West–East orientations. Since these radiometric distortions were systematic in all the acquired images, they must be the result of a diffraction effect caused by exposure of the protective glass of the aircraft under the sensor to sunlight (Fig. 2–4B). This diffraction effect seems to take place only when the sun is located at the rear of the air-platform [cases (1) and (3)]. In the cases (2) and (4), most probably the glass is shaded by the fuselage or the wings, thus, preventing the diffraction effect. Based on this, it was concluded that for this specific installation of the sensor, the image acquisition should be carried out in racetrack pattern with the air-platform flying toward the sun in order to reduce illumination variations of the incoming radiation.

Effects on the recorded radiance are obvious in the airborne HSIs (images A,B,C,D,E, and F). Images (a), (c), and (e) with flight direction South–North (a and c) and West–East (e) present a brightness gradient in the across-track direction (Fig. 2–5).

The graph of the mean column values for the 763.7 nm band in the across-track direction for the O_2-A absorption feature of image (c) shows the brightness variation of this image caused by the specific illumination conditions during the image capture (Fig. 2–6B). On the contrary, this is not observed in the graph of image (f) (Fig. 2–6A).

The *D*-value indicators of spectral smile for images (f) and (c) are shown in Fig. 2–7.

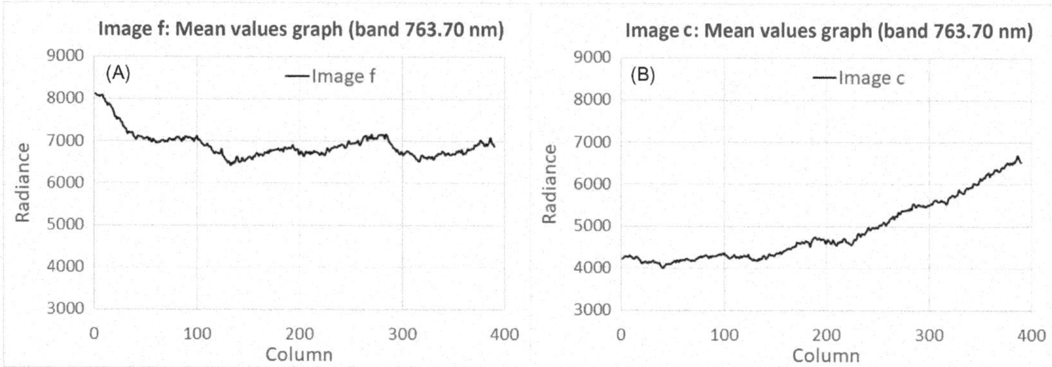

FIGURE 2–6 Illumination effects on the radiance of (A) image f and (B) image c.

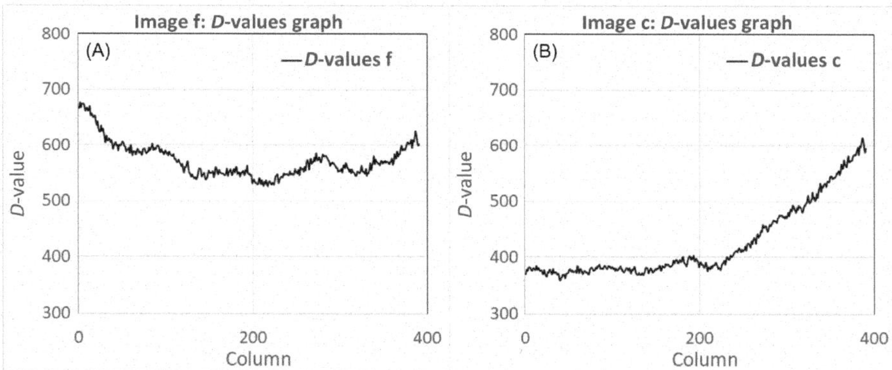

FIGURE 2–7 The *D*-value indicator (A) for image f and (B) for image c.

2.3 The modified trend line smile correction method

This method includes two main stages, namely the detection and quantification stage and the correction stage. During the first stage, the amount of spectral smile effect in the image is detected and quantified. For that purpose, the *D*-values indicator is calculated for every pixel of the radiance image using the values of the bands near the O_2-A absorption feature (around 760 nm). The *D*-values represent the right-shoulder slope between the respective bands for every pixel (Fig. 2–8). Assuming that the gases are well-mixed and relatively evenly distributed across the image, then the slope between the O_2 absorption band and the following band should be relatively the same for every pixel. Specifically, the *D*-values are calculated using Eq. (2.1):

$$D\text{-values} = \frac{\text{B2} - \text{B1}}{\overline{\text{FWHM}}} \quad (2.1)$$

where B2 stands for the band at the right of the O_2 absorption feature, B1 for the nearest band at the absorption feature, and $\overline{\text{FWHM}}$ for the mean full width at half maximum (FWHM) of the two bands.

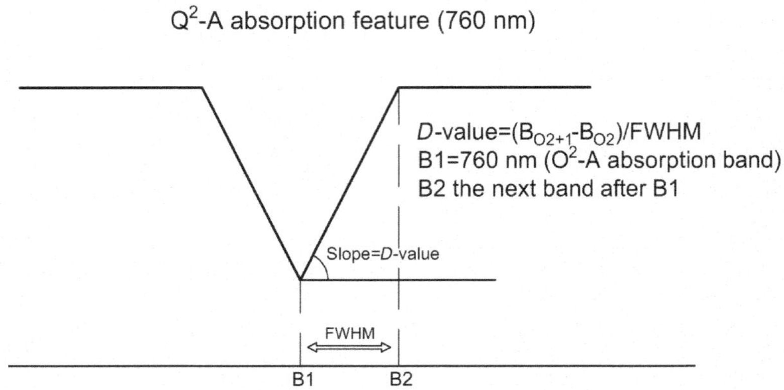

FIGURE 2–8 The *D*-values indicator at O_2-A absorption feature.

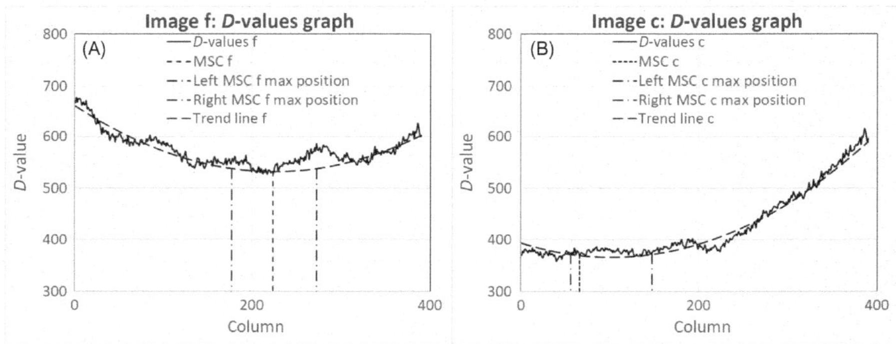

FIGURE 2–9 Definition of the minimum scale column (A) for image f, which is affected by spectral smile, and (B) for image c, which is affected by spectral smile and across-track illumination variation.

In order to locate the column that is the least affected by the smile effect (MSC), first a second order polynomial TL is fitted to the mean column *D*-value. Then two criteria are applied for the MSC retrieval:

1. The MSC should be located near the inflection point of the smile curve (Fig. 2–9). The spectral smile is described approximately by the TL function. By considering only the TL values, it is assumed that the inflection point of the TL, the TL_{infl}, is the least affected column by the smile, and the columns at the edges are those affected the most. Then the shift, ΔS_x, for every column x can be expressed as:

$$\Delta S_x = |TL_x - TL_{infl}| \tag{2.2}$$

Usually ΔS_x is maximized for one of the edge columns. For every column x, the ratio D_x of its shift to the maximum shift, ΔS_{max}, is calculated using Eq. (2.3).

$$D_x = \frac{\Delta S_x}{\Delta S_{max}} \tag{2.3}$$

A maximum ratio value of 1% was chosen in order to determine the maximum acceptable distance from the MSC to the inflection point. By replacing this value in equation Eq. (2.3), the maximum left and right position from the inflection point are defined on the TL (Eq. 2.4). Only columns with TL values less than TL + ΔS_{xmd} are MSC candidates.

$$\Delta S_{xmd} = \frac{\Delta S_{max}}{100} \tag{2.4}$$

2. At the interval defined, MSC is the column for which the TL and the graph of the mean column *D*-value are almost perfectly matching each other (Fig. 2–9). This criterion is important in order not to choose a column that is greatly affected by other parameters as MSC.

During the second stage, the graph and the TL are vertically shifted in order to assign the value of zero to $TL_{(MSC)}$ and then the correction values for the image are calculated. To achieve that, the TL function calculated earlier is scaled to the range of the MNF-1 mean column values. Then, the correction is implemented at the MNF-1 pixel values through the equation:

$$MNF - 1_{TLSC} = \Sigma(X_{MNF} + (-TL_{CX}) \pm \Delta MSC) \tag{2.5}$$

where MNF-1_{TLSC} is the corrected MNF-1 pixel value, X_{MNF} is the original pixel value of the MNF-1, TL_{CX} is the column value of the scaled TL for the column that the X_{MNF} belongs to, and ΔMSC is the difference between the value of the TL at the MSC and the *D*-value of the MSC, both scaled to the MNF-1 value range.

Finally, the corrected radiance image is reconstructed. Specifically, the corrected MNF-1 and the rest of the original MNF eigenimages are used in the inverse MNF transformation procedure for reconstructing the corrected radiance image. Fig. 2–10 shows a flowchart of the modified TLSC method.

2.4 Implementation and results

2.4.1 Data

For the study, six airborne HSI were captured using a CASI-550 hyperspectral imager, over Taxiarchis forest, Chalkidiki, Greece (Fig. 2–5). The area presents rough relief, whereas the sky was clear during the air campaign. The images were captured in a zigzag pattern with different aircraft orientations (South–North, North–South, West–East, and East–West) because one of the initial objectives of this campaign was the study of the bidirectional

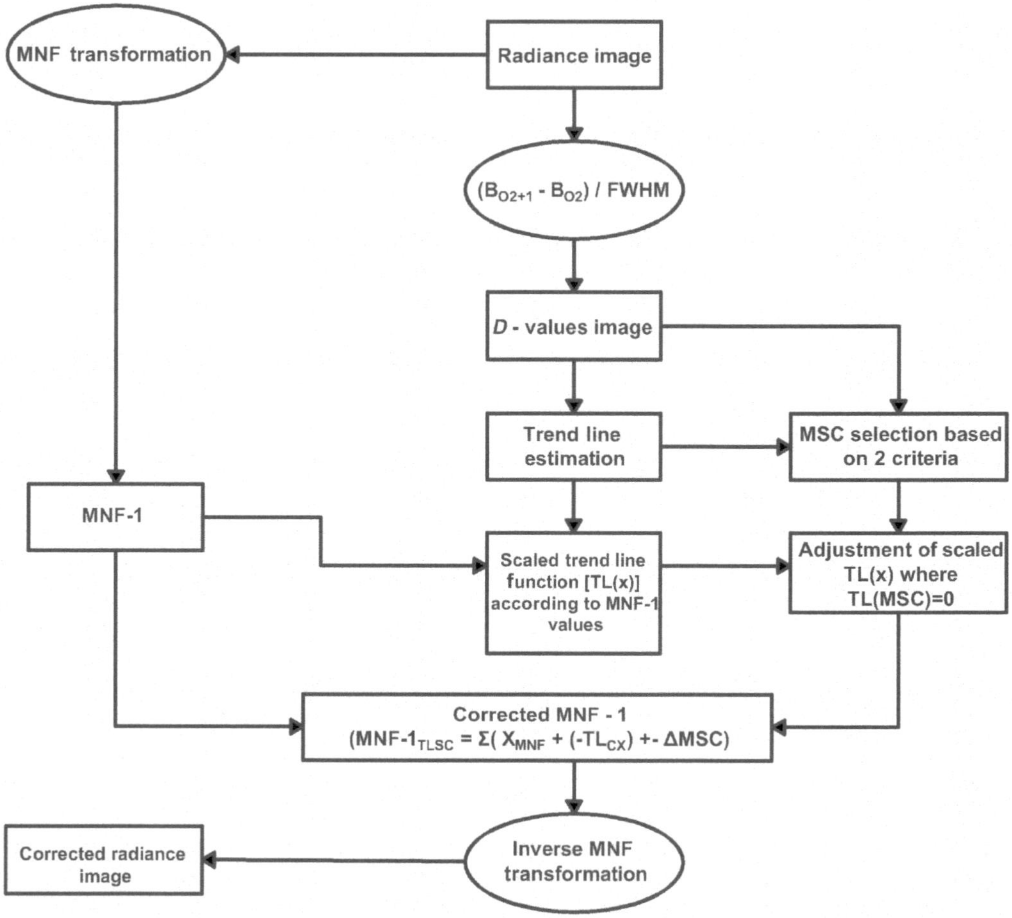

FIGURE 2–10 Flowchart of the modified trend line smile correction method. *MNF*, Minimum noise fraction.

reflectance for the various vegetation types. Images with South–North and North–South directions were obtained in the morning (approximately at 11:30 a.m.), whereas images with West–East and East–West directions in the afternoon (approximately at 15:00 p.m.) since they were the last images among the 17 images in total, that were captured within this campaign. All images have 72 spectral bands with 2 m spatial resolution and 3.8 nm spectral resolution. Spectral bands cover the VNIR spectrum from 421.8 to 975.7 nm. Each image presents an area with a 10 km length in the along-track direction and an 810 m swath in the across-track direction. Illumination gradient caused by platform orientation is observed in images (a), (c), and (e) (Fig. 2–5). The first MNF eigenvector (MNF-1) images present various radiometric distortions mainly caused by spectral smile and diffraction effects during the air campaign (Fig. 2–11).

FIGURE 2–11 The MNF-1 images present radiometric distortions.

2.4.2 Implementation and results

The modified TLCS method was implemented using ENVI software, some simple ANSI C in-house developed programs, and spreadsheet tools. The correction process was divided into two main stages; the first being the quantification of the radiometric distortions and the second being the correction stage.

The first step of the smile quantification process is the calculation of the *D*-values (derivative values) for every pixel of images (a), (b), (c), (d), (e), and (f) using Eq. (2.1):

$$D\text{-values} = \frac{\text{B2} - \text{B1}}{\overline{\text{FWHM}}}$$

where $\overline{\text{FWHM}}$ is 3.8 nm, B2 is band 47 (771.3 nm), and B1 is band 46 (763.7 nm), which is the nearest band to the oxygen absorption feature.

For every image, a *D*-value image has been created. In the second step, the calculation of the mean value for every column of the *D*-value images and the respective graphs, that is, the *D*-values indicator, was implemented (Fig. 2–12).

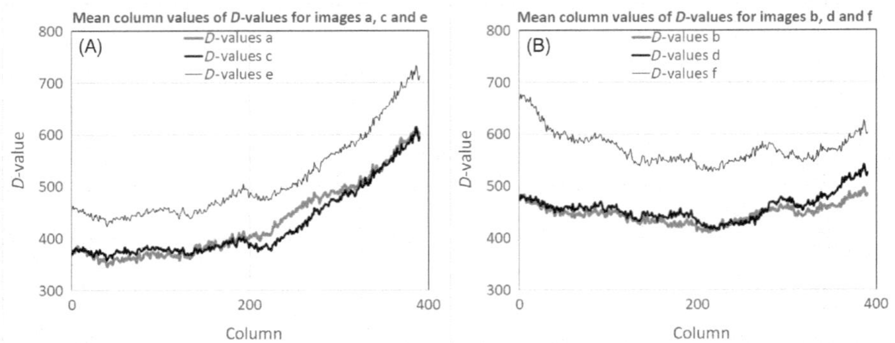

FIGURE 2–12 The *D*-values indicator (A) for images a, c, and e and (B) for images b, d, and f. As it is expected, the correlation between the shape of the *D*-values indicator and the flight direction is obvious. Images (a) and (c) that were obtained with a South–North flight direction present similar graphs to image (e), which was obtained with a West–East flight direction. The impact of both spectral smile and illumination conditions (protective glass diffraction effect) are observed in this indicator. On the contrary, less radiometric distortions are observed in images (b), (d), and (f). Only spectral smile seems to be the cause of the radiometric distortions in these images.

Table 2–1 MSC retrieval.

Image	Potential columns for MSC location	MSC
a	8–112	60
b	133–240	200
c	56–148	66
d	129–223	216
e	39–137	67
f	177–272	223

MSC, Minimum scale column.

The third step of the quantification process was the fitting of a TL to the *D*-value graphs. This step is important because the correction values for each column are calculated based on the TL values. The TLs were calculated as second order polynomials, which fitted the data well. However, higher order polynomials could also be used for more flexibility if more complex patterns were observed in the graphs, unrelated to the radiometry of the images.

The last step of this stage was the retrieval of the MSC. For every image, the two criteria explained in the previous chapter were applied in order to determine the MSC. Potential columns for the MSC location and MSC are presented in Table 2–1.

For images (a), (c), and (e), the MSC interval positions along the column axis are quite similar. The same observation can be made for the MSC location of images (b), (d), and (f). Using the MSC, the TL and the *D*-values indicator radiometric distortion are quantified for every image. The result is a correction value for every column equal to the variance of the respective TL value from the TL value of the MSC (Fig. 2–13).

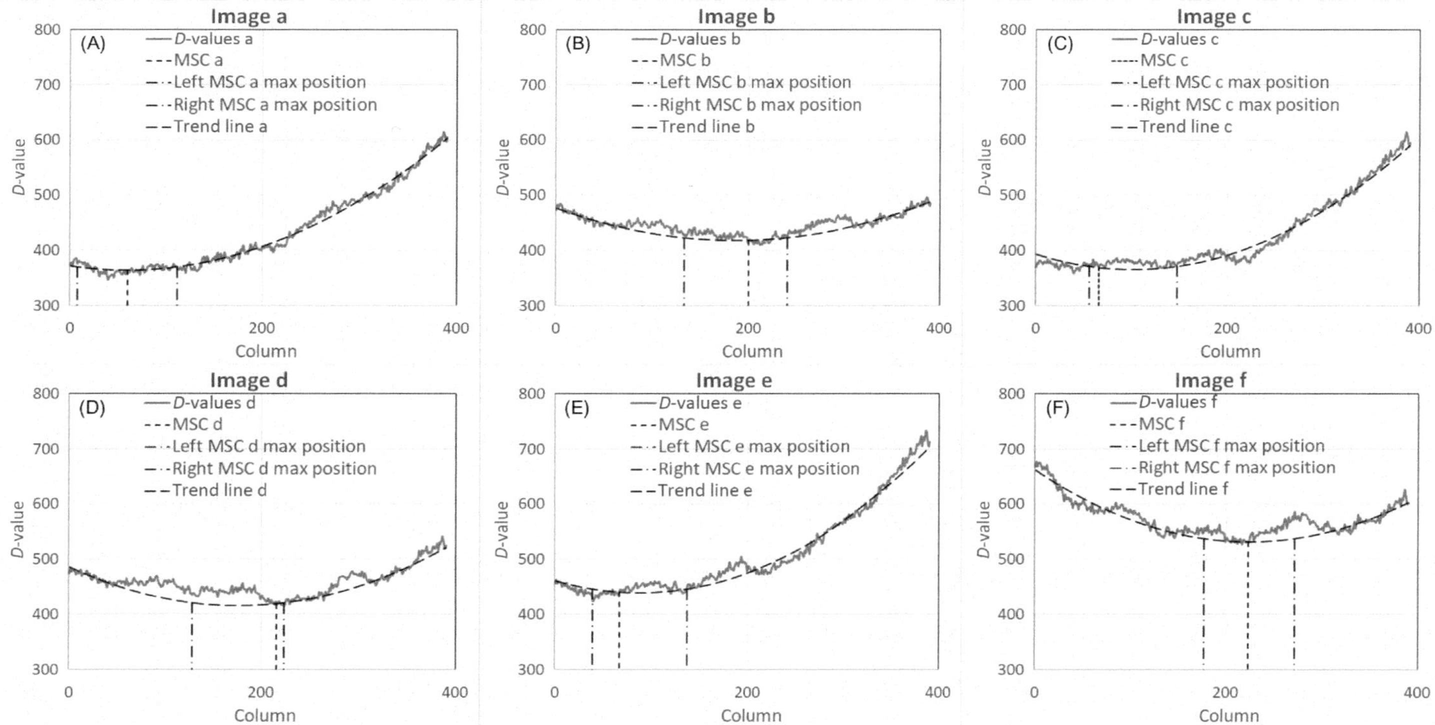

FIGURE 2–13 The retrieval of the minimum scale column.

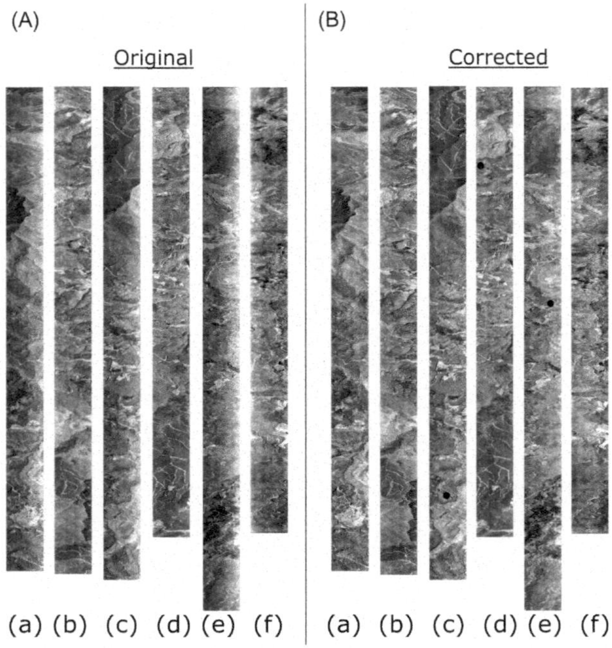

FIGURE 2–14 The original (A) and the corrected images (B). Black points present the pixel, which was used for extracting soil spectral signature.

The first step of the correction process was the vertical shift of the graphs and TL in order to assign the value of zero to the $TL_{(MSC)}$. The next step was the forward MNF transformation of the original images and the calculation of the mean value for each column of the MNF-1 images. Then, the TL values were scaled to the range of the corresponding MNF-1 mean values and a correction value was calculated for each column. For each pixel, the correction value was inserted into the MNF-1 pixel value using Eq. (2.5). Then, the corrected radiance images were reconstructed through the inverse MNF transformation (Fig. 2–14).

2.5 Evaluation and discussion

After the implementation of the modified TLSC method, spectral smile effect and radiometric variations caused by the illumination conditions during the flight were eliminated in the radiance images (Fig. 2–14). Images (a) (South–North direction), (c) (South–North direction), and (e) (West–East direction), which were characterized by nonuniform brightness in the across-track direction, do not exhibit such brightness gradient after the correction. Characteristic parts of the radiance images (a) and (c) before and after the correction for the band centered at 763.70 nm are shown in Figs. 2–15 and 2–16.

The mean column value graphs for the same band for images (a) and (c) are shown in Fig. 2–17. It can be observed that corrected image (a) presents a brightness slope after the

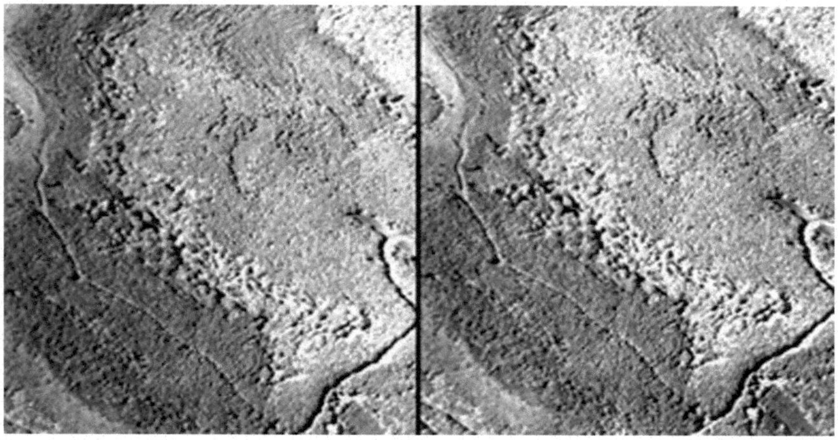

FIGURE 2–15 Image a before (left) and after the proposed correction (right). Band 763.70 nm.

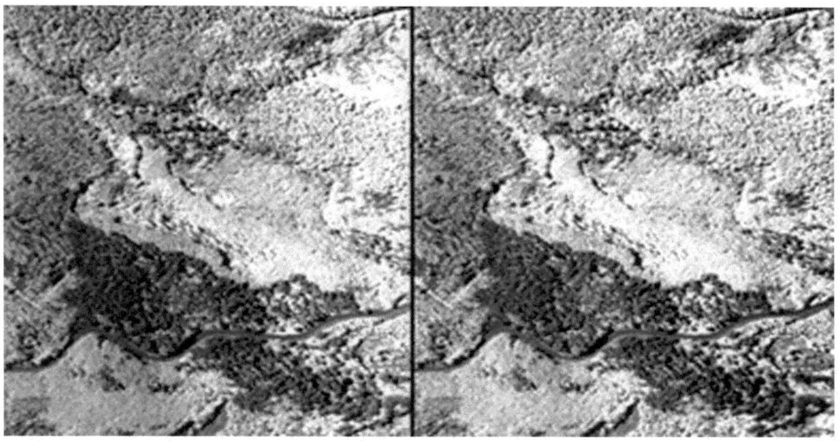

FIGURE 2–16 Image c before (left) and after the proposed correction (right). Band 763.70 nm.

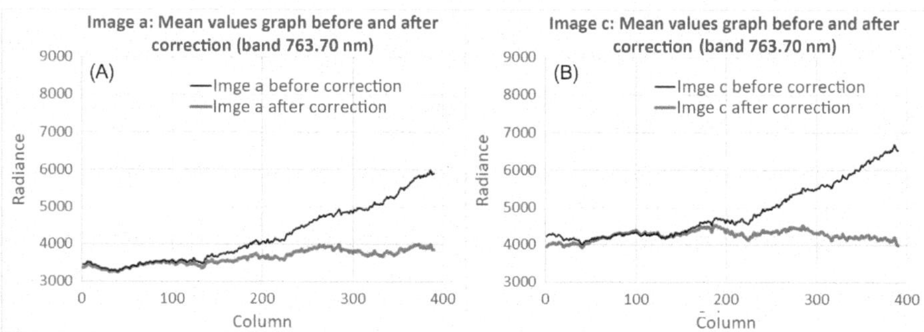

FIGURE 2–17 Mean column values for (A) image a and (B) image c.

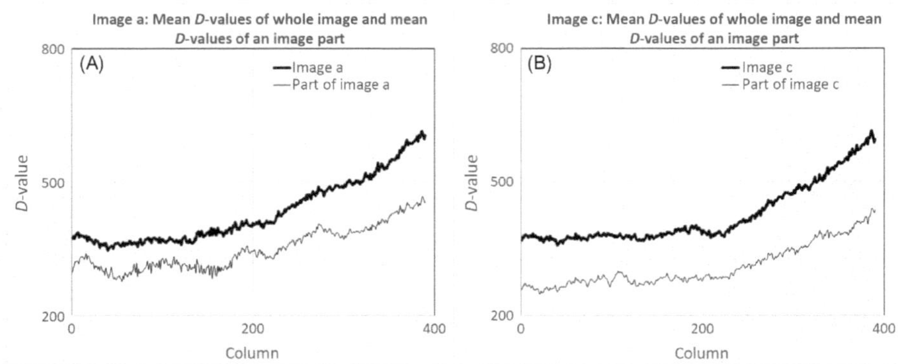

FIGURE 2–18 The *D*-values indicator for (A) the entire and a part of image a and (B) image c.

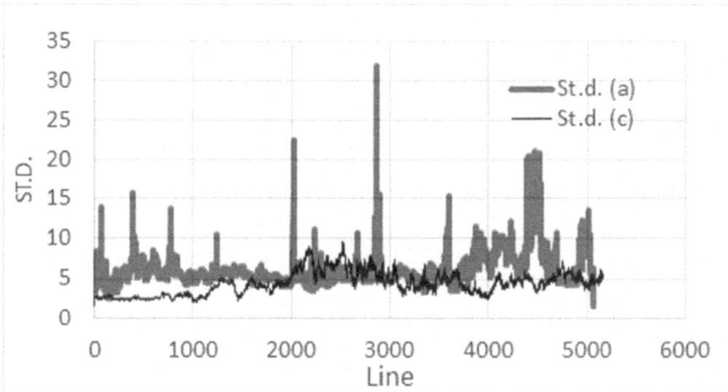

FIGURE 2–19 Standard deviation values for the lines of the MNF-1 for images a and c.

correction possibly due to variances of the land cover and the respective radiative energy rather than to radiometric distortions.

The wide difference in the radiance values of the various land cover categories for image (a) affects the *D*-values indicator. This is shown in Fig. 2–18A, where mean *D*-values are calculated for the entire image (a) as well as for a smaller part of the same image, which depicts intense land cover variances. Although the *D*-values indicator is assumed to be mainly affected by spectral smile distortions and illumination variances, this only occurs when a homogeneous area is presented in the image. Otherwise, the indicator is also influenced by land cover and presents irregularities (Fig. 2–18A, light line).

Furthermore, MNF-1 values are also influenced by radiance variances with consequences on the scaling factor, which is applied on the TL values before the calculation of the correction value of each column. In Fig. 2–19, the standard deviation for each line of the MNF-1 image is shown. The presence of high standard deviation values in some lines implies nonuniform land cover.

For comparison purposes, the *D*-values indicator is also calculated for part of image (c) (Fig. 2–18B). Although this image suffers from similar radiometric distortions with image

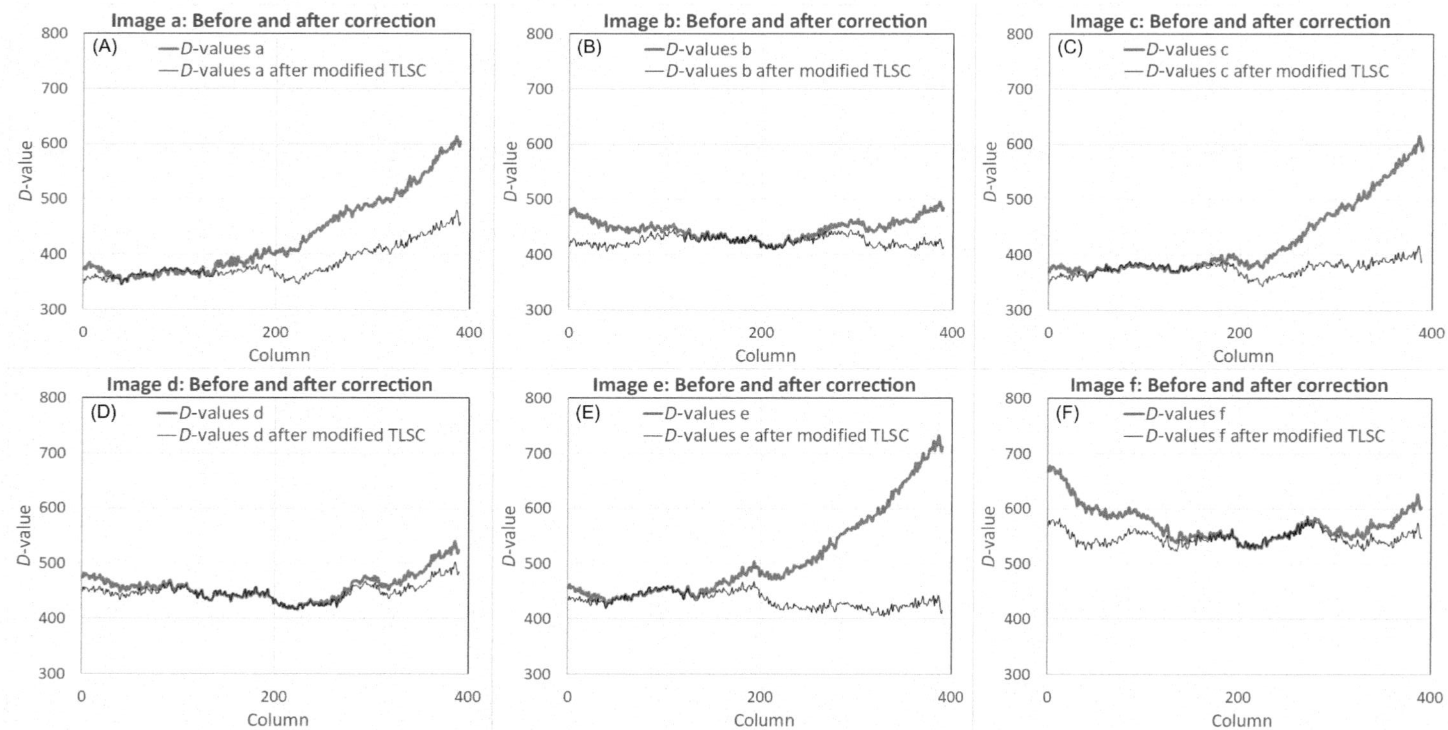

FIGURE 2–20 The *D*-values indicator before and after the correction. Graphs (A–F) correspond to the respective images. *TLSC*, Trend line smile correction.

Table 2–2 Standard deviation of mean *D*-values before and after correction.

Image	Standard deviation of mean *D*-values before correction	Standard deviation of mean *D*-values after correction	% Reduction
a	72.8	30.8	57.69
b	17.7	8.7	50.57
c	67.5	12.6	81.37
d	24.8	15.8	36.28
e	80.4	13.6	80.29
f	31.5	12.8	59.37

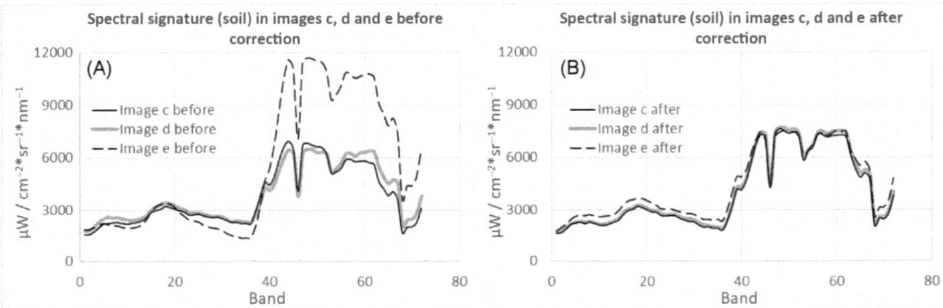

FIGURE 2–21 (A) Spectral signature of soil derived from images c, d, and e before the correction. (B) Spectral signature of soil derived from images c, d, and e after the correction.

(a), due to the fact that land cover is more uniform in the depicted area, the *D*-values indicator does not present any irregularities in the across-track direction (Fig. 2–18B, light line). Moreover, standard deviations for this image present similar values for each line (Fig. 2–19).

By comparing the *D*-values indicator for all the images before and after the implementation of the correction method (Fig. 2–20), it is safe to conclude that the radiometry of the images did not alter. Furthermore, smile curve as well as radiometric distortions have been significantly reduced in every image. This is further demonstrated by the reduction of the standard deviation of the mean *D*-values (Table 2–2). However, the *D*-values indicator still presents a slope after the correction for image (a), which presents various land cover categories and consequently variances in the radiative energy along the across-track direction. To address this issue, unsupervised classification and masking of images prior to spectral smile correction is advised. Then correction could be carried out differentially for each land cover category. However, further investigation is needed in order to build a more robust methodology.

Graphs in Fig. 2–21 show the spectral signatures for bear soil derived by the radiance values of a pixel (black points in Fig. 2–14), for which the corresponding ground area was depicted in the overlapping parts of images (c), (d), and (e). Spectral signatures before and after the correction are presented. Although the images have been captured under relatively the same atmospheric conditions, radiometric variations are observed before the correction, especially for image (e), caused by variations of the illumination at the sensor position. After spectral smile correction, spectral signatures almost coincide.

2.6 Conclusion

In this study, a modified TLSC method for spectral smile correction in airborne HSIs was proposed. For this scope, the spectral smile indicators, and especially that of *D*-values, were thoroughly examined. It was found that three factors affect spectral smile indicators, namely lack of sensor calibration, radiometric distortions caused by illumination variances in the across-track direction, and land cover variations in the across-track direction.

A method based on the TLSC method has been developed for the correction of spectral smile misregistrations. The method is applied on radiance values and does not require the use of any radiative transfer model. Instead, it applies two criteria on the TL for finding the MSC in the graph of the *D*-values indicator. Furthermore, it employs the MNF-1 eigenimage for the spectral smile correction. The method yielded satisfactory results for all the images. After the correction, spectral smile effects and radiometric distortions were effectively eliminated. However, strong radiance variances resulting from various land cover types affected the quality of the corrected images. As a result, in images with intense land cover changes, the mean column values diagram and *D*-values indicator still present a slope after the correction. This drawback is more pronounced in airborne images with high spatial resolution and can be overpassed by differentially applying the proposed methodology over homogeneous areas. Land cover unsupervised classification and image masks prior to spectral smile correction could contribute in this direction. However, further investigation is needed in order to build a more robust smile correction methodology.

List of abbreviations

FWHM full width at half maximum
HSI hyperspectral image
MNF minimum noise fraction
MSC minimum scale column
NCC normalized cross-correlation
PSF point spread function
SRF spatial response function
TL trend line
TLSC trend line smile correction

References

Aktaruzzaman, Md., 2008. Simulation and Correction of Spectral Smile Effect and Its Influence on Hyperspectral Mapping (M.Sc. thesis). International Institute for Geo-Information Science and Earth Observation, Enschede, The Netherlands. Available from: <https://webapps.itc.utwente.nl/librarywww/papers_2008/msc/gem/aktaruzzaman.pdf>.

Busetto, L., Meroni, M., Crosta, G.F., Guanter, L., Colombo, R., 2011. SpecCal: novel software for in-field spectral characterization of high-resolution spectrometers. Comput. Geosci. 37, 1685–1691. Available from: https://doi.org/10.1016/j.cageo.2010.12.005.

Dadon, A., Ben-Dor, E., Karnieli, A., 2010. Use of derivative calculations and minimum noise fraction transform for detecting and correcting the spectral curvature effect (smile) in Hyperion images. IEEE Trans. Geosci. Remote Sens. 48 (6), 2603–2612. Available from: https://doi.org/10.1109/TGRS.2010.2040391.

Duan, Y., Yan, L., Jing, X., 2013. A novel method of destriping for airborne hyperspectral image. In: 2013 IEEE International Geoscience and Remote Sensing Symposium—IGARSS, Melbourne, VIC, Australia, pp. 4447–4450.

Green, A., Berman, M., Switzer, P., Craig, M.D., 1988. A transformation for ordering multispectral data in terms of image quality with implications for noise removal. IEEE Trans. Geosci. Remote Sens. 26 (1), 65–74. Available from: https://doi.org/10.1109/36.3001.

Guanter, L., Richter, R., Moreno, J., 2006. Spectral calibration of hyperspectral imagery using atmospheric absorption features. Appl. Opt. 45, 2360–2370. Available from: https://doi.org/10.1364/AO.45.002360.

Guanter, L., Segl, K., Sang, B., Alonso, L., Kaufmann, H., Moreno, J., 2009. Scene-based spectral calibration assessment of high spectral resolution imaging spectrometers. Opt. Express 17, 11594–115606. Available from: https://doi.org/10.1364/OE.17.011594.

Hong, J., Kim, Y., Choi, B., Hwang, S., Jeong, D., Lee, J.H., et al., 2017. Efficient method to measure the spectral distortions using periodically distributed slit in hyperspectral imager. Opt. Express 25 (17), 20340–20351. Available from: https://doi.org/10.1364/OE.25.020340.

Høye, G., Løke, T., Fridman, A., 2015. Method for quantifying image quality in push-broom hyperspectral cameras. Opt. Eng. 54 (5), 053102. Available from: https://doi.org/10.1117/1.OE.54.5.053102.

Meroni, M., Busetto, L., Guanter, L., Cogliati, S., Crosta, G.F., Migliavacca, M., et al., 2010. Characterization of fine resolution field spectrometers using solar Fraunhofer lines and atmospheric absorption features. Appl. Opt. 49 (15), 2858–2871. Available from: https://doi.org/10.1364/AO.49.002858.

Mouroulis, P., 1999. Spectral and spatial uniformity in pushbroom imaging spectrometers. In: SPIE's International Symposium on Optical Science, Engineering, and Instrumentation, Denver, CO, United States, Proceedings, vol. 3753, Imaging Spectrometry V. Available from: https://doi.org/10.1117/12.366313.

Nieke, J., Schläpfer, D., Dell'Endice, F., Brazile, J., Itten, K.I., 2008. Uniformity of imaging spectrometry data products. IEEE Trans. Geosci. Remote Sens. 46 (10), 3326–3336. Available from: https://doi.org/10.1109/TGRS.2008.918764.

Oskouei, M.M., Babakan, S., 2016. Role of smile correction in mineral detection on hyperion data. J. Min. Environ. 7 (2), 261–272. Available from: https://doi.org/10.22044/jme.2016.567.

Richter, R., Schlapfer, D., Muller, A., 2011. Operational atmospheric correction for imaging spectrometers accounting for the smile effect. IEEE Trans. Geosci. Remote Sens. 49 (5), 1772–1780. Available from: https://doi.org/10.1109/TGRS.2010.2089799.

Saenz, K., Secker, J., Gao, B.C., Davis, C., Nadeau, C., 2002. Radiative transfer codes applied to hyperspectral data for the retrieval of surface reflectance. ISPRS J. Photogrammetry Remote Sens. 57 (3), 194–203. Available from: https://doi.org/10.1016/S0924-2716(02)00121-1.

Xu, Y., Wang, R., Liu, S., Yang, S., Yan, B., 2008. Atmospheric correction of hyperspectral data using MODTRAN model. In: Remote Sensing of the Environment: 16th National Symposium on Remote Sensing of China, vol. 7123(712306), pp. 1–7. Available from: https://doi.org/10.1117/12.815552.

Yokoya, N., Miyamura, N., Iwasaki, A., 2010. Detection and correction of spectral and spatial misregistrations for hyperspectral data using phase correlation method. Appl. Opt. 49, 4568–4575. Available from: https://doi.org/10.1364/AO.49.004568.

3

Anomaly detection in hyperspectral remote sensing images

Przemysław Głomb, Michał Romaszewski

INSTITUTE OF THEORETICAL AND APPLIED INFORMATICS, POLISH ACADEMY OF SCIENCES, GLIWICE, POLAND

3.1 Introduction

The objective of the anomaly detection (AD) is to identify patterns in a data signal that do not correspond to a well-defined notion of normal or typical behavior (Chandola et al., 2009). In general, AD generalizes or shares a lot of similarities with the typical data processing issues including:

Noise removal: The removal of the data points (or components) that were introduced, perhaps consistently, outside of the main data source.
Outlier detection: The detection of data points that are rare and do not conform to the basic data patterns.
Novelty detection: The detection of patterns distinct from historical patterns.

There are various sources of anomalies, for example, measurements or data handling errors (device error), malicious intent (e.g., credit card fraud), and rare event observation. The motivation for detecting anomalies is that they relate to important events that require intervention (e.g., faults or problems); also, they significantly influence data distribution, making it more complex, violating assumptions (e.g., Gaussianity), and requiring more complex processing methods. The two main challenges in AD are the definition of the typical/normal data for complex data sources and the limited number or lack of reference (training) data or exact anomaly definition. The field of AD is complex, and many methods and models are proposed and investigated, and many problems are examined (see reviews Chandola et al., 2009; Goldstein and Uchida, 2016; Hodge and Austin, 2004; Animesh and Park, 2007).

Traditional photography records two-dimensional images with pixels containing red, green, blue (RGB) color information. These sensors correspond to certain wavelength ranges, for example, green usually corresponds to the 495−570 nm range. Hyperspectral images (HSIs) increase the wavelength range and number of recorded measurements, while decreasing their width. The main advantage of hyperspectral over traditional imagery is the ability to isolate a specific narrow frequency range, where the interaction of light with

Hyperspectral Remote Sensing. DOI: https://doi.org/10.1016/B978-0-08-102894-0.00004-8

molecules is highly specific to certain chemical compounds. This gives the possibility to identify materials or analyze chemical process states (e.g., decay).

This chapter presents an introductory overview of the hyperspectral AD. In the scope of the introduction, an illustrative example is discussed, followed by an in-depth state of the art review. Section 3.2 presents a brief derivation of the theory of select methods. Section 3.3 describes the result of a reference AD scenario on several datasets. The aim of this chapter is to familiarize the reader with the problem and current approaches taken both in general and in particular examples, and finally to present a case study of how this problem is handled in practice.

3.1.1 An example of hyperspectral anomaly detection

Consider the case presented in Fig. 3–1, which explores a fragment of a well-known Cuprite dataset.[1] The ground truth, outlined in Fig. 3–1C, verified by onsite inspection (Swayze et al., 2014), shows a mostly uniform fragment of terrain with a number of artefacts. The main feature is the separation of two areas, to the north [above the dotted line (5) in Fig. 3–1C] jarosite and goethite minerals mixed with kaolinite, white mica, and alunite or halloysite and to the south, Fe minerals with white mica (cf. Fig. 3–5A and B in Swayze et al., 2014). Together they form a good example of a hyperspectral background; simpler than some example scenes (cf. Indian Pines and Pavia University[2]), while still being a real-life scenery. Within the area, there are several hills of different mineral structures corresponding to various alunite minerals [(1a–d) in Fig. 3–1C]. There is also a small chalcedony and/or opal region, a fragment of an interstate highway, and two white patches of ground [(2), (3), and (4) in Fig. 3–1C, respectively]. In this example, the wavelength range was limited to approximately 2000–2500 nm as in the work of Dobigeon et al. (2009), to "zoom in" on a region of mineral spectral features, especially alunite.

Consider now the application of three simple anomaly detectors, namely a global RX detector, RX with a BACON vector preselection, and a HMM. These are described in depth in the next section 3.2, but for this example, the most important feature is the difference between the RX and HMM detectors. The former uses a simple background model of a multivariate Gaussian model, while the latter, having many more free parameters, can accommodate much more complex models. The "naïve" application of a global RX detector is presented in Fig. 3–1D. This figure shows a map of detector responses; the blue values signify pixels conforming to the background model, while the white and red show deviations or potential anomalies. The global RX detector uses sample mean and covariance matrix estimation, which are biased based on the terrain artefacts. The result is visible in some parts of the alunite hills having strong and some weak responses. The BACON algorithm (Fig. 3–1E) uses an iterative selection of vectors, throwing away those that do not correspond well to the initial Gaussian. The result is a much clearer separation between the basic terrain and the

[1] Available online: www.spectir.com/contact#free-data-samples.

[2] Available online: www.ehu.eus/ccwintco/index.php/Hyperspectral_Remote_Sensing_Scenes.

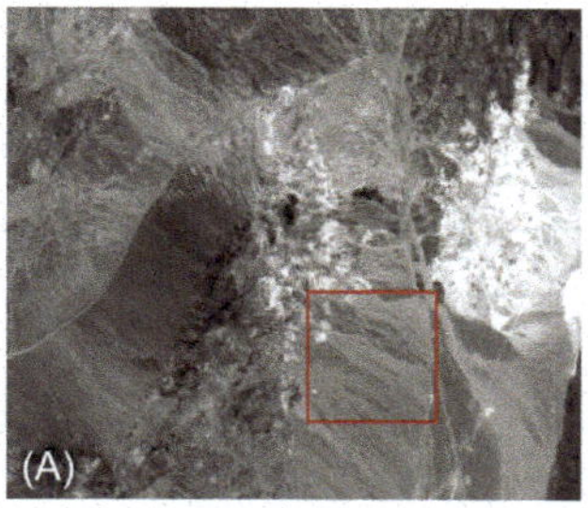

Aerial photograph of collection site at Cuprite, Nevada. Red square marks the position of an image fragment selected for this experiment. The wavelength range was taken as described by Dobigeon et al. (2009).

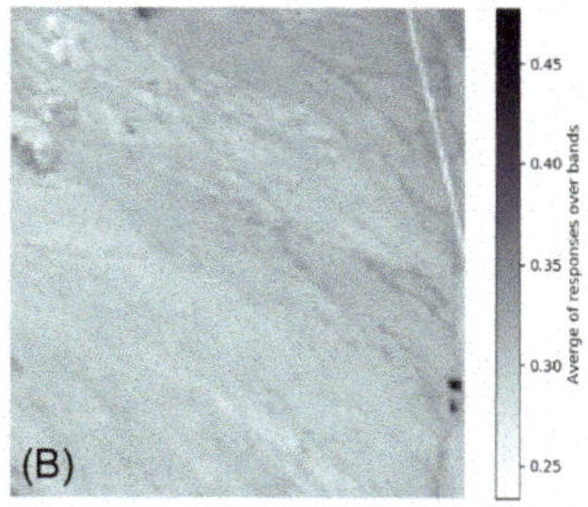

Mean of the band response of the selected image fragment in the wavelength range. The area contains a patch of a relatively uniform land cover with a number of terrain artefacts.

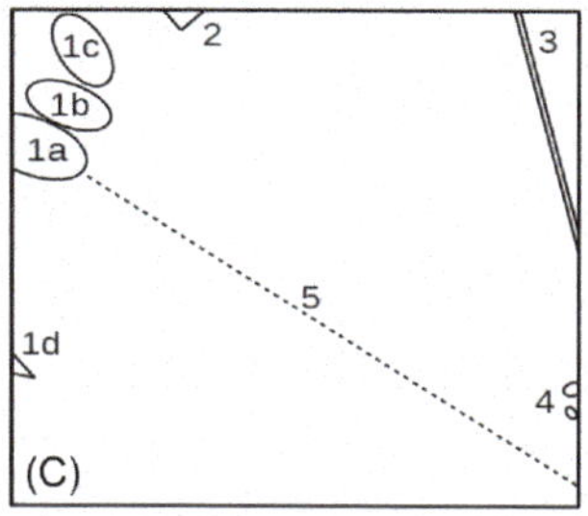

Terrain legend based on the work of Swayze et al. (2014). (1a–d) alunite deposits, (2) chalcedony and/or opal area, (3) highway, (4) white ground patches, (5) approximate division between different kinds of surface cover.

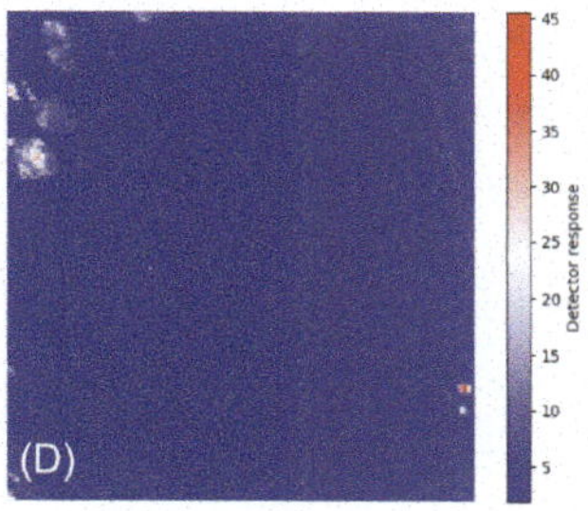

Global RX detector response map. The single Gaussian model favors the surface background, signaling alunite deposits as anomalies. However, global estimation of Gaussian parameters is biased in this situation, hence, the responses are not pronounced.

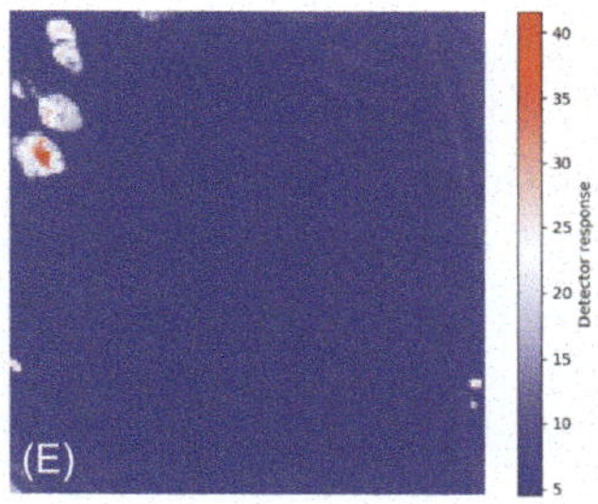

Reed–Xiaoli (RX) with BACON vector preselection response map. The BACON algorithm aims to remove from estimation vectors that do not conform to the model. As a result, the background corresponding to Gaussian is more reliably estimated, and nonconforming vectors have strong responses.

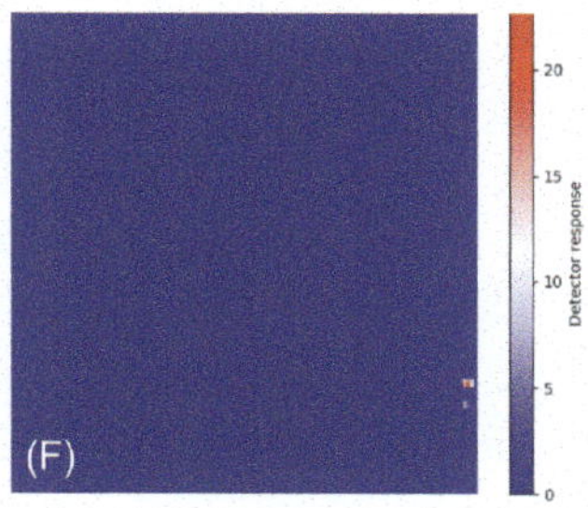

Hidden Markov model (HMM) detector response map. The state space model allows for more complex background, which includes alunite deposits. Hence it produces much less pronounced responses in (1) area, but singles out (4) as the main anomaly.

FIGURE 3–1 Illustration of a typical anomaly detection problem. The selected image contains, among other fragments, a relatively uniform background with some standing out terrain elements (1) and point anomalies (4). Depending on the detector algorithm and parameter estimation method, a variation of responses can be expected. See text.

artefacts. The HMM model (Fig. 3–1F) learns a nonparametric background model, which treats the alunite hills as part of a background, producing mostly uniform output except for a strong response to white ground patches.

Which of these results is correct? Is the alunite hills region a part of a background or an anomaly? It largely depends on the model that is used or the desired application. This example shows that the problem of AD is more complex in the evaluation phase, and care should be taken when preparing and describing the detection experiments. The type of images, the complexity of a method, and the procedure of parameter estimation can strongly affect the scores. We argue that while in some cases there are grounds for quantitative evaluation of different methods, the qualitative investigation is also an important feature of an AD experiment.

3.1.2 Literature review of hyperspectral anomaly detection

The basic HSI AD methods and models including generalized likelihood ratio testing and its derivatives were subject to a number of review and survey works including the work of

Manolakis and Shaw (2002) and Stein et al. (2002) who present an overview of full-pixel, subpixel, and a linear mixture effect in detection; Chang (2003) who provides an in depth treatment of the foundations of hyperspectral detection; Manolakis et al. (2009) who provides a critical discussion and arguments against the complexity of the detector; and Matteoli et al. (2010) who present a tutorial overview of both classic approaches and some of the more current developments.

Arguably the most basic and well-studied method is the RX detector (Reed and Yu, 1990) derived from the generalized likelihood ratio test. The image is modelled as Gaussian; for complex real-life images it is approximated by estimating the local covariance matrix and removing the local mean. The decision about a pixel under test (PUT) is made based on test statistics computed as the Mahalanobis distance between it and the (normalized) background. The distribution of this test statistics is independent of the unknown parameters (depends only on signal dimension), thus, it has a constant false alarm rate property (Stein et al., 2002). However, in practical applications, this normalization may not be enough; various algorithms are used to support the estimation by selecting a subset of samples, for example, iterative thresholding (Billor et al., 2000), random selection (Du and Zhang, 2011), weighting (Stein et al., 2002), or saliency estimation (Liu et al., 2018). The paper by Guo et al. (2016) compares several approaches for mitigating the departure from normal distribution in the RX estimation; it concludes that different mitigation methods offer similar improvements to the classic RX estimation. A form of generalization of this support is using a mixture of Gaussians as a background model; while it requires the number of components be to estimated, it has much greater potential of representing complex scenes. The approaches used by Matteoli et al. (2010) either split hyperspectral vectors into clusters using a number of parameters or combine the whole Gaussian mixture model (GMM) likelihood. Chang and Chiang (2002) analyze several options of proposed normalized RX detector. Matteoli et al. (2013) extends the mixture approach by proposing models other than Gaussian distributions as the background model. Maryam (2018) uses an iterative approach with the first run of RX to estimate the probable background, then performs feature reduction.

A number of AD methods are based on observing the intrinsic dimensionality of the hyperspectral data space. Specifically, it has been noted that HSI or their fragments, lie in a subspace of much lower dimensionality than the original spectra size. In this subspace, they can often be described by a simplex with vertices being the basic materials present in the scene. The pixels are modelled as a mixture of those basic materials, usually a linear one. Wang and Xue (2018) combine the linear mixing model and the generalized likelihood ratio test to test the probability adjusted for target and background subspaces. The PUT coefficients within those subspaces are obtained with ℓ_2-norm regularized estimation. Sun et al. (2018) generalizes the robust principal component approach (RPCA). The proposed method, randomized subspace learning-based anomaly detector, assumes the anomaly matrix is sparsely populated with nonzero columns, and that these columns form a different subspace from the background. To estimate that, they propose a multistage algorithm including random sampling and random Hadamard projections. Zhang et al. (2016) assume an image model of a low-rank background image, outliers, and noise added together. They use the GoDec algorithm, based on the bilateral random projections, to estimate this low-ranking background component. The covariance matrix inverse needed for computing the Mahalanobis distance between a PUT and the background is estimated from eigenvectors of this recovered matrix. Qu et al. (2016) use a three-stage algorithm with the steps of endmember unmixing, clustering, and low rank decomposition. Some methods focus on extracting a sparse decomposition, that is, estimating a current spectral pixel as a combination of a low number of dictionary pixels. Those methods use a similar decision—classifying the PUT to background or target depending on the dictionary representation error—but focus explicitly on minimization of the dictionary size. A regular (Chen et al., 2011) and kernelized versions (Zhang et al., 2015) of sparse decomposition were studied. A more complex method (Li et al., 2015) creates a dictionary and sparse representation from neighborhood pixels, considering different scenarios of neighborhood that are nonanomalous and have single anomalies, an anomalous cluster, or a composition of the former.

Another class of methods focuses on the combination of spatial and spectral information; an anomaly is defined as a pattern that does not fit the spatial neighborhood. Yuan et al. (2015) uses a filtering approach to single out the differences. This multistep algorithm uses the center-surround approach and includes a high pass filter, binary thresholding, and discrimination function based on counting the threshold values. Gu et al. (2015) extends the hypothesis testing and subspace representation by processing the pixel neighborhood area. Additionally, they combine the processing with wavelet transform and independent component analysis. Liu and Li (2018) use the exponential weighting of the nearest neighbors from positive and negative training sets. After optimization, which includes the coefficient for description of local information, a decision function is estimated. Lili and Zhao (2017) combine the local subspace representations for spatial and spectral domains to get the function for measuring local similarity level. Li et al. (2018) consider a complex model of the background structure in spatial patches in a HSI to find a good pixel representation; pixels that cannot be represented are classified as anomalies. Wang et al. (2018) use a process where the spatial maps of detected anomalies are processed and put back in as additional data sources. This forms an iterative process with each iteration having distinct steps of AD and anomaly spatial mapping. The resulting algorithm is able to detect and classify anomalies into distinctive groups.

Due to their successes in applications in other data domains, the kernel methods and/or the maximum margin classifier [e.g., one-class support vector machine (SVM)] are also applied to the hyperspectral domain. Heesung Kwon and Nasrabadi (2005) extend the RX detector by introducing kernels; the proposed kernel–RX method is much better suited to dealing with non-Gaussian background distributions. Dong et al. (2015) propose a maximum-margin metric learning for target detection. The method estimates a data transformation matrix that provides maximum separation between training samples of background and targets. The cutting plane algorithm is used to reduce the number of constraints in the optimization and, thus, improve the performance. Zhao et al. (2016) use kernel expansion as a basic element of a robust background regression technique. The optimization includes terms for matching data labels, regularization to prevent overfitting, and a fit to the robust background average density. The decision about anomalies was made based on the distance of the nearest neighbors to the robust background samples and potential anomalies. Zhao et al. (2015) improve the sparse representation approach using kernel methods. This method uses "stacking" of spatial and spectral kernels, which amounts to a weighted average of the respective kernel matrices. Banerjee et al. (2006) explore linear and kernel support vector data description (SVDD) representations for deriving a data model. This allows a derivation similar to an RX detector, without any underlying assumptions about the distribution.

Various other approaches are being used. Li and Du (2015) propose a collaborative approach for pixel representation where pixels from the outer window provide representation for the PUT. Suitable regularization is added to provide stability to the solution. Several extensions including a kernel framework and RPCA are also investigated. The resulting algorithm is straightforward to apply and can be viewed as a generalization of the original RX approach. Ma et al. (2018) use a deep belief network for PUT area reconstruction. The deep learning approach has the potential to internalize the complex structure of a hyperspectral neighborhood. It then uses several scenarios of anomalies present in the PUT, the neighborhood, both, or neither to form a decision function, in a way similar to that of Li et al. (2015). Yuan et al. (2016) start from local linear embedding, which is a local data representation based on the neighboring pixels. The representation error allows different neighbors to be weighed in the neighborhood graph. Postprocessing of the graph is used to isolate anomalous pixels. Zheng et al. (2016) builds a noise model based on linear mixing model representation. Several detectors are proposed, which combine abundance coefficient analysis with hypothesis testing.

3.2 Methods

The Section 3.1 presented the general scope of current approaches for AD. The selection of representative methods is a hard problem, and arguably each choice leaves out some important examples of methods. Here, well-known and established methods will be focused on in order to provide a solid reference for hyperspectral AD practitioners. This naturally leads to the classic models of global and local RX detectors. Since their introduction, they have been endlessly applied, and while unsophisticated, their solid derivation and simplicity are a strong argument for their inclusion. To adequately represent their modifications and upgrades, three additional versions were selected, namely weighted RX, which uses weights to nonuniformly scale the estimation of mean and covariance matrix, BACON, which iteratively discards the vectors that are far from the initial Gaussian, and mixture of Gaussian, which extends the distribution model. As a representation of other methods, one-class support vector machine (OCSVM), a representative approach using kernel methods; HMM, which is suitable for building a spectral model; collaborative-based detector (CRD), which is an example of a dictionary approach; and high-order two-dimensional crossing filter-based detector (HO2DC), as an example of a detector that focuses on spatial processing are included.

3.2.1 Gaussian model: the RX detector

The basic approach of Matteoli et al. (2010) considers an HSI of l bands, and a random vector $\boldsymbol{X} = X_1, \ldots, X_l$ corresponding to the PUT with components $X_i \in \mathbb{R}^+$ being the values (e.g., reflectance) associated with each frequency. A realization of this vector is denoted as $x = [x_1, \ldots, x_l]^\top$. The problem considered is assigning the PUT a label—a background pixel or a target (a pattern to be located or an anomaly).

In the basic formulation of the detection problem, the PUT x is classified considering the two hypotheses H_0 and H_1 with associated conditional probability distributions $f_{\mathbf{X}|H_0}(x)$ and $f_{\mathbf{X}|H_1}(x)$:

$$\hat{H}(x) = \begin{cases} H_0: & x \sim f_{\mathbf{X}|H_0} \text{ is a background pixel} \\ H_1: & x \sim f_{\mathbf{X}|H_1} \text{ is an anomaly} \end{cases} \tag{3-1}$$

When the cost structure of the decisions (e.g., the penalty for mistakenly ignoring an anomaly) is unavailable, the optimum decision rule can be derived according to the Neyman–Pearson (NP) criterion (Vincent Poor). The NP decision rule maximizes the detection probability $[P_D = \mathbb{P}(\hat{H}(x) = H_1 | H_1)$, correctly assuming H_1 hypothesis] while placing a constant bound on false alarm probability $[P_F = \mathbb{P}(\hat{H}(x) = H_1 | H_0)$, falsely rejecting $H_0]$; it is derived as a likelihood ratio (LR) test (or probability ratio test):

$$LR(x) = \frac{f_{\mathbf{X}|H_1}(x)}{f_{\mathbf{X}|H_0}(x)} \lessgtr \tau \tag{3-2}$$

where τ is the detection threshold parameter.

In a typical situation, the conditional probability distributions $f_{\mathbf{X}|H_0}(x)$ and $f_{\mathbf{X}|H_1}(x)$ are unavailable. In that case, one may substitute a family of distributions parameterized by a set of parameters. A typical family of distributions used is the Gaussian or normal distribution. There are several arguments for this model. For example, it is known to work well in many practical applications, its properties are well known, and it approximates well at least some processes related to the hyperspectral acquisition (electronic noise). This produces a set of hypotheses in the form:

$$\hat{H}(x) = \begin{cases} H_0: x \sim \mathcal{N}\left(\mu_b, \sum_{\mathbf{b}}\right) \\ H_1: x \sim \mathcal{N}\left(\mu_t, \sum_{\mathbf{t}}\right) \end{cases} \tag{3-3}$$

where $\boldsymbol{\mu}_b$ is the background mean, $\boldsymbol{\Sigma}_b$ is the background covariance matrix, and $\boldsymbol{\mu}_t$ and $\boldsymbol{\Sigma}_t$ are the anomaly or target mean and covariance matrix, respectively.

With Gaussian distributions, the generalized likelihood ratio (GLR) test becomes (Manolakis et al., 2009):

$$GLR(x) = \frac{|\sum_{\mathbf{t}}|^2 \exp(-\frac{1}{2}(x - \mu_t)^\top \sum_{\mathbf{t}}^{-1} (\mathbf{x} - \boldsymbol{\mu}_{\mathbf{t}}))}{|\sum_{\mathbf{b}}|^2 \exp\left(-\frac{1}{2}(x - \mu_b)^\top \sum_{\mathbf{b}}^{-1} (\mathbf{x} - \boldsymbol{\mu}_{\mathbf{b}})\right)} \tag{3-4}$$

Taking the logarithm produces the detection statistics (after ignoring constant terms) (Manolakis et al., 2009):

$$D_1(x) = (x - \mu_b)^\top \sum\nolimits_{\mathbf{b}}^{-1} (\mathbf{x} - \mu_\mathbf{b}) - (\mathbf{x} - \mu_\mathbf{t})^\top \sum\nolimits_{\mathbf{t}}^{-1} (\mathbf{x} - \mu_\mathbf{t}) \tag{3-5}$$

In many cases, no information about the target is available. The objective is then to detect anomalies or pixels that are sufficiently distinct from the background. This can be realized by assuming the equality $\boldsymbol{\Sigma}_b = \boldsymbol{\Sigma}_t$ of the covariance matrix as in the previous step, but also, since the target mean is not known, it is substituted by the maximum likelihood estimate under the H_1 hypothesis, $\hat{\mu}_t = x$ (Manolakis et al., 2009). This leads to the detection statistic:

$$r_{\mathrm{RX}}(x) = (x - \mu_b)^\top \sum\nolimits_{\mathbf{b}}^{-1} (\mathbf{x} - \mu_\mathbf{b}). \tag{3-6}$$

This is the basic formulation of the RX detector (Reed and Yu, 1990), which is regarded as the reference anomaly detector for hyperspectral data. Note that this statistic computes the Mahalanobis distance between the PUT and the center of the background distribution. The threshold τ value for selecting anomalies can be computed from a χ^2 distribution with l degrees of freedom, given the expected probability of the anomaly.

3.2.2 Extension of the model for local, partial, or multivariate Gaussian

There are three major components of hyperspectral processing pipeline, namely lighting, material, and acquisition device. Under preferable practical conditions, the hyperspectral data comes from imaging a uniform surface of one type of material with small variations in intensity resulting from some nonregularities of an object, lighting, or recording noise. In that case, the Gaussian model of μ and covariance matrix $\boldsymbol{\Sigma}$ are applicable, and the parameters can be estimated from data using:

$$\mu = \frac{1}{n}\sum_{i=1}^{n} \boldsymbol{x}_i \quad \sum = \frac{1}{n}\sum_{i=1}^{n} (\boldsymbol{x}_i - \mu)(\boldsymbol{x}_i - \mu)^\top \tag{3-7}$$

In real life conditions, the assumption of Gaussianity is hard to satisfy, especially for complex scenes. However, a quantitative inspection of images reveals regions that most likely conform to the Gaussian distribution, and those that certainly do not (refer to Fig. 3–2). While a global Gaussian-based detector can't be expected to work reliably, a local one could work, at least in some situations. This "local estimation" assumption is the foundation of the proposition of the original RX detector (Reed and Yu, 1990); the background mean and covariance matrix are estimated from the local neighborhood (window) of the PUT. To protect against contamination of the background estimates with anomalous pixels, a smaller "guard" window is often subtracted from the local neighborhood window. Hence the basic

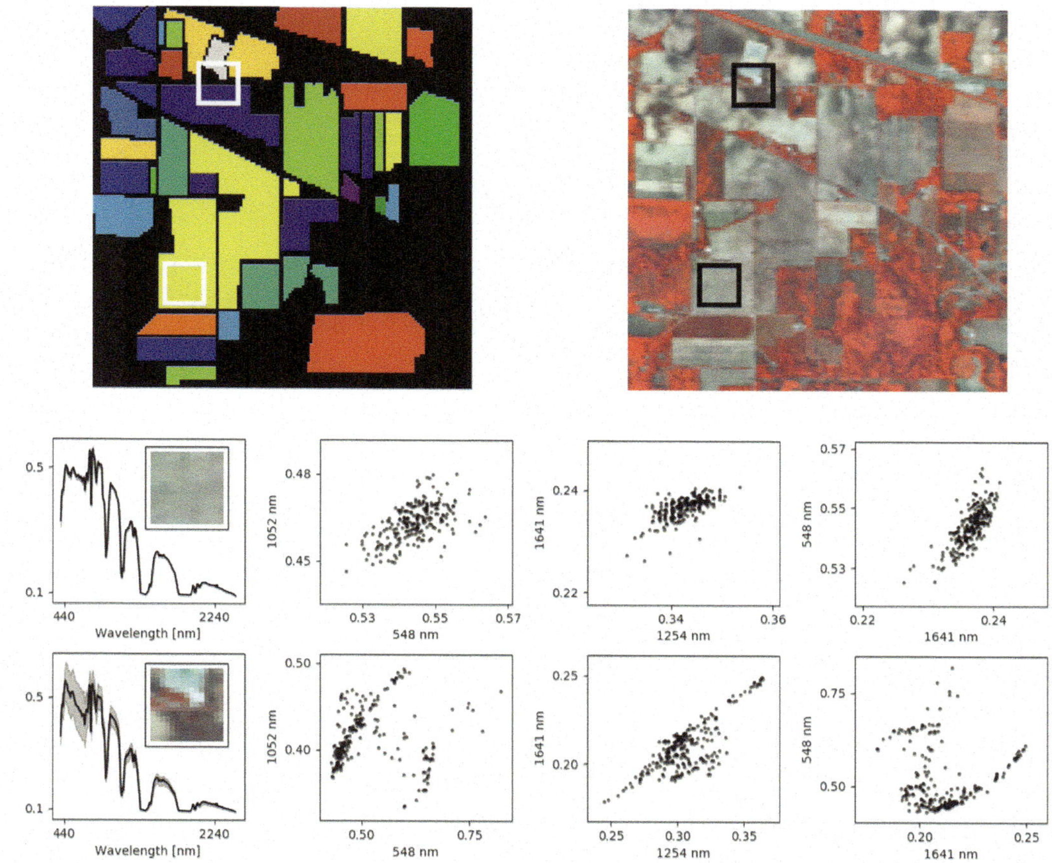

FIGURE 3–2 Examples of close to Gaussian and non-Gaussian hyperspectral data, from Indian Pines image. Two areas are selected, one inside and one outside a class. The scatterplots allow for simple inspection of distribution. Within the class, the structure often is Gaussian-like. If one looks outside the classes, it often is not.

local RX detector has three parameters for window sizes, namely mean, covariance, and guard window.

An approach to improve the performance of the global RX detector is to use a weighted estimation of the mean and covariance matrix (Guo et al., 2016). In this approach, dubbed "weighted RX," the initial global mean and covariance estimations are first used to compute probability for each pixel. Then, those probabilities are treated as weights in weighted estimation of the final statistics. This approach shifts the estimated mean and covariance matrix toward the "common ground" of the more probable samples. An approach the is similar in principle, but different in execution is used that combines the estimation with the BACON (Billor et al., 2000) outlier detection algorithm; the iterative procedure throws away vectors that do not conform to the representative set of the most probable samples. This approach is arguably more robust than the weight heuristic in the presence of large value anomalies.

Another way to improve the basic approach is to extend the distribution model beyond single Gaussian (Matteoli et al., 2010), using, for example, a mixture of Gaussians (MoG, or Gaussian Mixture Model, GMM) as a more general model (RXGMM):

$$r_{\mathrm{RXMoG}}(x) = -\log\left\{\sum_{i=1}^{g} w\mathcal{N}\left(\boldsymbol{x}; \boldsymbol{\mu}_i, \sum\nolimits_{\mathrm{i}}\right)\right\} \lessgtr \tau \tag{3-8}$$

3.2.3 Other approaches: non-Gaussian backgrounds

While the Gaussian approach is well-founded theoretically and various proposed upgrades provide robustness for some deviations from the model, there are cases when the data structure is either unknown or known to be non-Gaussian. This section presents some selected approaches for this case.

3.2.3.1 One-class support vector machine

The OCSVM (Schölkopf et al., 2000) detector aims to estimate a decision function that takes the value $+1$ in a "small" region capturing most of the data points and -1 elsewhere. This is done by mapping data points into a high-dimensional feature space (using a kernel function) and finding a maximal margin hyperplane separating the dataset from the origin. Thus the origin and its nearby vectors become anomalies.

The decision function for the tested vector corresponding to the PUT, $x \in \chi \subset \mathbb{R}^n$, is:

$$f(\boldsymbol{x}) = \left(\sum_i \alpha_i K(\boldsymbol{x}_i, \boldsymbol{x}) - \rho\right), \tag{3-9}$$

where $0 \leq \alpha_i \leq \frac{1}{\nu l}, \sum_i \alpha_i = 1$ and $\rho \in \mathbb{R}$ are coefficients computed through Lagrangian optimization (margin maximization on the training set), $\boldsymbol{x}_i$ and $\in \chi$ are training points, and K: $\chi \times \chi \rightarrow \mathcal{R}$ is a kernel function computing the similarity measure between the tested and training examples. Typically, by an application of a "kernel trick" (Scholkopf and Smola, 2001), a nonlinear mapping ϕ: $\chi \rightarrow \mathcal{H}$, $\boldsymbol{x} \mapsto \phi(\boldsymbol{x})$ to a special high-dimensional feature space $\mathcal{H}$ is introduced, and the kernel is computed as $K(\boldsymbol{x}, \boldsymbol{x}_i) = \langle \phi(\boldsymbol{x}), \phi(\boldsymbol{x}_i) \rangle$. The Gaussian radial basis function (RBF) $K(\boldsymbol{x}_i, \boldsymbol{x}_j) = \exp(-\gamma \|\boldsymbol{x}_i - \boldsymbol{x}_j\|^2)$ has been found to be the most versatile and effective for many different kinds of data. In addition, when using the RBF function, OCSVM gives similar results to another well-known anomaly detector, namely the SVDD (Tax and Duin, 2004). The OCSVM with RBF kernel has two parameters, namely the kernel parameter γ, which controls the influence of support vectors, while the parameter ν is both the upper bound on the fraction of outliers in the training set and a lower bound of the fraction of support vectors, thus, for AD tasks it should be small.

3.2.3.2 Collaborative-based detector

The CRD (Li and Du, 2015) uses pixels in a neighborhood window of the PUT as a dictionary to describe the PUT. For a PUT $\vec{y}$ a set of neighbouring pixels is formed. It is based on pixels from outer window (size $w_{\text{out}} \times w_{\text{out}}$ around $\vec{y}$) while excluding the inner window (size $w_{\text{in}} \times w_{\text{in}}$ around $\vec{y}$), $\vec{X_s} = \{\vec{x}_i\}_{i=1}^{s}(s = w_{\text{out}} \times w_{\text{out}} - w_{\text{in}} \times w_{\text{in}})$. The anomaly score is based on the optimum representation in the form:

$$r_{\text{CRD}} = \left\|\boldsymbol{y} - \boldsymbol{X}_s\hat{\alpha}\right\|_2^2.2 \tag{3-10}$$

The weight vector $\hat{\alpha}$ can be computed from:

$$\hat{\alpha} = (\tilde{\boldsymbol{X}}_s^\top \tilde{\boldsymbol{X}}_s + \lambda \boldsymbol{\Gamma}_y^\top \boldsymbol{\Gamma}_y)^{-1} \tilde{\boldsymbol{X}}_s^\top \tilde{\boldsymbol{y}} \tag{3-11}$$

where the Γy is the Tikhonov regularization matrix. This matrix prevents pixels that are different from the center pixel to have a large weight:

$$\boldsymbol{\Gamma}_y = \begin{bmatrix} \left\|\boldsymbol{y} - \boldsymbol{x}_1\right\|^2 & & 0 \\ & \ddots & \\ 0 & & \left\|\boldsymbol{y} - \boldsymbol{x}_s\right\|^2 \end{bmatrix} \tag{3-12}$$

and $\tilde{\boldsymbol{x}}_s, \tilde{\boldsymbol{y}}$ are $\boldsymbol{x}_s$ and $\boldsymbol{y}$ extended after last row with ones, to impose a sum-to-one constraint on the weight vector.

3.2.3.3 Hidden Markov model-based detector

The idea of HMM-based AD, to the best of the authors' knowledge, has not been applied before, but the HMM has been used for HSI processing (Chang and Du, 2001; Arabi et al., 2015). The HMM is a state-based model of a probabilistic sequence generator that can be described with the Markov process. Here a detector that uses a model of a HSI background is proposed. The model consists of a set of states, $Q = \{q_1, \ldots, qn\}$, where each associated with a Gaussian density is parameterized by μ_i, σ_i. The model contains a state transmission matrix, $\boldsymbol{A} = [a_{ij}]$, where $a_{ij} = \Pr(q_{t+1} = S_j | q_t = S_i)$ denotes the probability of transmission to state j if the current state is i and the initial probability vector of states is $\pi = [\pi_i]$, $\pi_i = \Pr(q_1 = S_i)$.

Given a set of m image vectors, $\boldsymbol{X} = \{x_i\}_{i=1}^{m}$ and the number of states n, one can use the Baum–Welch expectation maximization approach (Rabiner, 1989) to estimate the values $\gamma = \{a_{ij}, \pi_i, \mu_i, \sigma_i\}$ that correspond to a model of $\boldsymbol{X}$. Given the model γ and a PUT $\boldsymbol{y}$, one can use the forward algorithm to compute the probability $\Pr(\boldsymbol{y}|\gamma)$ of $\boldsymbol{y}$ being generated by the model γ and compute a simple anomaly score as:

$$r_{\text{HMM}} = 1 - \Pr(\boldsymbol{y}|\gamma) \tag{3-13}$$

Note that in practice, the logarithm of $\Pr(\boldsymbol{y}|\gamma)$ is often computed and used, to avoid overflow and underflow errors. The number of states n can be estimated, for example, by selecting a minimum value of an information criterion (e.g., Akaike or Bayesian IC), computed for a range of candidates with different n.

3.2.3.4 High-order two-dimensional crossing filter-based detector

The idea of the HO2DC (Yuan et al., 2015) is an example of a spatial processing method. In contrast to previous approaches, in this method, the spatial relationships between the pixels are the most important factor. The detector algorithm is a complex one, and will only be outlined here.

First, the k-th order difference images are obtained using:

$$l_i^{(k)} = \sum_{j \in \mathcal{N}_i} \left\| l_i^{(k-1)} - l_j^{(k-1)} \right\|^2 w_{ij} \tag{3-14}$$

where i is the pixel index, and for the first difference, the original hyperspectral pixels are used $l_i^0 = \boldsymbol{x}_i$. The set $\mathcal{N}_i$ is the subset of indexes forming the immediate neighborhood of i, and w_{ij} is a Gaussian-shaped distance weighting window centered at i.

Then, the high order 2D crossing count is obtained first using the binarized clip operator $b_i^{(k)}$:

$$b_i^{(k)} = \begin{cases} 1 & \text{if } l_i^{(k)} > F\left(l_i^{(k)}\right) \\ 0 & \text{otherwise} \end{cases} \tag{3-15}$$

where $F(\cdot)$ is a spatial smoothing (averaging) filter. Then the change monitor $d_i^{(k)}$ is computed:

$$d_i^{(k)} = \sum_{j \in \mathcal{N}_i} \left(b_i^{(k-1)} \oplus b_j^{(k-1)}\right) w_{ij} \tag{3-16}$$

where $\oplus$ is the exclusive or operation. Finally, the difference operator is computed:

$$\Delta_i^{(k)} = \begin{cases} d_i^{(k)} & k = 1 \\ \left| d_i^{(k)} - d_i^{(k-1)} \right| & \text{otherwise} \end{cases} \tag{3-17}$$

and the anomaly score is estimated using a function of $\Delta_i^{(k)}$ averaged in inner $\Delta_{inner}^{(k)}$ and outer $\Delta_{outer}^{(k)}$ windows:

$$r_{\mathrm{HO2DC}} = \sum_k \frac{\left\| \Delta_{inner}^{(k)} - \Delta_{outer}^{(k)} \right\|^2}{\Delta_{outer}^{(k)}} \tag{3-18}$$

3.3 Experiments

This section describes the datasets and methodologies used in experiments. A short summary of tested target detectors is provided in Table 3–1. In the following subsections 3.3.1–3.3.3 detectors will be referred to by their acronyms.

Table 3–1 A summary of anomaly detection algorithms.

Acronym	Full name	Comment	Reference
RXG[a]	Global RX	RX with estimated global covariance matrix	
RXL[a]	Local RX	RX with estimated local covariance matrix	Reed and Yu (1990)
RXB[a]	RX Bacon	RX with BACON vector preselection response map	Billor et al. (2000)
RXGMM[a]	RX GMM	RX with mixture of Gaussians background model	
RXW[a]	Weighted RX		Guo et al. (2016)
HMM	Hidden Markov model		
HO2DC		High-order 2D crossing filter based detector	Yuan et al. (2015)
CRD		Collaborative-based detector	Li and Du (2015)
OCSVM	One class SVM	Detector	Tax and Duin (2004)

[a]Implementation of Reed–Xiaoli detector, derived from generalized likelihood ratio test.

3.3.1 Detection datasets

This section describes three hyperspectral datasets used in experiments.

3.3.1.1 HYDICE urban

An image[3] was acquired using a HYDICE airborne sensor over an urban area. It has a spectral resolution of 10 nm covering a spectral range of 400–2500 nm, and a spatial resolution of ~ 1 m. The size of the original HSI is $307 \times 307 \times 210$. After removing the water absorption and low-Signal-to-noise ratio (SNR) bands (1–4, 76, 87, 101–111, 136–153, and 198–210), 162 bands are left. Experiments use a 90×80 fragment of the image located in its upper right part in coordinates (0,187). The scene and anomalies are presented in Fig. 3–3. Anomalies correspond to cars and manmade metallic objects. They were annotated manually based on reference annotations, for example, and analysis of endmember abundance maps (Zhu et al., 2014).

3.3.1.2 HyMap Cooke City

A HyMap airborne hyperspectral imaging sensor was used to capture an image dataset, covering one area of Cooke City, MT, United States, on July 4, 2006. It has a spectral resolution of ~ 15 nm covering a spectral range of 400–2500 nm, and a spatial resolution of ≈ 3 m. The image is publicly available as Target Detection Blind Test from the website of Chester F. Carlson Center for Imaging Science, RIT.[4] The size of the image is $200 \times 800 \times 126$. Experiments use a 100×100 fragment of the image located in its upper right part, in coordinates (0,700). Anomalies correspond to four types of fabric panels deployed in regions of interest annotated by the authors of the dataset and the bright white area in the upper right quarter of the image. Two versions of ground truth were used in these experiments. The first one corresponds to the original annotation provided by the authors, extended by the bright

[3] Available online: http://lesun.weebly.com/hyperspectral-data-set.html.

[4] Available online: http://dirsapps.cis.rit.edu/blindtest.

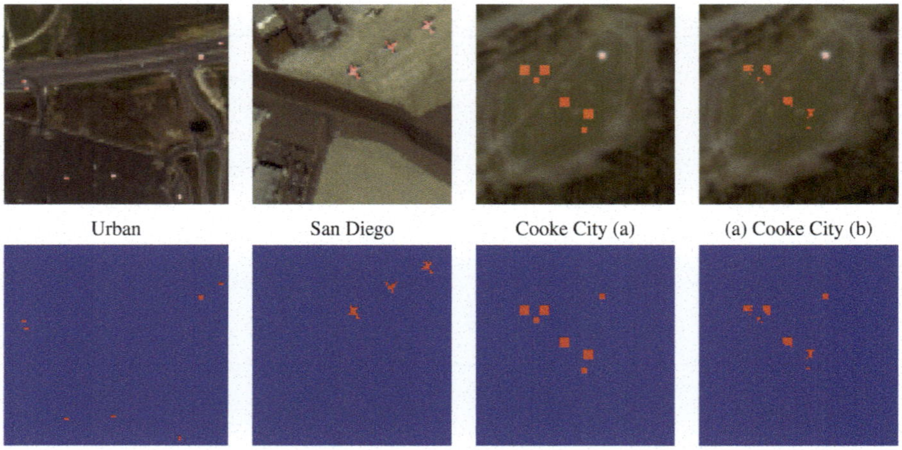

FIGURE 3–3 Datasets used in experiments along with ground truth maps of anomalies, denoted with red. HYDICE urban (A), AVIRIS San Diego (B), HyMap Cooke City (C and D) with two versions of the map (see Section 3.3.1.2).

anomaly in the upper right quarter of the image. However, since annotated regions of interest are significantly larger than the actual target sizes, an additional ground truth map was generated, aiming to remove pixels where the signature of the deployed target cannot be detected. This is done by:

1. For every pixel of the first ground truth map, and every target signature from the library provided by the authors of the dataset present in the image, responses of two target detectors, namely adaptive coherence estimator and matched filter (Manolakis et al., 2003), were generated.
2. Detection thresholds were selected in such way that target pixels were found in every region of interest.
3. Every pixel not located by at least one detector was removed from the ground truth map. Both ground truth maps are presented in Fig. 3–3.

3.3.1.3 AVIRIS San Diego

An image was acquired by an airborne visible/infrared imaging spectrometer (AVIRIS) from San Diego, CA, United States. It has a spectra range of 370–2510 nm with 224 spectral bands. Its spatial resolution is 3.5 m. After removing the water absorption and low-SNR bands (1–6, 33–35, 97, 107–113, 153–166, and 221–224). 189 bands are left. Experiments use a 100×100 fragment of the image. Anomalies correspond to three planes, annotated manually based on spectral similarity of pixels, located in the upper part of the image.

3.3.2 Experiment procedure

Parameters for every detector are described here. OCSVM: parameter γ of the RBF kernel was selected using the heuristics from Chapelle and Zien (2005). The ν parameter, which corresponds to the upper bound on the fraction of expected outliers was set to 0.1. *Local RX (RXL)*: a 3-pixel guard window was used in addition to a 20-pixel window to compute mean and a 25-pixel window to compute the covariance matrix. *HO2DC*: the number of L levels to use was set to 1, a 3-pixel inner window, a 5-pixel outer window, and a 7-pixel window for computing the neighborhood statistics were used. *CRD*: λ regularization parameter was set to 10^{-3}, a 5-pixel inner window and a 15-pixel outer window were used. *HMM*: 1000 randomly selected pixels were used for model selection, a maximum of 12 states with 10 components per state were used. *Global RX (RXG)*: the estimated number of components was between 3 and 7. *RX Bacon (RXB)*: a value of $c = 10$ was used, the number of iterations was capped at 50, and the expected ratio of outliers was $\alpha = 0.025$.

In the experiments, for every dataset and detector, the detector response was generated. The performance of the detectors was presented in the form of receiver operating characteristic (ROC) curves, and to quantify the detector performance, the area under curve measure was used. ROC curves present the true positive rate, corresponding to the probability of detecting an anomaly as a function of false positive rate (FPR), which corresponds to the probability of false detection.

3.3.3 Results and discussion

Detector responses are presented in Fig. 3–4 for the urban dataset, Fig. 3–5 for the San Diego dataset, and Fig. 3–6 for the Cooke City dataset. The corresponding ROC curves are presented in Fig. 3–7.

The approximate time of execution for detectors on the system with Intel Core i7–5820K CPU, 3.30 GHz, and 64 GB of RAM is presented in Table 3–2.

In the Urban and San Diego datasets, the majority of anomalies can be visually distinguished from their background even in the RGB images, but similar materials corresponding to (possibly painted) metallic surfaces are likely to be present in the images. In addition,

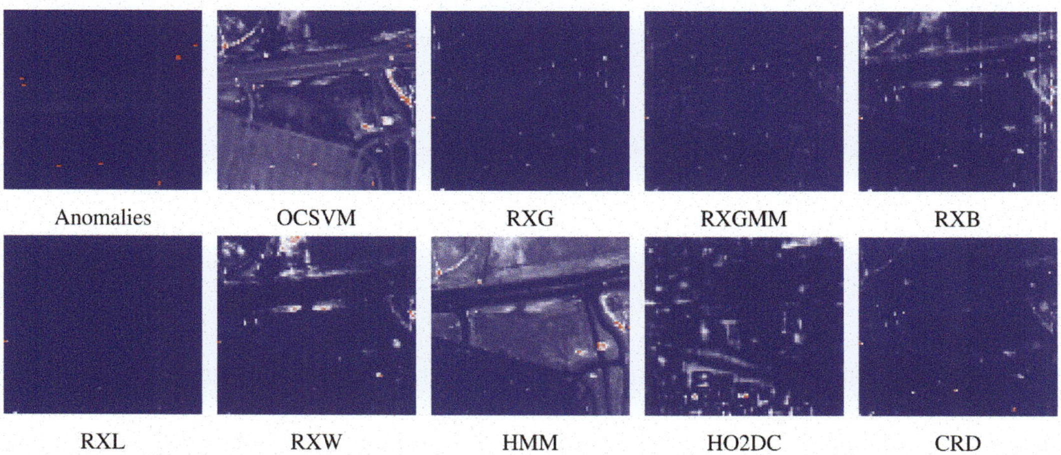

FIGURE 3–4 Ground truth of the Urban dataset and responses for every detector.

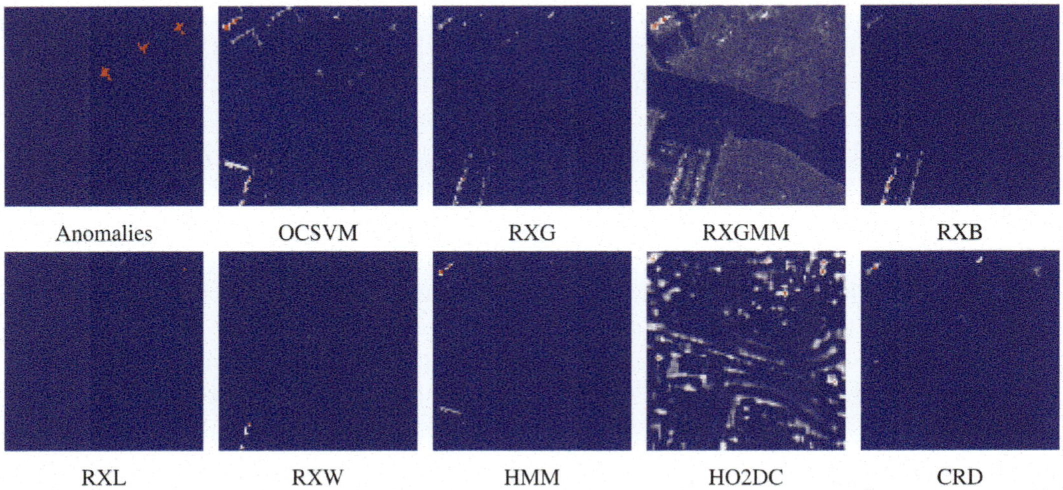

FIGURE 3–5 Ground truth of the San Diego dataset and responses for every detector.

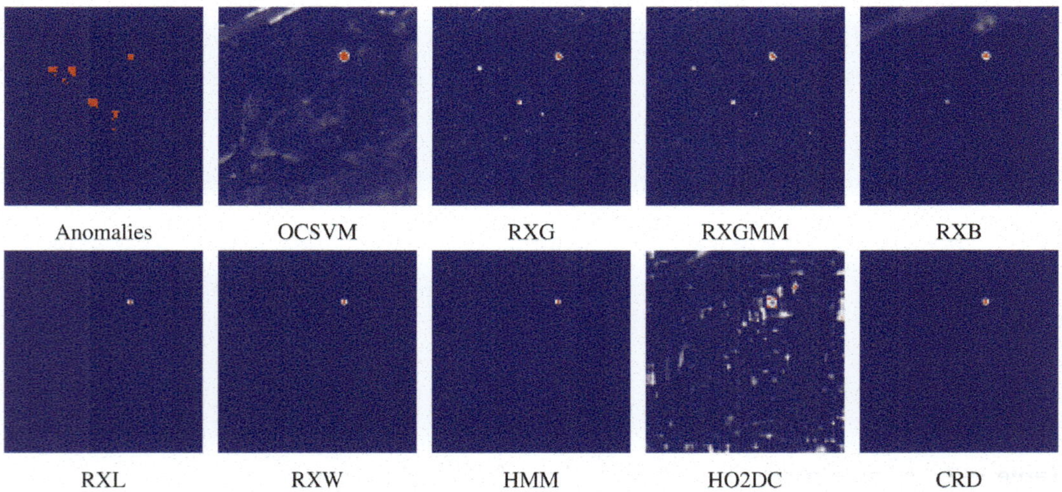

FIGURE 3–6 Ground truth of the Cooke City dataset and responses for every detector. Note that there are two versions of the ground truth for this dataset, both presented in Fig. 3–3, which will result in a set of scores when analyzing the detector performance.

anomalies in the San Diego dataset are relatively large. Most of the detectors can easily locate a subset of anomalies in these images, but they differ in the number of false positives returned when trying to locate all of them. In particular, the CRD detector seems to be able to detect anomalies with a low number of false positives.

In the Cooke City dataset, a large number of anomalies are subpixel targets with unique spectral characteristics (they are not present in the scene outside of the target areas).

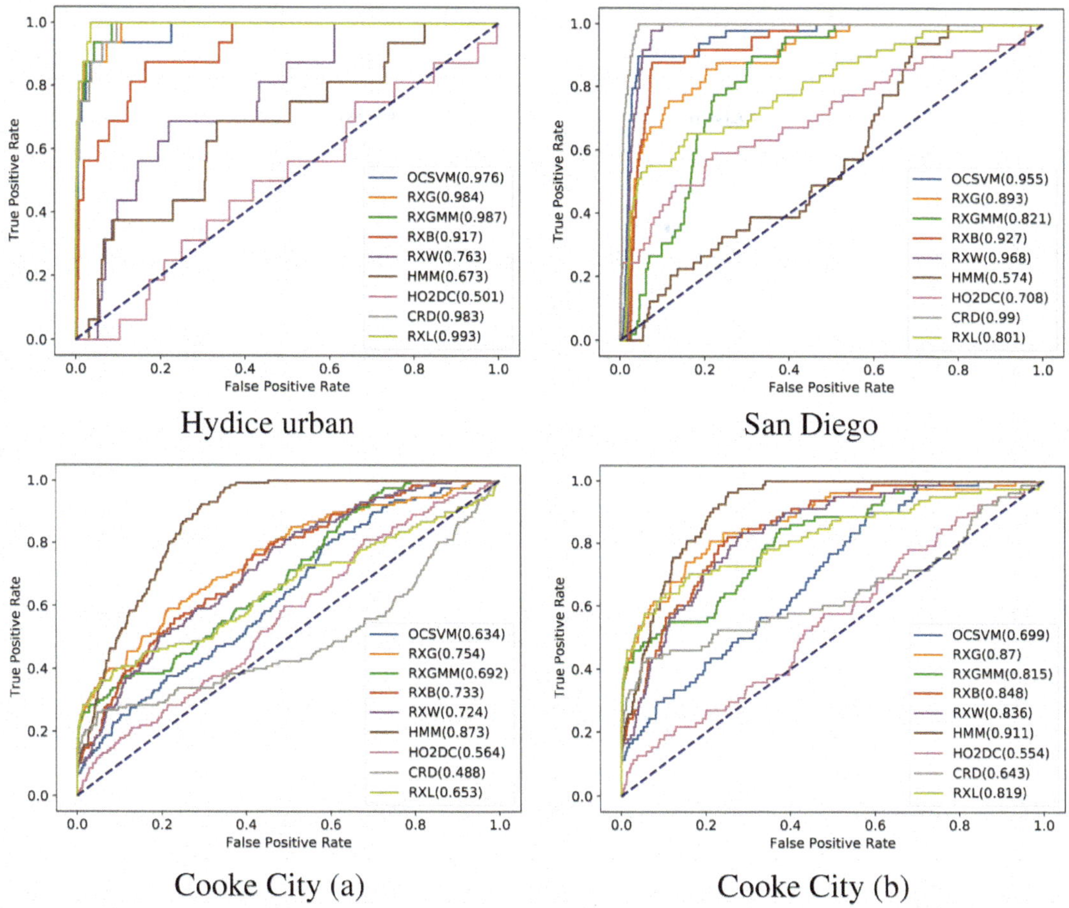

FIGURE 3–7 Detection performance in the form of receiver operating characteristic curves for tested detectors and datasets. Value in brackets near the detector's name is the area under curve. Two versions of results for the Cooke City dataset correspond to the two target maps, presented in Fig. 3–3.

Table 3–2 Approximate time of execution of tested detection algorithms in seconds.

Detector	Urban	San Diego	Cooke City
RXG	0.52	0.78	0.51
RXL	3450.98	6436.32	2645.27
RXB	6.37	10.45	29.91
RXGMM	35.96	26.46	17.66
RXW	27.64	44.86	53.28
HMM	5529.68	8523.70	3378.90
HO2DC	65.37	105.75	107.39
CRD	4773.68	7519.88	3553.05
OCSVM	3.49	7.64	5.40

RXG, Global RX; *RXL*, local RX; *RXB*, RX Bacon, *RXW*, weighted RX; *HMM*, hidden Markov model; *HO2DC*, high-order 2D crossing filter based detector; *CBD*, collaborative-based detector; *OCSVM*, one-class support vector machine.

The RXG seems to be highly sensitive to full pixel anomalous spectra, especially if they are visible on a small number of pixels in an image (e.g., Urban dataset). If the number of anomalous pixels is large, they may sufficiently influence the global covariance matrix to lower the detector sensitivity (e.g., San Diego dataset). The detector seems to be moderately sensitive to subpixel targets. Considering the low computational cost, stability, and predictability of the output and the fact that the method is nonparametric, RXG is a good reference when measuring AD performance.

The CRD detector seems to be highly sensitive to full pixel targets and to generate a small number of false positive results. On the other hand, it seems to be insensitive to subpixel targets. We can observe that both CRD and RXL seem to be less sensitive to background variability.

The OCSVM performance is good for full pixel targets, but mediocre for subpixel targets (Cooke City dataset). However, the output of the detector depends on its parameters. In particular, lowering the value of the γ parameter of the RBF kernel, thus, increasing the influence of support vectors seemed to reduce the performance for the Urban and San Diego datasets, but it increased the performance for the Cooke City dataset.

The HMM seems to perform poorly for the Urban and San Diego datasets, however, it gives the strongest responses for areas that are also favored by the OCSVM. On the other hand, for the Cooke City dataset, the HMM seems to be sensitive to anomalous subpixel targets, which suggests that it can be useful in applications where the searched signature is not present in the image.

The HO2DC positive response is structured in patches, separated by low response outputs. This structure seems to depend more on the background structure than anomalous targets, which corresponds to the ROC curve that is closest to the output of a random detector.

It is worth noting that since the number of anomalies used to test the detectors is typically small, the results, and in particular the FPR, may be biased (Manolakis et al., 2003). In addition, example difficulties related to measuring AD performance can be observed when analyzing detector responses for two versions of annotation for the Cooke City dataset, referred to as (a) and (b) in Fig. 3–3. Almost every detector is able to locate the white, anomalous arena in the upper right quarter of the images, but most of them have trouble in locating subpixel targets. Therefore initial detection rates are highly affected by this visible anomaly, which may produce "bumps" in the ROC curves when the detection results shift between pixels of the visible anomaly and the rest of the targets. On the other hand, the performance of the detectors for the ground truth map (b) generated using target detectors is, on average, better than for the (a) version. This suggests that anomaly detectors seem to be sensitive to the same subset of pixels that is detected by detectors used in ground truth estimation.

3.4 Conclusion

This chapter presents a basic overview of the AD problem and provides a representative review of state of the art target detection algorithms with different assumptions and approaches to the problem. Two main difficulties of AD are the inherent variability of background and the ambiguity of anomalies. Thus when no prior knowledge about both background and targets is available, nonparametric detectors based on statistical tests such as RXG are a natural choice. When assumptions about the nature of background are available (as, e.g., in satellite or aerial remote sensing), local detectors such as RXL or CRD or detectors based on the principle of high-density separation in the feature space such as OCSVM become viable. It is worth noting that when the information about targets becomes known, the problem may be reformulated as target detection (Manolakis et al., 2003). To sum it up, the task of finding an optimal target detector is difficult without additional assumptions relating to the qualitative properties of the results. The choice of detector should be determined by the application of prior knowledge and experimental analysis followed by an empirical assessment of the results. Achieving satisfactory performance of a detector is usually achieved by proper preprocessing of data and fine-tuning of the detector pipeline.

Acknowledgment

The authors would like to thank Lingxiao Zhu for making available the San Diego dataset.

List of abbreviations

AD anomaly detection
BACON blocked adaptive computationally efficient outlier nominator
CRD collaborative based detector
GMM Gaussian mixture model
HMM hidden Markov model
HO2DC high-order two-dimensional crossing filter-based detector
HSI hyper spectral images/imaging
LR likelihood ratio
PUT pixel under test (for anomaly detection)
OCSVM one-class support vector machine
RGB reed, green, blue imaging
RPCA robust principal component analysis
RX Reed-Xiaoli anomaly detector
SNR signal to noise ratio
SVM support vector machines

List of symbols

$\boldsymbol{X}$ random vector corresponding to the pixel under test
$\boldsymbol{x}$ a realization of the random vector X—a hyperspectral pixel
μ mean vector
Σ covariance matrix
N normal (Gaussian) distribution
$\oplus$ exclusive-or operation
$\boldsymbol{Pr}(\cdots)$ probability
$\{\ldots\}$ a set
$\boldsymbol{r}$ detector output

References

Animesh, P., Park, J.-M., 2007. An overview of anomaly detection techniques: existing solutions and latest technological trends. Comput. Netw. 51 (12), 3448–3470.

Arabi, S.Y.W., Fernandes, D., Pizarro, M.A., 2015. HMM for hyperspectral spectrum representation and classification with endmember entropy vectors. In: Proc. SPIE 9643, Image and Signal Processing for Remote Sensing XXI, 96430P. https://doi.org/10.1117/12.2194135

Banerjee, A., Burlina, P., Diehl, C., 2006. A support vector method for anomaly detection in hyperspectral imagery. IEEE Trans. Geosci. Remote Sens. 44 (8), 2282–2291.

Billor, N., Hadi, A.S., Velleman, P.F., 2000. Bacon: blocked adaptive computationally efficient outlier nominators. Computational Stat. Data Anal. 34 (3), 279–298.

Chandola, V., Banerjee, A., Kumar, V., 2009. Anomaly detection: a survey. ACM Comput. Surv. 41 (3), 15:1–15:58.

Chang, C.-I., 2003. Hyperspectral Imaging: Techniques for Spectral Detection and Classification. Springer, US.

Chang, C.-I., Chiang, S.-S., 2002. Anomaly detection and classification for hyperspectral imagery. IEEE Trans. Geosci. Remote Sens. 40 (6), 1314–1325.

Chang, C.-I., Du, Q., 2001. Hidden markov model approach to spectral analysis for hyperspectral imagery. Opt. Eng. 40 (10), 2277–2284. 8.

Chapelle, O., Zien, A., 2005. Semi-supervised classification by low density separation, AISTATS, vol. 2005. Citeseer, pp. 57–64.

Chen, Y., Nasrabadi, N.M., Tran, T.D., 2011. Sparse representation for target detection in hyperspectral imagery. IEEE J. Sel. Top. Signal. Process. 5 (3), 629–640.

Dobigeon, N., Moussaoui, S., Coulon, M., Tourneret, J.Y., Hero, A.O., 2009. Joint Bayesian endmember extraction and linear unmixing for hyperspectral imagery. IEEE Trans. Signal. Process. 57 (11), 4355–4368.

Dong, Y., Zhang, L., Zhang, L., Du, B., 2015. Maximum margin metric learning based target detection for hyperspectral images. ISPRS J. Photogram. Remote Sens. 108, 138–150.

Du, B., Zhang, L., 2011. Random-selection-based anomaly detector for hyperspectral imagery. IEEE Trans. Geosci. Remote Sens. 49 (5), 1578–1589.

Goldstein, M., Uchida, S., 2016. A comparative evaluation of unsupervised anomaly detection algorithms for multivariate data. PLOS One 11 (4), 1–31. 04.

Guo, Q., Pu, R., Cheng, J., 2016. Anomaly detection from hyperspectral remote sensing imagery. Geosciences 6 (4).

Gu, Y., Wang, Y., Zheng, H., Hu, Y., 2015. Hyperspectral target detection via exploiting spatial-spectral joint sparsity. Neurocomputing 169, 5–12.

Heesung Kwon, Nasrabadi, N.M., 2005. Kernel RX-algorithm: a nonlinear anomaly detector for hyperspectral imagery. IEEE Trans. Geosci. Remote Sens. 43 (2), 388–397.

Hodge, V.J., Austin, J., 2004. A survey of outlier detection methodologies. Artif. Intell. Rev. 22 (2), 85–126.

Lili, Z., Zhao, C., 2017. Hyperspectral anomaly detection based on spectral–spatial background joint sparse representation. Eur. J. Remote Sens. 50 (1), 362–376.

Liu, W., Feng, X., Wang, S., Hu, B., Gan, Y., Zhang, X., et al., 2018. Random selection-based adaptive saliency-weighted rxd anomaly detection for hyperspectral imagery. Int. J. Remote Sens. 39 (8), 2139–2158.

Liu, D., Li, J., 2018. Spectral-spatial target detection based on data field modeling for hyperspectral data. Chin. J. Aeronautics 31 (4), 795–805.

Li, W., Du, Q., 2015. Collaborative representation for hyperspectral anomaly detection. IEEE Trans. Geosci. Remote Sens. 53 (3), 1463–1474.

Li, F., Zhang, L., Zhang, X., Chen, Y., Jiang, D., Zhao, G., et al., 2018. Structured background modeling for hyperspectral anomaly detection. Sensors 18 (9), 3137.

Li, J., Zhang, H., Zhang, L., Ma, L., 2015. Hyperspectral anomaly detection by the use of background joint sparse representation. IEEE J. Sel. Top. Appl. Earth Obs. Remote Sens. 8 (6), 2523–2533.

Manolakis, D., Lockwood, R., Cooley, T., Jacobson, J., 2009. Is there a best hyperspectral detection algorithm? In: Proceedings of SPIE 7334, Algorithms and Technologies for Multispectral, Hyperspectral, and Ultraspectral Imagery XV, vol. 7334, 16 pp.

Manolakis, D., Marden, D., Shaw, G.A., et al., 2003. Hyperspectral image processing for automatic target detection applications. Linc. Lab. J. 14 (1), 79–116.

Manolakis, D., Shaw, G., 2002. Detection algorithms for hyperspectral imaging applications. IEEE Signal. Process. Mag. 19 (1), 29–43.

Maryam, I., 2018. Anomaly detection from hyperspectral images using clustering based feature reduction. J. Indian Soc. Remote Sens. 46 (9), 1389–1397.

Matteoli, S., Diani, M., Corsini, G., 2010. A tutorial overview of anomaly detection in hyperspectral images. IEEE Aerosp. Electron. Syst. Mag. 25 (7), 5–28.

Matteoli, S., Veracini, T., Diani, M., Corsini, G., 2013. Models and methods for automated background density estimation in hyperspectral anomaly detection. IEEE Trans. Geosci. Remote Sens. 51 (5), 2837–2852.

Ma, N., Peng, Y., Wang, S., Leong, P.H.W., 2018. An unsupervised deep hyperspectral anomaly detector. Sensors 18 (3), 693.

Qu, Y., Guo, R., Wang, W., Qi, H., Ayhan, B., Kwan, C., et al., 2016. Anomaly detection in hyperspectral images through spectral unmixing and low rank decomposition. In: 2016 IEEE International Geoscience and Remote Sensing Symposium (IGARSS), pp. 1855–1858.

Rabiner, L.R., 1989. A tutorial on hidden markov models and selected applications in speech recognition. Proc. IEEE 77 (2), 257–286.

Reed, I.S., Yu, X., 1990. Adaptive multiple-band CFAR detection of an optical pattern with unknown spectral distribution. IEEE Trans. Acoust. Speech Signal. Process. 38 (10), 1760–1770.

Scholkopf, B., Smola, A.J., 2001. Learning With Kernels: Support Vector Machines, Regularization, Optimization, and Beyond. MIT Press.

Schölkopf, B., Williamson, R.C., Smola, A.J., Shawe-Taylor, J., Platt, J.C., 2000. Support vector method for novelty detection. Adv. Neural Inf. Process. Syst. 582–588.

Stein, D.W.J., Beaven, S.G., Hoff, L.E., Winter, E.M., Schaum, A.P., Stocker, A.D., 2002. Anomaly detection from hyperspectral imagery. IEEE Signal. Process. Mag. 19 (1), 58–69.

Sun, W., Tian, L., Xu, Y., Du, B., Du, Q., 2018. A randomized subspace learning based anomaly detector for hyperspectral imagery. Remote Sens. 10 (3).

Swayze, G.A., Clark, R.N., Goetz, A.F.H., Livo, K.E., Breit, G.N., Kruse, F.A., et al., 2014. Mapping advanced argillic alteration at Cuprite, Nevada using imaging spectroscopy. Econ. Geol. 109 (5), 1179–1221.

Tax, D.M.J., Duin, R.P.W., 2004. Support vector data description. Mach. Learn. 54 (1), 45–66.

Vincent Poor, H., 1994. An Introduction to Signal Detection and Estimation. Springer-Verlag, New York.

Wang, Y., Lee, L., Xue, B., Wang, L., Song, M., Yu, C., et al., 2018. A posteriori hyperspectral anomaly detection for unlabeled classification. IEEE Trans. Geosci. Remote Sens. 56 (6), 3091–3106.

Wang, Z., Xue, J.-H., 2018. Matched shrunken subspace detectors for hyperspectral target detection. Neurocomputing 272, 226–236.

Yuan, Y., Ma, D., Wang, Q., 2016. Hyperspectral anomaly detection by graph pixel selection. IEEE Trans. Cybern. 46 (12), 3123–3134.

Yuan, Y., Wang, Q., Zhu, G., 2015. Fast hyperspectral anomaly detection via high-order 2-d crossing filter. IEEE Trans. Geosci. Remote Sens. 53 (2), 620–630.

Zhang, Y., Du, B., Zhang, L., Wang, S., 2016. A low-rank and sparse matrix decomposition-based mahalanobis distance method for hyperspectral anomaly detection. IEEE Trans. Geosci. Remote Sens. 54 (3), 1376–1389.

Zhang, Y., Zhang, L., Du, B., Wang, S., 2015. A nonlinear sparse representation-based binary hypothesis model for hyperspectral target detection. IEEE J. Sel. Top. Appl. Earth Obs. Remote Sens. 8 (6), 2513–2522.

Zhao, R., Du, B., Zhang, L., Zhang, L., 2016. A robust background regression based score estimation algorithm for hyperspectral anomaly detection. ISPRS J. Photogram. Remote Sens. 122, 126–144.

Zhao, C., Li, W., Li, X., Qi, B., 2015. Sparse representation based on stacked kernel for target detection in hyperspectral imagery. Opt. Int. J. Light Electron. Opt. 126 (24), 5633–5640.

Zheng, X., Yuan, Y., Lu, X., 2016. A target detection method for hyperspectral image based on mixture noise model. Neurocomputing 216, 331–341.

Zhu, F., Wang, Y., Fan, B., Xiang, S., Meng, G., Pan, C., 2014. Spectral unmixing via data-guided sparsity. IEEE Trans. Image Process. 23 (12), 5412–5427.

4

Atmospheric parameter retrieval and correction using hyperspectral data

Manoj K. Mishra, Bimal Bhattacharya

SPACE APPLICATIONS CENTRE, INDIAN SPACE RESEARCH ORGANIZATION, AHMEDABAD, INDIA

4.1 Introduction

Space/airborne imaging from hyperspectral optical spectrometers such as Hyperion and the Next Generation Airborne Visible/InfraRed Imaging Spectrometer (AVIRIS-NG) have a rich history of applications in a variety of fields such as mineral exploration, vegetation studies, ocean color, coastal monitoring, and atmospheric studies (Pearlman et al., 2003; Vane et al., 1993; Green et al., 1998; Hamlin et al., 2011; Gao and Goetz, 1990). The objective of remote sensing using hyperspectral sensors is to have quantitative measurements of components of the Earth System from calibrated spectra acquired as images for scientific research and applications. To achieve this, hyperspectral sensors measure the upwelling radiance in solar reflective spectrum from 400–2500 nm at a high spectral resolution on the order of around 5–10 nm and spatial resolution of 5–30 m. The measured constituent's absorption and scattering characteristics of the surface expressed in measured upwelling radiance spectrum are then exploited to accomplish various applications such as to detect/identify the distribution of surface materials and to monitor their change through periodic imaging. Major research or application areas that are connected to the distribution of materials are:

- vegetation (species type, chemistry, and absorbing and scattering constituents, etc.);
- open ocean, coastal, and inland water (in water constituents, ocean color, ocean biology, water quality, etc.);
- human generated minerals, snow and ice, hazards, and rocks and soils (minerals, altered minerals, etc.); and
- atmosphere (absorbers, scatters, aerosols, trace gases, air quality, etc.).

Most of the remote sensing applications require surface reflectance (the level-2 product) as primary input data, while radiance/reflectance spectrum (the level-1 radiometrically calibrated radiance/reflectance product)measured from space/airborne sensors is significantly

Hyperspectral Remote Sensing. DOI: https://doi.org/10.1016/B978-0-08-102894-0.00005-X

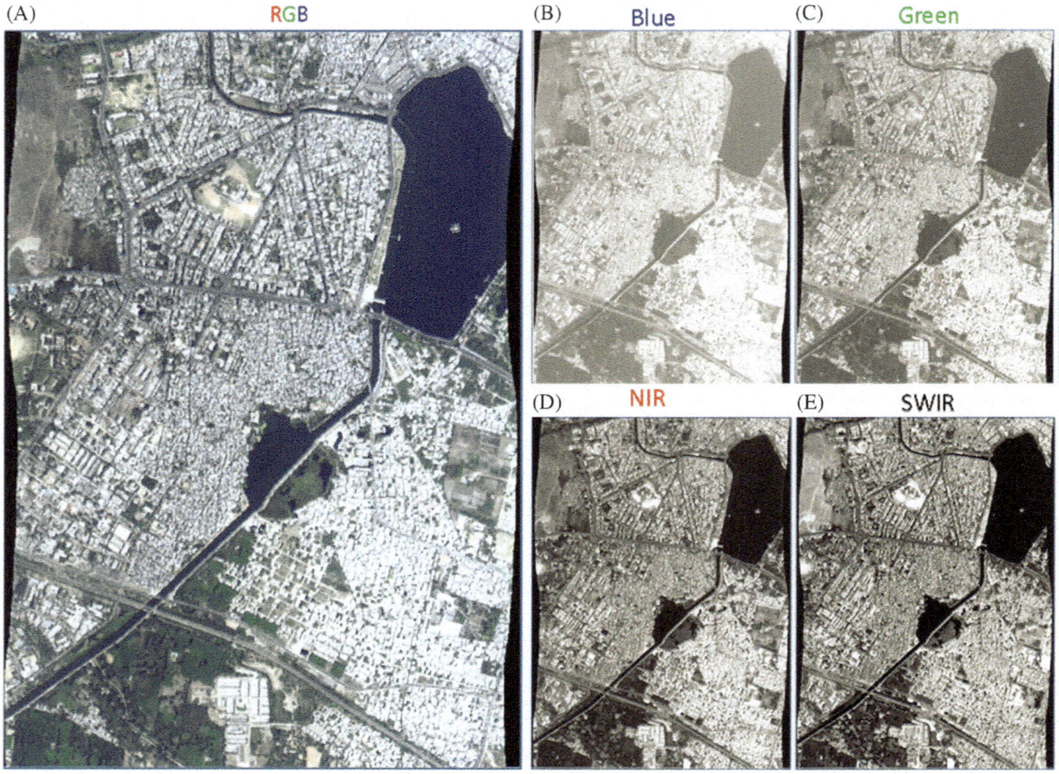

FIGURE 4–1 The effect of urban pollution (soot fine particles) at Kota. (A) Natural color composite image (RGB) made using sensor level apparent reflectance in red band (650 nm), green band (550 nm), and blue band (445 nm). (B) Blue band image. (C) Green band image. (D) Near-infrared (NIR) band image. (E) Shortwave infrared (SWIR) band image. The low contrast and haziness in RGB, blue, green, and NIR bands is due to the aerosol scattering effect. Due to the insensitivity of SWIR wavelengths to aerosol, the SWIR image shows high contrasting surface features.

contaminated by the absorption and scattering from atmospheric constituents such as aerosol and various atmospheric gases. As an example, Fig. 4–1 shows the impact of aerosol scattering on the natural color of a composite image remotely sensed by an airborne AVIRIS-NG sensor at an altitude of about 8 km over Kota city, India. The low contrast and haziness in RGB, blue, green, and near-infrared (NIR) bands are associated with significant aerosol scattering because of heavy urban pollution, especially due to an operational super thermal power station at Kota that emits a lot of carbonaceous aerosol particles into the atmosphere. Unlike visible and NIR images where prominent aerosol scattering makes images hazy, the high contrasting surface features in shortwave infrared (SWIR) band images are visible, which is due to the fact that aerosol scattering becomes negligibly small in SWIR wavelengths. Moreover, high Rayleigh scattering by gaseous molecules also decreases the visibility of contrasting surface features in visible wavelengths. Fig. 4–2 shows simulated atmospheric

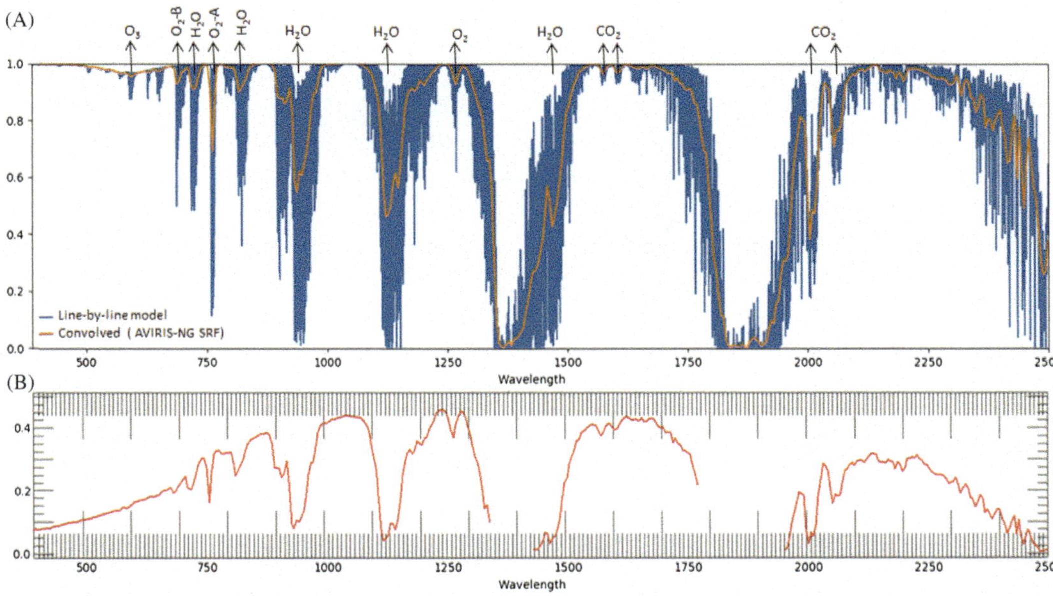

FIGURE 4–2 (A) Blue curve shows line-by-line simulated downward atmospheric transmittance for standard tropical atmosphere with water vapor of 1 cm. Red curve shows transmittance convolved with Next Generation Airborne Visible/InfraRed Imaging Spectrometer (AVIRIS-NG) spectral response function. Upward arrows show major gas absorption regions. (B) A sensor level reflectance spectrum acquired by the AVIRIS-NG over a bare soil pixel. The impact of gas absorption on the measured spectrum, leading to sharp dips in regions represented by the upward arrows (panel A), are clearly apparent.

transmittance for standard tropical atmosphere with water vapor equal to 1 cm and the effect of gas absorption on sensor level reflectance spectrum due to radiation absorption from water vapor and other trace gases such as carbon dioxide, ozone, oxygen, etc. Almost half of the 0.4–2.5 μm spectral region is affected by atmospheric gas absorptions and the shorter wavelength region below 1.0 μm is significantly affected by molecular and aerosol scattering. In order to study the distribution and quantification of surface materials using hyperspectral data, the accurate removal of atmospheric effects is required. The removal of atmospheric effects and the conversion of the radiance measured by a sensor to the reflectance of surface materials is necessary and the method to achieve this is called atmospheric correction.

4.2 Atmospheric correction techniques

Since the mid-1980s, atmospheric correction algorithms have evolved from the earlier empirical line method, dark object subtraction, and the flat-field method to more current methods based on rigorous radiative transfer modeling such as the atmospheric removal (ATREM) algorithm and the fast line-of-sight atmospheric analysis of hypercubes (FLAASH) algorithm (Gao and Goetz, 1990; Adler-Golden et al., 1999).

4.2.1 Nonphysical models for atmospheric correction

The empirical line method for atmospheric correction is just the scaling of measured image spectra that forces the image spectra to match reflectance spectra collected from the field (Karpouzli and Malthus, 2003). This method is capable of producing the accurate surface reflectance product, but it requires ground truth information. Moreover, highly heterogeneous atmosphere may require ground truth information at many numbers of pixels. The method of dark object subtraction assumes that dark objects reflect no light; therefore, any nonzero reflectance value must result from atmospheric scattering (Teillet and Fedosejevs, 1995). The method searches each band for the darkest pixel value. The atmospheric effects due to scattering are removed by subtracting the darkest pixel value from every pixel in the band. This simple technique is effective for haze correction in multispectral data, but it should not be used for hyperspectral data. The flat-field calibration method produces relative reflectance by dividing the mean spectrum of a user-defined region of interest into the spectrum of each pixel in the image (Roberts et al., 1986). The region of interest should be a spectrally flat material within the wavelength range of the sensor. Beach sand and concrete are popular choices. Materials with spectral features such as vegetation are poor choices. Since the mean spectrum of the region of interest (ROI) is divided into each pixel, the relative reflectance for pixels within the ROI will be flat and have a value around 1.0. All these methods require at least some ground information and are statistical in nature. Most of these methods may give good results for data from multispectral sensors designed to work in atmospheric windows free from gas absorption effects. However, for hyperspectral data in solar reflective spectrum, the gaseous absorption is unavoidable and truly no material is perfectly flat and dark; therefore, these methods should not be used for atmospheric correction of hyperspectral data.

4.2.2 Physics-based models for atmospheric correction

For these reasons, in recent past, new methods based on rigorous radiative transfer modeling such as ATREM and FLAASH have been developed (Gao and Goetz, 1990; Adler-Golden et al., 1999). Due to the high temporal and spatial dynamicity of aerosol and water vapor, the most difficult task in atmospheric correction of any hyperspectral or multispectral remote sensing data is to derive an accurate aerosol optical depth (AOD) and water vapor content. Atmospheric correction models such as ATREM and FLAASH retrieve the water vapor content for each pixel of a given scene and a single value of AOD for the entire scene before atmospheric correction. Apart from the requirement of accurate aerosol and water vapor information for atmospheric correction, these parameters are also useful for many applications such as determining aerosol and water vapor radiative forcing, Earth's radiation budget, etc. Atmospheric carbon dioxide (CO_2) is another significant greenhouse gas, which is increasing at a rapid rate (Etheridge et al., 1996; Leung et al., 2014), and CO_2 monitoring is executed with ground-based and satellite-borne spectroscopic sensors (Wunch et al., 2017; Bovensmann et al., 1999; Hamazaki et al., 2004, Frankenberg et al., 2015). The spatial and temporal variability of atmospheric CO_2 modulates the sensor level signal and, therefore, the retrieval of CO_2 is important for accurate atmospheric correction, especially in the SWIR region ranging from 1.93 to 2.12 μm.

The upwelling reflectance viewed at z altitude by any sensor, for a Lambertian surface, can be approximated as (Kotchenova et al., 2006; Lenoble, 1985; Tanre et al., 1990; Vermote et al., 1997):

$$\rho_\lambda^*(\theta_o,\theta,\varphi,z) = T^{gas}(\theta_o,\theta,\varphi,z)\times\left[\rho_\lambda^{atm}(\theta_o,\theta,\varphi,z)+\frac{T(\theta_o)T(\theta,z)\rho_\lambda^s(\theta_o,\theta,\varphi)}{1-S_\lambda\rho_\lambda^s(\theta_o,\theta,\varphi)}\right]. \tag{4.1}$$

Here $\rho_\lambda^*(\theta_o,\theta,\varphi,z)=\pi L_\lambda^*(\theta_o,\theta,\varphi,z)/E_s,\cos(\theta_o)$, where $L_\lambda^*(\theta_o,\theta,\varphi,z)$ is the upwelling radiance at-sensor level and $E_{s,\lambda}$ is the solar flux at the top of the atmosphere (TOA). The terms on the right side of Eq. (4.1) are the surface reflectance $\rho_\lambda^s(\theta_o,\theta,\varphi)$ and the atmospheric quantities joined to the radiative field in the coupled system. The at-sensor radiance like all of the radiative quantities of Eq. (4.1), depends on the radiation line-sight expressed by the angles θ_o, θ, φ, and z, for the solar zenith (with respect to nadir), sensor zenith (with respect to nadir), relative azimuth, and the sensor altitude respectively. The radiative quantities of Eq. (4.1) are the intrinsic reflectance of the molecule and aerosol layer, $\rho_\lambda^{atm}(\theta_o,\theta,\varphi,z)$, called path reflectance; the spherical albedo, S_λ, and the components of the flux transmission. $T^{gas}(\theta_o,\theta,\varphi,z)$ is the gaseous transmittance, whereas $T(\theta_o)=\exp\left(-\tau/\cos(\theta_o)+t_d(\theta_o)\right)$ and $T(\theta)=\exp\left(-\tau_z/\cos(\theta)+t_d(\theta)\right)$ are the summed direct and diffuse components respectively of the total transmittance for the illumination (descending) and view (ascending) directions. τ is the total optical thickness and τ_z is the optical thickness of the layer under the aircraft. The steps involved in atmospheric correction are shown in Fig. 4–1.

The objective of atmospheric correction is to derive surface reflectance, $\rho_\lambda^s(\theta_o,\theta,\varphi)$, from at-sensor radiance, $L_\lambda^*(\theta_o,\theta,\varphi,z)$. To do this, all the quantities on the right hand side (RHS) of Eq. (4.1) except for $\rho_\lambda^s(\theta_o,\theta,\varphi)$ should be known. Among all the quantities on the RHS of Eq. (4.1), the estimation of $\rho_\lambda^{atm}(\theta_o,\theta,\varphi,z)$ and $T^{gas}(\theta_o,\theta,\varphi,z)$ are the most difficult. The quantity of $\rho_\lambda^{atm}(\theta_o,\theta,\varphi,z)$ depends on the AOD, while for $T^{gas}(\theta_o,\theta,\varphi,z)$, it depends on atmospheric gases (especially water vapor). Once AOD and water vapor are known, these can be used as input to run a radiative transfer code like "Second Simulation of the Satellite Signal in the Solar Spectrum" (6S) or MODTRAN to estimate all other quantities except $\rho_\lambda^s(\theta_o,\theta,\varphi)$ in Eq. (4.1) to complete the task of atmospheric correction (Vermote et al., 1997; Adler-Golden et al., 1999). Therefore the main problem in atmospheric correction is to derive an accurate AOD and water vapor. Fig. 4–3 shows the steps involved in atmospheric correction based on radiative transfer calculations. In this method, for every image pixel, radiative transfer simulation is required. However, to reduce the processing time, precomputed look-up tables (LUTs) of various radiative quantities such as transmittance due to atmospheric gases as a function of concentration, aerosol-molecular path reflectance, transmittance and spherical albedo as a function of AOD and aerosol model are saved separately in the form of LUTs, which are used while processing data. These LUTs are also the function of wavelength and sun-sensor geometry.

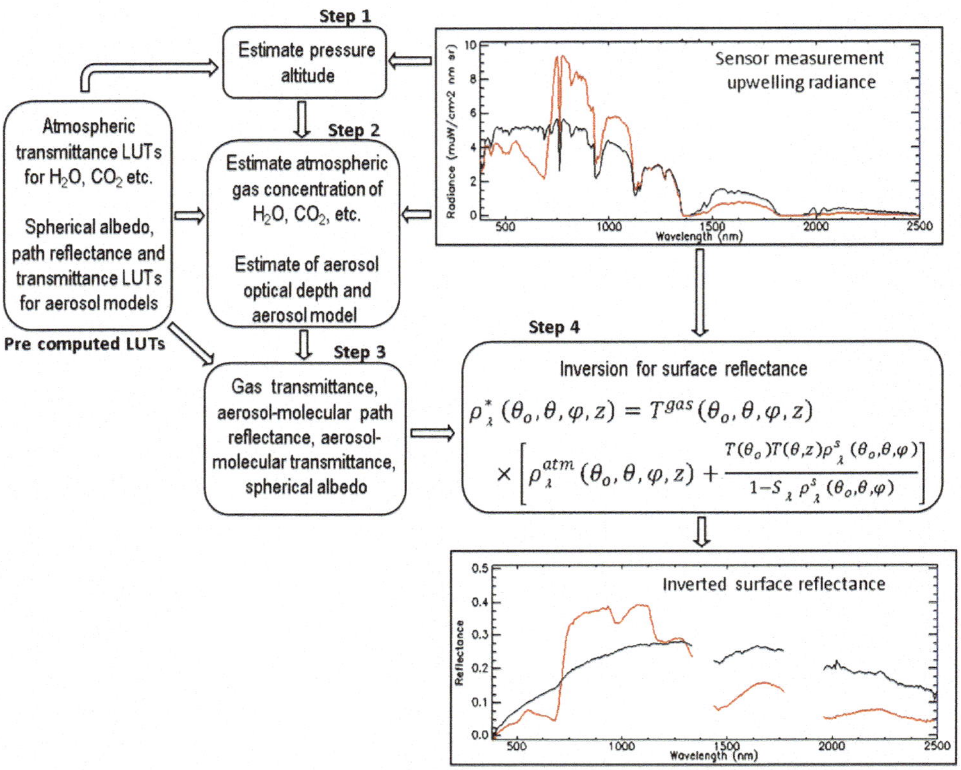

FIGURE 4–3 Steps involved in atmospheric correction approach based on radiative transfer calculation. The variables used in the equation in Step 4 are similar to the variables described in Eq. (4.1) in text. *LUTs*, Look-up tables.

4.3 Aerosol retrieval method

Unlike over water surfaces where AOD can be easily retrieved for each pixel by approximating the water surface as dark in NIR wavelengths due to absorption, aerosol retrieval methods over land are necessarily different and more complex due to the high reflectivity of land surfaces in the entire spectrum from 0.4 to 2.5 μm (Mishra et al., 2019). The dark dense vegetation (DDV) method is the most widely used aerosol retrieval method over land and it gives the AOD and best-fit aerosol model over reference vegetated pixels. In the DDV method, the correlation between the surface reflectance of the spectral channels in the visible and shortwave infrared (2.1 μm) are used to decouple the sensor-measured signal into contributions from the surface and atmosphere over reference vegetated pixels (Kaufman et al., 1997, 2002; Mishra et al., 2019). The atmospheric contribution over these reference pixels is then inverted to AOD at 0.55 μm based on a LUT approach, that is, radiative transfer calculations are precomputed for a set of aerosol models and AOD values.

Models such as ATREM and FLAASH assume a single value of AOD for the atmospheric correction of an entire scene; however, in reality, over regions with challenging atmospheres,

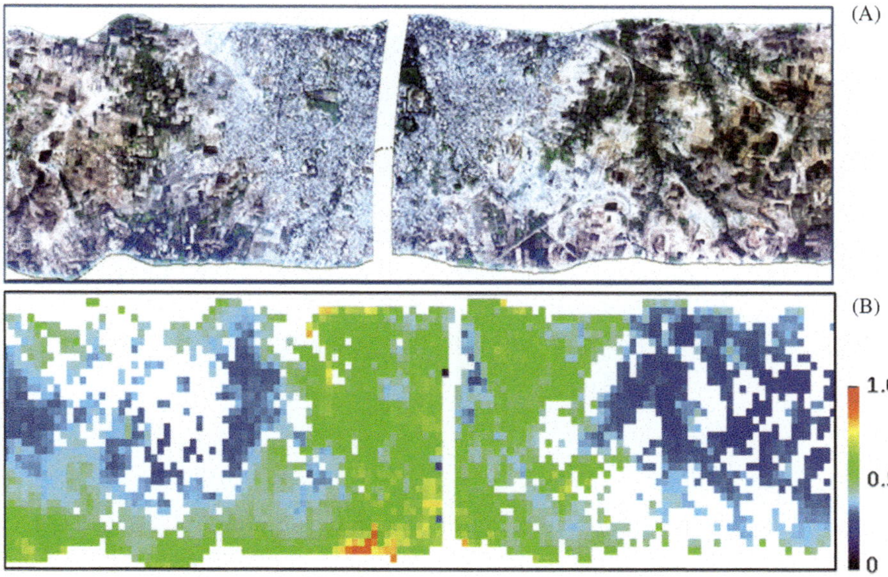

FIGURE 4–4 (A) An AVIRIS-NG true color image acquired on December 18, 2015, over Kalaburgi, Karnataka, at 10:42 UTC. (B) Aerosol optical depth map derived using the dark dense vegetation method and dividing the whole image into 20 × 20 pixel boxes. The AOD map shows significant variability of aerosol optical depth over the scene.

the assumption of a single AOD value for atmospheric correction of the whole scene does not always hold due to the heterogeneous nature of anthropogenic aerosol sources and high aerosol loading. Mishra et al. (2019) used the DDV method to derive a spatially variable AOD field over an AVIRIS-NG scene by dividing the entire scene into several boxes of 20 × 20 pixels. Fig. 4–4 shows the AOD map derived by dividing a whole AVIRIS-NG scene into 20 × 20 pixel boxes using the DDV method. The AOD map shows significant spatial variability of aerosol concentration, which must be taken into account when processing the data for atmospheric correction. Fig. 4–5A and B shows the effect of using a single AOD value and a spatially variable AOD for atmospheric correction on natural color composite images to derive surface reflectance respectively. It is clear that when spatially variable AOD is used, the localized haze (upper left corner of the image in Fig. 4–5A and B) is significantly reduced after atmospheric correction relative to when only one value of AOD for an entire scene is used.

AOD inversion is extremely sensitive to the selection of aerosol models. In most available models such as ATREM and FLAASH, a few predefined aerosol models such as continental, maritime, and desert dust urban aerosols models are available; however, if the actual aerosol type is considerably different from the assumed one, it may lead to high uncertainty in inverted AOD. The DDV method for AOD retrieval primarily uses the relationship between the apparent reflectance (the reflectance measured by the sensor at the TOA) data from the 2.1 μm channel and the surface albedo of the red or blue channels in the dark target area to remove the contribution of surface reflection in the red or blue channels. Since this

(A) (B)

FIGURE 4–5 Effect of aerosol correction on visual appearance of true color image made using (A) the top of atmosphere reflectance and (B) the atmospherically corrected surface reflectance. The AVIRIS-NG data used was acquired on December 18, 2015, over Kalaburgi, Karnataka, at 10:42 UTC.

relationship is not appreciably good for high aerosol loadings or for high optical paths, it leads to appreciable uncertainty in inverted AOD in such cases. For improving the AOD retrieval, more aerosol models representing the region of interest should be used.

4.4 Water vapor and other trace gas retrieval

Many atmospheric gases affect solar radiation while it moves from the top of the atmosphere to the surface and from the surface to the sensor. However, every gas does not produce observable absorption features in the spectral range of 0.4–2.5 μm under typical atmospheric conditions and at 5 nm AVIRIS-NG spectral resolution. Only about seven gases, namely water vapor, CO_2, ozone, nitrous oxide, carbon monoxide, methane, and oxygen show appreciable absorption of solar radiation in the spectral range of 0.4–2.5 μm under interest. CO_2, nitrous oxide, methane, and oxygen are uniformly mixed gases in the troposphere. Approximately 90% of the ozone is located in the stratosphere and the ozone concentration changes with season and latitude. Therefore, transmittance spectra for all gaseous absorption except for water vapor can be modeled with an assumed atmospheric model and climatology

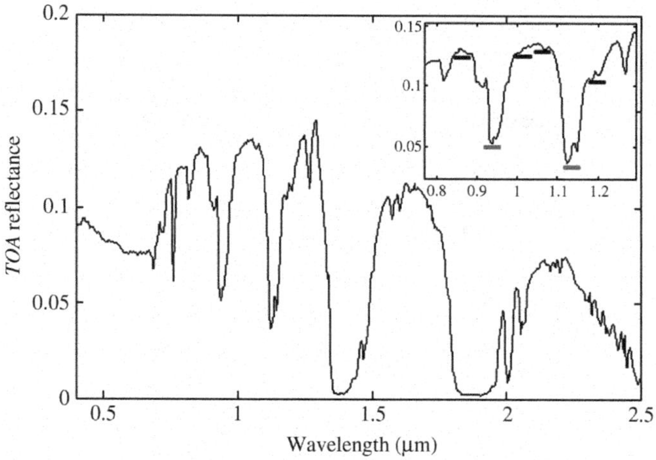

FIGURE 4–6 Retrieved water vapor map (unit: millimeter) using the continuum interpolated band ratio method. The AVIRIS-NG data used was acquired on December 18, 2015, over Kalaburgi, Karnataka. *TOA*, Top of the atmosphere.

database. The water vapor concentration varies rapidly with altitude. Moreover, the spatial and temporal variability of water vapor can be dramatic (Teillit, 1989), therefore, it is not possible to completely remove the water vapor features from hyperspectral data using standard atmospheric models. The most widely used water vapor retrieval method, namely continuum interpolated band ratio (CIBR) method, utilizes the strong water vapor absorption features at 0.94 and 1.14 μm (Bruegge et al., 1990; Frouin et al., 1990). Water vapor retrieval using the CIBR method is based on two assumptions:

1. The atmospheric transmittance of the 0.94 and 1.14 μm water vapor bands are appreciably sensitive to the column water vapor amount (Fig. 4–6).
2. The surface reflectance commonly varies linearly with wavelength in these two water vapor bands.

The steps involved in water vapor retrieval are:

Step 1: Calculate the apparent at-sensor or TOA reflectance observed by AVIRIS-NG and find the average apparent reflectance in predefined AVIRIS-NG channels centered at 0.865, 0.94, 1.03, 1.05, 1.14, and 1.23 μm (shown by horizontal lines in Fig. 4–5 inset).
Step 2: Calculate channel ratios at 0.94 and 1.14 μm regions defined as:

$$\begin{aligned} \tau_g(\lambda = 0.94) &= \frac{\rho(0.94)}{0.5[\rho(0.85) + \rho(1.03)]} \\ \tau_g(\lambda \approx 1.14) &= \frac{\rho(1.13)}{0.5[\rho(1.05) + \rho(1.23)]} \end{aligned} \tag{4.2}$$

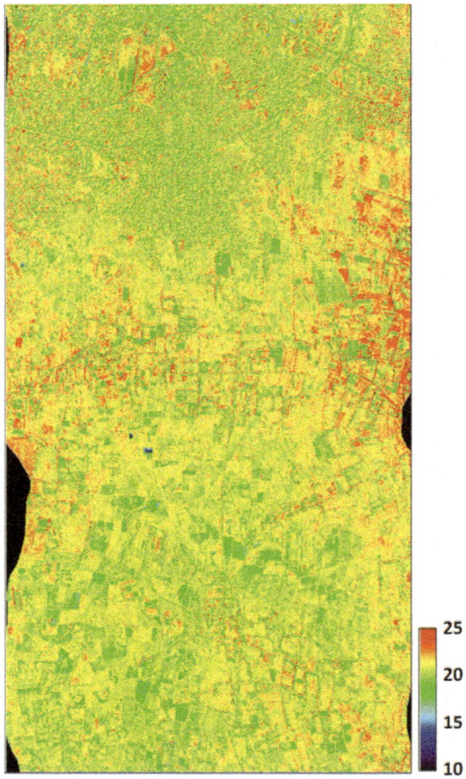

FIGURE 4–7 Example of an AVIRIS-NG at-sensor reflectance spectrum. Inset image shows the two water vapor absorption bands used for water vapor retrieval using the continuum interpolated band ratio method. The horizontal bars show the range for which AVIRIS reflectance is averaged to find band ratios.

Step 3: Model the theoretical transmittance spectrum for a given sensor-solar geometry for different values of water vapor amount using a radiative transfer model. Here, 6S code was used for radiative transfer computation. Fig. 4–7 gives the simulated gas transmittance plot.

Step 4: Compare band ratios at two water vapor bands with the theoretical transmittance spectrum for different water vapor contents and derive water vapor by interpolation. Repeat for every pixel.

Fig. 4–6 shows the water vapor map derived using the CIBIR method and AVIRIS-NG data acquired on December 18, 2015, over Kalaburgi, Karnataka. It should be noted that within one AVIRIS-NG flight path line with dimensions on the order of a few square kilometers, the variation of water vapor is significant, for example, in the range of 15–25 mm in this particular case (Fig. 4–6). For a detailed description and validation of water vapor derived using AVRIS-NG data with the aid of the CIBR method, see Mishra et al. (2019).

4.5 Atmospheric correction results and discussion

Using the retrieved AOD map, water vapor and the standard atmospheric values for well-mixed trace gaseous, CO_2, CH_4, O_2, and O_3, in a radiative transfer model, the aerosol-dependent radiative quantities, path reflectance $\rho_\lambda^{atm}(\theta_o, \theta, \varphi, z)$, spherical albedo S_λ downward transmittance $T(\theta_o)$ and upward transmittance $T(\theta)$, and the gaseous transmittance $T^{gas}(\theta_o, \theta, \varphi, z)$ can be estimated for each pixel of hyperspectral image cube. Having all these unknowns on the RHS of Eq. (4.1) and putting sensor-measured reflectance on the left hand side of Eq. (4.1), the solution for surface reflectance $\rho_\lambda^s(\theta_o, \theta, \varphi)$ can be obtained. Fig. 4–8 shows the spectra for derived surface reflectance (*blue curves*) after atmospheric correction and TOA apparent reflectance curve (*red curves*) over different types of target pixels for AVIRIS-NG data acquired over Jawar, Madhya Pradesh, India. It is clear from Fig. 4–8 (*red curves*) that almost half of the spectrum is contaminated by water vapor absorption with strong and broad absorption features around 0.82, 0.94, and 1.14 μm and numerous narrow absorption features at various wavelengths in the SWIR region (2.0–2.5 μm). After atmospheric correction, the absorption features due to water vapor absorption are faithfully removed (*blue curves* in Fig. 4–8). The dips in apparent reflectance due to other trace gases such as oxygen (near 0.76 μm) and CO_2 (1.9–2.12 μm) are also corrected. It is seen that due to aerosol scattering, the visible–NIR spectrum is modulated where increased reflectance values are observed due to which the shape of the spectrum deviates from the shape of the feature spectrum. For example, for vegetation target, the *red curve* in Fig. 4–8 shows a continuous increase in reflectance as we move from red to blue wavelength due to aerosol scattering. However, after atmospheric correction, it is seen that the spectrum shape of the vegetated targets (the *blue curve*) are faithfully recovered and becomes similar to the vegetation spectrum. It should be noted that over water surfaces, instead of using the DDV

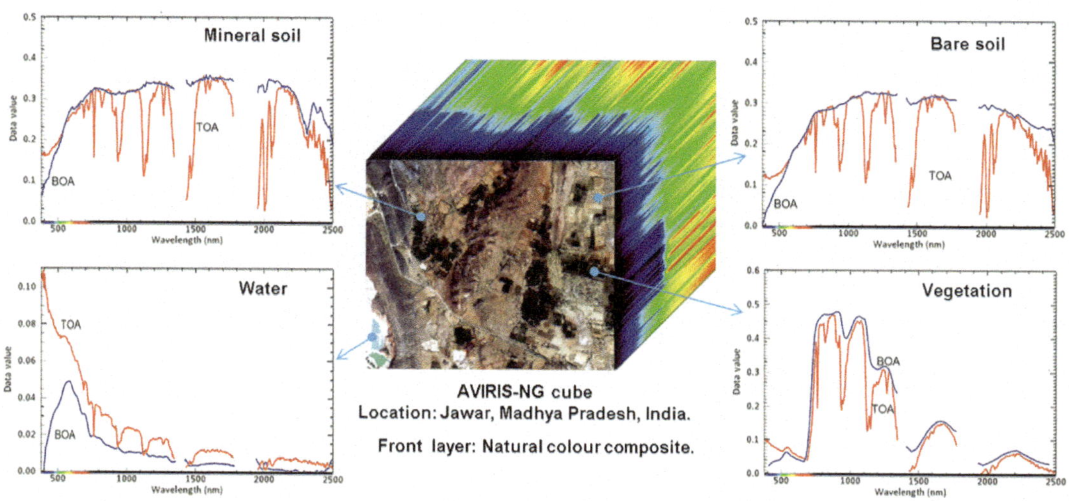

FIGURE 4–8 The spectra for derived surface reflectance (*blue curves*) after atmospheric correction and top of atmosphere apparent reflectance curve (*red curves*) over different types of target pixels for AVIRIS-NG data acquired over Jawar, Madhya Pradesh, India. *BOA*, Bottom of the atmosphere; *TOA*, top of the atmosphere.

algorithm, the aerosol retrieval is done using NIR bands assuming that the water surface is completely dark (Gordon et al., 1988; Gordon, 1997; Mishra et al., 2019). Over a water surface, the solution of Eq. (4.1) gives a surface reflectance that contains the contribution of sun glint, white foam, and water leaving radiance.

4.6 Conclusion

The space/airborne hyperspectral spectrometers are of prime importance in monitoring the Earth surface and understanding land or ocean surface processes. In order to fulfil such a goal, the retrieval of critical atmospheric parameters like aerosol and trace gases and atmospheric correction of sensor level radiance to derive surface reflectance is equally important. Various physical and non-physical atmospheric correction methods as well as water vapor and aerosol retrieval algorithms are described. The influence of the atmospheric constituents namely, columnar water vapor, aerosol optical depth and Rayleigh molecules and trace gases such as ozone content on hyperspectral data is quantified using radiative transfer models. The results show significant influence of atmosphere on surface signal when observed from hyperspectral sensor onboard space/airborne platforms. It is concluded that the atmospheric correction algorithms based on accurate radiative transfer models are capable of generating surface reflectance as well as atmospheric parameters (aerosol, water vapor, etc.) product of required quality.

List of abbreviations

AOD	aerosol optical depth
ATREM	atmospheric removal
AVIRIS-NG	Airborne Visible/InfraRed Imaging Spectrometer
CIBR	continuum interpolated band ratio
CZCS	coastal zone colour scanner
DDV	dark dense vegetation
FLAASH	fast line-of-site atmospheric analysis of spectral hypercubes
HICO	hyperspectral imager for the coastal ocean
LUTs	look-up table
NIR	near-infrared
OAC's	optically active constituents
RGB	red-blue-green
ROI	region of interest
SPM	suspended particulate matter
SSC	suspended sediment concentrations
SWIR	shortwave infrared
TOA	top of the atmosphere
UTC	coordinated universal time

List of symbols

λ	wavelength
$E_{s,\lambda}$	solar flux at the top of the atmosphere

$L^*_\lambda(\theta_o, \theta, \varphi, z)$	upwelling radiance at-sensor level
S_λ	spherical albedo
$T^{gas}(\theta_o, \theta, \varphi, z)$	total atmospheric gas transmittance
$t_d(\theta_o)$	diffused transmittance for the illumination (descending) direction
$t_d(\theta)$	diffused transmittance for the view (ascending) direction
θ_o	solar zenith angle
$\rho^*_\lambda(\theta_o, \theta, \varphi, z)$	upwelling reflectance at-sensor level
$\rho^{atm}_\lambda(\theta_o, \theta, \varphi, z)$	path reflectance due to aerosol and molecular interaction with radiation
$\rho^s_\lambda(\theta_o, \theta, \varphi)$	surface reflectance
τ_z	optical thickness of the layer under the aircraft
6S	Second Simulation of the Satellite Signal in the Solar Spectrum
CH_4	Methane
Chl-a	Chlorophyll-a
CO_2	Carbon dioxide
O_2	Oxygen
O_3	Ozone
z	sensor altitude
$T(\theta_o)$	total (direct and diffuse) transmittance for the illumination (descending) direction
$T(\theta)$	total (direct and diffuse) transmittance for the view (ascending) direction
θ	sensor zenith angle
τ	total optical thickness
φ	solar-sensor relative azimuth angle

References

Adler-Golden, S.M., Matthew, M.W., Bernstein, L.S., Levine, R.Y., Berk, A., Richtsmeier, S.C., et al., 1999. Atmospheric correction for short-wave spectral imagery based on MODTRAN4, In: SPIE Proceeding, Imaging Spectrometry V, vol. 3753, pp. 61–69.

Bovensmann, H., Burrows, J.P., Buchwitz, M., Frerick, J., Noel, S., Rozanov, V.V., 1999. SCIAMACHY: mission objectives and measurement modes. J. Atmos. Sci. 5, 127–150.

Bruegge, C.J., Conel, J.E., Margolis, J., Green, R.O., 1990. In-situ atmospheric water-vapour retrieval in support of AVIRIS validation, Imaging Spectroscopy of the Terrestrial Environment, vol. 1298. SPIE, pp. 150–163.

Etheridge, D.M., Steele, L.P., Langenfelds, R.L., Francey, R.J., Barnola, J.M., Morgan, V.I., 1996. Natural and anthropogenic changes in atmospheric CO_2 over the last 1000 years from air in Antarctic ice and firn. J. Geophys. Res. 101, 4115–4128.

Frankenberg, C., Pollock, R., Lee, R.A.M., Rosenberg, R., Blavierm, J.F., Crisp, D., et al., 2015. The orbiting carbon observatory (OCO-2): spectrometer performance evaluation using pre-launch direct sun measurements. Atmos. Meas. Tech. 8, 301–313.

Frouin, R., Deschamps, P.Y., Lecomte, P., 1990. Determination from space of atmospheric total water vapour amounts by differential absorption near 940 nm: theory and airborne verification. J. Appl. Meteorol. 29, 448–460.

Gao, B.C., Goetz, A.F., 1990. Column atmospheric water vapour and vegetation liquid water retrievals from airborne imaging spectrometer data. J. Geophys. Res. – Atmos. 95 (D4), 3549–3564.

Gordon, H.R., 1997. Atmospheric correction of ocean color imagery in the Earth observing system era. J. Geophys. Res. 102, 17081–17106.

Gordon, H.R., Brown, J.W., Evans, R.H., 1988. Exact Rayleigh scattering calculations for use with the Nimbus 7 coastal zone color scanner. Appl. Opt. 27, 862–871.

Green, R.O., Eastwood, M.L., Sarture, C.M., Chrien, T.G., Aronsson, M., Chippendale, B.J., 1998. Imaging spectroscopy and the Airborne Visible/Infrared Imaging Spectrometer (AVIRIS). Remote Sens. Environ. 65 (3), 227–248.

Hamazaki, T., Kaneko, Y., Kuze, A., 2004. Carbon dioxide monitoring from the GOSAT satellite. In: Proceedings of XXth ISPRS Conference, Istanbul, Turkey, 12–13 July 2004, pp. 3–5.

Hamlin, L., Green, R.O., Mouroulis, P., Eastwood, M., Wilson, D., Dudik, M., et al., 2011. Imaging spectrometer science measurements for terrestrial ecology: AVIRIS and new developments. In: IEEE Aerospace Conference Proceedings, pp. 1–7.

Karpouzli, E., Malthus, T., 2003. The empirical line method for the atmospheric correction of IKONOS imagery. Int. J. Remote Sens. 24 (5), 1143–1150.

Kaufman, Y.J., Wald, A., Remer, L.A., Gao, B.C., Li, R.R., Flynn, L., 1997. The MODIS 2.1-μm channel—correlation with visible reflectance for use in remote sensing of aerosol. IEEE Trans. Geosci. Remote Sens. 35 (5), 1286.

Kaufman, Y.J., Gobron, N., Pinty, B., Widlowski, J.L., Verstraete, M.M., 2002. Relationship between surface reflectance in the visible and mid-IR used in MODIS aerosol algorithm—theory. Geophys. Res. Lett. 29 (23), 2116.

Kotchenova, S.Y., Vermote, E.F., Matarrese, R., Klemm Jr, F.J., 2006. Validation of a vector version of the 6S radiative transfer code for atmospheric correction of satellite data. Part I: Path radiance. Appl. Opt. 45 (26), 6762.

Lenoble, J., 1985. Radiative Transfer in Scattering and Absorbing Atmospheres: Standard Computational Procedures. A. Deepak Publishing.

Leung, D.Y.C., Caramanna, G., Maroto-Valer, M.M., 2014. An overview of current status of carbon dioxide capture and storage technologies. Renew. Sustain. Energy Rev. 39, 426–443.

Mishra, M.K., Gupta, A., John, J., Shukla, B.P., Dennison, P., Srivastava, S.S., et al., 2019. Retrieval of atmospheric parameters and data-processing algorithms for AVIRIS-NG Indian campaign data. Curr. Sci. 116 (7), 1089–1100.

Pearlman, J.S., Barry, P.S., Segal, C.C., Shepanski, J., Beiso, D., Carman, S.L., 2003. Hyperion, a space-based imaging spectrometer. IEEE Trans. Geosci. Remote. Sens. 41 (6), 1160–1173.

Roberts, D.A., Yamaguchi, Y., Lyon, R., 1986. Comparison of various techniques for calibration of AIS data. In: Vane, G., Goetz, A.F.H. (Eds.), Proceedings of the 2nd Airborne Imaging Spectrometer Data Analysis Workshop. JPL Publication, Pasadena, CA, pp. 21–30. 86-35.

Tanre, D., Deroo, C., Duhaut, P., Herman, M., Morcrette, J.J., Perbos, J., et al., 1990. Description of a computer code to simulate the satellite signal in solar spectrum: the 5S code. Int. J. Remote Sens. 11 (4), 659–668.

Teillit, P.M., 1989. Surface reflectance retrieval using atmospheric correction algorithms. In: International Geoscience and Remote Sensing Symposium (IGARSS), vol. 2, pp. 864–867.

Teillet, P.M., Fedosejevs, G., 1995. On the dark target approach to atmospheric correction of remotely sensed data. Can. J. Remote Sens. 21 (4), 374–387.

Vane, G., Green, R.O., Chrien, T.G., Enmark, H.T., Hansen, E.G., Porter, W.M., 1993. The Airborne Visible/Infrared Imaging Spectrometer (AVIRIS). Remote Sens. Environ. 44 (2–3), 127–143.

Vermote, E.F., Tanre, D., Deuze, J.L., Herman, M., Morcrette, J.J., 1997. Second simulation of the satellite signal in the solar spectrum, 6S: an overview. IEEE Trans. Geosci. Remote. Sens. 35 (3), 675.

Wunch, D., 44 co-authors, 2017. Comparisons of the orbiting carbon observatory-2 (OCO-2) XCO2 measurements with TCCON. Atmos. Meas. Tech. 10, 2209–2238.

5 Hyperspectral image classifications and feature selection

Mahesh Pal

DEPARTMENT OF CIVIL ENGINEERING, NATIONAL INSTITUTE OF TECHNOLOGY, KURUKSHETRA, INDIA

5.1 Introduction

Land use/cover changes induced by human and natural processes play a major role in global as well as regional-scale patterns of the climate and the hydrology and biogeochemistry of the Earth system. Land use/cover mapping using moderate to coarse resolution datasets have been reported from different parts of the world at different scales. Land use/cover mapping using high-resolution data including multi- and hyperspectral data has several advantages since it aids in mapping using detailed information about select targets. Land use/cover mapping using satellite remote sensing data can help to establish the baseline data from which resource monitoring and management can be performed. These kinds of databases are important for national accounting of natural resources and planning at regular intervals.

The analysis and interpretation of data from hyperspectral sensors presents new possibilities for applications in areas such as crop area classification as these sensors provide greater detail on the spectral variation of targets than that provided by conventional multispectral systems, thus, providing the potential to derive more information about different objects in the area imaged. Hyperspectral sensors acquire tens to hundreds of images of any area in contiguous wavelengths spanning from visible to infrared regions (i.e., 350–2500 nm) of electromagnetic radiation. This technique of acquiring images makes it possible to accurately identify vegetation, soils, and building materials from the physical/chemical compositions of these objects (Chang, 2007). Images acquired using hyperspectral sensors are able to provide more accurate and detailed information about select targets as compared to multispectral sensors; a reason this is the efficient and fast-developing trend toward Earth remote sensing.

Research into the problems of land use/cover classification using hyperspectral data has been gaining momentum since 1990 with the availability of airborne visible/infrared imaging spectrometer (AVIRIS) data. Classification techniques that are now considered as standard such as the parallelepiped, minimum distance to mean, and maximum likelihood methods were developed at this time, and were mainly used for identifying and monitoring the type

Hyperspectral Remote Sensing. DOI: https://doi.org/10.1016/B978-0-08-102894-0.00010-3

and the extent of agricultural crops. The launch of Landsats-4 and -5 in 1982 and 1983, respectively, and SPOT-1 in 1986, made available datasets with a higher spatial resolution of 20–30 m compared with the 57 × 79 m pixel size of the Landsat MSS. The Landsat-4/5 Thematic Mapper produced data in six bands rather than four, while the SPOT HRV produced data in three bands. Perhaps surprisingly, the accuracy of land cover classifications was not necessarily improved (and, in some cases, it was reduced) by the use of higher spatial resolution data and by the availability of additional bands in the near and mid-infrared wavebands (Cushnie, 1987). Woodcock and Strahler (1987) suggested that this phenomenon was the consequence of an increase in within-class spectral variability as spatial resolution increased (i.e., pixel size decreases). Citing Hughes (1968), some researchers have reported that classification accuracy decreased as additional features were included. This effect has been termed "the curse of dimensionality."

In the late 1980s and early 1990s, research into alternative methods of image classification was largely driven by the perception that the classification techniques available at that time were not sufficiently powerful to identify patterns in multispectral data. More complex and powerful methods were sought, and those based on artificial neural networks, evidential reasoning, and decision trees were studied in detail. Related areas of research involving the evaluation of contextual information and texture measures were also stimulated by the need to improve the performance of image classifiers in order to provide adequate and timely data to environmental planners and government statisticians as well as to scientists engaged in climate change research. These developments are summarized by Tso and Mather (2016).

Over the past few decades, the use of several advanced classification algorithms using spectral information were proposed for hyperspectral data classification. This includes support vector machines, relevance vector machines, and other advanced classifiers performing well with limited training data in the presence of a large amount of features (Pal and Foody, 2012; Ghamisi et al., 2017). Spatial information has become an important part of hyperspectral data classification (Fauvel et al., 2008) suggesting the use of spatial–spectral classification methods that were found to perform well in terms of improved classification accuracy. Several approaches to include the spatial variability of spectral signatures have been proposed for the classification of hyperspectral data. A review of spatial–spectral classification algorithms used for the classification of hyperspectral data is provided by Ghamisi et al. (2018). Deep learning algorithms that use multiple stages of image processing like the human brain are found to work well for the classification of hyperspectral data (Chen et al., 2014; Petersson et al., 2016; Pan et al., 2018). Deep learning algorithms are found to perform well in comparison to shallow classifiers (e.g., support vector machine; SVM) because of their capabilities to extract more abstract and invariant features from data through multiple layers of processing. In spite of the improved performance of deep learning algorithms, some of issues, especially (1) the requirement for several user-defined parameters, (2) the requirement for a large amount of training data, (3) their black box type architecture, and (4) the need to adapt a deep learning architecture to work with remote sensing data acquired at different dates and times, still need serious attention (Ball et al., 2017).

Despite continuing research efforts over the past three decades, there is still no clear consensus of opinion on which is the most appropriate methodology for land cover classification using hyperspectral data. Considerations other than the nature and properties of classification algorithms that affect the accuracy of land cover classification include the spatial resolution of an image relative to the areal extent of the target classes, the degree of fragmentation of the target classes (i.e., whether they cover a large, contiguous, area or are small and dispersed), and the number of sample data/images during the training of a classifier.

However, the availability of a large number of features/bands in hyperspectral data represents a challenge in terms of storing and handling of large datasets, data redundancy because of the fact that the information contained in many adjacent bands are similar, and classification analysis. For example, the use of too many features may require a large amount of training data to avoid the Hughes effect and the estimation of parameters during the classification process, especially with statistical and neural classifiers. The size of a training set required for accurate parameter estimation may exceed that available during classification. Given that training data acquisition is difficult and costly, some means to accommodate the negative issues associated with hyperspectral data is required. Given that ground reference data are expensive and difficult to acquire, many strategies have been adopted to reduce the ground data requirements. This includes the use of semisupervised learning (the use of unlabeled training cases) (Marconcini et al., 2009), the adoption of dimensionality reduction approaches (feature reduction and selection) to reduce training dataset requirements (Pal and Foody, 2010), and the use of strategies that focus on the intelligent acquisition of informative training samples through the use of classification algorithms that are relatively insensitive to the Hughes effect and can effectively be applied for the analysis of hyperspectral data (Pal and Foody, 2012).

Feature extraction and feature selection are examples of commonly used dimensionality reduction techniques for remote sensing image classifications (Mather and Koch, 2011). Feature extraction techniques such as principal components analysis and minimum noise fraction transform original remote sensing data into a smaller set of new features containing the majority of the original dataset's information. On the other hand, the feature selection process aims to identify and select a subset of the original features in which class separability is maximized while excluding highly correlated and redundant features from the classification process. Feature selection procedures are dependent on the properties of the input data as well as on the classifier used (Guyon and Elisseeff, 2003). These approaches require a criterion to be defined by which it is possible to judge the quality of each feature in terms of its discriminating power (Dash and Liu, 1997). A computational procedure is then required to search through the range of potential subsets of features and select the "best" subset of features based upon some predefined criterion. The search procedure could simply consist of an exhaustive search through all possible subsets of features since this is guaranteed to find the optimal subset. In a practical application, however, the computational requirements of this approach are unreasonably large, and generally, a nonexhaustive search procedure is used. A wide range of feature selection approaches have been used for dimensionality reduction of remotely sensed data (Kavzoglu and Mather, 2002; Serpico and Bruzzone, 2001; Pal, 2006). Based on whether or not they use classification algorithms to evaluate subsets, the different methods can be grouped into three categories, namely filters, wrappers, and embedded approaches (Pal and Foody, 2010). These approaches may select different subsets and these in turn may vary in suitability for use as a preprocessing algorithm for different classifiers. Because of these differences and the wide range of reasons for undertaking a feature selection as well as the numerous issues that influence the outputs of a given feature selection approach, feature selection still remains a topic for research (Loughrey and Cunningham, 2004). Keeping in view of the issues related to the choice of classification and feature selection algorithms for hyperspectral data, this chapter reports the results of a modified radial basis function (RBF) network for classification and a Bayesian theory-based filter feature selection method.

5.2 Modified radial basis function neural network

A RBF network is a special type of neural network consisting of input, hidden, and output layers. This neural network is considered to be a good alternative to a back-propagation neural network and can be trained in one step (Venkatesan and Anitha, 2006). The input data is passed through the input layer to the hidden layer nodes. Each node of hidden layer produces an activation based on the associated RBF. Output layer nodes then compute the linear combination of the activation of the hidden nodes. The way in which the RBF network reacts to the given input data is determined by the activation function of the hidden nodes and the connecting weights between the hidden and the output layers. The output function of the RBF network corresponding to the input data vector $\boldsymbol{x}$ can be written as:

$$f_j(x) = \sum_{i=1}^{p} w_{ji}\Phi(||x - c_i||; \rho_i) \tag{5.1}$$

where $f_j(x)$ is the function corresponding to the jth output unit and is a linear combination of p number of neurons in the hidden layer with c_i as the center vector of the ith neuron, ρ_i is the bandwidth of the ith neuron, and w_{ji} is the weight of the ith neuron and the jth output. The norm is usually taken to be the Euclidean distance and the RBF is also taken to be a Gaussian function.

The design of an RBF neural network involves the training of the hidden layer by identifying a suitable value for the center vector (c_i) and kernel bandwidth (ρ_i) and fixing the weights between the hidden and output layers (Bruzzone and Prieto, 1999). Several techniques have been proposed in the literature for the selection of the centers of the kernel functions (Bishop, 1995), which includes applying a clustering technique for the use of the whole training set. After selecting the centers of the kernel functions, the selection of the bandwidth of the kernel functions is carried out. Moody and Darken (1989) suggested that the choice of bandwidth controls the generalization capabilities of an RBF network and, thus, it needs to be chosen in a way that does not affect the generalization ability of the classifier. Finally, a sum of squares error function (Bishop, 1995) is generally used to select the weights corresponding to the connections between the hidden nodes and the output units. Several approaches, which include a two-phase approach, pruning methods, constructive methods, and evolutionary computation techniques, are proposed to select centers, bandwidth, and weights, and these are discussed in detail in Yu et al. (2009).

This chapter discusses the use of a modified RBF neural network concentrating on the calculation of the weights (Oyang et al., 2005) for hyperspectral image classification. The works of Li et al. (2016) and Camps-Valls et al. (2004) also used different forms of RBF network for hyperspectral image classification. The conventional cluster-based learning algorithm for RBF networks works by placing one RBF at the center of each cluster. Several studies related to land cover classification suggest that samples lying near the boundaries between different classes carry more information than those at the center of a cluster (Pal and Foody, 2012). Because of this, conventional RBF networks, in general, are not able to achieve the same level of accuracy as those of algorithms like SVM, which uses samples near the class boundaries.

The modified RBF network used in this work uses a Cholesky decomposition approach and the least mean square error method to determine the weights associated with the links between the hidden layer and the output layers. As the selection of centers for RBF network is still a research problem, a simple random method, which may have selected any number of training datasets, was used to select the centers in this work. For this study, a fixed value of kernel bandwidth (i.e., 5) was employed.

Assuming O to be the output of the hidden layer of the RBF network defined as:

$$O = [\Phi_1(\boldsymbol{x})\ \Phi_2(\boldsymbol{x})\ldots\ldots\ldots\Phi_p(\boldsymbol{x})]^{\mathrm{T}} \tag{5.2}$$

where $\Phi_1(\boldsymbol{x})$ is the output value of the first kernel function with input value $\boldsymbol{x}$, then, the discriminant function $f_j\ (\boldsymbol{x})$ of class j can be expressed as:

$$f_j(\boldsymbol{x}) = w_j^{\mathrm{T}} O, \quad j = 1, 2, \ldots, k \tag{5.3}$$

where k is the number of classes, and w_j is the weight vector of class j. Oyang et al. (2005) propose the use of the least mean square method to calculate w_j for a classification problem. If $\boldsymbol{V}_i$ is the ith column vector of $k \times k$ identity matrix and $\boldsymbol{W}$ *is* a $p \times k$ matrix of weights, the objective function to be minimized is defined as:

$$J(\boldsymbol{W}) = \sum_{j=1}^{k} P_j\ E_j\ \{||\boldsymbol{W}^{\mathrm{T}}\ O - \boldsymbol{V}_j||^2\} \tag{5.4}$$

where P_j and E_j {} are a priori probability and the expected value of class j, respectively.

To find the optimal value of $\boldsymbol{W}$, Oyang et al. (2005) propose to set the gradient of $J\ (\boldsymbol{W})$ to zero. If $\boldsymbol{A}$ is a nonsingular class-conditional matrix of the second-order moments of O, the optimal value of $\boldsymbol{W}$ can be calculated by:

$$\boldsymbol{W}^* = \boldsymbol{A}^{-1}\ M \tag{5.5}$$

where $M = \sum_{j=1}^{k} P_j\ E_J\ \{O\}\ V_j^{\mathrm{T}}$ and $\boldsymbol{A}_i = E_i(O\ O^{\mathrm{T}})$.

If $\boldsymbol{A}$ is a singular matrix, Oyang et al. (2005) propose to add a regularization term (l) to make the matrix invertible. The optimal value of weight vector can then be calculated by:

$$\boldsymbol{W}^{*} = (\boldsymbol{A} + \lambda \boldsymbol{I})^{-1} M \tag{5.6}$$

where $(\boldsymbol{A} + \lambda\boldsymbol{I})$ is a positive definite matrix and nonsingular. Oyang et al. (2005) propose the use of Cholesky decomposition to decompose $(\boldsymbol{A} + \lambda\boldsymbol{I})$ as:

$$(\boldsymbol{A} + \lambda \boldsymbol{I}) = \boldsymbol{L}\boldsymbol{L}^{\mathrm{T}} \tag{5.7}$$

where $\boldsymbol{L}$ is a lower triangular matrix. These optimal weights can then be used to calculate the optimal discriminant function $f_j(\boldsymbol{x})$.

5.3 Bayesian framework for feature selection

The Bayesian framework for feature selection (BFFS) is a filter-based feature selection algorithm based on Bayesian theory and the receiver operating characteristic (ROC) analysis (Yang and Hu, 2006). It selects combinations of features based on the area under the curve (AUC) of their combined ROC curve. The AUC of the ROC curve is an important measure of discriminability between the two classes described by their likelihood distributions. The equivalent statistical metric to the AUC is the Wilcoxon statistics, which was designed to estimate the probability of the rank of two random variables. In the Bayesian framework, the probability for each class and their priors can be estimated independently and used to eliminate both irrelevant and redundant features.

Since removing or adding an irrelevant feature does not change the expected AUC, both backward and forward greedy selection (filter) algorithms can be designed to use the expected AUC as an evaluation function. A backward elimination approach provides a greedy algorithm for feature selection. It starts with the full feature set and removes one feature at each iteration. A feature $f_j \in G^k$ to be removed is determined using the equation:

$$f_j = \underset{f_j \in G^k}{\operatorname{argmin}} (E_{\mathrm{AUC}}(G^k) - E_{\mathrm{AUC}}(G^k - \{f_j\}))$$

where G^k is the feature set after the kth iteration and E_{AUC} is the expectation of the AUC, which is the evaluation function used by the BFFS approach. Due to the large computational demand in the estimation of the AUC in a high-dimensional space, an approximation algorithm is used to calculate the AUC in a lower-dimensional space in BFFS (Singh and Provan, 1996). Yang and Hu (2006) suggested that the decrease of the total AUC after the removal of a feature f_i is related to the overlap of the discriminability of the feature with other features. Thus in the approximation algorithm, a feature subset H^k is selected out of the current feature set G^k and the degree of discriminability overlap in H^k is used to approximate that in G^k (Yang and Hu, 2006). A heuristic approach is used to select k_s features from G^k that have the largest overlap with feature f_i and it is assumed that the discriminability overlap of feature f_i with other features in G^k is dominated by this feature subset. Further detail of this algorithm is provided by Yang and Hu (2006).

5.4 Study area and data sources

Two study areas including La Mancha Alta, covering an area of approximately 8000 km^2 to the south of Madrid, Spain and the Anand district of Gujarat state, India covering an area of about 320 km^2 were used. The area in Spain is a Mediterranean semiarid wetland that supports the rain-fed cultivation of wheat, barley, and other crops such as vines and olives, whereas the area in Anand district covers the research farms of Anand Agricultural University and the adjoining agricultural area. The climate of the study area is semiarid. The datasets available include a Digital Airborne Imaging Spectrometer (DAIS) hyperspectral image collected on June 29, 2000, for the area in Spain and data from a next generation AVIRIS (AVIRIS-NG) campaign that was carried out on February 6–7, 2016, over Anand. The DAIS 7915 instrument used is a 79-channel imaging spectrometer developed and operated by the German Space Agency (DLR). This instrument has a spatial resolution of 5 m. It operates in the wavelength range of 0.4–12.5 μm. Bands 1–72, covering the optical region of the spectrum were used in this study. The data in these bands show moderate to severe striping problems, especially in the near-infrared region (bands 41–72). The bands that were most severely affected by striping were identified by visual inspection. As a result, seven bands with severe striping problems (bands 41, 42, and 68–72) were removed from the dataset. Striping in the remaining 65 bands was reduced using a Fourier-based filtering technique. It was not possible to collect information on land cover types at the time of the DAIS overflights. Field observations were undertaken in late June 2001, exactly one year after the image data were acquired, to generate a reference dataset. Only those land cover types that were most likely to have remained unaltered since the previous year were used. The use of the ground reference data collected one year after the date of image acquisition was justified on the grounds that, in an area of Mediterranean climate, the weather patterns and soil conditions tend to be quite similar from one year to the next. For the AVIRIS-NG sensor, data was collected in 425 wavebands covering the 350–2500 nm range. Major crops in this area during the campaign period were wheat at different stages, mustard, chickpea, oat, amaranth, potato, tomato, brinjal, chili, different cultivars of tobacco at different stages, fodder maize at different stages, pearl millet, cabbage, fodder sorghum, and fennel. A total of 13 classes were used to classify the AVIRIS-NG data. Out of the total 425 bands, bands 1–5, 196–207, and 285–320 were removed due to stripping problems in the reflectance data. Thus a total of 372 bands covering a study area of 566 columns and 1101 rows were used for classification and feature selection.

5.5 Results

The aim of this chapter is to evaluate the performance of the modified RBF classifier for land cover classification using hyperspectral data. For the DAIS dataset, the results of the modified RBF network were compared with an SVM classifier. Table 5–1 provides the results obtained using the modified RBF and SVM classifier using the DAIS hyperspectral dataset. The results indicate that the modified RBFN classifier works equally well in comparison to SVM. A classification accuracy of 92.2% was achieved by the modified RBF classifier in comparison to 91.8%

Table 5–1 Results with DAIS hyperspectral dataset.

Classifier	Classification accuracy	Computational cost (s)
Modified RBF network	92.2%	3.78 + 110 (test time)
SVM	91.98% $C = 5000$, Gamma = 2	7.35

DAIS, Digital airborne imaging spectrometer; *SVM*, support vector machine; *RBF*, radial basis function.

Table 5–2 Results with reduced dataset using DAIS data.

Selected features	Accuracy with modified RBF	Accuracy with SVM
1, 6, 11, 15, 27, 31, 33, 39, 40, 46, 51, 61, 63, 64, 65	91.24%	90.05%

DAIS, Digital airborne imaging spectrometer; *SVM*, support vector machine; *RBF*, radial basis function.

Table 5–3 Results with AVIRIS-NG hyperspectral data.

Features	Accuracy with modified RBF
372	96.76%
30 selected features (60, 62, 63, 64, 67, 69, 74, 97, 109, 116, 136, 146, 148, 183, 186, 188, 189, 190, 192, 197, 210, 216, 220, 242, 272, 315, 369, 370, 371, 372)	96.14%

AVIRIS-NG, Next generation airborne visible/infrared imaging spectrometer; *RBF*, radial basis function.

achieved by the SVM. Table 5–1 also provides the computational cost of the modified RBF and SVM classifiers. The results (Table 5–1) suggest the suitability of the modified RBF classifier in terms of computational cost.

To compare the performance of the modified RBF on a reduced dataset, a BFFS-based feature selection approach was used to reduce the dimensionality of the DAIS dataset. This criterion filters out the irrelevant features (Yang and Hu, 2006) and provides a subset of selected features. A total of 15 selected features (Table 5–2) were then used to compare both classifiers in terms of classification accuracy with the reduced dataset. Table 5–2 provides the classification accuracy of the selected features using both classification algorithms. In comparison to an accuracy of 90.05% for SVM, the modified RBF provides an accuracy of 91.28% for the selected features.

Keeping in view of the improved performance of the modified RBFN and the BFFS-based feature selection approach with DAIS hyperspectral data, both approaches were used with the AVIRIS-NG dataset. The results presented in Table 5–3 suggest an encouraging performance of the modified RBFN classifier and BFFS-based feature selection approach with this dataset as well.

5.6 Conclusion

A modified RBF network classifier and BFFS-based filter feature selection approach were used for classification and feature selection using two hyperspectral datasets. The results suggest several conclusions. First, the modified RBF classifier works equally well when compared to an SVM classifier in terms of classification accuracy. However, the modified RBF classifier has a smaller computational cost in comparison to the SVM classifier. Second, the results also suggest that the modified RBF requires fewer user-defined parameters in comparison to the SVM classifier. Another conclusion from this work is the usefulness of a BFFS-based feature selection approach with both hyperspectral datasets. A total of 15 and 30 selected features with the DAIS and AVIRIS-NG datasets respectively provide a comparable value of classification accuracy to that achieved by the full dataset using the modified RBF classifier.

List of abbreviations

AVIRIS-NG	airborne visible/infrared imaging spectrometer new generation
AUC	area under the curve
BFFS	Bayesian framework for feature selection
DAIS	Digital *Airborne* Imaging Spectrometer
PCA	principal components analysis
RBFN	radial basis function neural network
ROC	receiver operating characteristic
SVM	support vector machine

List of symbols

c	centre vector
E	expected value
k	number of classes
K	matrix of second order moment
O	output of hidden layer
P	priori probability
p	number of neuron
w	weight
λ	regularisation parameter
ρ	kernel bandwidth
$\phi(x)$	output of kernel function

Acknowledgment

Mahesh Pal thanks Space Application centre (ISRO), Ahmedabad for providing AVIRIS-NG data used in this study under the research project "Classification and feature selection of AVIRIS-NG airborne hyperspectral data for crop cover mapping/urban mapping". This data was collected over Anand (Gujrat) site under AVIRIS-NG airborne campaign carried out in 2016.

References

Ball, J.E., Anderson, D.T., Chan, C.S., 2017. Comprehensive survey of deep learning in remote sensing: theories, tools, and challenges for the community. J. Appl. Remote Sens. 11 (4), 042609.

Bishop, C.M., 1995. Neural Networks for Pattern Recognition. Clarendon, Oxford.

Bruzzone, L., Prieto, D.F., 1999. A technique for the selection of kernel-function parameters in RBF neural networks for classification of remote-sensing images. IEEE Trans. Geosci. Remote Sens. 37 (2), 1179–1184.

Camps-Valls, G., Serrano-López, A.J., Gómez-Chova, L., Martín-Guerrero, J.D., Calpe-Maravilla, J., Moreno, J., 2004. Regularized RBF networks for hyperspectral data classification. International Conference Image Analysis and Recognition. Springer, Berlin, Heidelberg, pp. 429–436.

Chang, C.-I., 2007. Hyperspectral Data Exploitation: Theory and Applications. John Wiley and Sons, New Jersey.

Chen, Y., Lin, Z., Zhao, X., Wang, G., Gu, Y., 2014. Deep learning-based classification of hyperspectral data. IEEE J. Select. Top. Appl. Earth Obs. Remote Sens. 7 (6), 2094–2107.

Cushnie, J.L., 1987. The interactive effect of spatial resolution and degree of internal variability with land-cover types on classification accuracies. Int. J. Remote Sens. 8 (1), 15–29.

Dash, M., Liu, H., 1997. Feature selection for classification. Intell. Data Anal.: Int. J. 1 (3), 131–156.

Fauvel, M., Benediktsson, J.A., Chanussot, J., Sveinsson, J.R., 2008. Spectral and spatial classification of hyperspectral data using SVMs and morphological profiles. IEEE Trans. Geosci. Remote Sens. 46 (11), 3804–3814.

Ghamisi, P., Maggiori, E., Li, S., Souza, R., Tarablaka, Y., Moser, G., et al., 2018. New Frontiers in spectral–spatial hyperspectral image classification: the latest advances based on mathematical morphology, Markov random fields, segmentation, sparse representation, and deep learning. IEEE Geosci. Remote Sens. Mag. 6 (3), 10–43.

Ghamisi, P., Plaza, J., Chen, Y., Li, J., Plaza, A.J., 2017. Advanced spectral classifiers for hyperspectral images: a review. IEEE Geosci. Remote Sens. Mag. 5 (1), 8–32.

Guyon, I., Elisseeff, A., 2003. An introduction to variable and feature selection. J. Mach. Learn. Res. 3, 1157–1182.

Hughes, G.F., 1968. On the mean accuracy of statistical pattern recognizers. IEEE Trans. Inf. Theory IT-14, 55–63.

Kavzoglu, T., Mather, P.M., 2002. The role of feature selection in artificial neural network applications. Int. J. Remote Sens. 23 (15), 2787–2803.

Li, J., Du, Q., Li, Y., 2016. An efficient radial basis function neural network for hyperspectral remote sensing image classification. Soft Comput. 20 (12), 4753–4759.

Loughrey, J., Cunningham, P., 2004. Overfitting in wrapper-based feature subset selection: The harder you try the worse it gets. In: International Conference on Innovative Techniques and Applications of Artificial Intelligence, Springer, London, pp. 33–43.

Marconcini, M., Camps-Valls, G., Bruzzone, L., 2009. A composite semisupervised SVM for classification of hyperspectral images. IEEE Geosci. Remote Sens. Lett. 6 (2), 234–238.

Mather, P.M., Koch, M., 2011. Computer Processing of Remotely-Sensed Images: An Introduction. John Wiley & Sons.

Moody, J., Darken, C.J., 1989. Fast learning in networks of locally-tuned processing units. Neural Comput. 1 (2), 281–294.

Oyang, Y.-J., Hwang, S.-C., Ou, Y.-Y., Chen, C.-Y., Chen, Z.-W., 2005. Data classification with the radial basis function network based on a novel kernel density estimation algorithm. IEEE Trans. Neural Netw. 16 (1), 225–236.

Pal, M., 2006. Support vector machine-based feature selection for land cover classification: a case study with DAIS hyperspectral data. Int. J. Remote Sens. 27 (14), 2877–2894.

Pal, M., Foody, G.M., 2010. Feature selection for classification of hyperspectral data by SVM. IEEE Trans. Geosci. Remote Sens. 48 (5), 2297–2307.

Pal, M., Foody, G.M., 2012. Evaluation of SVM, RVM and SMLR for accurate image classification with limited ground data. IEEE J. Sel. Top. Appl. Earth Observations Remote. Sens. 5 (5), 1344–1355.

Pan, B., Shi, Z., Xu, X., 2018. MugNet: deep learning for hyperspectral image classification using limited samples. ISPRS J. Photogram. Remote Sens. 145, 108–119.

Petersson, H., Gustafsson, D., Bergstrom, D., 2016. Hyperspectral image analysis using deep learning—a review. In: 2016 6th IEEE International Conference on Image Processing Theory Tools and Applications (IPTA), pp. 1–6.

Serpico, S.B., Bruzzone, L., 2001. A new search algorithm for feature selection in hyperspectral remote sensing images. IEEE Trans. Geosci. Remote Sens. 39 (7), 1360–1367.

Singh, M., Provan, G. M., 1996. Efficient learning of selective Bayesian classifiers. In: Machine Learning: Proceedings of the Thirteenth International network Conference on Machine Learning. Morgan Kaufmann.

Tso, B., Mather, P., 2016. Classification Methods for Remotely Sensed Data. CRC Press.

Venkatesan, P., Anitha, S., 2006. Application of a radial basis function neural network for diagnosis of diabetes mellitus. Curr. Sci. 91 (9), 1195–1199.

Woodcock, C.E., Strahler, A.H., 1987. The factor of scale in remote sensing. Remote Sens. Environ. 21, 311–332.

Yang, G.Z., Hu, X., 2006. Multi-sensor fusion. In: Yang, G.Z. (Ed.), Body Sensor Networks. Springer, Heidelberg, pp. 262–280.

Yu, S., Zhu, K., Gao, S., 2009. A hybrid MPSO-BP structure adaptive algorithm for RBFNs. Neural Comput. Appl. 18 (7), 769–779.

SECTION II

Hyperspectral Remote Sensing Application in Vegetation

6

Identification of functionally distinct plants using linear spectral mixture analysis

Ramandeep Kaur M. Malhi[1], Prashant K. Srivastava[1], G. Sandhya Kiran[2]

[1]REMOTE SENSING LABORATORY, INSTITUTE OF ENVIRONMENT AND SUSTAINABLE DEVELOPMENT, BANARAS HINDU UNIVERSITY, VARANASI, INDIA

[2]THE MAHARAJA SAYAJIRAO UNIVERSITY OF BARODA, VADODARA, INDIA

6.1 Introduction

Over the past few years various global changes have occurred due to many processes like immense land use change, species invasion, climate change, and transformed biogeochemical cycles, which have adversely disturbed ecosystems worldwide at unprecedented scales (Dentener et al., 2006; Foster et al., 2003; McLauchlan, 2006; Muler, 2016). This has also driven a great rise in species extinctions (Bu et al., 2018). One of the major drivers behind ecosystem change is biodiversity loss (Hooper et al., 2012). The degradation of ecosystem services at a global level is stressing the need to understand the primary mechanisms of biotic communities by monitoring ecosystem processes (Liu et al., 2018). The functioning of an ecosystem is affected by several factors; plant functional trait (FT) composition is one of those factors (Liu et al., 2016). Many studies show the varying trends of FTs such as leaf economics spectrum (Osnas et al., 2013; Wright et al., 2004), wood economics spectrum (Chave et al., 2009), and the seed size–seed number tradeoff (Moles and Westoby, 2006; Venable, 1992). Thus the quantification of FTs is considered to be a suitable method by many ecologists for understanding the changes in ecosystem functions (Adler et al., 2014; Lavorel and Garnier, 2002; McGill et al., 2006).

6.2 Plant functional traits

Plant FTs are referred to as the morpho-physio-phenological traits of any plant species. These FTs are directly related to three performance components of individual plant species, namely growth, reproduction, and survival. This, thereby, will affect the vitality of individual species (Violle et al., 2007). In the past few years, FTs are increasingly recognized as having a

Hyperspectral Remote Sensing. DOI: https://doi.org/10.1016/B978-0-08-102894-0.00008-5

key role in understanding and predicting the adaptation of ecosystems to environmental changes. With the increased availability of trait data, the use of trait-based approaches is increasing to obtain insights into the functional aspects of plant communities. FTs, therefore, provide a promising basis for a more quantitative and predictive global change science. These FTs also address ecological questions like how a species responds to its surrounding environment (McGill et al., 2006). The influence of ecosystem processes through abiotic conditions, changes the ecosystem flux rate of energy or it alters the physiological functions of species. FT composition is also considered as an important driver of carbon sequestration in terrestrial ecosystems since plant species vary in their carbon storing and releasing capabilities (De Deyn et al., 2008). Thus many ecosystem functions are determined by FTs (Boyer et al., 2009; Geert Hiddink et al., 2009; Hajek et al., 2016; Lavorel et al., 2011).

Variations in FTs have been observed by many researchers among species (Kooyman and Westoby, 2009; Wardle et al., 1998) as well as among groups of species (Garnier, 1991). Multiple FTs effect ecosystem functions and every trait varies independently from one another (Eviner and Chapin, 2003). Multiple or single key traits can aid in monitoring an ecosystem's processes (de Bello et al., 2010). Thus the selection of a trait in any study depends upon the ecosystem process to be studied. Thus the quantification of plant FTs becomes imperative since these traits greatly respond to the dominant ecosystem processes (Gitay and Noble, 1997). Information about variations in these traits can aid in identifying functionally distinct plants. This can further be used for grouping these functionally distinct plants into different plant functional types (PFTs).

6.3 Plant functional types

The structural, physiological, and phenological characteristics of a plant, that is, FTs, can help in determining a plant's performance in three aspects, namely growth, reproduction, and survival. As discussed previously, the variations in traits among plant species can be determined by studying different environmental factors like climate, topography, or soil properties together with species interaction (Díaz et al., 2016; Grime et al., 2014; Wright et al., 2004). These variations in FTs can be taken into account and can aid in classifying different species into different PFTs based upon their similar FTs (Lavorel et al., 1997).

PFTs can, thus, be defined as groups of plant species sharing similar functioning, similar responses to environmental factors, and/or similar effects on ecosystems (Smith et al., 1997). Information on different PFTs in terms of their geographic distribution and abundance around the globe is essentially required for global climate change research (Sun et al., 2008). The land surface models traditionally used consider vegetation as discrete biomes such as evergreen broadleaf forests, shrubs, grasses, and savannas. These biomes then set surface biogeophysical variables such as albedo, leaf area index (LAI), fraction of photosynthetically active radiation (fPAR), canopy roughness, and stomatal physiology for each grid cell (Bonan, 1993; Prince and Goward, 1995; Running and Coughlan, 1988; Sellers et al., 1986). A demerit of these biome-based classifications is that biomes are outputs of classification and are not natural vegetation units. The parameters needed for these biome-based land models include leaf-level and whole-plant parameters. These parameters cannot be parameterized for mixed life-form biomes such as mixed forests and savannas (Bonan et al., 2002). To overcome this problem, scientists are moving away from the use of the traditional biome-based model to PFT-based models. For example, the National Center for Atmospheric Research land surface model (NCAR LSM) has shifted from using biome-based land cover information to using satellite-derived PFT maps (Bonan et al., 2002; Tian et al., 2004). Accurate specifications of PFTs are important inputs in the carbon models that scale carbon fluxes (Denning et al., 1996; Sellers et al., 1997). Using remote sensing techniques to extract reliable PFT information can, therefore, contribute to improved predictive capabilities of global and regional carbon cycle, climate, and ecosystem models.

6.4 Remote sensing in identification of plant functional types or functionally distinct plants

Remote sensing has proved its potential in determining temporal and spatial variations in land surface biogeophysical variables required by global change research (Liang, 2005; Myneni et al., 1997; Roughgarden et al., 1991; Sellers et al., 1995; Townshend et al., 1987). Several researchers have used remote sensing data for extracting land cover information or information about certain biogeophysical parameters (De Fries et al., 1998; Fonji and Taff, 2014; Kimes et al., 1991; Liping et al., 2018; Lloyd, 1990; Townshend and Justice, 1981; Townshend et al., 1987). In the past few years, scientists have shown increasing interest in exploring the potential of remote sensing in PFTs identification. Certain researches are noted on the identification of PFTs using remote sensing data (Bonan et al., 2002; Kattenborn et al., 2019; Sun et al., 2008; Townsend et al., 2017). The mapping of PFTs using coarser remote sensing has also proved to be a significant input for dynamic ecosystem models (Sitch et al., 2003; Smith et al., 2001) and Earth system models (Poulter et al., 2011). Certain studies have also been reported where the importance of the methods used to determine PFTs or the quality of PFT data is emphasized, which in turn effects the results of the effect on global climate modeling (Oleson and Bonan, 2000; Tian et al., 2004). Hyperspectral data have shown particular promise in providing high quality PFT data because of their discrimination ability of different species.

6.5 Hyperspectral remote sensing in plant functional types identification

The mapping of functionally distinct plants or PFTs using hyperspectral data that measure reflected solar radiance using a large set of narrow, contiguous spectral bands has shown great potential (Clark et al., 2005; Martin et al., 1998; Roberts et al., 1998). Hyperspectral Earth observation (EO) data have great capability in determining the spatial distribution of PFTs. The relationship between plant traits and electromagnetic radiation, that is, the absorption and scattering processes within a canopy, makes this feasible. Reflected electromagnetic radiation is measured by hyperspectral EO-sensors, which in turn helps to retrieve optically relevant FTs (Kattenborn et al., 2019). In biodiversity-rich areas like Shoolpaneshwar Wildlife Sanctuary where the forests are heterogeneous, the identification of functionally distinct plants or PFTs is feasible through hyperspectral remote sensing. Hyperspectral data prove their potential in highly heterogeneous areas, where pixels with mixed classes would be abundant and an entire area of interest with secondary forests of varying ages might occupy only a few pixels (Castro et al., 2003). In such heterogeneous areas, the measured spectral signal for every pixel of surface reflectance data will be a result of fractions in which various functionally distinct plants or PFTs and also the soil background occur. Spectral mixture analysis (SMA) is an image processing tool that proves useful in such cases. SMA models a pixel as a linear or nonlinear combination of its constituent spectral components or spectral endmembers weighted by their subpixel fractional cover. By model inversion, it provides subpixel endmember fractions.

6.6 Spectral mixture analysis

SMA is basically a physically-based image-processing tool aiding in precise repeated derivation of quantitative subpixel information (Roberts et al., 1999; Smith et al., 1990). SMA works under the assumption that a spectrum computed by a sensor is considered as a linear combination of the spectra of all the components within the pixel and the spectral proportions of the endmembers is considered to reflect proportions of the area occupied by definite features on the Earth surface (Adams et al., 1995; Lu et al., 2004). Mathematically SMA can be written as:

$$R_i = \sum_{k=1} f_k R_{ik} + \varepsilon_i \tag{6.1}$$

where i is the spectral band used; $k = 1, \ldots, n$ (number of endmembers); R_i is the spectral reflectance of band i of a pixel that contains one or more endmembers; f_k is the proportion of endmember k within the pixel; R_{ik} is known as the spectral reflectance of endmember k within the pixel on band i, and i is the error for band ε_i.

In this chapter, a case study carried out is also discussed in which functionally distinct plants of the Shoolpaneshwar Wildlife Sanctuary, Narmada, Gujarat are mapped using EO-1 Hyperion data combined with SMA, which accounts for endmember variability.

6.7 Study site

The study site selected for this case study is part of the Shoolpaneshwar Wildlife Sanctuary (Fig. 6–1), a protected area located in the western Satpura Range south of the Narmada River, Narmada District, Gujarat State, India. The sanctuary was established in 1989 and

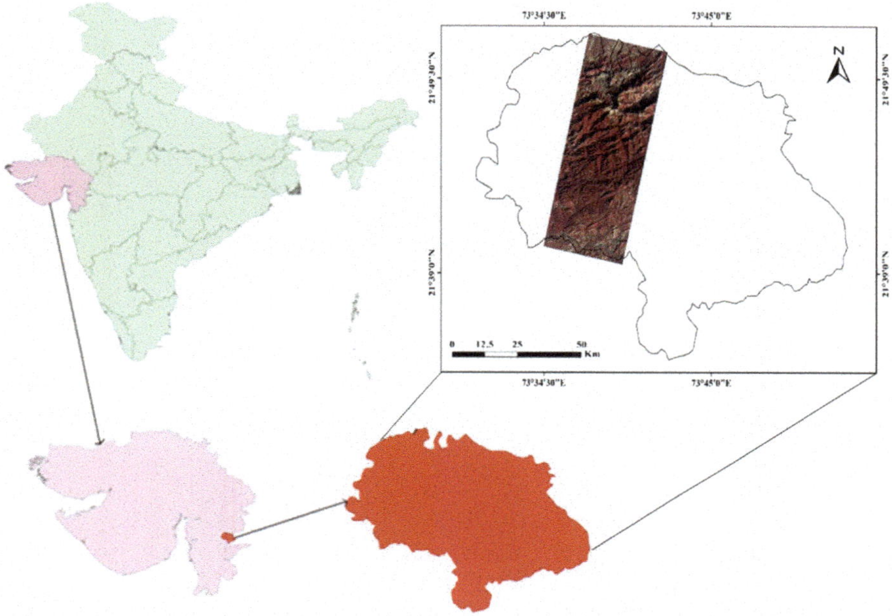

FIGURE 6–1 Map showing study area.

covers a 607.71 km^2 geographical area. It extends from 21°03′N to 21°59′N latitude and from 73°05′E to 74°10′E longitude at an altitude of 800–900 MSL.

6.8 Materials and methodology

6.8.1 Ground data collection and analysis

The ground location of homogenous patches of three forest species, namely *Tectona grandis, Butea monosperma,* and *Bambusa bambos* was recorded using a GPS eTrex 10. The plant samples were collected for laboratory estimations of their FTs. Their FTs, namely chlorophyll content (CC), LAI, and height and diameter at breast height, were measured using standard procedures. A standard acetone extraction method was used for determining the amount of chlorophyll in the leaf samples of selected plants. The homogenization of the leaf tissue in 80% acetone was carried out and then absorbance at 663 and 645 nm was measured. Then, specific absorption coefficients for chlorophyll a and b (Arnon, 1949) were used for the calculation of CC (Wu et al., 2008). The LAI of these species was estimated using a plant canopy analyzer, namely the LAI-2000 (LI COR Inc., Lincoln, NE, United States). Height was measured using a clinometer and diameter at breast height (DBH) using measuring tape. The values of the FTs of the three selected plants were compared to see whether they varied functionally or not.

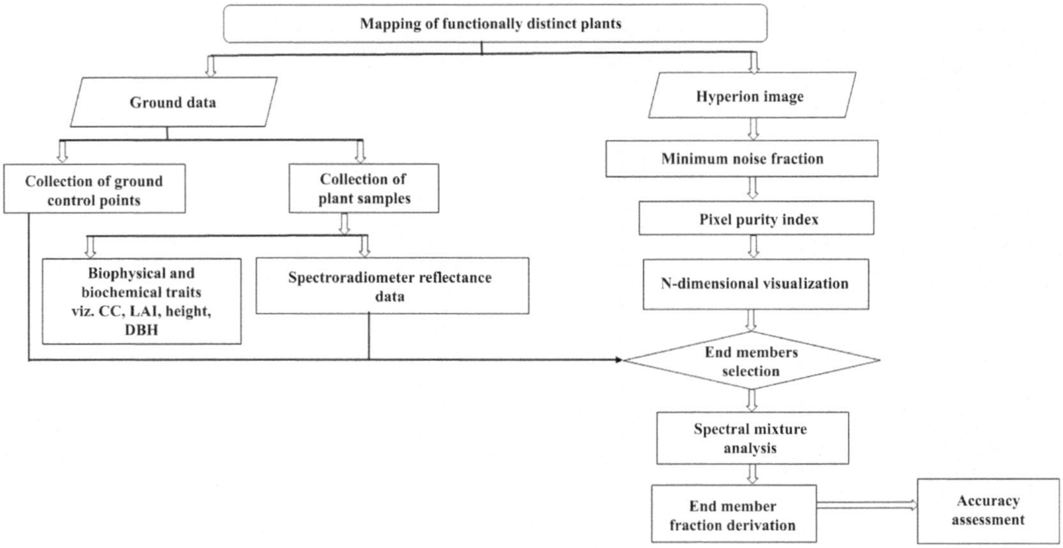

FIGURE 6–2 Flow chart showing entire methodology.

6.9 Satellite data and their analysis

Hyperspectral data namely EO-1 Hyperion was downloaded from United States Geological Survey (USGS) Earth explorer site (https://earthexplorer.usgs.gov/). The data are composed of 242 spectral bands and have a spatial resolution of 30 m. The procurement date of the data, that is, November 26, 2016 was selected corresponding to the field sampling period. The removal of the bad bands and pixels was executed manually using ENVI 5.3 and a total of 196 radiometrically calibrated bands remained for further processing. The ENVI fast line-of-sight atmospheric analysis of spectral hypercubes (FLAASH) plugin was employed for the atmospheric correction of the images. Minimum noise fraction (MNF) transformation was performed to eliminate any additional noise remaining in the scene. MNF bands occurring after an 80% variance threshold were discarded from further analysis. A manual approach based on field data was used for the endmember selection from the Hyperion images using a pixel purity index (PPI). Based on the available field information, candidate pixels were selected from locations where the functionally distinct plants appeared pure or had a relative homogeneous species composition. The computation of PPI was done by projecting *n*-D scatter plots repeatedly. The determined endmembers were used as inputs in SMA. The linear spectral unmixing algorithm was applied and the final product of the analysis was a set of fractional abundance images for each functionally distinct plant.

A flow chart depicting the entire methodology is shown in Fig. 6–2.

6.10 Results and discussion

Significant variations in the CC estimated for *T. grandis*, *B. monosperma*, and *B. bambos* were observed. *B. monosperma* had the lowest CC, whereas *B. bambos* showed the highest CC (Fig. 6–3A). *T. grandis* showed intermediate CC values. The opposite trend was observed in the case of height (Fig. 6–3B). Maximum height was observed in *B. monosperma* and the least was seen in *B. bambos*. Comparative values of LAI were observed for *T. grandis* and *B. bambos* and the highest average value for LAI was measured in *B. monosperma* (Fig. 6–3C). DBH also varied across these three species. *B. monosperma* had the highest DBH followed by *T. grandis* and then *B. bambos* (Fig. 6–3D). The significant variations in the values of the different FTs in these three forest species indicated that these are functionally distinct plants.

Hyperion image is subjected to SMA which helped in deriving the fractional composition measurements, even small basal area fractions of three functionally distinct plant species of Shoolpaneshwar forests for each pixel were measured. The product of SMA was obtained as a set of fractional abundance maps for each class of functionally distinct plants, namely

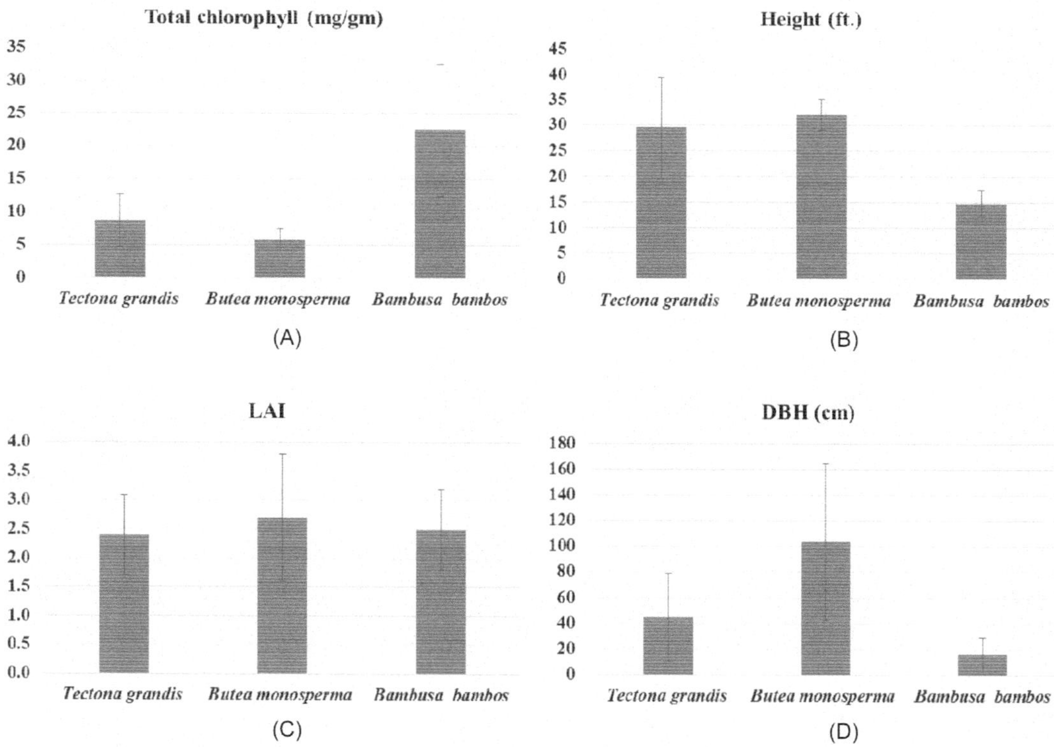

FIGURE 6–3 Functional traits estimated for *Tectona grandis*, *Butea monosperma*, and *Bambusa bambos*. (A) Total chlorophyll, (B) height, (C) LAI, and (D) DBH. *CC*, Chlorophyll content; *DBH*, diameter at breast height; *LAI*, leaf area index.

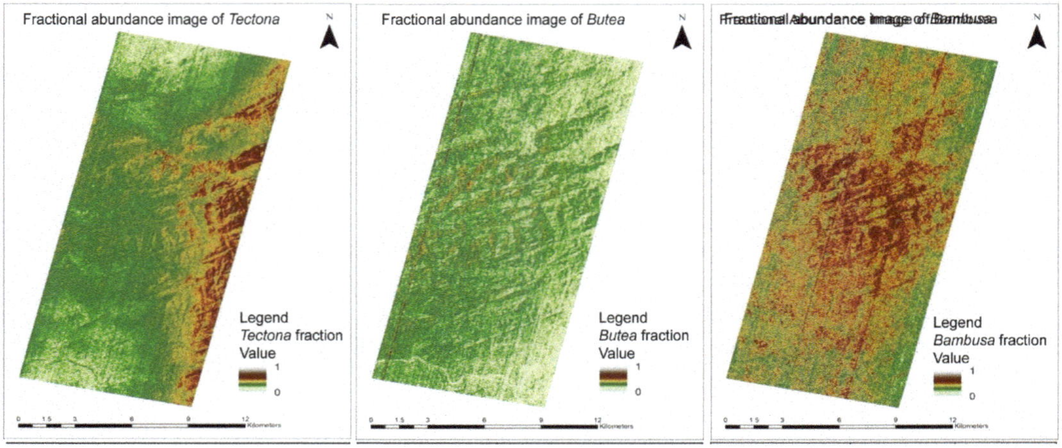

FIGURE 6–4 Fractional abundance images of (A) *Tectona grandis*, (B) *Butea monosperma*, and (C) *Bambusa bambos*.

T. grandis, *B. monosperma*, and *B. bambos*. Fig. 6–4A depicts the fractional abundance of *T. grandis* in the study area. Pale yellow to green areas in the map show lower fractions of *T. grandis*, while dark brown areas indicate higher fractions of *T. grandis* in those areas. Similarly Fig. 6–4B and C shows the fractional abundance images of *B. monosperma* and *B. bambos*, respectively. A comparison of the three abundance images highlights that the study area has higher *T. grandis* and *B. bambos* endmembers, while *B. monosperma* showed comparatively lower fraction values.

6.11 Conclusion

PFT-based classification should be adopted for by ecologists as it can be of great aid in global change research. PFT-based models are better options compared to biome-based models. Better mapping of PFTs can help in obtaining better simulations of carbon cycle, climate, and ecosystem change regionally as well as globally. Remote sensing techniques, particularly hyperspectral remote sensing, although being comparatively new, hold great potential to extract reliable PFT information and can, therefore, contribute to improved predictive capabilities of global and regional carbon cycle, climate, and ecosystem models. SMA combined with hyperspectral data gives promising results in identifying the endmembers of functionally distinct plants. The case study discussed in this chapter highlighted the capability of SMA to identify functionally distinct plants, in this case, *T. grandis*, *B. monosperma*, and *B. bambos*. Shoolpaneshwar was chosen as a study area due to its high diversity of plant species. Ground and laboratory estimations of the functional plant traits of three plant species were measured. SMA adequately characterized these functionally distinct plants.

Acknowledgments

A special thanks is extended to the Department of Science and Technology and Science and Engineering Research Board for funding *PDF/2017/002620*. The authors also offer thanks to the Ministry of Environment, Forest and Climate Change and Gujarat Forest Department for sanctioning the necessary permission for field sampling in the sanctuary. Local forest officials are also thanked for extending help during field sampling.

List of Abbreviation

CC	chlorophyll content
DBH	diameter at breast height
ENVI	environment for visulizing images
EO	Earth observation
FLAASH	fast line-of-sight atmospheric analysis of spectral hypercubes
FPAR	fraction of photosynthetically active radiation
FTs	functional traits
GPS	global positioning system
LAI	leaf area index
MNF	minimum noise fraction
NCAR LSM	National Center for Atmospheric Research land surface model
PFTs	plant functional types
PPI	pixel purity index
SMA	spectral mixture analysis
USGS	United States Geological Survey

References

Adams, J.B., Sabol, D.E., Kapos, V., Almeida Filho, R., Roberts, D.A., Smith, M.O., et al., 1995. Classification of multispectral images based on fractions of endmembers: application to land-cover change in the Brazilian Amazon. Remote Sens. Environ. 52 (2), 137–154.

Adler, P.B., Salguero-Gómez, R., Compagnoni, A., Hsu, J.S., Ray-Mukherjee, J., Mbeau-Ache, C., et al., 2014. Functional traits explain variation in plant life history strategies. Proc. Natl Acad. Sci. USA 111 (2), 740–745.

Arnon, D.I., 1949. Copper enzymes in isolated chloroplasts. Polyphenoloxidase in *Beta vulgaris*. Plant. Physiol. 24 (1), 1.

Bonan, G.B., 1993. Importance of leaf area index and forest type when estimating photosynthesis in boreal forests. Remote Sens. Environ. 43 (3), 303–314.

Bonan, G.B., Levis, S., Kergoat, L., Oleson, K.W., 2002. Landscapes as patches of plant functional types: an integrating concept for climate and ecosystem models. Glob. Biogeochem. Cycles 16 (2), 5-1-5-23.

Boyer, K.E., Kertesz, J.S., Bruno, J.F., 2009. Biodiversity effects on productivity and stability of marine macroalgal communities: the role of environmental context. Oikos 118 (7), 1062–1072.

Bu, W., Huang, J., Xu, H., Zang, R., Ding, Y., Li, Y., et al., 2018. Plant functional traits are the mediators in regulating effects of abiotic site conditions on aboveground carbon stock-evidence from a 30 ha tropical forest plot. Front. Plant. Sci. 9.

Castro, K., Sanchez-Azofeifa, G.A., Rivard, B., 2003. Monitoring secondary tropical forests using space-borne data: implications for Central America. Int. J. Remote Sens. 24 (9), 1853–1894.

Chave, J., Coomes, D., Jansen, S., Lewis, S.L., Swenson, N.G., Zanne, A.E., 2009. Towards a worldwide wood economics spectrum. Ecol. Lett. 12 (4), 351–366.

Clark, M.L., Roberts, D.A., Clark, D.B., 2005. Hyperspectral discrimination of tropical rain forest tree species at leaf to crown scales. Remote Sens. Environ. 96 (3-4), 375–398.

de Bello, F., Lavorel, S., Díaz, S., Harrington, R., Cornelissen, J.H., Bardgett, R.D., et al., 2010. Towards an assessment of multiple ecosystem processes and services via functional traits. Biodivers. Conserv. 19 (10), 2873–2893.

De Deyn, G.B., Cornelissen, J.H., Bardgett, R.D., 2008. Plant functional traits and soil carbon sequestration in contrasting biomes. Ecol. Lett. 11 (5), 516–531.

De Fries, R., Hansen, M., Townshend, J., Sohlberg, R., 1998. Global land cover classifications at 8 km spatial resolution: the use of training data derived from Landsat imagery in decision tree classifiers. Int. J. Remote Sens. 19 (16), 3141–3168.

Denning, A.S., Collatz, G.J., Zhang, C., Randall, D.A., Berry, J.A., Sellers, P.J., et al., 1996. Simulations of terrestrial carbon metabolism and atmospheric CO_2 in a general circulation model: Part 1: Surface carbon fluxes. Tellus B: Chem. Phys. Meteorol. 48 (4), 521–542.

Dentener, F., Stevenson, D., Ellingsen, Kv, Van Noije, T., Schultz, M., Amann, M., et al., 2006. The global atmospheric environment for the next generation. Environ. Sci. Technol. 40 (11), 3586–3594.

Díaz, S., Kattge, J., Cornelissen, J.H., Wright, I.J., Lavorel, S., Dray, S., et al., 2016. The global spectrum of plant form and function. Nature 529 (7585), 167.

Eviner, V.T., Chapin III, F.S., 2003. Functional matrix: a conceptual framework for predicting multiple plant effects on ecosystem processes. Annu. Rev. Ecol. Evol. Syst. 34 (1), 455–485.

Fonji, S.F., Taff, G.N., 2014. Using satellite data to monitor land-use land-cover change in North-eastern Latvia. Springerplus 3 (1), 61.

Foster, D., Swanson, F., Aber, J., Burke, I., Brokaw, N., Tilman, D., et al., 2003. The importance of land-use legacies to ecology and conservation. BioScience 53 (1), 77–88.

Garnier, E., 1991. Resource capture, biomass allocation and growth in herbaceous plants. Trends Ecol. Evol. 6 (4), 126–131.

Geert Hiddink, J., Wynter Davies, T., Perkins, M., Machairopoulou, M., Neill, S.P., 2009. Context dependency of relationships between biodiversity and ecosystem functioning is different for multiple ecosystem functions. Oikos 118 (12), 1892–1900.

Gitay, H., Noble, I., 1997. What are functional types and how should we seek them. Plant functional types: their relevance to ecosystem properties and global change. In Plant Functional Types(Smith, T.M. et al., eds), Cambridge University Press, 1, 3–19.

Grime, J.P., Hodgson, J.G., Hunt, R., 2014. Comparative Plant Ecology: A Functional Approach to Common British Species. Springer.

Hajek, P., Kurjak, D., von Wühlisch, G., Delzon, S., Schuldt, B., 2016. Intraspecific variation in wood anatomical, hydraulic, and foliar traits in ten European beech provenances differing in growth yield. Front. Plant. Sci. 7, 791.

Hooper, D.U., Adair, E.C., Cardinale, B.J., Byrnes, J.E., Hungate, B.A., Matulich, K.L., et al., 2012. A global synthesis reveals biodiversity loss as a major driver of ecosystem change. Nature 486 (7401), 105.

Kattenborn, T., Fassnacht, F.E., Schmidtlein, S., 2019. Differentiating plant functional types using reflectance: which traits make the difference? Remote Sens. Ecol. Conserv. 5 (1), 5–19.

Kimes, D.S., Harrison, P.R., Ratcliffe, P.A., 1991. A knowledge-based expert system for inferring vegetation characteristics. Remote Sens. 12 (10), 1987–2020.

Kooyman, R.M., Westoby, M., 2009. Costs of height gain in rainforest saplings: main-stem scaling, functional traits and strategy variation across 75 species. Ann. Bot. 104 (5), 987–993.

Lavorel, S., Garnier, É., 2002. Predicting changes in community composition and ecosystem functioning from plant traits: revisiting the Holy Grail. Funct. Ecol. 16 (5), 545–556.

Lavorel, S., McIntyre, S., Landsberg, J., Forbes, T., 1997. Plant functional classifications: from general groups to specific groups based on response to disturbance. Trends Ecol. Evol. 12 (12), 474–478.

Lavorel, S., Grigulis, K., Lamarque, P., Colace, M.P., Garden, D., Girel, J., et al., 2011. Using plant functional traits to understand the landscape distribution of multiple ecosystem services. J. Ecol. 99 (1), 135–147.

Liang, S., 2005. Quantitative Remote Sensing of Land Surfaces, vol. 30. John Wiley & Sons.

Liping, C., Yujun, S., Saeed, S., 2018. Monitoring and predicting land use and land cover changes using remote sensing and GIS techniques—a case study of a hilly area, Jiangle, China. PLoS One 13 (7), e0200493.

Liu, X., Swenson, N.G., Lin, D., Mi, X., Umaña, M.N., Schmid, B., et al., 2016. Linking individual-level functional traits to tree growth in a subtropical forest. Ecology 97 (9), 2396–2405.

Liu, X., Trogisch, S., He, J.-S., Niklaus, P.A., Bruelheide, H., Tang, Z., et al., 2018. Tree species richness increases ecosystem carbon storage in subtropical forests. Proc. R. Soc. B: Biol. Sci. 285 (1885), 20181240.

Lloyd, D., 1990. A phenological classification of terrestrial vegetation cover using shortwave vegetation index imagery. Remote Sens. 11 (12), 2269–2279.

Lu, D., Batistella, M., Moran, E., Mausel, P., 2004. Application of spectral mixture analysis to Amazonian land-use and land-cover classification. Int. J. Remote Sens. 25 (23), 5345–5358.

Martin, M., Newman, S., Aber, J., Congalton, R., 1998. Determining forest species composition using high spectral resolution remote sensing data. Remote Sens. Environ. 65 (3), 249–254.

McGill, B.J., Enquist, B.J., Weiher, E., Westoby, M., 2006. Rebuilding community ecology from functional traits. Trends Ecol. Evol. 21 (4), 178–185.

McLauchlan, K., 2006. The nature and longevity of agricultural impacts on soil carbon and nutrients: a review. Ecosystems 9 (8), 1364–1382.

Moles, A.T., Westoby, M., 2006. Seed size and plant strategy across the whole life cycle. Oikos 113 (1), 91–105.

Muler, A.L., 2016. Identifying plant functional traits to assist ecological intervention in a drying landscape, Theses: Doctorates and Masters, Edith Cowan University, pp.154.

Myneni, R.B., Keeling, C., Tucker, C.J., Asrar, G., Nemani, R.R., 1997. Increased plant growth in the northern high latitudes from 1981 to 1991. Nature 386 (6626), 698.

Oleson, K.W., Bonan, G.B., 2000. The effects of remotely sensed plant functional type and leaf area index on simulations of boreal forest surface fluxes by the NCAR land surface model. J. Hydrometeorol. 1 (5), 431–446.

Osnas, J.L., Lichstein, J.W., Reich, P.B., Pacala, S.W., 2013. Global leaf trait relationships: mass, area, and the leaf economics spectrum. Science 340 (6133), 741–744.

Poulter, B., Ciais, P., Hodson, E., Lischke, H., Maignan, F., Plummer, S., et al., 2011. Plant functional type mapping for earth system models. Geosci. Model. Dev. 4 (4), 993–1010.

Prince, S.D., Goward, S.N., 1995. Global primary production: a remote sensing approach. J. Biogeogr. 815–835.

Roberts, D.A., Gardner, M., Church, R., Ustin, S., Scheer, G., Green, R., 1998. Mapping chaparral in the Santa Monica Mountains using multiple endmember spectral mixture models. Remote Sens. Environ. 65 (3), 267–279.

Roberts, D., Batista, G., Pereira, J., Waller, E., Nelson, B., 1999. Change identification using multitemporal spectral mixture analysis: applications in eastern Amazonia, *Environ. Mon. Methods App.* 9, 1998, 137–159.

Roughgarden, J., Running, S.W., Matson, P.A., 1991. What does remote sensing do for ecology? Ecology 72 (6), 1918–1922.

Running, S.W., Coughlan, J.C., 1988. A general model of forest ecosystem processes for regional applications I. Hydrologic balance, canopy gas exchange and primary production processes. Ecol. Model. 42 (2), 125–154.

Sellers, P., Mintz, Y., Sud, Y.E.A., Dalcher, A., 1986. A simple biosphere model (SiB) for use within general circulation models. J. Atmos. Sci. 43 (6), 505–531.

Sellers, P., Meeson, B., Hall, F., Asrar, G., Murphy, R., Schiffer, R., et al., 1995. Remote sensing of the land surface for studies of global change: models–algorithms–experiments. Remote Sens. Environ. 51 (1), 3–26.

Sellers, P., Dickinson, R., Randall, D., Betts, A., Hall, F., Berry, J., et al., 1997. Modeling the exchanges of energy, water, and carbon between continents and the atmosphere. Science 275 (5299), 502–509.

Sitch, S., Smith, B., Prentice, I.C., Arneth, A., Bondeau, A., Cramer, W., et al., 2003. Evaluation of ecosystem dynamics, plant geography and terrestrial carbon cycling in the LPJ dynamic global vegetation model. Glob. Change Biol. 9 (2), 161–185.

Smith, M.O., Ustin, S.L., Adams, J.B., Gillespie, A.R., 1990. Vegetation in deserts: I. A regional measure of abundance from multispectral images. Remote Sens. Environ. 31 (1), 1–26.

Smith, T.M., Smith, T., Shugart, H.H., Woodward, F., 1997. Plant Functional Types: Their Relevance to Ecosystem Properties and Global Change, vol. 1. Cambridge University Press.

Smith, B., Prentice, I.C., Sykes, M.T., 2001. Representation of vegetation dynamics in the modelling of terrestrial ecosystems: comparing two contrasting approaches within European climate space. Glob. Ecol. Biogeogr. 10 (6), 621–637.

Sun, W., Liang, S., Xu, G., Fang, H., Dickinson, R., 2008. Mapping plant functional types from MODIS data using multisource evidential reasoning. Remote Sens. Environ. 112 (3), 1010–1024.

Tian, Y., Dickinson, R., Zhou, L., Myneni, R., Friedl, M., Schaaf, C., et al., 2004. Land boundary conditions from MODIS data and consequences for the albedo of a climate model. Geophys. Res. Lett. 31 (5).

Townsend, P.A., Chlus, A., Wang, Z., Singh, A., 2017. Remote sensing spectroscopy to discriminate plant functional types and physiological function. In: 2017 IEEE International Geoscience and Remote Sensing Symposium (IGARSS), IEEE, pp. 3043–3046.

Townshend, J., Justice, C., 1981. Information extraction from remotely sensed data. A user view. Remote Sens. 2 (4), 313–329.

Townshend, J.R., Justice, C., Kalb, V., 1987. Characterization and classification of South American land cover types using satellite data. Int. J. Remote Sens. 8 (8), 1189–1207.

Venable, D.L., 1992. Size-number trade-offs and the variation of seed size with plant resource status. Am. Nat. 140 (2), 287–304.

Violle, C., Navas, M.L., Vile, D., Kazakou, E., Fortunel, C., Hummel, I., et al., 2007. Let the concept of trait be functional!. Oikos 116 (5), 882–892.

Wardle, D., Barker, G., Bonner, K., Nicholson, K., 1998. Can comparative approaches based on plant ecophysiological traits predict the nature of biotic interactions and individual plant species effects in ecosystems? J. Ecol. 86 (3), 405–420.

Wright, I.J., Reich, P.B., Westoby, M., Ackerly, D.D., Baruch, Z., Bongers, F., et al., 2004. The worldwide leaf economics spectrum. Nature 428 (6985), 821.

Wu, C., Niu, Z., Tang, Q., Huang, W., 2008. Estimating chlorophyll content from hyperspectral vegetation indices: modeling and validation. Agric. For. Meteorol. 148 (8–9), 1230–1241.

7

Estimation of chengal trees relative abundance using coarse spatial resolution hyperspectral systems

Noordyana Hassan[1,2], Mazlan Hashim[1,2], Shinya Numata[3], Mohamad Zakri Tarmidi[1,2]

[1]GEOSCIENCE AND DIGITAL EARTH CENTRE (INSTEG), RESEARCH INSTITUTE OF SUSTAINABLE ENVIRONMENT, UNIVERSITI TEKNOLOGI MALAYSIA, UTM JOHOR, JOHOR, MALAYSIA
[2]DEPARTMENT OF GEOINFORMATICS, FACULTY OF BUILT ENVIRONMENT AND SURVEYING, UNIVERSITI TEKNOLOGI MALAYSIA, UTM JOHOR, JOHOR, MALAYSIA
[3]DEPARTMENT OF TOURISM SCIENCE, GRADUATE SCHOOL OF URBAN ENVIRONMENTAL SCIENCES, TOKYO METROPOLITAN UNIVERSITY, MINAMI-OSAWA 1-1, HACHIOUJI, TOKYO, JAPAN

7.1 Introduction

Forest degradation and deforestation monitoring is crucial as it is occurring at an alarming rate (Zaw Htun et al., 2011). Forest degradation and deforestation have a negative impact toward ecosystem function, biodiversity, and climate. Thus these problems have become key issues in reducing emission from deforestation and degradation (REDD +). Forest degradation and deforestation are caused by anthropogenic pressures such as unsustainable timber harvesting (Zaw Htun et al., 2011). Therefore to overcome this problem, the Food and Agriculture Organization (FAO) has introduced sustainable forest management (SFM) policies. The conventional measurement was employed in selective logging scheme where the abundance, distribution, and the volume of trees were measured in this scheme (Numata et al., 2006; Zaw Htun et al., 2011).

Studies at a global scale, especially for forest regions, can be carried out by utilizing a remote sensing technique due to its ability to provide spatial and temporal resolution data. Remote sensing techniques have been widely conducted over the rainforest in Malaysia (Sy et al., 2012). SFM is crucial to stop extreme forest degradation. Remote sensing techniques can provide access for monitoring the tropical rainforest due to the availability of high temporal resolution remote sensing data. Relative abundance studies on tree species in tropical

Hyperspectral Remote Sensing. DOI: https://doi.org/10.1016/B978-0-08-102894-0.00006-1

rainforests could essentially be one of the solutions to cope with forest degradation and deforestation.

Previous studies suggested that relative abundance of tree species may be estimated by using spectral unmixing approach such as through canopy fractional cover (CFC), which was developed from the linear mixture model (LMM) and mixture tuned matched filtering (MTMF) (Hassan and Hashim, 2011; Parker Williams and Hunt, 2002). Spectral unmixing is used to model each pixel problem as a linear combination of a finite number of spectrally distinct signatures or "endmembers," and a subpixel estimate of end member abundance can be obtained (Parker Williams and Hunt, 2002). In this study, the spectral unmixing approach used was MTMF and modified canopy fractional cover (mCFC) which has been modified by author from CFC developed by Wang et al. (2005) to estimate the relative abundance of chengal trees exist in the Pasoh Forest Reserve. The main goal of this study was to assess the modified spectral unmixing model for estimating the relative abundance of chengal trees. The best model to estimate relative abundance of chengal trees was identified based on accuracy evaluation. The accuracy of mCFC and MTMF were evaluated using census data provided by the Forest Research Institute of Malaysia (FRIM). As such, we can hypothesize that mCFC may give more accurate results than MTMF as it applies a vegetation index in the model which can eliminate soil background effects.

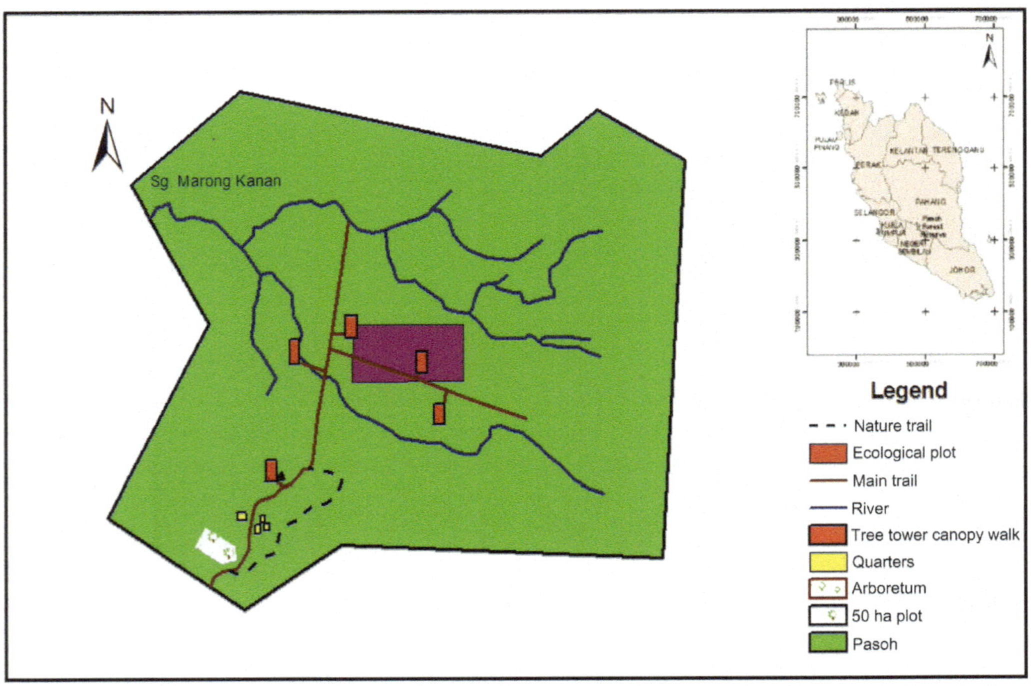

FIGURE 7–1 The study area which is 50 ha plot (purple box) in Pasoh Forest Reserve. *Adapted from Okuda, T., Manokaran, N., Matsumoto, Y., Niiyama, K., Thomas, S.C., Ashton, P.S., 2003. Pasoh Ecology of a Lowland Rain Forest in Southeast Asia. Springer, Japan (Okuda et al., 2003).*

7.2 Materials and methodology

7.2.1 Study area

The study area was located in Pasoh Forest Reserve, Negeri Sembilan, Malaysia (2°58′N latitude and 102°18′E longitude) (Fig. 7–1). The Pasoh Forest Reserve is mainly dominated by lowland, mixed-dipterocarp forest. Shorea and Dipterocarpus are the main species in this forest. The study was restricted to a 50 ha ecological plot, being 0.5 km wide and 1 km long. This ecological plot was established by FRIM. There are 338, 360 trees with diameter at breast height (DBH) ≥1 cm. This ecological plot also has 81 families, 259 genera, and 818 species. However, there are only 30 species of dipterocarpacea in the plot. The average height of the emergent tree is 46 m, while the main canopy height was 20–30 m (Hassan and Hashim, 2011). Various useful timber species exist in the 50 ha plot, representing the heterogeneous nature of species of the tropical rainforest in Malaysia. The study area was selected based on remotely sensed data availability of the area.

7.2.2 Hyperspectral system

Hyperion EO-1 was onboard an Earth observation satellite. The data was acquired from the official United State Geological Survey website (USGS) via Glovis. Data for 2003 with 20% cloud cover obtained from Glovis was downloaded. Hyperion EO-1 was used for estimating the distribution and composition of tree species in the Pasoh Forest Reserve.

7.2.3 Ancillary data (topographic map, census data, and spectral radiometer data)

Tree census data was used to perform an accuracy assessment of the spectral unmixing models. The census data used was obtained for 2005. This dataset usually involves information including the DBH of censused trees (>1 cm in DBH), species, genus, and family names, as well as the location of the trees. The distribution of trees were plotted to show the general pattern of distribution in the study area. DBH information was used to estimate the tree height and tree crown for plotting the trees in three-dimensional views. The plotted tree distribution was used as reference data in the accuracy assessment. In this study the main focus was on trees that have DBH of more than 40 cm because the top layer of these trees each the canopy sensed by the remote sensor.

7.2.3.1 Estimation of tree height from diameter at breast height

In order to use census data, tree height must be identified. According to Avery and Burkhart (2002) there is a strong relationship between DBH and tree height. The height of the tree can be predicted by using the allometry equation (Kato et al., 1978):

$$\frac{1}{H} = \frac{1}{(\alpha D)} + \frac{1}{H_{max}} \tag{7.1}$$

where H is tree height in meters, D is DBH in centimeters, H_{max} is the maximum tree height in study area, which is 61 m and is a regression coefficient. According to Kato et al. (1978), the correlation coefficient is 1.5 for primary forest and 2.0 for logged forest. In this study, the correlation coefficient used is 1.5 because Pasoh Forest Reserve consists primarily of a forested area.

7.2.3.2 Estimation of tree crown from tree height

Tree crown is an important factor to estimate the canopy size of each tree because bigger canopies are generally above other, smaller canopies Tree crown can be estimated based on the crown–diameter relationship model. The crown–diameter relationship model in this area was estimated using the below equation:

$$y = 3.5266 \ln(x) + 2.8283 \tag{7.2}$$

where y is the tree crown and x is the tree DBH. As the crown and height of the tree has been identified, the tree can be plotted in three-dimensional views as illustrated in Fig. 7–2.

7.2.4 Spectral radiometer data collection

Leaves of trees were randomly shuffled and separated into six piles per dipterocarp species. Each pile of leaves (top side up) was placed on top of black paper until the black paper could not been seen. Next, the spectral response of each leaf plate was recorded 10 times. The pistol grip was rotated 90° horizontally for each plate after fifth record to compensate for the bidirectional reflectance distribution function.

Then the means of the 10 records were calculated to construct a radiance curve. The radiance was converted into a reflectance curve using the reference panel including conducting the correction of the spectrometer internal current (dark current). The steps above were followed for the other leaf plates, resulting in six reflectance curves—one for each tree species.

Spectral measurement was conducted under laboratory conditions by using a spectroradiometer (FieldSpec Pro, Analytical Spectral Device, Inc.). This spectroradiometer is equipped with three spectrometers (i.e., visible near infrared, short-wave near infrared 1, and short-wave near infrared 2), covering 325–1075 nm, with sampling intervals of 1.4 nm between 325 and 1000 nm, and 2 nm between 1000 nm and 1075 nm. The spectral resolution of the spectrometer was 3 nm for the wavelength interval 350–1000 nm, and 10 nm for the wavelength interval 1000–1075 nm.

FIGURE 7–2 Plotted tree in three-dimensional views based on calculated tree height and tree crown. *Hassan, N., Hashim, M., 2011. Decomposition of mixed pixels of ASTER satellite data for mapping chengal (Neobalanocarpus heimii sp.) tree. In: 2011 IEEE International Conference on Control System, Computing and Engineering (ICCSCE), pp. 7479. https://doi.org/10.1109/ICCSCE.2011.6190499.*

The sensor, equipped with a field of view 25°, was mounted on a tripod and positioned 0.5 m above the leaf plate at the nadir position. A halogen lamp fixed on the tripod at the same position as the sensor of the spectrometer was used to illuminate the sample plate. The room was conditioned to be dark in order to avoid unwanted external energy sources.

7.2.5 Hyperspectral image preprocessing

To estimate the relative abundance of chengal trees, a spectral unmixing approach was employed to Hyperion EO-1 which was acquired from USGS via Glovis. Data downloaded was captured by Earth observation satellite on 2003 with 20% cloud cover. In order to remove atmospheric effects that existed in the Hyperion EO-1 scene, Fast Line-of-Sight Atmospheric Analysis of Spectral Hypercubes (FLAASH) was employed. FLAASH is the most commonly used method to transform radiance images into reflectance images, especially useful for hyperspectral imagery. The two main functions of FLAASH are: (1) it can be used as an atmospheric correction tool that corrects wavelengths in the visible through near-infrared and short-wave infrared regions (i.e., up to 3 μm); and (2) it can be used for converting radiance images into reflectance images. Geometric correction was carried out through the EO-1 Hyperion L1R image which had already been projected in Universal Transverse Mercator (UTM) with WGS 84 datum. This geometric correction was employed to the image in order to convert the UTM projection into RSO projection.

7.2.6 Canopy fractional cover

In the LMM, the spectral reflectance of ground cover is represented as a sum of each component that exists in a pixel, which becomes weighted by their percentage cover. These components as known as end members. The LMM can be represented by the equation (Defries et al., 2000):

$$R_i = \sum_{j=1}^{n} r_{ij} f_j + e_i \tag{7.3}$$

where R_i is the reflectance of the band i, r_{ij} is the reflectance end member j in band i, f_i is the percentage cover of end member j in such pixel and e_i is some insignificant remaining components within the pixel in band i.

If there is another end member in a pixel it will become a two-component mixture model where the reflectance of the components are independent of each other. The two-component linear model can be determined using Eq. (7.4) (Wang et al., 2005):

$$R = R_{canopy} fc + R_{open}(1 - fc) + \varepsilon \tag{7.4}$$

where fc is tree fractional cover in a pixel. Eq. (7.4) shows that the spectral response of a pixel at a certain wavelength is a linear combination of responses from the tree canopy and open areas which were weighted by the corresponding cover of each component, fc and $(1 - fc)$, respectively.

However, surface reflectance usually change significantly with wavelengths. Consequently, different spectral bands may result in different fractional cover. Furthermore, at certain wavelength, the spectral reflectance are influenced by the vegetation structure, soil and vegetation moisture, the texture and external factor like sensor geometry. Thus in order to overcome this problem, vegetation indices were used to identify forest fractional cover (Jasinski, 1990; Carlson and Ripley, 1997; Gutman and Ignatov, 1998; Zeng et al., 2000; Wang et al., 2005) where vegetation index was used by replacing spectral response, R in Eq. (7.4) and becomes:

$$VI = VI_{canopy} fc + VI_{open}(1 - fc) \tag{7.5}$$

where VI_{canopy} represents the vegetation index for the tree canopy and VI_{open} represents the vegetation index for open areas. Thus fractional cover can be expressed as (Wang et al., 2005):

$$fc = \frac{VI - VI_{open}}{VI_{canopy} - VI_{open}} \tag{7.6}$$

where VI_{canopy} and VI_{open} are two different end members that are obtained from the satellite's remotely sensed data. According to Wang et al. (2005), CFC was used to estimate the fractional cover of forest degradation. However, for this study, the CFC model was modified (i.e., mCFC) to estimate the relative abundance of specific tree species in the 50 ha plot.

7.2.7 Vegetation index used in canopy fractional cover

The most widely used vegetation index in remote sensing is distance and slope-based vegetation indices. The differential between near infrared and red reflectance and the ratio of these two spectral bands are used by the distance and slope-based approaches, respectively. Both approaches aim to be more sensitive to the vegetation component, while minimizing external effects due to reflectance of red and near-infrared variation at the sensor caused by solar irradiance, atmospheric conditions, canopy background, and vegetation canopy structure and composition (Qi et al., 1994). However, the normalized difference vegetation index (NDVI) can be transformed further to become several indices with the purpose to improve the vegetation performance and reduce the atmospheric and soil effects.

Modified soil adjusted vegetation index (MSAVI) is a modified version of the soil adjusted vegetation index (SAVI) developed by Huete (1988) and Qi et al. (1994). Further versions derived from SAVI include the transformed soil adjusted vegetation index (TSAVI) developed by Baret et al. (1989) and the second version of the modified soil adjusted vegetation index (MSAVI2) also developed by Qi et al. (1994). Meanwhile, the global environment monitoring index is newer vegetation index that can minimize atmospheric effects and increase the sensitivity in sensing vegetation.

However, it is vital to remember that the vegetation indices will suppress the spectral variation of reflectance when different spectral bands are combined (Jasinski, 1990) and, consequently, reduce the effects of those factors (Gutman and Ignatov, 1998). The L factor introduced in MSAVI is to make the adjustment between the level of vegetation cover being observed and on the product of NDVI and the weighted difference vegetation index. Thus the vegetation index could be more accurate.

The MSAVI model can be expressed as (Qi et al., 1994):

$$MSAVI_0 = \frac{\rho NIR - \rho R}{\rho NIR + \rho r + L_0}(1 + L_0) \tag{7.7}$$

where NIR is the reflectance at near-infrared band; R is the reflectance at red band; and L is the correction factor.

The correction L factor can be express as:

$$L = [(NIR - 0.5R) \times s + 1 + NIR + 0.5R]^2 - 8.0 \times s \times (NIR - 0.5R) \tag{7.8}$$

where s is the slope of the soil line calculated from the surface reflectance at deforestation areas.

In order to make the calculation easy, another L factor can be obtained to further minimize the soil effect:

$$L_1 = 1 - MSAVI_0 \tag{7.9}$$

Then $MSAVI_1$ will become:

$$MSAVI_1 = \frac{NIR - R}{NIR + R + MSAVI(n-1)}(2 - MSAVI_0) \tag{7.10}$$

By continuing this process n times, the equations for L and MSAVI will become:

$$L_n = 1 - MSAVI_{(n-1)} \tag{7.11}$$

$$MSAVI_n = \frac{NIR - R}{NIR + R + MSAVI(n-1)}(2 - MSAVI_{(n-1)}) \tag{7.12}$$

With this process, there exists an interaction time, N, such that $MSAVI_N$ will be equal to $MSAVI_{(N-1)}$, where the soil effects cannot be minimized further. Then the equation will be:

$$MSAVI_N = \frac{NIR - R}{NIR + R + MSAVI(N-1)}(2 - MSAVI_{(N)}) \tag{7.13}$$

The solution for this equation is:

$$MSAVI_N = \frac{-b - \sqrt{(b^2 - 4c)}}{2} \tag{7.14}$$

where b is $-(2NIR + 1)c$ is $2(NIR - R)$.

Therefore with an interactive L function of $L = 1 - MSAVI_2$, the resultant MSAVI by induction, $MSAVI_2$, becomes

$$MSAVI_2 = \frac{2NIR + 1\sqrt{(2NIR + 1)^2 - 8(NIR - R)}}{2} \tag{7.15}$$

where ρNIR is the reflectance at near-infrared band, and ρR is the reflectance at red band.

7.2.8 Mixture tuned matched filtering

Mixture analysis was carried out using MTMF with input features selected from the previous MNF transformation of the ASTER dataset. MTMF is the fraction of an image which allows false positives to be identified and eliminated from the abundance results (Boardman, 1998). Therefore MTMF can be used to identify tree species in the ASTER dataset as it can calculate the quantity of the target that is much smaller than pixel size.

There are two phases in the MTMF algorithm, namely a matched filter (MF) calculation for an abundance estimation and a mixture tuning (MT) calculation for false positive identification or rejection. In the remote sensing context, MF is a filtering process of input data for matching the target spectrum and eliminates the remaining background spectra. Meanwhile, the MT calculates a value of infeasibility for each MF classified pixel. The MTMF algorithm as described here implicitly requires zero mean, unit noise variance input data (such as MNF transformed data) for proper mixture tuning calculation.

The MF score was calculated for each pixel by using minimum noise fraction (MNF) transformed data as input to the MF vector. This vector transforms the target spectrum into MNF space and projects it into inverse covariance of MNF data. Inverse covariance of MNF data is normalized to the magnitude of the target spectrum. This ensures that the matched filter vector has a unit length that corresponds to target components ranging from 0% to 100%. This mechanism is summarized in Eq. (7.17) (Boardman, 1998):

$$\vec{v} = \frac{[C_{\mathrm{MNF}}]^{-1}\#\vec{t}_{\mathrm{MNF}}}{(\vec{t}_{\mathrm{MNF}})^{T}\#[C_{\mathrm{MNF}}]^{-1}\#\vec{t}_{\mathrm{MNF}}} \tag{7.16}$$

where $\vec{v}$ is the matched filter vector, $[C_{\mathrm{MNF}}]^{-1}$ is the MNF inverse covariance matrix (a diagonal matrix of eigenvalue reciprocals), and $\vec{t}_{\mathrm{MNF}}$ is the vector of the target spectra in MNF space. MF values are subsequently calculated for each pixel (creating an $i \times j$ target abundance image) by using the MNF transformed data as input to the MF vector, as expressed in Eq. (7.17) (Boardman, 1998):

$$[\mathrm{MF}] = \vec{v}\#[\mathrm{MNF}] \tag{7.17}$$

Output of MF scores represent a linear solution for the magnitude of the MF score, where MF scores are normally distributed and have a mean of zero. MF values of zero and lower represent background (no target component), and pixels with a MF score greater than zero are considered to contain a fractional target component equivalent to the MF score.

The MT step of the MTMF algorithm assesses the probability of an MF estimation error for each pixel based on the concept of mixing feasibility. The calculation of infeasibility value of each pixel takes place over three distinct steps: (1) determination of the target vector component of the pixel, (2) interpolation of variance eigenvalues respective to the target vector component, and (3) calculation of the standardized separation between a pixel and its ideal target vector component. The target vector component of a pixel is the scalar product of its MF score and the target vector, as illustrated in Eq. (7.18):

$$\vec{c}_i = \mathrm{MF}_i \times \vec{t}_{\mathrm{MNF}} \tag{7.18}$$

where $\rightarrow c_i$ is the target vector component for pixel i and MF_i is the matched filter value of the pixel. An ideal pixel, containing some fraction of the target, lies directly on the target vector. Actual pixels, however, likely contain some degree of noise variance and background mixing, resulting in a pixel that does not lie directly on the target.

The proximity of each pixel to its idealized location on the target vector was a conceptual measure of infeasibility. The units of MNF space are defined in terms of noise standard deviations. Thus the proximity of a pixel's actual location to its ideal projected location (on the target vector) is a statistical measure of its spectral variance from the target. Pixels with a high percentage target component are expected to have a low degree of mixing freedom (low variance) because their spectral characteristics should be dominated by the target. Alternatively, pixels with a low percentage target component are expected to have a higher degree of mixing freedom (high variance) because they mix with combinations of background components to form varied spectral signatures. Because pixels with small target vector components have higher degrees of mixing freedom than do pixels with large target vector components, it is necessary to establish variance thresholds to consistently assess acceptable degrees of data variance based on mixing freedom. The mathematical derivation of infeasibility requires eigenvalue interpolation and is determined by Boardman (1998) as:

$$\vec{e}_i = \left(\sqrt{\vec{e}_{\mathrm{MNF}}} - \mathrm{MF}_i \times \left(\sqrt{\vec{e}_{\mathrm{MNF}}} - \vec{e}_n\right)\right)^2 \tag{7.19}$$

where $\vec{e}_i$ is the interpolated vector of eigenvalues for pixel i, $\vec{e}_{\mathrm{MNF}}$ is the vector of MNF eigenvalues, and $\vec{e}_n$ is the vector of MNF noise eigenvalues (a vector of one). These standardizations (variance thresholds) were determined by the linear interpolation of eigenvalues between the 100% target distribution (eigenvalues are all equal to one, having no mixing freedom and therefore only unit noise variance) and the 100% background distribution (eigenvalues generally much greater than one, representative of high degrees of mixing freedom). The infeasibility value of a pixel is calculated as the geometric distance from the pixel to the target vector, normalized to the variance threshold magnitude for the respective MF value. The mathematical derivation of infeasibility values is illustrated in Eq. (7.20):

$$I_i = \frac{\vec{s}_i - \vec{c}_i}{\vec{e}_i} \tag{7.20}$$

where I_i is the resulting infeasibility value for pixel i, and $\vec{s}_i$ is the MNF spectra for pixel i.

7.2.9 Relative abundance assessment

As the relative abundance per pixel basis has been estimated by using mCFC, the next step is to estimate the relative abundance of tree species per hectare in each compartment that been built by FRIM. The percentage of relative abundance of tree species obtained using mCFC was then evaluated using a three-dimensional plot of census data on a Hyperion EO-1 30 m × 30 m grid. The three-dimensional census data on Hyperion grid was produced based on the calculated tree crown cover.

7.3 Results and analysis

7.3.1 Chengal trees relative abundance estimation using mixture tuned matched filtering

MTMF was utilized to Hyperion EO-1 which allows false positive to be identified and eliminated from abundance results (Mitchell and Glenn, 2009). The results obtained from MTMF were in fraction with the subpixel abundance values of chengal trees and feasibility image that ranging from 2 to 20. Eight points which represent chengal trees were figured out based on MTMF fraction and the infeasibility scale. The highest percentage of chengal trees had low infeasibility value (≤6.00). The low infeasibility value showing the correct MTMF fraction of the chengal trees. Meanwhile, a higher MTMF fraction value with higher infeasibility value (≥6.00) shows that the pixel was dominated by other tree species. Fig. 7–3A represents the relative abundance of chengal and other tree species where chengal was classified as red and other tree species as blue. Species was classified based on MTMF fractional value as shown in Table 7–1. From the results of regression analysis, the *P*-value indicates that the MTMF fraction is not significant compared with the census fraction value. Despite insignificance with the census fractional value, the correlation of determination obtained was also relatively low.

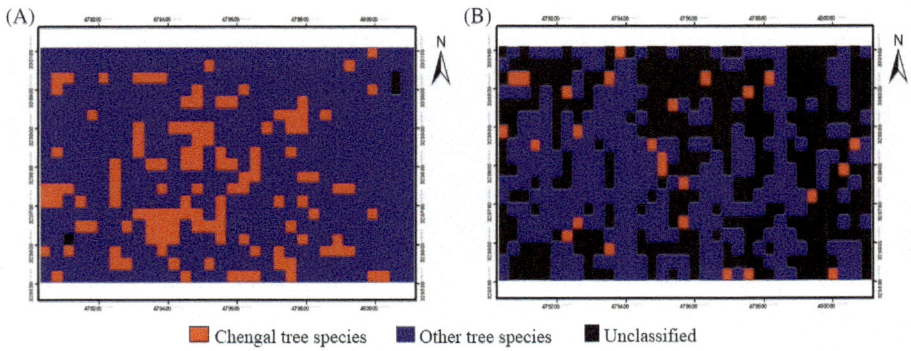

FIGURE 7–3 Chengal trees relative abundance estimation using the (A) MTMF model and (B) mCFC in the 50 ha plot. *Hassan, N., 2014. Relative abundance estimations of chengal tree in a tropical rainforest by using modified Canopy Fractional Cover (mCFC). In: IOP Conference Series: Earth and Environmental Science. doi:10.1088/1755-1315/18/1/012189 (Hassan, 2014).*

Table 7–1 Calculated fractional cover of trees in a 50 ha plot.

Tree species	Fractional cover	
	MTMF	**mCFC**
Chengal tree	0.2–0.9	0.01–0.999
Other tree species	0.94	1.000–10.000
Unclassified		−0.900 to 0.000

mCFC, Modified canopy fractional cover; *MTMF*, mixture tuned matched filtering.

Source: Hassan, N., 2014. Relative abundance estimations of chengal tree in a tropical rainforest by using modified Canopy Fractional Cover (mCFC). In: IOP Conference Series: Earth and Environmental Science. doi:10.1088/1755-1315/18/1/012189.

7.3.2 Relative abundance of chengal trees estimation by modified canopy fractional cover

mCFC was applied to Hyperion EO-1 and the relative abundance of chengal trees was estimated. The result obtained was represented in a fractional image with a fractional value of chengal trees and other tree species that exist in the Hyperion EO-1 image. Fig. 7–3B shows the chengal and other tree species fractional cover in the 50 ha plot. Red, blue, and black areas represent the existence of chengal, other tree species, and unclassified species, respectively. The species that were classified based on fractional values are listed in Table 7–1. Table 7–1 shows that chengal trees were found in the range 0.000–0.999 of fraction values. Meanwhile, other tree species were detected within 1.000–10.000 of fraction value and an unclassified end member was found in the range of −0.9000 to 0.000. The existence of an unclassified end member may be due to haze in the image and the course spatial resolution of Hyperion data. The overall performance of mCFC was good with $r_2 = 0.667$. There is no significant difference between mCFC fractional values and census data fractional values ($P < .05$). The standard error of y-estimates was ± 0.16, hence we accept our hypothesis that mCFC gives better results of relative abundance as it applies a vegetation index in the model that can eliminate soil background effects.

7.4 Discussion

From the findings, the mCFC model explained the relative abundance of chengal trees better than the MTMF model. The MTMF model may not explain the estimated relative abundance of chengal trees in 50 ha plot well due to the high contribution of background effects (i.e., haze existed in the image, soil background, target species, and another species). Background effects may be due to the gap between tree canopies or degradation. Moreover, the limited number of ground sample data may also affect the results obtained. The number of ground sample data chosen was small due to the small number of tall chengal trees in 50 ha plot. However, this result showed better performance than a previous study conducted by Mitchell and Glenn (2009) that applied MTMF on Landsat TM 5 with 30 m spatial resolution and low-spectral resolution where the accuracy obtained was only 35%. In this study, the mCFC model shows better performance than MTMF as it applies a vegetation index that can eliminate soil background effects that exist in Hyperion EO-1 image. However, the results obtained from mCFC are relatively low due to low-spatial resolution of data as well as existence of haze in the data which inevitably obscures the clarity of the dataset despite the removal of atmospheric effects (Kochummen et al, 1990).

7.5 Conclusion

In this chapter, mCFC and MTMF models were compared for estimation of relative abundance of useful timber species. From the accuracy evaluation, mCFC is better than MTMF to estimate relative abundance of chengal trees in a 50 ha plot of a Malaysian lowland rainforest. Results of the accuracy test also suggest that the mCFC model would be better for estimating relative abundance of tree species especially for lowland dipterocarp forests where focal forests are relatively degraded or ground sample data are limited (i.e., census data). Therefore in order to estimate the relative abundance of tree species for the whole area of the Pasoh Forest Reserve (2450 ha), a scaled-up technique is required. Further study is needed to focus on relative abundance of tree species for the whole Pasoh Forest Reserve by applying a scaled-up mCFC model.

List of abbreviations

CFC canopy fractional cover
DBH diameter at breast height
FAO Food and Agriculture Organization
FLAASH fast line-of-sight atmospheric analysis of spectral hypercubes
FRIM Forest Research Institute of Malaysia
Hyperion EO-1 Hyperion Earth Observation-1
L1R level 1R
LMM linear mixture model
mCFC modified canopy fractional cover
MF matched filtering
MNF minimum noise fraction
MSAVI2 modified soil adjusted vegetation index 2
MTMF mixture tuned matched filtering
NDVI normalized difference vegetation index
REDD reducing emission from deforestation and degradation
SAVI soil adjusted vegetation index
SFM sustainable forest management
TSAVI transformed soil adjusted vegetation index
USGS State Geological Survey

References

Avery, T.E., Burkhart, H.E., 2002. Forest Measurements, fifth ed. McGraw Hill, New York, p. 456.

Baret, Frédéric, G. Guyot, Major, D.J., 1989. TSAVI: a vegetation index which minimizes soil brightness effects on LAI and APAR estimation. In: 12th Canadian symposium on remote sensing geoscience and remote sensing symposium, vol. 3, pp. 1355–1358. IEEE.

Boardman, J.W., 1998. Leveraging the high dimensionality of AVIRIS data for improved sub-pixel target unmixing and rejection of false positives: mixture tuned matched filtering. In: 7th JPL Airborne Earth Science Workshop.

Carlson, T.N., Ripley, D.A., 1997. On the relation between NDVI, fractional vegetation cover, and leaf area index. Remote Sens. Environ. Available from: https://doi.org/10.1016/S0034-4257(97)00104-1.

Defries, R.S., Hansen, M.C., Townshend, J.R.G., 2000. Global continuous fields of vegetation characteristics: a linear mixture model applied to multi-year 8 km AVHRR data. Int. J. Remote Sens. Available from: https://doi.org/10.1080/014311600210236.

Gutman, G., Ignatov, A., 1998. The derivation of the green vegetation fraction from NOAA/AVHRR data for use in numerical weather prediction models. Int. J. Remote Sens. Available from: https://doi.org/10.1080/014311698215333.

Hassan, N., 2014. Relative abundance estimations of chengal tree in a tropical rainforest by using modified Canopy Fractional Cover (mCFC). In: IOP Conference Series: Earth and Environmental Science. https://doi.org/10.1088/1755-1315/18/1/012189.

Hassan, N., Hashim, M., 2011. Decomposition of mixed pixels of ASTER satellite data for mapping chengal (*Neobalanocarpus heimii* sp.) tree. In: 2011 IEEE International Conference on Control System, Computing and Engineering (ICCSCE), pp. 74–79. https://doi.org/10.1109/ICCSCE.2011.6190499.

Huete, A., 1988. A soil-adjusted vegetation index (SAVI). Remote Sensing of Environment. Remote Sensing of Environment, 25, 295–309.

Jasinski, M.F., 1990. Sensitivity of the normalized difference vegetation index to subpixel canopy cover, soil albedo, and pixel scale. Remote Sens. Environ. Available from: https://doi.org/10.1016/0034-4257(90)90016-F.

Kochummen, K.M., LaFrankie, J.J.V., Manokaran, N., 1990. Floristic composition of Pasoh Forest Reserve, a lowland rain forest in Peninsular Malaysia. J. Trop. For. Sci. 3 (1), 1–13. Available from: <http://www.cabdirect.org/abstracts/19930667132.html>.

Kato, R., Tadaki, Y., Ogawa, I.I., 1978. Plant biomass and growth increment studies in Pasoh Reserve Forest. National Journal, 30, 211–224.

Mitchell, J.J., Glenn, N.F., 2009. Subpixel abundance estimates in mixture-tuned matched filtering classifications of leafy spurge (*Euphorbia esula* L.). Int. J. Remote Sens. 30 (23), 6099–6119. Available from: https://doi.org/10.1080/01431160902810620. Taylor & Francis.

Numata, S., et al., 2006. Canopy gap dynamics of two different forest stands in a Malaysian lowland rain forest. J. Trop. Forest Sci. 18 (2), 109–116.

Okuda, T., Manokaran, N., Matsumoto, Y., Niiyama, K., Thomas, S.C., Ashton, P.S., 2003. Pasoh Ecology of a Lowland Rain Forest in Southeast Asia. Springer, Japan.

Parker Williams, A., Hunt, E.R., 2002. Estimation of leafy spurge cover from hyperspectral imagery using mixture tuned matched filtering. Remote Sens. Environ. 82 (2–3), 446–456. Available from: https://doi.org/10.1016/S0034-4257(02)00061-5.

Qi, J., et al., 1994. A modified soil adjusted vegetation index. Remote Sens. Environ. Available from: https://doi.org/10.1016/0034-4257(94)90134-1.

Sy, V., et al., 2012. Synergies of multiple remote sensing data sources for REDD + monitoring. Curr. Opin. Env. Sustain. 4 (6).

Wang, C., Qi, J., Cochrane, M., 2005. Assessment of tropical forest degradation with canopy fractional cover from Landsat ETM + and IKONOS imagery. Earth Interact. Am. Meteorol. Soc. 9 (22), 1–18. Available from: http://dx.doi.org/10.1175/EI133.1.

Zaw Htun, N., Mizoue, N., Yoshida, S., 2011. Classifying tropical deciduous vegetation: a comparison of multiple approaches in Popa Mountain Park, Myanmar. Int. J. Remote Sens. 32 (24), 8935–8948. Available from: https://doi.org/10.1080/01431161.2010.531779. Taylor & Francis.

8

Hyperspectral remote sensing in precision agriculture: present status, challenges, and future trends

Prachi Singh[1], Prem Chandra Pandey[2], George P. Petropoulos[3], Andrew Pavlides[4], Prashant K. Srivastava[1,5], Nikos Koutsias[6], Khidir Abdala Kwal Deng[7], Yangson Bao[7]

[1]REMOTE SENSING LABORATORY, INSTITUTE OF ENVIRONMENT AND SUSTAINABLE DEVELOPMENT, BANARAS HINDU UNIVERSITY, VARANASI, INDIA
[2]CENTER FOR ENVIRONMENTAL SCIENCES & ENGINEERING, SCHOOL OF NATURAL SCIENCES, SHIV NADAR UNIVERSITY, GREATER NOIDA, INDIA
[3]DEPARTMENT OF GEOGRAPHY, HAROKOPIO UNIVERSITY OF ATHENS, ATHENS, GREECE
[4]SCHOOL OF MINERAL RESOURCES ENGINEERING, TECHNICAL UNIVERSITY OF CRETE, CHANIA, GREECE
[5]DST-MAHAMANA CENTRE FOR EXCELLENCE IN CLIMATE CHANGE RESEARCH, BANARAS HINDU UNIVERSITY, VARANASI, INDIA
[6]DEPARTMENT OF ENVIRONMENTAL ENGINEERING, UNIVERSITY OF PATRAS, AGRINIO, GREECE
[7]COLLABORATIVE INNOVATION CENTER ON FORECAST AND EVALUATION OF METEOROLOGICAL DISASTERS, NANJING, UNIVERSITY OF INFORMATION SCIENCE & TECHNOLOGY, NANJING, P.R. CHINA

8.1 Introduction

Precision agriculture (PA) is the science of improving crop yields and assisting management decisions using high technology sensor and analysis tools. PA is a new concept adopted throughout the world to increase production, reduce labor time, and ensure the effective management of fertilizers and irrigation processes. It uses a large amount of data and information to improve the use of agricultural resources, yields, and the quality of crops (Mulla, 2013). PA is an advanced innovation and optimized field level management strategy used in agriculture that aims to improve the productivity of resources on agriculture fields. Thus PA is a new advanced method in which farmers provide optimized inputs such as water and fertilizer to enhance productivity, quality, and yield (Gebbers and Adamchuk, 2010). It requires a huge

Hyperspectral Remote Sensing. DOI: https://doi.org/10.1016/B978-0-08-102894-0.00009-7

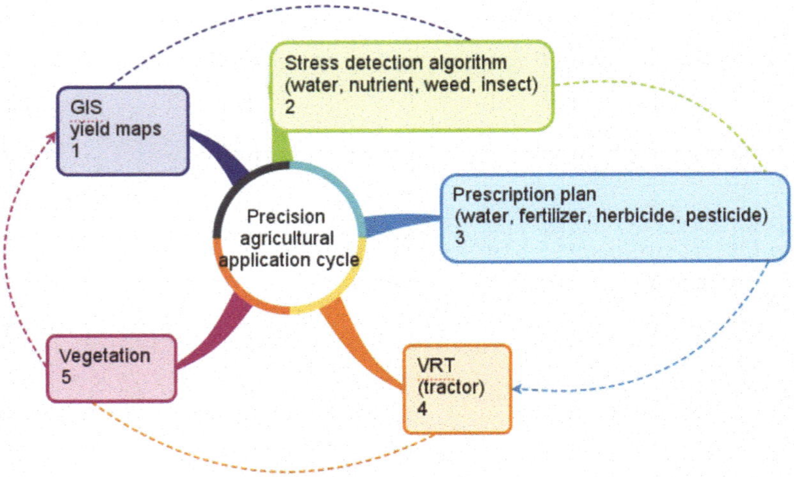

FIGURE 8–1 Cycle of precision agriculture. *GIS*, Geographic information system; *VRT*, variable rate technology. *Adopted from Abdullahi, H., Sheriff, R., 2017. Case study to investigate the adoption of precision agriculture in nigeria using simple analysis to determine variability on a maize plantation. J. Agric. Econ. Rural Dev. 3 (3), 279–292 (Abdullahi and Sheriff, 2017), p. 284. Under CCBY license—Creative Commons-licensed research.*

amount of information about the crop condition or crop health in the growing season at high spatial resolution. Independently of the data source, the most crucial objective of PA is to provide support to farmers in managing their business. Such support comes in diverse ways, but the end result is typically a decrease of the necessary resources (Fig. 8–1).

Modern agricultural production relies on monitoring crop status by observing and measuring variables such as soil condition, plant health, fertilizer and pesticide effect, irrigation, and crop yield. Managing all of these factors is a considerable challenge for crop producers. The rapid enhancement of precise monitoring of agricultural growth and its health assessment is important for sensible use of farming resources and as well as in managing crop yields (Nigam et al., 2019). Such challenges can be addressed by implementing remote sensing (RS) systems such as hyperspectral imaging to produce precise biophysical indicator maps across the various cycles of crop development.

RS is a rapidly expanding technology implemented in various agricultural applications. In particular, imaging spectroscopy in large continuous narrow bands provides significant information for understanding the biophysical and biochemical properties of agricultural plants. It is also useful to identify the changes in various physical processes, which can be better identified using multispectral RS (Sahoo et al., 2015). Advanced techniques of RS have been used for large scale crop inventory and yield predictions (Mulla, 2013). RS applications are used in agriculture studies that are based on the interaction between electromagnetic radiation and soil or plant material on the Earth's surface (Atzberger, 2013). RS combined with geographic information systems (GISs) and/or global positioning systems (GPSs) are often used in PA. This allows farmers and other agriculture producers to reduce inputs and maximize cost benefits using modern technologies rather than traditional field approaches. Nowadays, variable rate technology (VRT) is introduced to increase precision farming practices. VRT is a vital component for PA and is becoming more prevalent for large land holders. In VRT, collections of field variable information and other input data are helpful in defining suitable quantities of chemical inputs required for the fields. Hence the demand of precision agricultural techniques, valuable products, fine RS information as well as VRT has grown tremendously (Brisco et al., 1998).

This chapter describes the latest developments in Earth Observation (EO) techniques and platforms for PA with particular emphasis on the use of hyperspectral sensors for this purpose. As part of this, it provides useful information regarding the identification of research challenges, limitations, and advantages of different platforms and sensors for PA with specific emphasis placed on hyperspectral sensors.

8.2 Multispectral remote sensing in precision agriculture

RS is a key component of PA. Multispectral satellites play a vital role in collecting high spatial resolution data for agricultural practices. Multispectral imaging camera sensors embedded in agriculture drones permit farmers to manage crops, soil, parasites, and fertilizer and water demand more accurately. Therefore such drones have proven to be beneficial in term of increasing yields and other benefits. Multispectral sensors use four bands, namely red, green, red-edge, and near infrared (NIR) bands to capture images of crops and vegetation in visible and invisible regions (Liaghat and Balasundram, 2010) (Fig. 8–2).

8.2.1 Advantages

Modern PA is useful to maximize yields and resources while reducing environmental influences such as over-fertilization and the use of pesticides. There are a lot of benefits to using multispectral data or imaging including cost reductions, their simplicity to use, high accuracy, and broad range of applications. Advanced multispectral imaging equipment offers

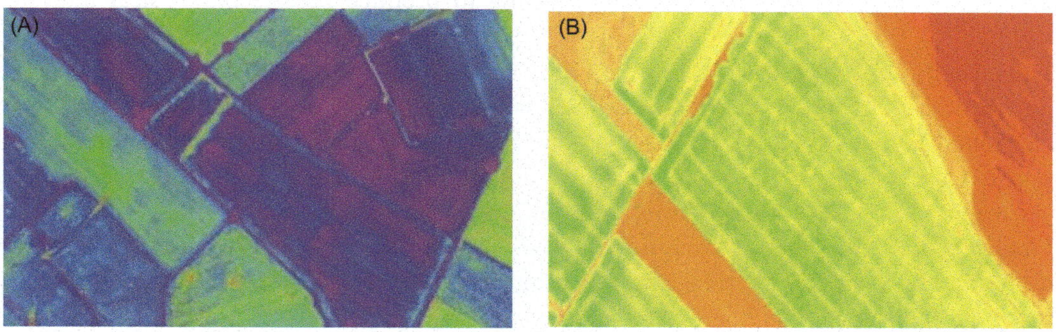

FIGURE 8–2 (A) Multispectral image of the field and (B) normalized difference vegetation index image of the agricultural field generated from unmanned aerial vehicles.

new innovations in the practical application of PA techniques. This is useful for the valuation of crop stresses, quality of soils, and vegetative cover as well as for yield estimation (Bagheri et al., 2013; Wójtowicz et al., 2016). The visible range is from 0.4 to 0.7 μm. Multispectral cameras and sensors are able to see in the visible and NIR regions. There are several multispectral sensors like Sentinel 2 with a 10 m resolution, LANDSAT ETM + with a 30 m resolution, SPOT 5 with a 10 m resolution, IKONOS with a 1 m resolution, and QuickBird with a 0.60 m resolution. The use of unmanned aerial vehicles (UAVs) can provide a spatial resolution ranging from 0.10 to 2 m depending upon the flight height. There are various types of multispectral camera sensors useful for agricultural practices such as (Corrigan, 2018; Deng et al., 2018):

1. Sentera Quad
2. Parrot Sequoia
3. Tetracams AC lite sensor
4. MicaSense Rededge Sensor
5. Airinov multiSPEC 4C Agronomic Sensor

8.2.2 Multispectral data limitations in precision agriculture

In the past few years, several multispectral broadband sensors have been utilized for agricultural crop yield and its regional assessment (e.g., Sebastian et al., 2015). Due to broad spectral channels, multispectral sensors are capable of providing information on vegetation, like vegetation structure, coverage area, and crop greenness. However, these sensors have limitations in terms of spectral resolution and bandwidth. As such, they are not adequate for detecting subtle changes in crop types, stages, and their biochemical parameters (Lamine et al., 2018). Therefore multispectral sensors are unable to provide information about crop biophysical and biochemical variables, which are important parts of PA research (Meerdink et al., 2016; Nigam et al., 2019). To counter these limitations, hyperspectral remote sensing (HRS) is being employed in agriculture research for the identification of crop types, mapping, monitoring health, and biochemical parameters.

8.2.3 Advantages of hyperspectral over multispectral data

Hyperspectral imaging sensor (HIS) data have more advantages than multispectral data for the identification and discrimination of target features or objects. They provide detailed information about any object because of narrowband information acquisition. Hyperspectral sensors have fine spectral resolution. For example, AVIRIS with 224 spectral bands provides 4 m resolution data, while Hyperion and CHRIS Proba have spatial resolutions of 15–30 m (Pandey et al., 2018, 2019). The high spectral resolution of hyperspectral data has the benefit of PA capturing and monitoring, but it also includes redundant data, which affect accuracy level. These informative bands are useful to discriminate vegetation types or species (Mulla, 2013). HIS provides a nondestructive method of measuring plant growth parameters and nutrient levels in crop plants and is useful for the implementation of RS-based PA.

The narrowband contiguous bands of hyperspectral data are useful to estimate biophysical (leaf area index, biomass) and biochemical (chlorophyll, leaf nitrogen) variables. Using various band combinations, many types of vegetation indices such as normalized difference vegetation index, soil adjusted vegetation index, ratio vegetation index, and various biophysical properties can be estimated. These indices are useful to estimate chlorophyll content, plant stress, water content, and biomass in plants. Hyperspectral indices are useful to detect the differences among crops under different irrigation treatments. The estimation of such indices is possible with multispectral data (Manjunath et al., 2011). HIS is a vital tool for researchers and farmers and is carried out using low-cost, portable devices with high-quality accurate data. The capability of these devices allows for an improved quality and accuracy in information extraction. Hyperspectral imaging allows for day-to-day monitoring and promises to create a new model of agricultural efficiency (Gevaert et al., 2015).

8.2.4 Precision farming requirement

RS with UAVs is a game-changer in PA. Maes and Steppe (2018) demonstrated the implementation and use of UAVs, a RS technology, as a game-changer in PA research due to its unprecedented spectral, spatial, and temporal resolution product delivery. UAVs can be used to deliver different optical data such as multispectral as well as hyperspectral data of the field in observation. These data can be obtained in multiangular directions and have managed to provided height data as well (Maes and Steppe, 2018). The everyday practices of agriculture can be observed by UAVs, for soil monitoring, early weed detection, pathogen infestation and pest control, nutrient assessment, fertilizer applications, and surveying for crop and horticulture management at different spatial scales.(Laliberte et al., 2010; Hruska et al., 2012; Zarco-Tejada et al., 2012; Uto et al., 2013). These innovative technologies have been implemented and are in demand for new opportunities in PA by agronomists and farmers to provide stress, pest infestation, soil condition, disease control, and yield maps (Laliberte et al., 2010; Zarco-Tejada et al., 2012; Uto et al., 2013; Stehr, 2015; Von Bueren et al., 2015).

In addition to its unprecedented spectral, spatial, and temporal resolution, it also provides detailed vegetation height data and multiangular observations. In this chapter, the progress of RS with UAVs in drought stress, weed and pathogen detection, nutrient status and growth vigor assessment, and in yield prediction is discussed. To transfer this knowledge to everyday practice in PA, future research should focus on exploiting the complementarity of hyperspectral or multispectral data with thermal data in integrated datasets.

For PA, a set of rules or steps is required that provides better yield and less effort when performing several works related to agricultural activities such as:

1. Before heading toward farming, field survey is important according to crop type. Therefore soil sampling can be performed using UAVs having a hyperspectral camera for the analysis of soil conductivity, salinity, pH, soil moisture, and soil type according to chosen crop plant.
2. Early weed detection can help to remove unwanted plants growing with crops and prevent unnecessary competition among them.

3. Nutrient status assessment for nitrogen, potassium, and phosphorus, which are essential for plant growth and health.
4. After crops are planted, there is an important factor that should not be ignored, that is, the detection of pathogens or any disease that might infect the crops.
5. Even during the growth of crops, water is required and there should be major steps to assess the drought stress detection of crops.
6. The detection of lodging is also an important part of precision farming, which is carried out in order to determine and treat the cause of any yield reduction, and this step will help to prevent reductions in yield.
7. Finally, it helps in the yield estimation of crops, which may be assessed for the area acquired for farming.

The spatial coverage provides a wide-area view at low to very low spatial resolutions. Different indices can be calculated using the narrow bands of hyperspectral data and correlated with sampling points to archive a greater accuracy for a wider area and different land-use types. This is a highly cost-effective technique that requires little man-power and can produce temporal change analyses for a given region. It can also help in obtaining significant growth in the production rate within agricultural fields. There are numerous techniques and UAV-based hyperspectral cameras useful for the early detection of weeds, which includes the identification of the color and texture of the leaves and their chlorophyll content (Andreo 2013; Nawar et al., 2017).

The early detection of weeds is vital in PA for crop protection and yield enhancement (Rumpf et al., 2010). A number of researches have been carried out using hyperspectral data for weed detection using RS methods from traditional to advanced techniques. Earlier attempts have been made to detect weeds among crops using colored film image texture (Burks et al., 2000a,b; El-Faki et al., 2000) using inputs of red, green, and blue (RGB) color indices into a discriminant analysis and neural network for weed detection. Later on, with the advancement of RS techniques, researchers have implemented HRS for early weed detection in agricultural fields (Haapala, 2003; Slaughter et al., 2004; Okamoto et al., 2007) using a visible/NIR (400–2500 nm) spectrophotometer. Hyperspectral images can provide significant data that can help in finding weeds early using different vegetation indices and other computational classification techniques. With its fine resolution and wide range of spectral bands, it can give an accurate estimation and is important in calculating the total crop loss and for risk assessment in crop fields.

Hyperspectral data contain narrow bands and there are different spectral indices that can generate the status for different nutrient levels in plants. Plant health is a vital issue and it has to be monitored on a continuous basis, in which spatial datasets play an important role (Haboudane et al., 2002; Oerke et al., 2012; Zhan et al., 2014). A field-based plant spectrum can be used as an endmember to generate a classified scenario of different plant types and their health status.

In the same way, the early detection of disease condition is important for yield protection. HRS techniques are employed to discriminate between diseased and nondiseased crop leaves (Rumpf et al., 2010). This differentiation helps in the protection of crops and, thus, ultimately increasing yields. This works in the way that it differentiates diseased leaves from nondiseased crops (such as cercospora leaf spot, leaf rust, and powdery mildew disease), thus, providing information about the condition of crop plants even before the appearance of specific symptoms. Hyperspectral imaging can distinguish between healthy leaves and leaves inoculated with pathogens in an agricultural field timeously in contrast to traditional methods of physically examining each crop as visual monitoring of diseases at early stages in the field, which is time-consuming and expensive (Steddom et al., 2005; Steiner et al., 2008; Mahlein et al., 2012). Early stage identification, detection, and differentiation of any diseases allow for an efficient application of agro-chemicals (Hillnhuetter and Mahlein, 2008). This will prevent the deliberate use of chemcical sprays on crop plants, thus, reducing chemical exposure to crops. One of the aim of PA is to reduce and optimize the pesticide applications due to increasing cost of chemical, fertilizers and its associated ecological impact on soil, water and air (Rumpf et al., 2010). Continuous monitoring of crops is important as many crops are susceptible to climatic conditions, especially humidity. Different conditions allow pathogens to attack plants and can lead to the destruction of crops as well as degrading the quality of soil in a particular area. Continuous monitoring of crop health and soil moisture can reduce the risk. Soil moisture is an important aspect as it can be useful for detecting the correct time for irrigation processes (Sutherst et al., 2011). There are several criteria for assessing the water requirements for maintaining crop health, which include field-based sampling methods as well as spatial modeling techniques.

Soil moisture is crucial for these kinds of issues and there are several EO satellites that provide frequent analyses of the status of soil moisture with a global coverage and a high spatial resolution. Drought is one of the major issues with the changing climate and frequent changes in the weather system and certain advancements have been made in the field of drought stress detection using different spatial maps and machine learning algorithms under simple GIS environments. These methods can help in the detection of drought stress in any part of globe with a continuous temporal coverage (Lee et al., 2010).

Lodging in cereals is often a result of the combined effects of the inadequate standing power of crops and conditions such as rain, wind, hail, topography, soil, previous crops, and others. Lodging affects wheat, rice, and other cereals, and reducing it is a major goal of agricultural research (Khobra et al., 2019). Lodging is the loss of the standing capability of crop stems in the field or it can be due to combined effects of inadequate standing power of the crop. It is the collapse of the crop stem when it can no longer support its own weight in the field. All cereal crops and all varieties are susceptible to varying degrees of lodging due to root lodging and stem failure, and it has a range of causes such as rain, wind, hail, topography, soil, and previous crops. It generally affects wheat, rice, and other cereals, and reduces yields. Nowadays, preventing lodging is a major goal of agricultural practices.

Yield estimation is a crucial process and its accurate estimation depends upon the accuracy of the spectral and other sampling results of the endmembers and the GIS environment under which it has been calculated. Overall, it can be said that the spatial database including the hyperspectral data and other layers plays an important role in the identification, monitoring, and mapping of different crop yields with high accuracy at very fine resolution level (Fauvel et al., 2012; Khan et al., 2018).

8.2.5 Spaceborne remote sensing: advantages and disadvantages

Due to some limitations in the data provided by airborne sensors, various types of spaceborne sensors are becoming more popular for agricultural practices. Spaceborne RS measurements can provide more regular and accurate information on agricultural and hydrological conditions of the land surface of large areas (Brown et al., 2018). These sensors also have the capability to identify and monitor crop growth and other biophysical and biochemical parameters at large spatial and temporal scales, which was not possible with traditional and conventional methods. As spaceborne sensors are well placed on stable platforms they have less problems with distortion than airborne sensors. However, spaceborne sensors have their own limitations. Because of height, they usually provide coarse resolution data affecting the result accuracy. Furthermore, cloud cover limits the effectiveness of these sensors, while sensors mounted on UAVs or airplanes can be used below cloud height (Ali et al., 2015).

8.3 Hyperspectral sensors: present status

NASA launched the first successful civilian hyperspectral satellite sensor, Hyperion EO-1, into the Earth's orbit in November, 2000 (Chatziantoniou et al., 2017; Lamine et al., 2019). The Hyperion sensor is considered as the first main "real and genuine" spaceborne hyperspectral instrument in satellite orbit. It was mounted on board the Earth Observer-1 (EO-1) satellite, launched by NASA's New Millennium Program around 2000. It has a total of 242 spectral channels that acquire images at a 30 m spatial resolution and at a 10 nm spectral resolution [70 spectral channels are part of the visible and near infrared (VNIR) and 172 channels form part of the short-wave infrared (SWIR) spectrum] (Han et al., 2002). The Hyperion EO-1 sensor was mainly launched to prepare a mineral spectral library and mineral mapping by United States Geological Survey (Petropoulos et al., 2012).

CHRIS (compact high resolution imaging spectrometer) followed this successful launch in 2001 on ESA's PROBA platform (Barnsley et al., 2004). ESA's CHRIS PROBA was used to gather the bidirectional reflectance distribution (BRDF) function information for enhanced knowledge of spectral reflectance. The Hyperion EO-1 sensor has ground coverage field of view providing a 7.5 km swath, while the CHRIS sensor is a high spatial resolution hyperspectral spectrometer (18 m at nadir) with 14 km swath (Ben-Dor et al., 2013).

Since 2013, the development of the Italian PRISMA (PRecursore IperSpettrale della Missione Applicativa; Hyperspectral Precursor of the Application Mission) by ASI, Germany's EnMAP (Environmental Mapping and Analysis Program), and NASA's HyspIRI (Hyperspectral Infrared Imager) concepts has enabled the commercial and research themes progress of hyperspectral community compared to multispectral themes. HyspIRI measures the spectral range 380–2500 nm from visible to infrared in 10 nm contiguous bands, while a second multispectral instrument measures from 3 to 12 μm in mid infra-red (MIR)–thermal infrared (TIR). The second multispectral image have a spatial resolution of 60 m at nadir with revisit period of 19 days for visible-shortwave infraRed and 5 days revisit time for TIR.

ISRO's (Indian Space Research Organization) Moon mission, Chandrayan-1, which launched in 2008, carried a HySI hyperspectral sensor that was useful in delivering information about the mineral composition of the lunar surface, its mineral mapping, and the presence of water molecules on the surface of the Moon. ISRO also launched "HySIS." The satellite has 55 spectral bands that can be used for a range of research themes such as crop and environment monitoring, which are essential for applications of PA.

On June 23, 2017, a small and lightweight hyperspectral camera was successfully launched into space on the Aalto-1 nanosatellite measuring a wavelength range of 500–900 nm. It is a unique hyperspectral tunable spectral imager operating in space due to its miniature size, representing half a CubeSat unit (0.5 U) in size or 5 × 10 × 10 cm. This scalable sensing technology will offer opportunities for new nanosatellite-based services for land cover, crop yield predictions, or other research domains.

8.4 Hyperspectral data in agriculture

HRS is the acquisition of images in hundreds of narrow contiguous spectral bands to obtain high-resolution data for each pixel of a particular scene (Thenkabail and Lyon, 2016). Extracted spectral signatures from hyperspectral images are used to identify and classify the feature objects. Many applications of hyperspectral and multispectral imaging are being verified in various types of farming techniques including in quality control, classifying, and sorting of agricultural products, and in the identification of insects and contaminants as well as in food safety.

8.4.1 Recent approaches

8.4.1.1 Analytical spectral device field radio spectrometer

Field spectroradiometry is an integral part of RS science since both use the Sun's radiation as the primary light source (Maid and Deshmukh, 2018). The key variable in spectroradiometry is spectral reflectance. This is emphasized as opposed to imaging spectroscopy, which primarily analyses the causes and exhibition of spectral absorbance from imagery and scattering processes that occur when light strikes the target (Milton et al., 2009; Manevski et al., 2011). Field spectroradiometry spectral reflectance is not a natural characteristic of materials. It is derived as "spectral reflectance factor" R(l), which is a ratio of the radiance reflected by a surface to that reflected into the same beam geometry standard surface irradiated under the same conditions of illumination and observation (Milton et al., 2009). It is dimensionless and ranges from 0 to values beyond 1 (for highly reflective surfaces, for instance, snow). Spectral reflectance depends on the wavelength of the incident radiation, the properties of the target, and the incident radiation angle. Field spectroradiometry has progressively evolved in both design and mobility aspects. Modern measurement devices—called spectroradiometers—are presently available, measuring spectra in a wide spectral range (300–2500 nm) with high precision and accuracy and that are portable and easy to use (Lamine et al., 2019).

NIR spectroscopy has been used as a logical tool in PA (Finch et al., 2014). Analytical spectral devices (ASD) field radio spectrometer collects a range of spectral radiance data between 350 and 2500 nm in visible, NIR, and SWIR spectral regions, which is reflected by Earth objects. The design of this instrument is more flexible than other laboratory instruments. The ASD field spectroradiometer provides accurate spectral field measurements for any type of agricultural research. The collected data are exported in ASCII text, after which they can be analyzed using different specialized software. Field spectroradiometry and spectral library data led to improved outcomes thanks to the improved relationships investigated and the reliability of the results. For instance, Rao et al. (2007) illustrated the use of field spectroradiometry information from agricultural sites in India, for Hyperion data classification at different spatial scales (canopy level and pixel level). Outstanding outcomes were achieved for overall accuracy (86.5% and 88.8%, respectively) proving the efficiency of these methods in regards to PA (Petropoulos et al., 2015).

8.4.1.2 Global positioning system-guided unmanned aerial vehicles employing hyperspectral data

Progress in UAVs has advanced traditional satellite-based capabilities, providing a capacity for high spatial, spectral, and temporal responses. However, while some HIS sensors have been developed for use onboard UAVs, significant investment is required to develop a system and data processing workflow that accurately retrieves georeferenced mosaics. UAVs are rapidly being used for various applications like investigation, the acquisition of images, surveying, and the collection of spatial information. The developments in UAVs present advanced solutions for crop management and monitoring. They can provide high-resolution images for small fruitful areas and support PA (Laliberte et al., 2010). GPS in UAVs has a vital role in operating UAVs safely. GPS navigation techniques can increase the accuracy of information and are used to detect the position of the vehicle. Along with inertial measurement unit data, GPS delivers highly accurate information (Laliberte et al., 2007). Several pushbroom hyperspectral sensors have been employed in UAVs (Hruska et al., 2012; Zarco-Tejada et al., 2012). These innovative technologies are used to assist with environmental and RS related tasks. Point-based spectrometers are also used in UAVs by researchers (Uto et al., 2013; Von Bueren et al., 2015). GPS-guided UAVs based drones give farmers the ability to inspect crops from perspectives that were never possible before. Using this technology, they are able to identify problem areas in the field, and also to remedy diseases and defects. Day by day this technology becomes more useful for agronomics and farmers for understanding the usefulness of PA, decreasing costs and environmental impact, and increasing crop yield.

8.4.2 Case studies

8.4.2.1 Field spectroradiometry

The flexibility of UAV systems along with the number of additional features means that there is something for every agricultural application. Once guidelines are employed, drones could become a commonly used device for agronomists and farmers (Stehr, 2015).

According to a case study from Florida, an ASD FieldSpec 4 spectroradiometer was used for PA. In this study, HRS techniques were used for PA. All team members of Highland Precision Ag visualized the ground truth data in order to prepare predictive analytics and also focused on developing a digital recording system that would permit them to aggregate this data and information for future use (Kenaston and Crockett, 2018) (Fig. 8–3).

In a different study, Virginia Tech University performed an experiment using UAVs to monitor airborne pathogens. The researchers used drones for collecting sample data and flied the drones in the lower atmosphere. Using these samples and according to weather patterns, researchers predicted the condition of bacteria, their origin, and their destination. These aerial images were combined with yield maps to inspect for patterns.

Another study by Lanthier et al. (2008) was been done in an agricultural region near St Jean-sur-Richelieu, in Southwest Quebec. In this study, a CASI (compact airborne spectroscopic imager) sensor was used with different resolutions of 1, 2, 4 m. Pixel-oriented and objected-oriented classifications were used for understanding the accuracy level of the raw data obtained. For the pixel-oriented classification, a maximum likelihood algorithm was used and for the object-oriented classification, a nearest neighbor classifier was used. Raw reflectance data were converted into reflectance data and a CAM5 transfer code was used for atmospheric corrections. In an accuracy assessment, the object-oriented classification showed greater accuracy in all types of resolution data. In a coarser resolution (4 m), the kappa coefficient was estimated at 0.8268, and in a higher resolution (1 m), the kappa coefficient was found to be 0.7730. The object-oriented classification showed a higher accuracy than that of the pixel-oriented classification and was useful in the task of delineating agricultural classes (Lanthier et al., 2008).

FIGURE 8–3 Chris Crockett of Highland Precision Ag uses the ASD FieldSpec 4 spectroradiometer to measure canopy leaf-level constituents. *Adopted from Kenaston, B., Crockett, C., 2018. ASD—highland precision agriculture case study: improving precision agriculture with visible and near-infrared spectroscopy. NIR News 29 (2), 7–9. Permission granted under the STM Guidelines.*

8.4.2.2 Crop characterization and discrimination

Hyperspectral (narrow band) indices have been discovered to be essential for providing biochemical and biophysical attribute quantification and monitoring. As such, they are significant for crop studies. Leaf biochemical characteristics estimation such as the levels of chlorophyll and nitrogen help in the assessment of available nutrients as well as plant productivity status and stress. In a comparison between RS techniques and direct field techniques, RS techniques were found to be more efficient in terms of time, they are nondestructive, and deliver spatial quantification and monitoring attributes of agricultural crops. Jain et al. (2007) recommended that in the red edge and NIR regions, the optimum bandwidth is 5–10 nm. For crop stress studies, the optimum bandwidth is 25 nm in the regions of 500–700 and 800–900 nm. Kumar et al. (2016) performed a field hyperspectral data analysis of tea plantations for discriminating spectral behavior in regards to type, plant age, stage growth, pruning status, light conditions, and disease incidence (Fig. 8–4). To recognize the most suitable bands for retrieving these factors, a stepwise discriminant examination and a primary component analysis were conducted. The most appropriate band for discriminating different plant types of tea for both in sunlit and nonsunlit conditions was found to be the green region. For plant age, plant growth stage, and bush health, the most appropriate was the blue region. For pruned and unpruned tea, discrimination in the red and NIR regions were found to give the best results.

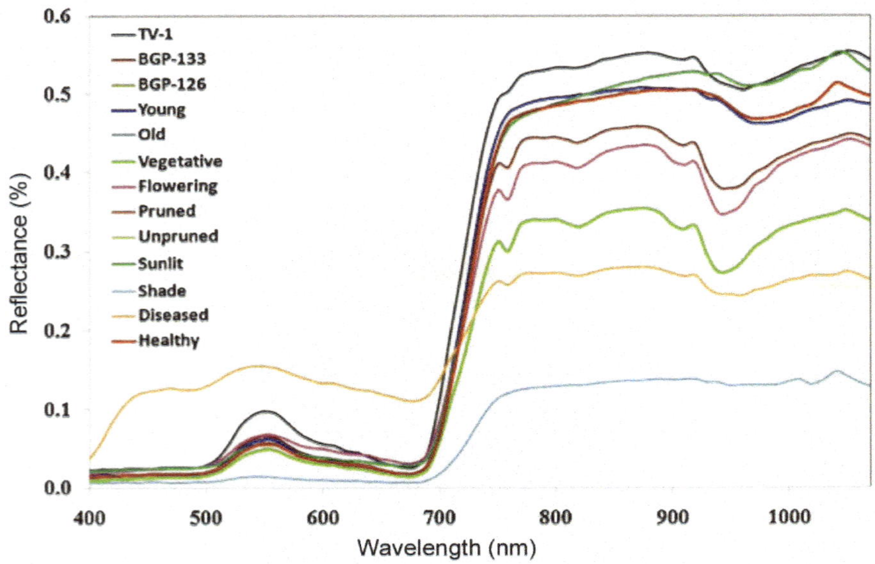

FIGURE 8–4 Field hyperspectral data analysis of tea plantations for discriminating spectral behavior. *Adopted from Amit Kumar et al., 2013. Field hyperspectral data analysis for discriminating spectral behavior of tea plantations under various management practices. Int. J. App. Earth Obs. 23, 352–359.*

8.4.2.3 Land Use Land Cover (LULC) mapping

Land cover is an essential variable of the Earth's system, and is intimately connected with anthropogenic and physical environments (Otukei and Blaschke, 2010; Chatziantoniou et al., 2017). Landscape information data acquired from field sites are reliable, hence, they can be linked with the RS datasets acquired at different spatial and temporal scales of the spectral characteristics of the target features of a landscape (McCoy, 2004).

Vegetation land cover, both agricultural and natural, remains the focus of most studies as the spectral characteristics of plants are dynamically influenced by many internal and external factors. The capacity of hyperspectral systems to better segregate and identify several ground features over traditional multispectral systems has been shown by several researchers (Zhang and Ma, 2009; Otukei and Blaschke, 2010; Chatziantoniou et al., 2017). These days, hyperspectral RS systems are viewed as one of the most noteworthy EO information sources (Du et al., 2010) and are, therefore, are utilized for different applications covering land cover assessments. Generally, HIS data are preferred over multispectral data sources for digital image classification for land cover thematic maps (Shah et al., 2004; Manevski et al., 2011). Petropoulos et al. (2015) evaluated the combined use of Hyperion imagery with support vector machines (SVMs) and artificial neural network (ANN) classifiers for a heterogeneous region in Greece. Their results (Fig. 8–5) showed close classification accuracy between the

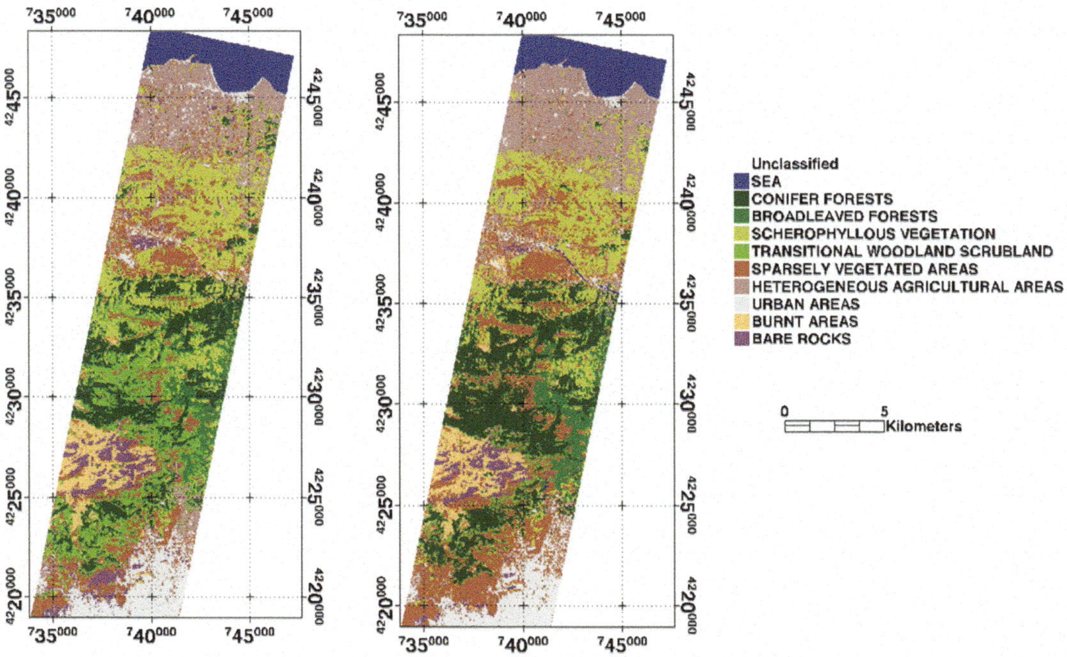

FIGURE 8–5 Hyperion pixel-based classification using support vector machine classifier (left) and artificial neural networks (right). *Adopted from Petropoulos G.P., Arvanitis K. and Sigrimis N., 2012, Hyperion hyperspectral imagery analysis combined with machine learning classifiers for land use/cover mapping, Expert Syst. Appl.* ***39*** *(3), 3800–3809. https://doi.org/10.1016/j.eswa.2011.09.083.*

two classifiers (higher than 85% in both cases) with the SVMs outperforming the ANN classifier by 3.31% in overall accuracy and by 0.038 in kappa coefficient. In another study, the same authors also compared SVMs and object-based classification for another region in Greece representative of typical Mediterranean conditions. The findings from their work showed that both classifiers were able to produce comparatively accurate land cover maps of the studied area. The overall accuracy and kappa for the object-based classification were 81% and 0.779, respectively, whereas for SVMs, these were 76% and 0.719, respectively.

Numerous studies demonstrated that HIS data from airborne and satellite platforms are capable of producing accurate maps of land cover. Validation methods based on error matrix statistics have generally shown overall accuracies and kappa coefficients of 85% and 0.800 or higher respectively. All in all, to improve the estimation of land cover from hyperspectral data it is important to integrate knowledge on the spectral properties of the land cover targets and the factors affecting the spectral variations across scales (Petropoulos et al., 2013; Elatawneh et al., 2014).

8.4.2.4 Insect, invasive plant species, and plant disease monitoring

Hyperspectral instruments have the ability to acquire images with narrow spectral bands (<20 nm) that are continuously measured. This enables the extraction of the spectral signature for the observed objects or materials. HRS encompasses more than 200 bands. HRS is applied in many fields such as disaster and ecological monitoring as well as agriculture, forestry, and geology. The anticipated population growth of the world makes the use of HRS technologies a matter of crucial importance (Jadhav and Patil, 2014).

Hyperspectral image processing in plant disease monitoring, insect pest detection, and monitoring invasive plant species is of great importance to farmers to reduce the economic losses caused by these threats. Studies have displayed that invasive plants pose a serious threat to forest environments and other plant species (Evangelista et al., 2008; Wang, 2008). For example, *Tamarix* (salt cedar) absorbs the limited source of soil moisture, therefore, it increases soil salinity. In the work of Evangelista et al. (2008), HRS was proven to be suitable to study the spatial distribution of tamarisk as well as the other invasive species. In this study, six satellite images of Landsat 8/ETM + obtained at different times steps for the duration of the growing season were used for vegetation indices computation. After considering the changes of the values over time, these indices were used in combination with the maximum entropy model for detecting and mapping the tamarisk distribution.

8.4.2.5 Drought mapping

Since water is an inevitable factor in agriculture, a sufficient supply of nutrients, pesticides, and water to all crops in the field can be ensured by rationalizing water use. Study conducted by authors (WUSTL 2014) examined drought and water stress to quickly identify water concerned stresses in crop fields to spot irrigation specific areas for appropriate energy use, time saving, and water supply. In this study, the SOC-700 hyperspectral imager showed remarkable ability in tracking the water stress development up to four days before stress hindrances were visible to the human eye. It continues to be an indispensable tool in evaluating both leaf and landscape levels of water content. Changes in photosynthetic pigments indicate whether stress has surpassed a certain level. For example, in chlorosis, the reflected red wavelengths increase thereby producing a typical yellow color. HIS display such changes before they are visible to the naked eye. A hyperspectral imager was applied to determine drought stress in leaves and was highly adequate in monitoring the consequence of dryness over time on the individual green leaves of a tree. Authors demonstrated the level of dehydration characteristics and loss of water content from the leaves when kept for 12 h before the imaging as well as a freshly cut leaf at room temperature. The color is a darker green and the margins are curling for the fresh leaf while red-orange tone for drought stressed leaf (as shown in Fig. 8–6).

From Fig. 8–6, demonstrated through another example to assess the drought stress on leaves. As indicated in Fig. 8–6, both upper leaves look very similar in visible range (upper left corner leaf in normal weather conditions and upper right leaf from a plant suffering from drought). But a scanner for infrared radiation, left and right bottom leaves reveals significant differences that may help scientists to take corrective action as soon as possible. These results clearly indicate the change in the leaf showing the stress of drought. It is evident from above examples that hyperspectral imaging is useful in investigating plant drought tolerance. This technology is also fully applicable in crops fields to assess water stress, fire scars and assist in irrigation programming (Fig. 8–7).

To determine the spectral behavior of agricultural areas and compare them with other vegetative areas and burned surfaces in the hyperspectral image, representative sampling

FIGURE 8–6 Hyperspectral spectroscopy applied to determine drought stress in leaves to monitor the consequence of dryness over time with special permission from Prof. Mikhial Berezin.

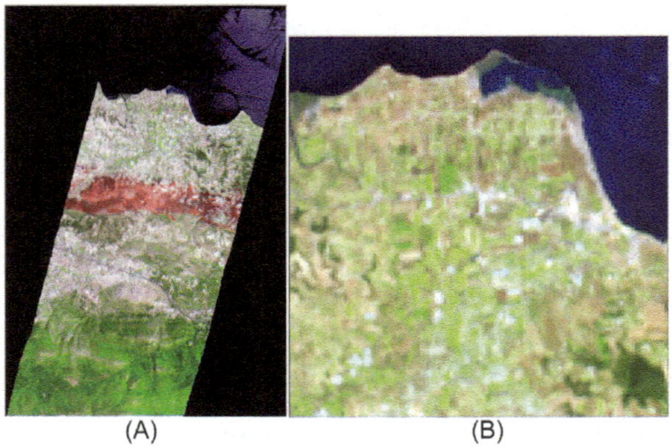

FIGURE 8–7 (A) Hyperion image. Several landscape components exist in the image, the sea in the upper part of the image, then bare land and agricultural land exactly below the sea in light green, then a fire scar in red, then urban features, and in the bottom, forests and forested areas in light and dark green. (B) Zoom of an area with agricultural parcels where a sampling plot has been established to show the variability in reflectance values (data available from the US Geological Survey).

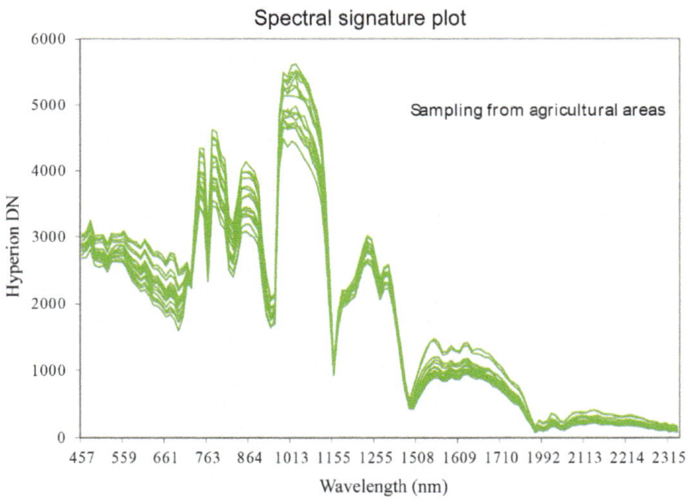

FIGURE 8–8 Spectral signatures of the agricultural areas (actually one agricultural parcel) as recorded by EO-1 Hyperion sensor (original digital Number (DN) raw values).

plots were allocated over the satellite imagery manually by photointerpretation of the various color composites and the use of very high spatial resolution imagery. Radiometric values of the pixels found in these sampling plots were recorded to characterize the spectral signal of agricultural areas (Fig. 8–8), vegetation areas (Fig. 8–9), and burned surfaces (Fig. 8–10). As can be observed, there is a high variability of the spectral signatures, associated with

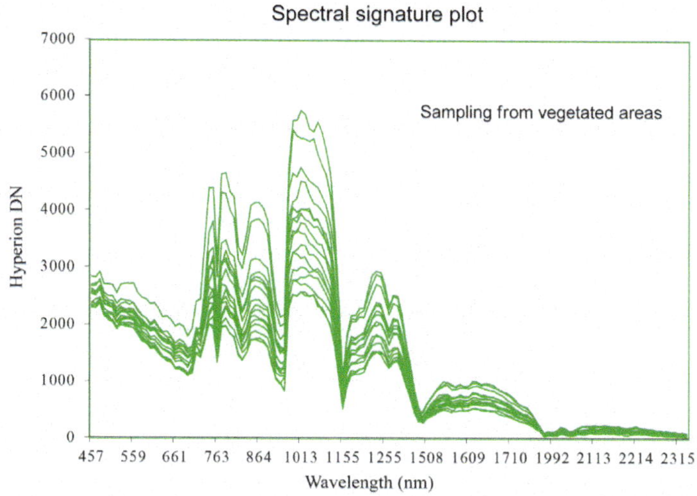

FIGURE 8–9 Spectral signatures of the vegetated areas as recorded by EO-1 Hyperion sensor (original DN raw values).

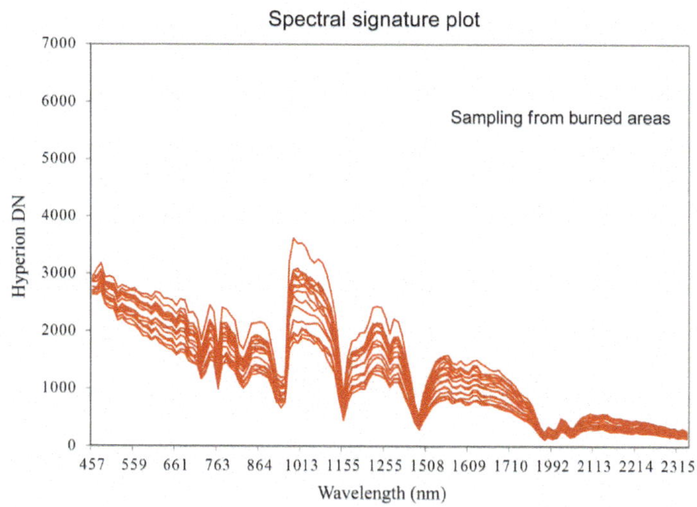

FIGURE 8–10 Spectral signatures of the burned surfaces as recorded by EO-1 Hyperion sensor (original DN raw values).

many different aspects of the landscape, within each landscape component (Figs. 8–8 to 8–10) and among the landscape components (Fig. 8–11).

This example demonstrated the usefulness of HRS and its diverse field of applications. The spectral signatures of agricultural areas and vegetated areas, used here to differentiate between them and burned areas, could be used in PA using similar methods with the one detailed here.

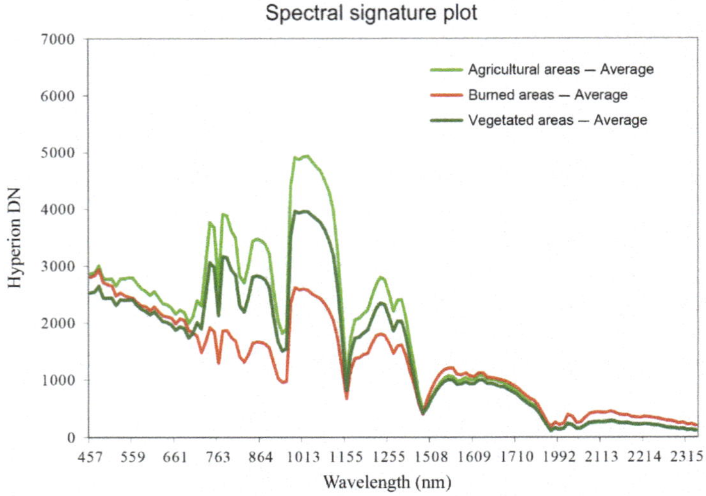

FIGURE 8–11 Spectral signatures (average values) of the agricultural areas, vegetated areas, and burned surfaces as recorded by EO-1 Hyperion sensor (original DN raw values).

8.5 Hyperspectral sensors: future missions

Recognizing the importance of HIS, several countries have planned space missions in the near future. Israel and Italy are joining the elite group to launch their HIS into the Earth's orbit. SHALOM (Space-borne Hyperspectral Applicative Land and Ocean Mission) is a joint mission between the Israel Space Agency (ISA) and the Italy Space Agency (ASI) that was agreed upon on June 16, 2009. Its purpose is to support research themes in both countries (Feingersh and Dor, 2015). The SHALOM HRS satellite will be launched in order to perform a joint study of the feasibility of development, launch, and operation of commercial satellites. SHALOM will have spatial resolution of 10 and 2.5 m for hyperspectral images and panchromatic (PAN) images, respectively. Hyperspectral images will have a 2-day revisit time covering 200 km^2 daily with 10 m spectral resolution and 241 spectral bands at the range of 400–2500 nm. The future use of SHALOM will be for environmental quality, crisis monitoring, exploration for mineral and natural resources, monitoring waterbodies, and assisting PA activities in Israel and Italy (Ben Dor et al., 2014).

The PRISMA (PRecursore IperSpettrale della Missione Applicativa; Hyperspectral Precursor of the Application Mission) satellite, of the ASI, launched on May 30, 2018. PRISMA has integrated a hyperspectral sensor with a medium resolution PAN photo camera sensitive to all colors. PRISMA is a pushbroom sensor covering a spectral range of 400–2500 nm with a swath of 30–60 km and ground sampling distance of 20–30 m (2.5–5 m for PAN images). This feature enables PRISMA to detect the chemical–physical composition of the geometric features of observed objects while distinguishing them.

The European Space Agency (ESA) has planned to launch the FLEX (FLuorescence Explorer) satellite as the eighth Earth explorer mission in space in 2022 (Middleton et al., 2017; Colombo et al., 2018; ESA 2018, 2019) to observe vegetation fluorescence (Bovensmann et al., 2015; Pandey et al., 2018a,b). FLEX will be comprised of three instrument arrays such as fluorescence, hyperspectral reflectance, and canopy temperature for the measurement of the interrelated features related to them. The FLEX mapper with 300 m resolution will be used to monitor croplands and forests due to its unique capability of fluorescence detection (an indicator of photosynthesis in both healthy and physiologically perturbed vegetation) (Jonathan, 2015). The FLEX mapper will provide an opportunity to researchers to assess canopy fluorescence at a global level from space to monitor global steady-state chlorophyll fluorescence in terrestrial vegetation (Colombo et al., 2018).

The Chinese Centre for Resources Satellite Data and Application (CRESDA) has developed two small satellites to monitor environmental resources. HJ-1A is a hyperspectral sensor, while HJ-1B is an infrared camera. The HJ-1A and HJ-1B (*"Huan Jing"* meaning environment) are mounted on the Huan Jing satellite constellation consisting of two more instruments HJ-1C and a radar satellite. These two sensors deliver 3–100 m spatial resolution imagery.

In the past few years, commercial HIS have been miniaturized and their performance has been demonstrated on UAVs. Indeed, a mounting body of studies over the past few years deals with the potential of using hyperspectral images acquired from UAV to classify various land covers and the feasibility of applying different spectral parameters (vegetation indices) to classify vegetation (Huang et al., 2017). As these data are characterized by very high spatial resolution, they should be more capable of covering the relatively large spectral variability within and also between plots/fields as demonstrated by some studies (Yuan et al., 2017). Other studies report limited classification results, despite increased spectral discrimination between various land use/cover targets from UAV image-derived endmember pixels. This has been largely attributed to the extensive processing of these data such as corrections for radiative and geometric distortions and noise removal (Mitchell et al., 2012).

8.6 Conclusion

The agriculture industry is only beginning to harness the advantages of advanced RS techniques such as GPS-guided drones, HRS, and there is already a large impact on yields that collected data can bring to operations (Singh et al., 2017; Pandey et al., 2018a,b). Farmers are currently using advanced agricultural technologies for increasing yields and sustainability, mitigating climate change, and improving their bottom line. In particular, hyperspectral imaging is capable of providing accurate information about the features of Earth. Hyperspectral sensors are becoming a vital tool for PA. Although a novel technology, it has led to remarkable achievements and has proven the significant capability of this technology in PA over the past decade or so.

UAVs are becoming increasingly popular in HRS for PA. UAVs can improve accuracy and enhance information about agricultural fields. Also, GPS-based UAVs have a vital role in the context of PA when used in combination with HIS. Such systems can identify crop type, growth stage, and crop health status as well as for provide assistance to optimize fertilizer use. UAV-based HRS will be used further for various types of agricultural practices in the future for increasing crop productivity (Pandey et al., 2018a,b). Immense opportunities remain to be exploited in the implementation of UAV-based hyperspectral sensing (and its combination with other imaging systems) to provide a transferable and scalable integrated framework for crop growth monitoring and yield prediction.

In this chapter, some of the challenges and issues in translating the available technological capacity into a useful and useable image collection were explored. The future launch of more hyperspectral satellites and the rapid developments in UAVs and sensor technologies with enhanced spatial, spectral, and temporal resolution will, in turn, require sophisticated and accurate classification algorithms for land cover mapping. The expected results are of great relevance to regional and global scale mapping of land use changes that will affect agricultural areas and also the use of PA. This remains to be seen.

Acknowledgments

Authors wish to thank the United States Geological Survey (USGS) for providing at no cost the Hyperion imagery used in the present study. The authors also wish to express their gratitude to the anonymous reviewers for their useful and constructive comments, which resulted in an overall improvement of the originally submitted manuscript. Dr. Pandey is thankful to Shiv Nadar University for help and support. Dr. Petropoulos's contribution was supported by the EU Marie Curie Project ENViSIoN-EO (project ID: 334533). Ms Prachi Singh and Dr. Prashant K Srivastava would like to thank the Space Application Centre, Indian Space Research Organisation and National Mission on Himalayan Studies, Ministry of Environment, Forest & Climate Change (MoEF&CC), Government of India for supporting this work.

List of abbreviations

ANN artificial neural network
ASCII American Standard Code for Information Interchange
ASD analytical spectral devices
ASI Italy Space Agency
AVIRIS airborne visible/infrared imaging spectrometer
BRDF bidirectional reflectance distribution functions
CAM5 Canadian advanced modified – simulation du signal satellitaire dans le spectre solaire
CASI compact airborne spectroscopic imager
CHRIS compact high resolution imaging spectrometer
CRESDA Chinese Centre for Resources Satellite Data and Application
DN digital number
EnMAP Environmental Mapping and Analysis Programmer
EO Earth Observation
EO-1 Earth Observer-1
ESA European Space Agency
ETM + Enhanced Thematic Mapper Plus

FLEX	Fluorescence Explorer
GIS	geographical information system
GPS	geographical positioning system
h	hours
HIS	hyperspectral imaging sensor
HRS	hyperspectral remote sensing
HySI	hyperspectral imaging
HySIS	hyperspectral imaging satellite
HyspIRI	hyperspectral infrared imager
IKONOS	Name of Mission Launched in 1999 by Europe
IR	infrared
ISA	Israel Space Agency
ISRO	Indian Space Research Organization
LAI	leaf area index
MIR	mid infrared
MSI	Moisture Stress Index
NASA	National Aeronautics and Space Administration
NDVI	normalized difference vegetation index
NIR	near infrared
PA	precision agriculture
PAN	panchromatic
PRISMA	Precursore IperSpettrale della Missione Applicativa
RGB	red, green, and blue
RS	remote sensing
RVI	ratio vegetation index
SAVI	soil adjusted vegetation index
SHALOM	Space-Borne Hyperspectral Applicative Land and Ocean Mission
SOC	system on chip
SPOT	(French: Satellite Pour l'Observation De La Terre, Lit. "Satellite for Observation of Earth")
STM	scanning tunneling microscopy
SVM	support vector machine
SWIR	short-wave infrared
TNCs	total nitrogen contents
TIR	thermal infrared
UAV	unmanned aerial vehicle
USGS	United States Geological Survey
VNIR	visible and near-infrared
VRT	variable rate technology
VSWIR	visible shortwave infrared
WBI	Water band Index
WUSTL	Washington University in St. Louis

References

Abdullahi, H., Sheriff, R., 2017. Case study to investigate the adoption of precision agriculture in nigeria using simple analysis to determine variability on a maize plantation. J. Agric. Econ. Rural Dev. 3 (3), 279–292.

Ali, I., Greifeneder, F., Stamenkovic, J., Neumann, M., Notarnicola, C., 2015. Review of machine learning approaches for biomass and soil moisture retrievals from remote sensing data. Remote Sens. 7 (12), 16398–16421.

Andreo, V., 2013. Remote Sensing and Geographic Information Systems in Precision Farming. Instituto de Altos Estudios Espaciales "Mario Gulich"-CONAE/UNC Facultad de Matematica Astronomia y Física–UNC.

Atzberger, C., 2013. Advances in remote sensing of agriculture: context description, existing operational monitoring systems and major information needs. Remote Sens. 5 (2), 949–981.

Bagheri, N., Ahmadi, H., Alavipanah, S.K., Omid, M., 2013. Multispectral remote sensing for site-specific nitrogen fertilizer management. Pesqui. Agropecuária Brasileira 48 (10), 1394–1401.

Barnsley, M.J., Settle, J.J., Cutter, M.A., Lobb, D.R., Teston, F., 2004. The PROBA/CHRIS mission: a low-cost smallsat for hyperspectral multiangle observations of the earth surface and atmosphere. IEEE Trans. Geosci. Remote Sens. 42 (7), 1512–1520.

Ben-Dor, E., Schläpfer, D., Plaza, A.J., Malthus, T., 2013. Hyperspectral remote sensing. Airborne Meas. Environ. Res. Methods Instrum. 413–456.

Ben Dor, E., Kafri, A., Giancarlo, V., 2014. SHALOM: Spaceborne Hyperspectral Applicative Land and Ocean Mission: A Joint Project of ASI-ISA an Update for 2014. Quebec IGARSS July 2014. <https://www.grss-ieee.org/wp-content/uploads/2014/12/2014_07_ISIS_Session1_Mission/shalom_IGARSS_2014_2.pdf> (accessed 25.10.17).

Bovensmann, H., Bösch, H., Brunner, D., Ciais, P., Crisp, D., Dolman, H., et al., 2015. Report for Mission Selection: CarbonSat—An Earth Explorer to Observe Greenhouse Gases.

Brisco, B., Brown, R., Hirose, T., McNairn, H., Staenz, K., 1998. Precision agriculture and the role of remote sensing: a review. Can. J. Remote Sens. 24 (3), 315–327.

Brown, A.R., Petropoulos, G.P., Ferentinos, K.P., 2018. Appraisal of the Sentinel-1 and 2 use in a large-scale wildfire assessment: a case study from Portugal's fires of 2017. Appl. Geogr. 100, 78–89. Available from: https://doi.org/10.1016/j.apgeog.2018.10.004.

Burks, T., Shearer, S., Gates, R., Donohue, K., 2000a. Backpropagation neural network design and evaluation for classifying weed species using color image texture. Trans. ASAE 43 (4), 1029.

Burks, T., Shearer, S., Payne, F., 2000b. Classification of weed species using color texture features and discriminant analysis. Trans. ASAE 43 (2), 441.

Chatziantoniou, A., Psomiadis, E., Petropoulos, G., 2017. Co-orbital Sentinel 1 and 2 for LULC mapping with emphasis on wetlands in a mediterranean setting based on machine learning. Remote Sens. 9 (12), 1259. Available from: https://doi.org/10.3390/rs9121259.

Colombo, R., Celesti, M., Bianchi, R., Campbell, P.K., Cogliati, S., Cook, B.D., et al., 2018. Variability of sun-induced chlorophyll fluorescence according to stand age-related processes in a managed loblolly pine forest. Glob. Change Biol. 24 (7), 2980–2996.

Corrigan, F., 2018. Multispectral Imaging Camera Drones in Farming Yield Big Benefits. DroneZon. <https://www.dronezon.com/learn-about-drones-quadcopters/multispectral-sensor-drones-in-farming-yield-big-benefits/> (accessed 01.06.19).

Deng, L., Mao, Z., Li, X., Hu, Z., Duan, F., Yan, Y., 2018. UAV-based multispectral remote sensing for precision agriculture: a comparison between different cameras. ISPRS J. Photogramm. Remote Sens. 146, 124–136.

Du, P., Tan, K., Xing, X., 2010. Wavelet SVM in reproducing kernel Hilbert space for hyperspectral remote sensing image classification. Opt. Communications. 283 (24), 4978–4984.

El-Faki, M.S., Zhang, N., Peterson, D., 2000. Weed detection using color machine vision. Trans. ASAE 43 (6), 1969.

Elatawneh, A., Kalaitzidis, C., Petropoulos, G.P., Schneider, T., 2014. Evaluation of diverse classification approaches for land use/cover mapping in a Mediterranean region utilizing Hyperion data. Int. J. Digital Earth 7 (3), 194–216.

ESA, 2018. Missions. Europe: European Space Agency. <https://earth.esa.int/web/guest/missions> (accessed 10.01.18).

ESA, 2019. EO-Sharing Earth Observation Resources. <https://directory.eoportal.org/web/eoportal/satellite-missions/h/hysis> (accessed).

Evangelista, P.H., Kumar, S., Stohlgren, T.J., Jarnevich, C.S., Crall, A.W., Norman III, J.B., et al., 2008. Modelling invasion for a habitat generalist and a specialist plant species. Divers. Distrib. 14 (5), 808–817.

Fauvel, M., Tarabalka, Y., Benediktsson, J.A., Chanussot, J., Tilton, J.C., 2012. Advances in spectral–spatial classification of hyperspectral images. Proc. IEEE 101 (3), 652–675.

Feingersh, T., Dor, E.B., 2015. SHALOM—a commercial hyperspectral space mission. In: Qian, S.-E. (Ed.), Optical Payloads for Space Missions. John Wiley & Sons Ltd, Chichester, p. 247. Available from: http://dx.doi.org/10.1002/9781118945179.ch11.

Finch, S., Samuel, A., Lane, G.P., 2014. Lockhart and Wiseman's Crop Husbandry Including Grassland. Elsevier.

Gebbers, R., Adamchuk, V.I., 2010. Precision agriculture and food security. Science. 327 (5967), 828–831.

Gevaert, C.M., Suomalainen, J., Tang, J., Kooistra, L., 2015. Generation of spectral–temporal response surfaces by combining multispectral satellite and hyperspectral UAV imagery for precision agriculture applications. IEEE J. Sel. Top. Appl. Earth Obs. Remote Sens. 8 (6), 3140–3146.

Haapala, H.E., 2003. Discriminating weeds from processing tomato plants using visible and near infrared reflectance. In: 2003 ASAE Annual Meeting; 2003: American Society of Agricultural and Biological Engineers.

Haboudane, D., Miller, J.R., Tremblay, N., Zarco-Tejada, P.J., Dextraze, L., 2002. Integrated narrow-band vegetation indices for prediction of crop chlorophyll content for application to precision agriculture. Remote Sens. Environ. 81 (2–3), 416–426.

Han, T., Goodenough, D.G., Dyk, A., Love, J., 2002. Detection and correction of abnormal pixels in Hyperion images. IEEE International Geoscience and Remote Sensing Symposium. IEEE.

Hillnhuetter, C., Mahlein, A.-K., 2008. Early detection and localisation of sugar beet diseases: new approaches. Gesunde Pflanzen 60 (4), 143–149.

Hruska, R., Mitchell, J., Anderson, M., Glenn, N.F., 2012. Radiometric and geometric analysis of hyperspectral imagery acquired from an unmanned aerial vehicle. Remote Sens. 4 (9), 2736–2752.

Huang, J., Sun, Y.-H., Wang, M.-Y., Zhang, D.-D., Sada, R., Li, M.-C., 2017. Juvenile tree classification based on hyperspectral image acquired from an unmanned aerial vehicle. Int. J. Remote Sens. 38 (8–10), 2273–2295. Available from: https://doi.org/10.1080/01431161.2016.1219076.

Jadhav, B., Patil, P., 2014. Hyperspectral remote sensing for agricultural management: a survey. Int. J. Comput. Appl. 106 (7).

Jain, N., Ray, S.S., Singh, J., Panigrahy, S., 2007. Use of hyperspectral data to assess the effects of different nitrogen applications on a potato crop. Precis. Agric. 8 (4–5), 225–239.

Jonathan, A., 2015. Flex satellite will map Earth's plant glow. BBC News, Retrieved 2019 11–20.

Kenaston, B., Crockett, C., 2018. ASD—highland precision agriculture case study: improving precision agriculture with visible and near-infrared spectroscopy. NIR News 29 (2), 7–9.

Khan, M.J., Khan, H.S., Yousaf, A., Khurshid, K., Abbas, A., 2018. Modern trends in hyperspectral image analysis: a review. IEEE Access 6, 14118–14129.

Khobra, R., Sareen, S., Meena, B.K., Kumar, A., Tiwari, V., Singh, G., 2019. Exploring the traits for lodging tolerance in wheat genotypes: a review. Physiol. Mol. Biol. Plants 25, 1–12.

Kumar, A., Saxena, S., Shrivastava, S., Bharti, V., Kumar, U., Dhama, K., 2016. Hyperspectral imaging (HSI): applications in animal and dairy sector. J. Exp. Biol. Agric. Sci. 4 (4), 448–461.

Laliberte, A.S., Herrick, J.E., Rango, A., Winters, C., 2010. Acquisition, orthorectification, and object-based classification of unmanned aerial vehicle (UAV) imagery for rangeland monitoring. Photogramm. Eng. Remote Sens. 76 (6), 661–672.

Laliberte, A.S., Rango, A., Herrick, J., 2007. Unmanned aerial vehicles for rangeland mapping and monitoring: a comparison of two systems. In: ASPRS Annual Conference Proceedings.

Lamine, S., Petropoulos, G.P., Brewer, P.A., Bachari, N.-E.-I., Srivastava, P.K., Manevski, K., et al., 2019. Heavy metal soil contamination detection using combined geochemistry and field spectroradiometry in the United Kingdom. Sensors 19 (4), 762. Available from: https://doi.org/10.3390/s19040762.

Lamine, S., Petropoulos, G.P., Singh, S.K., Szabó, S., Bachari, N.E.I., Srivastava, P.K., et al., 2018. Quantifying land use/land cover spatio-temporal landscape pattern dynamics from Hyperion using SVMs classifier and FRAGSTATS. Geocarto Int. 33 (8), 862–878. Available from: https://doi.org/10.1080/10106049.2017.1307460.

Lanthier, Y., Bannari, A., Haboudane, D., Miller, J.R., Tremblay, N., 2008. Hyperspectral data segmentation and classification in precision agriculture: a multi-scale analysis. In: IGARSS 2008–2008 IEEE International Geoscience and Remote Sensing Symposium. IEEE.

Lee, W., Alchanatis, V., Yang, C., Hirafuji, M., Moshou, D., Li, C., 2010. Sensing technologies for precision specialty crop production. Comput. Electron. Agric. 74 (1), 2–33.

Liaghat, S., Balasundram, S.K., 2010. A review: the role of remote sensing in precision agriculture. Am. J. Agric. Biol. Sci. 5 (1), 50–55.

Maes, W.H., Steppe, K., 2018. Perspectives for remote sensing with unmanned aerial vehicles in precision agriculture. Trends Plant Sci. 24 (2), 152–164. Available from: https://doi.org/10.1016/j.tplants.2018.11.007.

Mahlein, A.-K., Oerke, E.-C., Steiner, U., Dehne, H.-W., 2012. Recent advances in sensing plant diseases for precision crop protection. Eur. J. Plant Pathol. 133 (1), 197–209.

Maid, M., Deshmukh, R., 2018. Hyperspectral analysis of wheat leaf rust (WLR) disease: a review. Int. J. Comput. Sci. Eng. 6 (1).

Manevski, K., Manakos, I., Petropoulos, G.P., Kalaitzidis, C., 2011. Discrimination of common Mediterranean plant species using field spectroradiometry. Int. J. Appl. Earth Obs. Geoinf. 13 (6), 922–933.

Manjunath, K., Ray, S.S., Panigrahy, S., 2011. Discrimination of spectrally-close crops using ground-based hyperspectral data. J. Indian Soc. Remote Sens. 39 (4), 599–602.

McCoy, R.M., 2004. Field Methods in Remote Sensing. Guilford Publications.

Meerdink, S.K., Roberts, D.A., King, J.Y., Roth, K.L., Dennison, P.E., Amaral, C.H., et al., 2016. Linking seasonal foliar traits to VSWIR–TIR spectroscopy across California ecosystems. Remote Sens. Environ. 186, 322–338.

Middleton, E., Rascher, U., Huemmrich, K., Cook, B., Noormets, A., Schickling, A., et al., 2017. The 2013 FLEX—US airborne campaign at the Parker tract loblolly pine plantation in North Carolina, USA. Remote Sens. 9 (6), 612.

Milton, E.J., Schaepman, M.E., Anderson, K., Kneubühler, M., Fox, N., 2009. Progress in field spectroscopy. Remote Sens. Environ. 113, S92–S109.

Mitchell, J.J., Glenn, N.F., Anderson, M.O., Hruska, R.C., Halford, A., Baun, C., et al., 2012. Unmanned aerial vehicle (UAV) hyperspectral remote sensing for dryland vegetation monitoring. In: 2012 4th Workshop on Hyperspectral Image and Signal Processing: Evolution in Remote Sensing (WHISPERS), 4–7 June 2012.

Mulla, D.J., 2013. Twenty five years of remote sensing in precision agriculture: key advances and remaining knowledge gaps. Biosyst. Eng. 114 (4), 358–371.

Nawar, S., Corstanje, R., Halcro, G., Mulla, D., Mouazen, A.M., 2017. Delineation of soil management zones for variable-rate fertilization: a review. Advances in Agronomy. Elsevier, pp. 175–245.

Nigam, R., Tripathy, R., Dutta, S., Bhagia, N., Nagori, R., Chandrasekar, K., et al., 2019. Crop type discrimination and health assessment using hyperspectral imaging. Curr. Sci. (00113891) 116 (7).

Oerke, E.-C., Dehne, H.-W., Schönbeck, F., Weber, A., 2012. Crop Production and Crop Protection: Estimated Losses in Major Food and Cash Crops. Elsevier.

Okamoto, H., Murata, T., Kataoka, T., Hata, S.I., 2007. Plant classification for weed detection using hyperspectral imaging with wavelet analysis. Weed Biol. Manag. 7 (1), 31–37.

Otukei, J.R., Blaschke, T., 2010. Land cover change assessment using decision trees, support vector machines and maximum likelihood classification algorithms. Int. J. Appl. Earth Obs. Geoinf. S12, S27–S31.

Pandey, P.C., Koutsia, N., Petropoulos, G.P., Srivastava, P.K., Ben Dor, E., 2019. Land use/land cover in view of earth observation: data sources, input dimensions and classifiers—a review of the state of the art. Geocarto Int. Available from: https://doi.org/10.1080/10106049.2019.1629647.

Pandey, P.C., Manevski, K., Srivastava, P., Petropoulos, G., 2018a. The use of hyperspectral earth observation data for land use/cover classification: present status, challenges and future outlook. In: Thenkabail, P. (Ed.), Hyperspectral Remote Sensing of Vegetation, first ed. pp. 147–173.

Pandey, P.C., Manevski, K., Srivastava, P.K., Petropoulos, G.P., 2018b. The use of hyperspectral earth observation data for land use/cover classification: present status, challenges and future outlook. In: P. Thenkabail (ed.), Hyperspectral Remote Sensing of Vegetation (1st ed.), pp. 147–173.

Petropoulos, G.P., Arvanitis, K., Sigrimis, N., 2012. Hyperion hyperspectral imagery analysis combined with machine learning classifiers for land use/cover mapping. Expert Syst. Appl. 39 (3), 3800–3809. Available from: https://doi.org/10.1016/j.eswa.2011.09.083.

Petropoulos, G.P., Kalivas, D.P., Georgopoulou, I.A., Srivastava, P.K., 2015. Urban vegetation cover extraction from hyperspectral imagery and geographic information system spatial analysis techniques: case of Athens, Greece. J. Appl. Remote Sens. 9 (1), 096088. Available from: https://doi.org/10.1117/1.JRS.9.096088.

Petropoulos, G.P., Vadrevu, K.P., Kalaitzidis, C., 2013. Spectral angle mapper and object-based classification combined with hyperspectral remote sensing imagery for obtaining land use/cover mapping in a Mediterranean region. Geocarto Int. 28 (2), 114–129.

Rao, N.R., Garg, P.K., Ghosh, S.K., 2007. Development of an agricultural crops spectral library and classification of crops at cultivar level using hyperspectral data. Precis. Agric. 8 (4–5), 173–185. Available from: https://doi.org/10.1007/s11119-007-9037-x.

Rumpf, T., Mahlein, A.-K., Steiner, U., Oerke, E.-C., Dehne, H.-W., Plümer, L., 2010. Early detection and classification of plant diseases with support vector machines based on hyperspectral reflectance. Comput. Electron Agric. 74 (1), 91–99.

Sahoo, R.N., Ray, S., Manjunath, K., 2015. Hyperspectral remote sensing of agriculture. Curr. Sci. 108 (5), 848–859.

Sebastian, C., Remondino, F., De Giglio, M., Dubbini, M., Gattelli, M., Lucieer, A., et al., 2015. Evaluating multispectral images and vegetation indices for precision farming applications from UAV images. Remote Sens. 7 (4), 4026–4047.

Shah, C.A., Arora, M.K., Varshney, P.K., 2004. Unsupervised classification of hyperspectral data: an ICA mixture model based approach. Int. J. Remote Sens. 25 (2), 481–487.

Singh, S.K., Srivastava, P.K., Szabó, S., Petropoulos, G.P., Gupta, M., Islam, T., 2017. Landscape transform and spatial metrics for mapping spatiotemporal land cover dynamics using Earth Observation data-sets. Geocarto Int. 32 (2), 113–127.

Slaughter, D., Lanini, W., Giles, D., 2004. Discriminating weeds from processing tomato plants using visible and near-infrared spectroscopy. Trans. ASAE 47 (6), 1907.

Steddom, K., Bredehoeft, M., Khan, M., Rush, C., 2005. Comparison of visual and multispectral radiometric disease evaluations of Cercospora leaf spot of sugar beet. Plant Dis. 89 (2), 153–158.

Stehr, N.J., 2015. Drones: The newest technology for precision agriculture. Nat. Sci. Educ. 44 (1), 89–91.

Steiner, U., Buerling, K., Oerke, E.-C., 2008. Sensor use in plant protection. Gesunde Pflanzen 60 (4), 131–141.

Sutherst, R.W., Constable, F., Finlay, K.J., Harrington, R., Luck, J., Zalucki, M.P., 2011. Adapting to crop pest and pathogen risks under a changing climate. Wiley Interdiscip. Rev.: Clim. Change 2 (2), 220–237.

Thenkabail, P.S., Lyon, J.G., 2016. Hyperspectral Remote Sensing of Vegetation. CRC Press.

Uto, K., Seki, H., Saito, G., Kosugi, Y., 2013. Characterization of rice paddies by a UAV-mounted miniature hyperspectral sensor system. IEEE J. Sel. Top. Appl. Earth Observations Remote. Sens. 6 (2), 851–860.

VetriDeepika, K., Johnson, A., LafrulHudha, M., Ramya, M., 2016. Enhancement of Precision Agriculture Using Hyperspectral Imaging. National Conference on Emerging Trends in Electronics, Instrumentation, Automation & Control (ETEIAC) organized by Karpagam College of Engineering, 11th Mar, Special Issue Published in International Journal of Trend in Research and Development (IJTRD), 2016, 19–21.

Von Bueren, S., Burkart, A., Hueni, A., Rascher, U., Tuohy, M., Yule, I., 2015. Deploying four optical UAV-based sensors over grassland: challenges and limitations. Biogeosciences. 12 (1), 163–175.

Wang, L., 2008. Invasive species spread mapping using multi-resolution remote sensing data. Int. Arch. Photogramm. Remote. Sens. Spat. Inf. Sci. 37, 135–142.

Wójtowicz, M., Wójtowicz, A., Piekarczyk, J., 2016. Application of remote sensing methods in agriculture. Commun. Biometry Crop Sci. 11 (1), 31–50.

WUSTL, 2014. How does climate change impact you [Online]. USA: Washington University in St. Louis. Available from: https://engineering.wustl.edu/our-school/leadership/offices/marketing-communications/magazine/Documents/Engineering_Momentum_Fall_2014.pdf (Accessed 10 May 2020).

Yuan, H., Yang, G., Li, C., Wang, Y., Liu, J., Yu, H., et al., 2017. Retrieving soybean leaf area index from unmanned aerial vehicle hyperspectral remote sensing: analysis of RF, ANN, and SVM regression models. Remote Sens. 9 (4), 309.

Zarco-Tejada, P.J., González-Dugo, V., Berni, J.A., 2012. Fluorescence, temperature and narrow-band indices acquired from a UAV platform for water stress detection using a micro-hyperspectral imager and a thermal camera. Remote Sens. Environ. 117, 322–337.

Zhang, R., Ma, J., 2009. Feature selection for hyperspectral data based on recursive support vector machines. Int. J. Remote Sens. 30 (14), 3669–3677.

Zhan, J., Thrall, P.H., Burdon, J.J., 2014. Achieving sustainable plant disease management through evolutionary principles. Trends Plant Sci. 19 (9), 570–575.

9

Discriminating tropical grasses grown under different nitrogen fertilizer regimes in KwaZulu-Natal, South Africa

Rowan Naicker, Onisimo Mutanga, Mbulisi Sibanda, Kabir Peerbhay

DEPARTMENT OF GEOGRAPHY, SCHOOL OF AGRICULTURAL, EARTH AND ENVIRONMENTAL SCIENCE, UNIVERSITY OF KWAZULU-NATAL, PIETERMARITZBURG, SOUTH AFRICA

9.1 Introduction

Grassland biomes occupy 37% of the Earth's surface and are pivotal in ensuring the smooth operation of the biosphere (Egoh et al., 2011). For instance, grasslands regulate global land-surface processes and Earth's water and carbon cycles, they improve soil fertility, and maintain biodiversity as well as regional hydrological water balance (Egoh et al., 2011; Dzerefos and Witkowski, 2001; Naicker et al., 2016). Furthermore, grasslands play a critical role as a source of livelihood to rural communities. In South Africa, for example, grasslands have an economic value of about R9.7 billion with about 9.2 million South Africans benefiting directly from these ecosystems (de Wit et al., 2006). Grasslands are generally used for livestock grazing purposes and are also a source of medicinal plants (Naicker et al., 2016; Egoh et al., 2011).

Despite these numerous roles, grasslands still remain prone to degradation through overgrazing and unsuitable agronomic practices (Kowaljow et al., 2010). Grassland degradation usually occurs when their productivity and quality is compromised, and they cannot adequately fulfill their ecological role. This has severe implications on the regional water balance, biodiversity, and agricultural/farm productivity as well as the livelihoods of rural communities. To improve rangeland productivity, inorganic nitrogen-based fertilizers are often applied (Omaliko et al., 1984; Kowaljow et al., 2010). For example, the application of nitrogenous fertilizers has been found to effectively restore the productivity of degraded grasslands and has, therefore, been adopted as a general management practice. For example, Muir et al. (2001), in their study of biomass production of Alamo switchgrass, noted that at least 168 kg N/ha/year of nitrogen-based fertilizers was required to bolster biomass productivity.

Hyperspectral Remote Sensing. **DOI: https://doi.org/10.1016/B978-0-08-102894-0.00014-0**

However, the application of nitrogen fertilizers to bolster the growth rate of plants was noted by Khan et al. (2008) and Bakht et al. (2010) to be highly dependent on the phenological stage. For instance, early grass growth such as during the vegetative or boot stage tends to have a higher nutrient intake as opposed to the maturity stage (Kilcher, 1981). Excess nitrogen availability, which is often associated with maturity, can lead to nutrient imbalances and cause reductions in photosynthetic nitrogen use efficiency and net photosynthetic activity (Aber et al., 1995). The elevated deposition of nitrogen can hamper the health and functioning of many ecosystems, especially grasslands (Bonanomi et al., 2009). Increased nitrogen deposition can cause eutrophication, which results in the loss of species diversity, stunted growth, and increased mortality (Socolow, 1999; Cassman et al., 2002; Hoegberg et al., 2006). For instance, Bobbink and Willems (1991) documented a reduction of species within a calcareous grassland that had undergone nitrogen enrichment. In addition to the variation of nutrients between phenological stages, both soil nitrogen supply and plant nitrogen demand can vary temporally and spatially, highlighting the importance of optimum fertilizer application (Scharf et al., 2002; Knoblauch et al., 2017). Furthermore, monitoring the effects of fertilizer treatment on grass quality and quantity has been a long-standing challenge for rangeland managers (Sibanda et al., 2015). In this regard, there is a need for comprehensive frameworks for monitoring and characterizing the productivity of these rehabilitated native grasslands (Sibanda et al., 2015).

Remote sensing techniques offer practical alternatives through the noninvasive discrimination of vegetation characteristics (Asner et al., 1998; Ramoelo et al., 2012). Foliar biochemicals such as nitrogen have been extensively researched within the remote sensing community (Curran, 1989; Curran et al., 2001; Serrano et al., 2002; Mutanga and Skidmore, 2004). For instance, hyperspectral data, which provide information within numerous spectral channels, can characterize vegetation characteristics through their unique spectral signatures and have proven particularly useful in discriminating and predicting nitrogen concentrations (Hadoux et al., 2012; Ramoelo et al., 2013; Ling et al., 2014). Mutanga et al. (2004) used in situ hyperspectral data to successfully predict macronutrients within a savanna rangeland. Similarly, Ling et al. (2014) demonstrated the use of *in situ* hyperspectral data in extracting canopy nitrogen information from a tallgrass prairie that had undergone varying levels of fire and grazing treatments. Whilst in a different study, Sibanda et al. (2015) proved the potential of *in situ* hyperspectral data in discriminating between different fertilizer treatments in a fertilized grassland; although they did not assess the impact of different nitrogen fertilizers on the grass at different phenological stages. Despite the plethora of remote sensing research on the characterization of the foliar biochemical properties of grass, few studies have attempted to discriminate native tropical grasses grown under different nitrogen fertilizer treatments at different stages of the phenological cycle.

Although hyperspectral datasets have been associated with a lot of success, they often present the problems of multicollinearity and multidimensionality (Ling et al., 2014; Sibanda et al., 2015). The high correlation between the contiguous bands within a hyperspectral dataset causes matrix inversion and hampers the attainment of accurate parameter estimates (Huang et al., 2004a; Shao and Li, 2012; Wolter et al., 2009; Boulesteix and Strimmer, 2006).

To address the complexities associated with multidimensional datasets, Sibanda et al. (2015) employed the use of the analysis of variance prior to the use of a classification algorithm. Several researchers (Huang et al., 2004a; Boulesteix and Strimmer, 2006; Peerbhay et al., 2013; Amigo et al., 2013; Delalieux et al., 2007) have documented partial least squares discriminate analysis (PLS-DA) as a robust classification algorithm owing to its ability to exploit the variance of the input variables whilst maximizing the covariance between the input variables and the target (Pérez-Enciso and Tenenhaus, 2003).

In this context, the prime objective of this study was to discriminate tropical grasses grown under different nitrogen ranges [(1) 0% (control), (2) 20%–30%, (3) 40%–50%, and (4) 60%–80%] using *in situ* hyperspectral data and multivariate techniques such as PLS-DA at different phenological stages. However, to ascertain the robustness of PLS-DA, it was compared against the partial least squares linear discriminant analysis (PLS-LDA) algorithm.

9.2 Materials and methods

9.2.1 Study area

The investigation was undertaken at the Ukulinga Research Farm. The farm, which is located in Pietermaritzburg, South Africa, is home to a long-term grassland fertilizer treatment trial (Fig. 9–1). The grass growing season occurs between October and April, during which, Pietermaritzburg experiences most of its annual rainfall of approximately 694 mm (Sibanda et al., 2015). The site consists of several dominant grass species, namely *Themeda triandra* (red grass), *Heteropogon contortus* (black speargrass), *Eragrostis plana* (canegrass), *Panicum maximum* (guinea grass), *Setaria nigrirostris* (black-seed bristle grass), and *Tristachya leucothrix* (trident grass) (Fynn and O'connor, 2005). The grass species are grown on acidic and infertile Westleigh form soils (Fynn and O'connor, 2005; Morris and Fynn, 2001). Despite several grass species prevailing within the site, the objective of this study was not to pursue species-specific effects due to the highly interconnected, mixed-species nature of the grassland.

9.2.2 Experimental design

This study sought to discriminate grasses grown under different levels of ammonium nitrate (NH_4NO_3) and ammonium sulfate ($(NH_4)_2SO_4$). The NH_4NO_3 was applied at 21.0, 42.1, and 63.2 g/m^2, and the $(NH_4)_2SO_4$ was applied at 33.6, 67.2, and 100.8 g/m^2. Each of these ammonium fertilizers were applied twice yearly. This allowed for varying ranges of nitrogen concentrations to be measured, namely (1) 0% (control), (2) 20%–30%, (3) 40%–50%, and (4) 60%–80%. These nitrogen fertilizers were applied on native grasses growing in 78 plots, each characterized by a dimension of 3 m × 3 m. The fertilizer treatments were randomly assigned to plots within three replicate blocks as illustrated in the work of Sibanda et al. (2017). Fertilization took place in October 2016 (prior to seed germination) and December 2016 (during the boot stage).

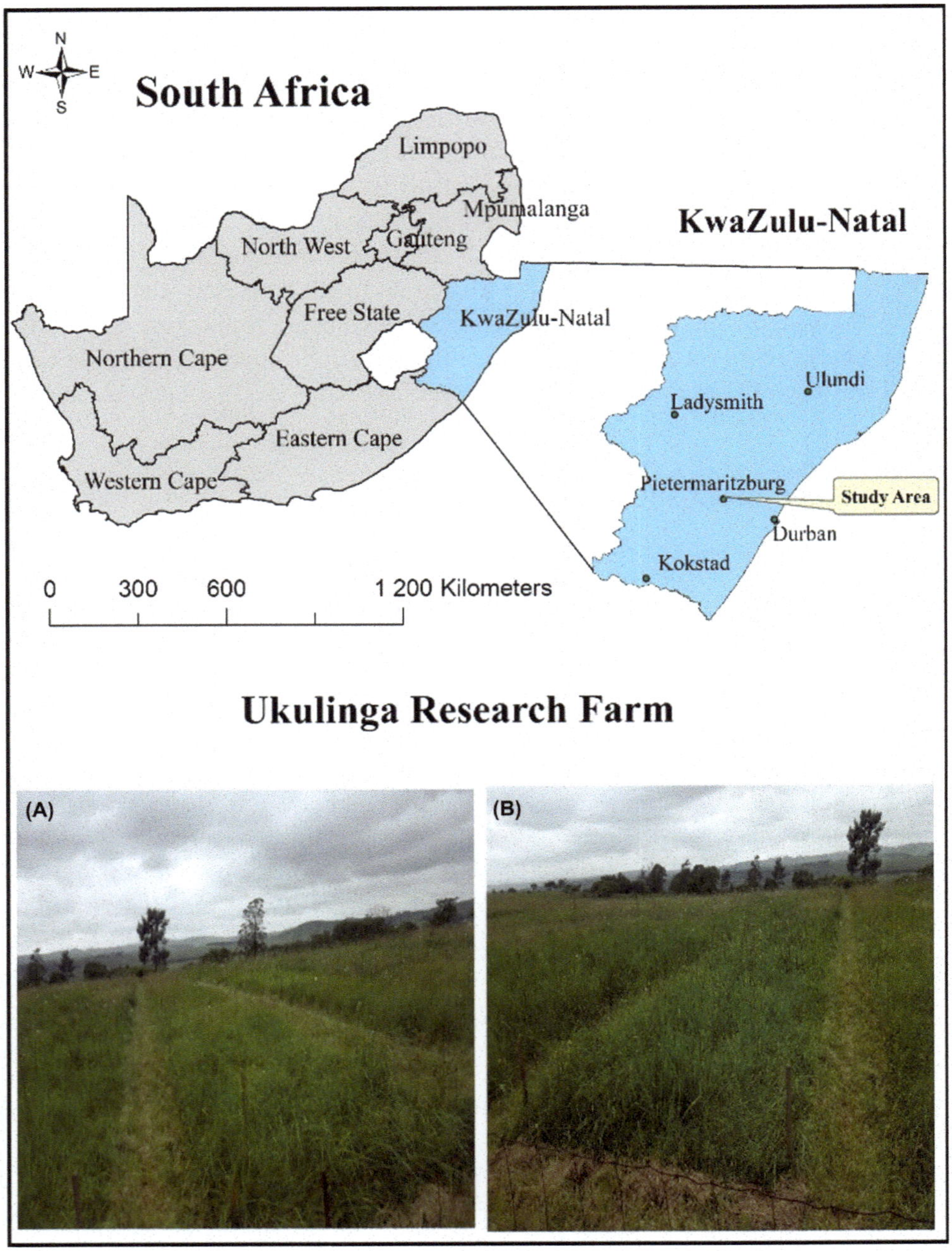

FIGURE 9–1 Long-term grassland fertilizer treatment trials located within the Ukulinga Research farm, where (A) depicts the boot stage and (B) depicts the maturity or ripening stage of the grass phenological cycle.

9.2.3 Field data collection and laboratory analyses

Three stages of the grass phenological cycle were explored in this study (Haggar, 1976; Moore et al., 1991) and these were the seed germination stage, the boot stage, and the maturity stage. However, the seed germination stage of the grass phenological cycle was excluded within this study because the plant nitrogen could not be measured. Spectral reflectance measurements, grass clippings, and soil samples were collected for both the boot stage (December 2016) and the maturity stage (April 2017). The grasses had an average height of 90 cm during the boot stage and reached approximately 1.2 m at maturity. Spectral reflectance measurements were performed in the spectral range of 350–2500 nm using an analytical spectral device FieldSpec 3 spectrometer (Sibanda et al., 2015; Mutanga et al., 2012). 10 spectral measurements were randomly taken from each plot at nadir and 1 m above the grass canopy with a 5 degrees field of view (Ferwerda et al., 2005). This yielded a ground diameter coverage of 8.7 cm. These readings were taken between 10h00 and 14h00 on a cloudless and sunny day to ensure maximum irradiance (Sibanda et al., 2015). The spectral measurements at each plot were averaged to obtain the final spectral measurement of the grass canopy within each of the 78 experimental plots (Adam et al., 2014).

To establish the impact of nitrogen fertilizer assimilation by native grasses at different levels, soil samples were collected and grass samples were clipped from the plots administered with different levels of these two nitrogenous fertilizers ((NH_4NO_3) and $(NH_4)_2SO_4$). Specifically, 78 grass samples were dried and milled prior to being taken for analysis at the Cedara Agricultural College laboratory for plant nitrogen concentration (Balota et al., 2014). Meanwhile, soil samples were collected at 0–10 cm for inorganic nitrogen content ($n = 78$) and at 10–20 cm for soil fertility ($n = 78$), and these were analyzed at the Cedara Agricultural College (Balota et al., 2014). Site factors, namely pH, nitrates (NO_3–N), and ammonium (NH_4–N) from each soil sample, were derived in the laboratory.

9.2.4 Statistical data analyses

To establish whether the grasses grown under varying nitrogen fertilizer treatments could be characterized using remotely sensed data, various stages of statistical analyses were performed. Prior to any statistical analysis, known regions (350–399, 1355–1420, 1810–1940, and 2462–2500 nm) that are associated with noise were removed from the grass reflectance data (Abdel-Rahman et al., 2014). A Shapiro–Wilk's test was conducted to ascertain that all data did not significantly deviate from the normal distribution curve (Shapiro and Wilk, 1965) ($\alpha = 0.05$) as an exploratory data analysis step for the boot and the mature grass growth stages.

Thereafter, PLS-DA and PLS-LDA were conducted in the Tanagra data mining statistical software package to discriminate the grasses grown under different nitrogen concentrations (Rakotomalala, 2005). The datasets for the boot and the mature grass growth stages were split into two, that is, 70% of the samples ($n = 45$) were used for training and 30% of the samples ($n = 24$) were used to validate the accuracy of the algorithms used. Thereafter, 100 iterations were conducted in running each algorithm using the datasets for the boot and the mature grass growing stages as illustrated in literature (Boulesteix and Strimmer, 2006; Huang et al., 2004b, Wolter et al., 2009; Peerbhay et al., 2013). Afterwards, the optimal variables selected during the boot and the mature stages were combined and used to discriminate grasses across the entire growing season. Thereafter, to understand the differences in classification between the grasses grown under varying nitrogen fertilizer treatments, the study investigated the soil–plant nitrogen assimilation relationship during the boot and the mature grass growing stages. Therefore the Spearman's rank correlation coefficient was used to test whether there was a significant relationship between any of the site factors [namely pH, nitrates (NO_3–N), and ammonium (NH_4–N)] and foliar nitrogen (Iman and Conover, 1982). Subsequently, a Kruskal–Wallis test was used to establish whether there were any significant differences in the reflectance of grass grown under different nitrogen levels (Vargha and Delaney, 1998).

9.2.4.1 Partial least squares classification ensembles

The PLS-DA and PLS-LDA methods were chosen to assist in discriminating between the grasses grown under varying nitrogen concentrations. PLS-DA is designed for regression analyses, but performs well for classification and discrimination problems (Huang et al., 2004a; Dorigo et al., 2007). This technique maximizes the covariance between the predictor and the predicted for each component, which results in the modeling of relationships between X and Y through a series of local least-square fits (Pérez-Enciso and Tenenhaus, 2003). The PLS model creates multiple sums of squares (eigenvalues) of spectral matrices. This produces scores that explain the variance of the spectral reflectance data as well as the high correlation with the response variable. The PLS-DA algorithm aims to find a straight line that divides a space into two regions and to find the discriminator or decision function. Furthermore, the model seeks to provide dimension reduction through an application where the response variable (Y) is related to the predictor variables (X) (Huang et al., 2004a; Boulesteix and Strimmer, 2006; Wolter et al., 2009; Wold et al., 2001). The classification capacity of the PLS-DA algorithm improves with increases in the number of latent variables (Wold et al., 2001). However, to avoid the problem of overfitting, the ideal number of latent variables must be determined using cross validation (CV) (Wold et al., 2001). 12 latent variables, which culminated in a lower CV error, were used in discriminating between nitrogen concentrations within the fertilizer groups. For comparison purposes, a PLS-LDA method was implemented. PLS-LDA combines the PLS method and linear discriminant analysis to circumvent the problem of a multiclass application (Amigo et al., 2013). The model solves this problem by maximizing the variance between the classes whilst minimizing the variance within the classes (Amigo et al., 2013; Malec and Királová, 2018). PLS-LDA attempts to find the linear subspace and allows for the easiest discrimination.

The fundamentals of the PLS equations are expressed in Eqs. (9.1) and (9.2):

$$X = TP + E \tag{9.1}$$

where X represents the matrix of the wavebands ($n = 1868$), T is a factor score matrix, P is the X loadings, and E is the residuals.

$$Y = Tq + f \tag{9.2}$$

where Y is a matrix of the response variable (nitrogen ranges), T is the scores for Y, q is the Y loadings, and f is the residuals.

9.2.4.2 *Variable importance in the projection*

In order to produce noteworthy classification accuracies, the preselection of important variables and the elimination of redundant variables are vital (Gomez et al., 2008). The variable importance in the projection (VIP) calculates the importance of each waveband in the discrimination model by computing a ranked score for each waveband within the hyperspectral dataset (Chong and Jun, 2005). Wavebands with a VIP score greater than one are considered important (Peerbhay et al., 2013). Common wavebands, which were identified as important within the boot stage and the maturity stage, were used to derive new PLS-DA and PLS-LDA models. Thereafter, to optimize the models using these important wavebands, each component was added progressively to reduce the cross-validation error. A total of 100 iterations were also conducted at this stage. Once stabilized, the selected components were used to classify the test dataset.

9.2.4.3 *Accuracy assessment*

The dataset was divided into 70% training ($n = 45$) and 30% testing data ($n = 24$) (Peerbhay et al., 2014). The training data, through a 10-fold cross-validation, were used to determine the accuracy of the models for discriminating grasses growing under different levels of nitrogen fertilizer treatments (Peerbhay et al., 2013). Thereafter, a confusion matrix was developed, and user, producer, and overall accuracies were calculated (Stehman, 1996; Congalton and Green, 1999). Last, to determine if a significant difference existed between the different error matrices, a discrete multivariate technique, namely the Kappa analysis, was used. The Kappa analysis uses the k (KHAT) statistic to measure the level of agreement between the classified data and the reference data (Congalton and Green, 1999). The coefficient ranges from zero to one, with one equating to a perfect level of agreement (Stehman, 1996).

9.3 Results

9.3.1 The classification of fertilizer treatments across different phenological stages

The classification outcomes of the PLS-DA and PLS-LDA models for the boot and the mature grass growing stages are depicted in Fig. 9–2. The boot stage produced better classification results with higher user, producer, and overall accuracies as compared to the maturity stage. The PLS-LDA model yielded a user accuracy of 92%, a producer accuracy of 94%, and an overall accuracy of 91% for the boot stage (Fig. 9–2A), whereas during maturity, the user

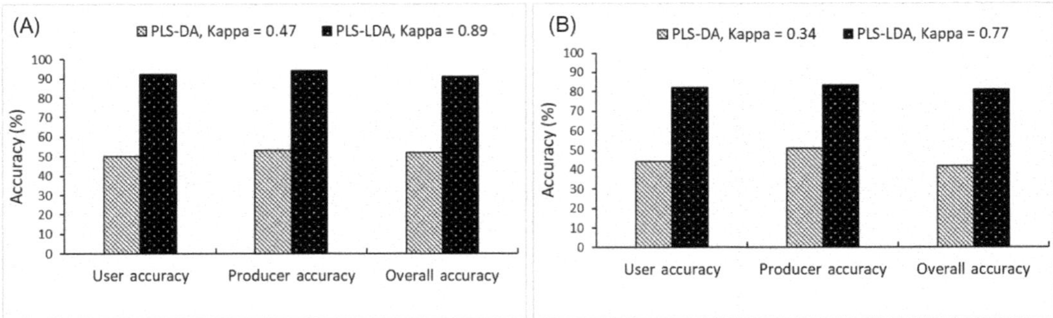

FIGURE 9–2 Classification results for (A) the partial least squares discriminate analysis (PLS-DA) and partial least squares linear discriminate analysis (PLS-LDA) models in the boot stage and (B) the PLS-DA and PLS-LDA models in the maturity stage, detailing Kappa, user, producer, and overall model accuracies.

accuracy declined to 82%, the producer accuracy decreased to 83%, and the overall accuracy decreased to 81%. The PLS-DA models exhibited a similar trend with overall model accuracy decreasing from 52% (boot stage) to 42% (mature grass stage) (Fig. 9–2). Furthermore, the PLS-LDA model proved to be a more robust model in classifying grasses grown under different fertilizer treatments with Kappa values of 0.89 and 0.77, compared to the 0.47 and 0.34 for the PLS-DA models (Fig. 9–2).

9.3.2 Comparing the performance of PLS-DA and PLS-LDA after optimization

The VIP identified a total of 233 common wavebands with VIP scores greater than one. In total, 82 bands were located within the visible part of the spectrum (390–700 nm), 22 bands were situated within the near-infrared (750–950 nm) part of the spectrum, and 77 bands were positioned within the shortwave infrared (1100–2500 nm) region. The highest VIP scores were located within the shortwave infrared region with wavebands 1923, 1925, and 1926 nm identified as the most important in characterizing nitrogen treatments with VIP scores of 3.1, 3.0, and 3.5, respectively. Thereafter, using the commonly identified wavebands, the PLS-DA and PLS-LDA models were optimized. The results showed a decrease in classification accuracies for both models derived using both the boot and the mature stages. The Kappa values decreased from 0.47, 0.89, 0.34, and 0.77 to 0.40, 0.81, 0.24, and 0.65, respectively. However, the general trends identified in the previous stage were observed again when the PLS-LDA model outperformed the PLS-DA model and when the boot stage produced higher classification accuracies as compared to the maturity stage.

9.4 Discussion

This study demonstrated the potential of remote sensing and multivariate statistical techniques in discriminating between tropical grasses grown under different nitrogen treatments. Moreover, the PLS-LDA model, which maximized the variance between treatment types whilst minimizing the variance within each class, proved to be the more robust model in characterizing nitrogen variability within the grassland.

9.4.1 Discrimination of different nitrogen fertilizer treatment regimes and characterization of the soil–plant nitrogen relationship

The results of this study showed that grasses grown under different levels of nitrogenous fertilizer treatments could be optimally discriminated during the boot growth stage. The PLS-DA model produced an overall classification accuracy of 52% and 42% for the boot and maturity stages respectively. In comparison, the PLS-LDA model substantially outperformed the PLS-DA model with overall accuracies of 91% for the boot stage and 81% for the maturity stage. The variations in classification accuracies could be attributed to soil and plant nitrogen assimilation within the phenological cycle. In assessing soil–plant nitrogen assimilation, plant nitrogen was found to be significantly ($P < .05$) correlated to soil nitrates only. During the boot stage, plant nitrogen showed a strong positive linear correlation to nitrates ($<.05$) (Fig. 9–3A). However, at the maturity stage, plant nitrogen showed a weak correlation to nitrates (Fig. 9–3B). In addition, fertilizer treatment was found to affect plant nitrogen assimilation. Significant differences in foliar nitrogen content, for the boot and maturity stages, were observed between the grasses grown under NH_4NO_3 and $(NH_4)_2SO_4$ (Fig. 9–4). The control group (level 1) depicted a low amount of plant nitrogen, whilst at level 2, both NH_4NO_3 and $(NH_4)_2SO_4$ showed minimal differences. At level 3, NH_4NO_3 displayed a greater effect on plant nitrogen concentration as compared to $(NH_4)_2SO_4$. However, at level 4, which had a high nitrogen concentration (60%–80%), $(NH_4)_2SO_4$ appeared to supply the grasses with a greater amount of nitrogen (Fig. 9–4). Furthermore, during the maturity stage, all the increments of plant nitrogen concentrations were higher (Fig. 9–4B).

These results highlight that plant nitrogen assimilation was still relatively incremental at the boot stage, whilst at maturity, nitrogen uptake had reached saturation levels. This demonstrates

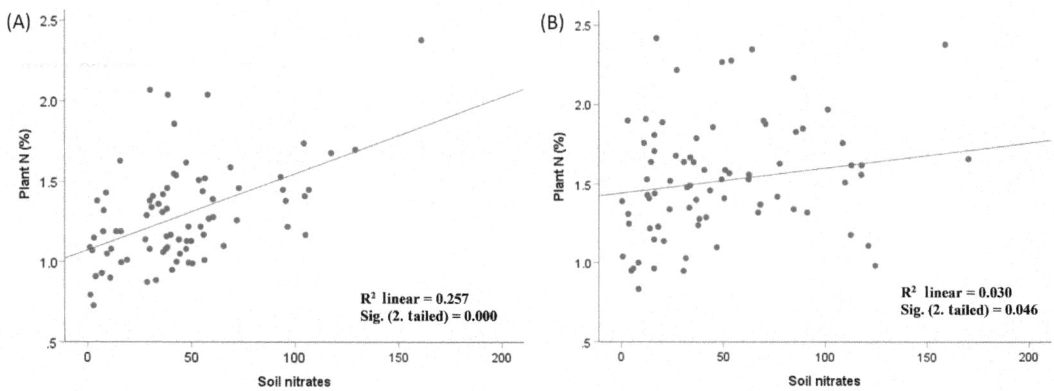

FIGURE 9–3 Correlation of plant nitrogen to nitrates at the (A) boot and (B) maturity stages.

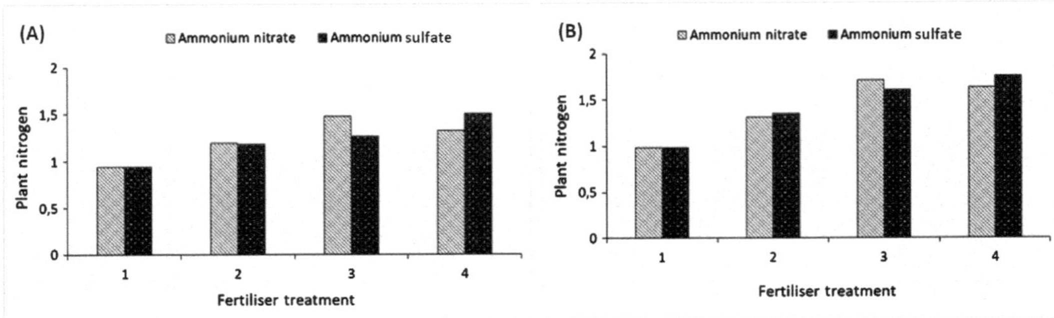

FIGURE 9–4 Effects of treatment type (ammonium sulfate and ammonium nitrate) and nitrogen addition on the amount of nitrogen absorbed by grasses across the phenological cycle for (A) the boot stage and (B) the maturity stage (where fertilizer treatments represents nitrogen ranges of (1) 0% (control), (2) 20%–30%, (3) 40%–50%, and (4) 60%–80%).

that as the amount of nitrates in the soil increases, the amount of nitrogen assimilated within the grasses also increases. Furthermore, there are studies that have confirmed a significant relationship between nitrogen concentrations and spectral reflectance variability (Sibanda et al., 2015). Therefore the optimal discrimination of grasses grown under different levels of fertilizer treatments in this study was determined by the variability in nitrogen content, which, in turn, facilitated high chlorophyll variabilities amongst the grasses, but at a specific stage of the phenological cycle.

During the maturity phase of the grass phenological cycle, results indicated a decrease in the overall classification accuracies in discriminating grasses grown under different treatments. This could be explained by a weak correlation between the foliar nitrogen concentration and nitrates in the soil (Fig. 9–3B). This indicates that at this stage, the grass no longer required added nitrogen to support net primary productivity and was starting to reach an early state of saturation. At this point, the introduction of additional fertilizer would cause the nitrogen saturated soil to discard surplus nitrogen, resulting in eutrophication within the soil. As a result, despite variable additions of nitrogen in different treatments, there is no significant difference in the plant assimilation process, and, thus, the discrimination using spectral data is also relatively weak at maturity. Furthermore, the morphological transition from being erectophile (with predominately vertical leaves) during the boot stage to being weak planophile (with predominately horizontal leaves) during the mature stage, significantly influences the overall classification between the two stages as well as across the entire growing season. Pinter Jr et al. (1985), in a study of wheat canopies, demonstrated that planophiles are positioned to receive greater total irradiance than erectophiles and exhibit higher reflectance in the visible and near-infrared regions, which should culminate in better classification during maturity when planophiles are abundant. However, in a later study, Wang et al. (2009) showed that certain erectophile types still had a higher canopy reflectance in the near-infrared region than that of planophile types. In addition, it was documented that the spectral reflectance of a sparse planophile canopy is similar to that of a dense erectophile canopy (Wang et al., 2009). This may have significantly influenced overall classification accuracies.

A total of 233 common wavebands were identified within both morphological stages. These wavebands ranged from 82 bands within the visible spectrum to 22 and 77 bands within the near-infrared and shortwave infrared regions, respectively. In particular, wavebands 1923, 1925, and 1926 nm were identified as the most important in characterizing nitrogen treatments. These results correlate with previous studies such as those of Curran (1989), Kumar et al. (2002), and Mutanga et al. (2004), who have documented several wavebands within the shortwave infrared spectrum that are synonymous with nitrogen absorption features. Moreover, these important wavebands are located within 15 nm of known nitrogen absorption features that were previously documented by Curran (1989). However, the use of these 233 wavebands reduced the classification accuracies of the original PLS models. Again, this could be attributed to the fact that at different regions within the phenological cycle (such as during the boot stage), net primary productivity and chlorophyll concentration may be higher as compared to the maturity and ripening stages of plant growth. This will result in a difference in light absorption between the stages (Lucas et al., 2008). As such, the increase in chlorophyll concentration and variability across treatments during the boot stage would result in proportional variability in light absorption and improved classification accuracy. Similarly, Peerbhay et al. (2013) documented a reduction in user and producer accuracies with the use of wavebands selected using the VIP method. This reduction in both user and producer accuracy, was attributed to an explanatory compromise between the spectral variance within each group and the spectral variance for the entire data set (Peerbhay et al., 2013).

9.4.2 Comparing the performance of PLS-DA and PLS-LDA classification ensembles

The PLS-LDA model outperformed the PLS-DA model with overall accuracies of 91% and 81% in comparison to the 52% and 42% of the PLS-DA model. This difference in classification is uncommon, however, several studies have documented the discriminatory potential of the PLS-LDA model, stressing that the PLS-LDA model outperforms the PLS-DA model (Hadoux et al., 2012; Boulesteix and Strimmer, 2006; Amigo et al., 2013). This was demonstrated by Shao and Li (2012), whose classification of sweet corn using PLS-LDA resulted in an overall classification accuracy of 94.3%. Similarly, Lottering and Mutanga (2016) used Worldview-2 imagery and PLS-LDA in successfully discriminating forest species at a subspecies level with an overall accuracy of 90%. Moreover, studies such as that by Malec and Királová (2018) have highlighted that the PLS-LDA model is particularly suited for small sample classification and in instances of collinearity or near-collinearity within datasets (which is exhibited in hyperspectral data). Although some chemometrics experts have characterized the PLS-DA model as a mere step within a full classification procedure (Brereton and Lloyd, 2014) and despite the poor performance of the model within this study, numerous authors (Peerbhay et al., 2013; Amigo et al., 2013; Wang et al., 2018; Shi et al., 2018) have proven the value of the PLS-DA model within the remote sensing and chemometrics fields.

9.5 Conclusion

Based on the findings of this study, it was concluded that:

1. The rate of nitrogen assimilation varies between the boot and maturity stages of the phenological cycle of plants with the maturity stage reaching saturation.
2. Hyperspectral remotely sensed data and multivariate techniques can successfully characterize grasses grown under different fertilizer treatments at various stages of the phenological cycle.
3. The boot stage of plant growth is the optimal phase for plant discrimination using hyperspectral data.
4. The PLS-LDA model is a more robust model, as compared to the PLS-DA model, for characterizing grasses grown under different levels of nitrogen fertilizers across the whole phenological cycle using spectral data from the shortwave infrared region of the electromagnetic spectrum.

Overall, the findings of this study will allow rangeland managers to accurately discriminate between different levels of fertilizer application and ensure the health and correct functioning of grasslands. However, this only provides the first step in ensuring native grassland health. Future studies should investigate the extent to which current nitrogen variability within grasslands can be adequately discriminated and predicted. This will provide a framework to establish a nitrogen threshold within a given region. Moreover, the use of nitrogen fertilizers is common practice in ensuring the productivity of rangelands. As such, future research may wish to document the point at which nitrophilous grassland species begin to outcompete native grassland species. This will assist rangeland managers in ensuring the health and correct functioning of tropical grasslands. Last, the influence of additional site factors (such as temperature, rainfall, and altitude) on the soil–plant nitrogen relationship should be explored.

Acknowledgment

Thank you to the staff at Ukulinga Research Farm and Cedara Agricultural College for their assistance during data collection.

Author contributions

Conceptualization, Onisimo Mutanga and Mbulisi Sibanda; Data curation, Rowan Naicker; Formal analysis, Rowan Naicker and Mbulisi Sibanda; Funding acquisition, Onisimo Mutanga; Investigation, Rowan Naicker, Mbulisi Sibanda, and Kabir Peerbhay; Methodology, Rowan Naicker, Mbulisi Sibanda, and Kabir Peerbhay; Project administration, Onisimo Mutanga and Kabir Peerbhay; Resources, Onisimo Mutanga; Software, Mbulisi Sibanda and Kabir Peerbhay; Supervision, Onisimo Mutanga, Mbulisi Sibanda, and Kabir Peerbhay; Validation, Onisimo Mutanga and Kabir Peerbhay; Writing of original draft, Rowan Naicker; Writing, review, and editing, Onisimo Mutanga, Mbulisi Sibanda, and Kabir Peerbhay.

Funding

This study was supported by eThekwini Municipality through the Durban Research Action Partnership: KwaZulu-Natal Global Environmental Change Programme. The South African Research Chairs Initiative of the Department of Science and Technology and the National Research Foundation of South Africa (grant no 84157) financially supported the research.

Conflict of Interest

The authors declare no conflict of interest.

List of abbreviations

CV cross validation
N nitrogen
NIR near-infrared
PLS partial least squares
PLS-DA partial least squares discriminate analysis
PLS-LDA partial least squares linear discriminant analysis
VIP variable importance in the projection

List of symbols

k KHAT statistic
n total number of cases
P probability value
R Rands
α Alpha—"significance level"

References

Abdel-Rahman, E.M., Mutanga, O., Odindi, J., Adam, E., Odindo, A., Ismail, R., 2014. A comparison of partial least squares (PLS) and sparse PLS regressions for predicting yield of Swiss chard grown under different irrigation water sources using hyperspectral data. Comput. Electron. Agric. 106, 11–19.

Aber, J.D., Ollinger, S.V., Federer, C.A., Reich, P.B., Goulden, M.L., Kicklighter, D.W., et al., 1995. Predicting the effects of climate change on water yield and forest production in the northeastern United States. Clim. Res. 5, 207–222.

Adam, E., Mutanga, O., Odindi, J., Abdel-Rahman, E.M., 2014. Land-use/cover classification in a heterogeneous coastal landscape using RapidEye imagery: evaluating the performance of random forest and support vector machines classifiers. Int. J. Remote Sens. 35, 3440–3458.

Amigo, J.M., Martí, I., Gowen, A., 2013. Hyperspectral imaging and chemometrics: a perfect combination for the analysis of food structure, composition and quality. Data Handling in Science and Technology. Elsevier.

Asner, G.P., Wessman, C.A., Schimel, D.S., 1998. Heterogeneity of savanna canopy structure and function from imaging spectrometry and inverse modeling. Ecol. Appl. 8, 1022–1036.

Bakht, J., Shafi, M., Yousaf, M., Shah, H.U., 2010. Physiology, phenology and yield of sunflower (autumn) as affected by NPK fertilizer and hybrids. Pak. J. Bot. 42, 1909–1922.

Balota, E.L., Yada, I.F., Amaral, H., Nakatani, A.S., Dick, R.P., Coyne, M.S., 2014. Long-term land use influences soil microbial biomass P and S, phosphatase and arylsulfatase activities, and S mineralization in a Brazilian oxisol. Land Degrad. Dev. 25, 397–406.

Bobbink, R., Willems, J., 1991. Impact of different cutting regimes on the performance of Brachypodium pinnatum in Dutch chalk grassland. Biol. Conserv. 56, 1–21.

Bonanomi, G., Caporaso, S., Allegrezza, M., 2009. Effects of nitrogen enrichment, plant litter removal and cutting on a species-rich Mediterranean calcareous grassland. Plant. Biosyst. 143, 443–455.

Boulesteix, A.-L., Strimmer, K., 2006. Partial least squares: a versatile tool for the analysis of high-dimensional genomic data. Brief Bioinf. 8, 32–44.

Brereton, R.G., Lloyd, G.R., 2014. Partial least squares discriminant analysis: taking the magic away. J. Chemom. 28, 213–225.

Cassman, K.G., Dobermann, A., Walters, D.T., 2002. Agroecosystems, nitrogen-use efficiency, and nitrogen management. AMBIO: J. Hum. Environ. 31, 132–140.

Chong, I.-G., Jun, C.-H., 2005. Performance of some variable selection methods when multicollinearity is present. Chemom. Intell. Lab. Syst. 78, 103–112.

Congalton, R.G., Green, K., 1999. Assessing the Accuracy of Remotely Sensed Data: Principles and Applications. Lewis Publishers, Boca Raton, FL.

Curran, P.J., 1989. Remote sensing of foliar chemistry. Remote Sens. Environ. 30, 271–278.

Curran, P.J., Dungan, J.L., Peterson, D.L., 2001. Estimating the foliar biochemical concentration of leaves with reflectance spectrometry: testing the Kokaly and Clark methodologies. Remote Sens. Environ. 76, 349–359.

De Witt, M.P., Blignaut, J.N., 2006. Using monetary valuation results with specific reference to grasslands in South Africa. *Report prepared for SANBI as part of the project "Making the case for the value of ecosystem goods and services provided by the Grassland Biome". Report No. SO6002. De Witt Sustainable Options (Pty) Ltd.*

Delalieux, S., Van aardt, J., Keulemans, W., Schrevens, E., Coppin, P., 2007. Detection of biotic stress (*Venturia inaequalis*) in apple trees using hyperspectral data: non-parametric statistical approaches and physiological implications. Eur. J. Agron. 27, 130–143.

Dorigo, W.A., Zurita-Milla, R., De wit, A.J., Brazile, J., Singh, R., Schaepman, M.E., 2007. A review on reflective remote sensing and data assimilation techniques for enhanced agroecosystem modeling. Int. J. Appl. Earth Obs. Geoinf. 9, 165–193.

Dzerefos, C.M., Witkowski, E., 2001. Density and potential utilisation of medicinal grassland plants from Abe Bailey Nature Reserve, South Africa. Biodivers. Conserv. 10, 1875–1896.

Egoh, B.N., Reyers, B., Rouget, M., Richardson, D.M., 2011. Identifying priority areas for ecosystem service management in South African grasslands. J. Environ. Manag. 92, 1642–1650.

Ferwerda, J.G., Skidmore, A.K., Mutanga, O., 2005. Nitrogen detection with hyperspectral normalized ratio indices across multiple plant species. Int. J. Remote Sens. 26, 4083–4095.

Fynn, R.W., O'connor, T.G., 2005. Determinants of community organization of a South African mesic grassland. J. Veg. Sci. 16, 93–102.

Gomez, C., Rossel, R.A.V., Mcbratney, A.B., 2008. Soil organic carbon prediction by hyperspectral remote sensing and field vis–NIR spectroscopy: an Australian case study. Geoderma 146, 403–411.

Hadoux, X., Gorretta, N., Rabatel, G., 2012. Weeds–wheat discrimination using hyperspectral imagery. In: CIGR-Ageng 2012, International Conference on Agricultural Engineering, 6 p.

Haggar, R., 1976. The seasonal productivity, quality and response to nitrogen of four indigenous grasses compared with Lolium perenne. Grass Forage Sci. 31, 197–207.

Hoegberg, P., Fan, H., Quist, M., Binkley, D., Tamm, C.O., 2006. Tree growth and soil acidification in response to 30 years of experimental nitrogen loading on boreal forest. Glob. Change Biol. 12, 489–499.

Huang, W., Wang, J., Wang, Z., Zhaochun, J., Liu, L., Wang, J., 2004a. Inversion of foliar biochemical parameters at various physiological stages and grain quality indicators of winter wheat with canopy reflectance. Int. J. Remote Sens. 25, 2409–2419.

Huang, Z., Turner, B.J., Dury, S.J., Wallis, I.R., Foley, W.J., 2004b. Estimating foliage nitrogen concentration from HYMAP data using continuum removal analysis. Remote. Sens. Environ. 93, 18–29.

Iman, R.L., Conover, W.-J., 1982. A distribution-free approach to inducing rank correlation among input variables. Commun. Stat. Simul. Comput. 11, 311–334.

Khan, A., Jan, M.T., Arif, M., Marwat, K.B., Jan, A., 2008. Phenology and crop stand of wheat as affected by nitrogen sources and tillage systems. Pak. J. Bot. 40, 1103–1112.

Kilcher, M., 1981. Plant development, stage of maturity and nutrient composition. Rangel. Ecol. Manag./ J. Range Manag. Arch. 34, 363–364.

Knoblauch, C., Watson, C., Berendonk, C., Becker, R., Wrage-Mönnig, N., Wichern, F., 2017. Relationship between remote sensing data, plant biomass and soil nitrogen dynamics in intensively managed grasslands under controlled conditions. Sensors 17, 1483.

Kowaljow, E., Mazzarino, M.J., Satti, P., Jiménez-Rodríguez, C., 2010. Organic and inorganic fertilizer effects on a degraded Patagonian rangeland. Plant Soil 332, 135–145.

Kumar, L., Schmidt, K., Dury, S., Skidmore, A., 2002. Imaging spectrometry and vegetation science. Imaging Spectrometry. Springer.

Ling, B., Goodin, D.G., Mohler, R.L., Laws, A.N., Joern, A., 2014. Estimating canopy nitrogen content in a heterogeneous grassland with varying fire and grazing treatments: Konza Prairie, Kansas, USA. Remote Sens. 6, 4430–4453.

Lottering, R., Mutanga, O., 2016. Optimizing the spatial resolution of WorldView-2 imagery for discriminating forest vegetation at subspecies level in KwaZulu-Natal, South Africa. Geocarto Int. 31, 870–880.

Lucas, R., Bunting, P., Paterson, M., Chisholm, L., 2008. Classification of Australian forest communities using aerial photography, CASI and HyMap data. Remote Sens. Environ. 112, 2088–2103.

Malec, L., Kiráľová, A., 2018. Evaluating competencies of graduates in tourism as a prerequisite for future employability. Prague Econ. Pap. 2018, 196–214.

Moore, K., Moser, L.E., Vogel, K.P., Waller, S.S., Johnson, B., Pedersen, J.F., 1991. Describing and quantifying growth stages of perennial forage grasses. Agron. J. 83, 1073–1077.

Morris, C., Fynn, R., 2001. The Ukulinga long-term grassland trials: reaping the fruits of meticulous, patient research. Bull. Grassl. Soc. South Afr. 11, 7–22.

Muir, J.P., Sanderson, M.A., Ocumpaugh, W.R., Jones, R.M., Reed, R.L., 2001. Biomass production of 'Alamo' switchgrass in response to nitrogen, phosphorus, and row spacing. Agron. J. 93, 896–901.

Mutanga, O., Skidmore, A.K., 2004. Integrating imaging spectroscopy and neural networks to map grass quality in the Kruger National Park, South Africa. Remote Sens. Environ. 90, 104–115.

Mutanga, O., Skidmore, A.K., Prins, H.H.T., 2004. Predicting in situ pasture quality in the Kruger National Park, South Africa, using continuum-removed absorption features. Remote Sens. Environ. 89, 393–408.

Mutanga, O., Adam, E., Cho, M.A., 2012. High density biomass estimation for wetland vegetation using WorldView-2 imagery and random forest regression algorithm. Int. J. Appl. Earth Obs. Geoinf. 18, 399–406.

Naicker, R., Rouget, M., Mutanga, O., 2016. Assessing habitat fragmentation of the KwaZulu-Natal Sandstone Sourveld, a threatened ecosystem. Bothalia-African Biodivers. Conserv. 46, 1–10.

Omaliko, C., Mammah, O., Agbakoba, A., 1984. Some aspects of rangeland improvement in a derived Savanna ecosystem. J. Range Manag. 415–419.

Peerbhay, K.Y., Mutanga, O., Ismail, R., 2013. Commercial tree species discrimination using airborne AISA Eagle hyperspectral imagery and partial least squares discriminant analysis (PLS-DA) in KwaZulu-Natal, South Africa. ISPRS J. Photogramm. Remote Sens. 79, 19–28.

Peerbhay, K.Y., Mutanga, O., Ismail, R., 2014. Investigating the capability of few strategically placed worldview-2 multispectral bands to discriminate forest species in KwaZulu-Natal, South Africa. IEEE J. Sel. Top. Appl. Earth Obs. Remote Sens. 7, 307–316.

Pérez-Enciso, M., Tenenhaus, M., 2003. Prediction of clinical outcome with microarray data: a partial least squares discriminant analysis (PLS-DA) approach. Hum. Genet. 112, 581–592.

Pinter Jr, P.J., Jackson, R.D., Elaine ezra, C., Gausman, H.W., 1985. Sun-angle and canopy-architecture effects on the spectral reflectance of six wheat cultivars. Int. J. Remote Sens. 6, 1813–1825.

Rakotomalala, R., 2005. Tanagra: a free software for research and academic purposes. Proc. EGC 2, 697–702.

Ramoelo, A., Skidmore, A.K., Cho, M.A., Schlerf, M., Mathieu, R., Heitkonig, I.M.A., 2012. Regional estimation of savanna grass nitrogen using the red-edge band of the spaceborne RapidEye sensor. Int. J. Appl. Earth Obs. Geoinf. 19, 151–162.

Ramoelo, A., Skidmore, A.K., Schlerf, M., Heitkonig, I.M.A., Mathieu, R., Cho, M.A., 2013. Savanna grass nitrogen to phosphorous ratio estimation using field spectroscopy and the potential for estimation with imaging spectroscopy. Int. J. Appl. Earth Obs. Geoinf. 23, 334–343.

Scharf, P.C., Schmidt, J.P., Kitchen, N.R., Sudduth, K.A., Hong, S.Y., Lory, J.A., et al., 2002. Remote sensing for nitrogen management. J. Soil Water Conserv. 57, 518–524.

Serrano, L., Penuelas, J., Ustin, S.L., 2002. Remote sensing of nitrogen and lignin in Mediterranean vegetation from AVIRIS data: decomposing biochemical from structural signals. Remote Sens. Environ. 81, 355–364.

Shao, X., Li, Y., 2012. Classification and prediction by LF NMR. Food Bioprocess Technol. 5, 1817–1823.

Shapiro, S.S., Wilk, M.B., 1965. An analysis of variance test for normality (complete samples). Biometrika 52, 591–611.

Shi, Y., Huang, W., Ye, H., Ruan, C., Xing, N., Geng, Y., et al., 2018. Partial least square discriminant analysis based on normalized two-stage vegetation indices for mapping damage from rice diseases using planetscope datasets. Sensors 18, 1901.

Sibanda, M., Mutanga, O., Rouget, M., Odindi, J., 2015. Exploring the potential of in situ hyperspectral data and multivariate techniques in discriminating different fertilizer treatments in grasslands. J. Appl. Remote Sens. 9, 096033.

Sibanda, M., Mutanga, O., Rouget, M., Kumar, L., 2017. Estimating biomass of native grass grown under complex management treatments using worldview-3 spectral derivatives. Remote Sens. 9, 55.

Socolow, R.H., 1999. Nitrogen management and the future of food: lessons from the management of energy and carbon. Proc. Natl Acad. Sci. U.S.A. 96, 6001–6008.

Stehman, S., 1996. Estimating the kappa coefficient and its variance under stratified random sampling. Photogramm. Eng. Remote Sens. 62, 401–407.

Vargha, A., Delaney, H.D., 1998. The Kruskal–Wallis test and stochastic homogeneity. J. Educ. Behav. Stat. 23, 170–192.

Wang, J., Wang, Z., Uchida, S., Huang, W., 2009. Relative discrimination of planophile and erectophile wheat types using multi-temporal spectrum measurements. Jpn. Agric. Res. Quart. JARQ 43, 157–166.

Wang, R., Gamon, J.A., Schweiger, A.K., Cavender-Bares, J., Townsend, P.A., Zygielbaum, A.I., et al., 2018. Influence of species richness, evenness, and composition on optical diversity: a simulation study. Remote Sens. Environ. 211, 218–228.

Wold, S., Sjöström, M., Eriksson, L., 2001. PLS-regression: a basic tool of chemometrics. Chemom. Intell. Lab. Syst. 58, 109–130.

Wolter, P.T., Townsend, P.A., Sturtevant, B.R., 2009. Estimation of forest structural parameters using 5 and 10 meter SPOT-5 satellite data. Remote Sens. Environ. 113, 2019–2036.

SECTION

III

Hyperspectral Remote Sensing Application in Water, Snow, Urban Research

10

Effect of contamination and adjacency factors on snow using spectroradiometer and hyperspectral images

P.K. Garg

CIVIL ENGINEERING DEPARTMENT, INDIAN INSTITUTE OF TECHNOLOGY, ROORKEE, INDIA

10.1 Remote sensing of snow

Snow is composed of small crystalline ice particles consisting of a multitude of snowflakes that fall from clouds. Snow is a form of precipitation, but in the study of hydrology, it is different because of the lag between when snow is produced, when it falls, groundwater recharge, and other involved hydrologic processes. Approximately 40%–50% of the Northern Hemisphere is covered in snow during midwinter (Hall et al., 1995; Pepe et al., 2005), making snow cover the most prevalent landcover type during the season. Snow cover is important for local water availability, river runoff, and groundwater recharge, especially in middle and high latitudes on a regional scale (Akyurek and Sorman, 2002; Jain et al., 2008). Precipitation falling as snow in cold regions is temporarily stored in snowpack or icepack until the beginning of the melt season. Using a remote sensing perspective, snow cover is one of the most easily identified measures of water resources from satellite imagery.

Snow is easily identified by remote sensing because its spectral signature is different from other surrounding features. The reflectance of snow is high when it is small grain size snow or dry snow. In remote sensing, snow is identified using the normalized differential snow index (NDSI) method. Remote sensing is a valuable tool for modeling and predicting snowmelt runoff. Present operational remote sensing satellite systems are limited to determining the snow depth, snow quality, and snow–water equivalent, and snow physical parameters are not directly measured by these systems. Most research in snow cover mapping is being done using multispectral remote sensing or hyperspectral images.

Hyperspectral Remote Sensing. DOI: https://doi.org/10.1016/B978-0-08-102894-0.00016-4

10.2 Snow spectra

A snow spectrum has typically high reflectance values in the visible part of the spectrum with the highest values around 500 nm where the reflectance ranges between 90% and 100% (Dozier, 1989). It decreases in the near-infrared region, and is most steep between 1200 and 1500 nm. There are several peaks around 1800 and 2250 nm, however, the reflectance in the shortwave infrared varies greatly with grain size. A grain size of 50 μm has a reflectance of about 40% with a peak at 1800 nm, while in contrast, it decreases to 3% for a 1000 μm grain size. So, a reduction of wavelength in the visible spectrum is observed for a reduced grain size. There is only 5% reflectance for 500 nm, which confirms the fact. The presence of dark particles and impurities in the snow (like wind-transported soil particles) may be the main reason for the reduction in reflectance in the visual part of the wavelength. This phenomenon can be observed in the late snowmelt season with a mixture of bare ground and snow cover, and will generally increase with time until the snow has totally melted.

Field measurements of snow properties provide point observations, and are restricted to limited locations due to rough terrain and harsh weather conditions. The reflectance of fresh snow is approximately 90% in the visible region and this is reduced at longer wavelengths. Freshly fallen snow almost immediately begins to compact and metamorphose, and this changes the snowpack characteristics. The albedo of fresh snow in the visible region of the spectrum remains high, but decreases slowly with age, while in the near-infrared region, the albedo of aging snow decreases considerably as compared to fresh snow (Singh et al., 2011).

Within wavelengths from 0.4 to 1.35 μm, snow shows very high reflectance in a range of 0.7 to 1.0 μm. Maximum reflectance, which varies in snowbound regions, ranges from 450 to 550 nm. The spectral response of snow depends on the orientation and elevation of the Sun, topographic position of snow in terms of slope, orientation, health of snow, and the atmosphere. The spectral reflectance curve is also affected by factors such as soil nutrient status, snow grain size, spectral albedo, and color of the soil.

Vane and Goetz (1988) published a detailed comprehensive analysis of imaging spectroscopy using an airborne imaging spectrometer. There was also a need to define algorithms and develop software that could be used to analyze this data. Further, advancements in sensor technology enabled a change of platforms from spaceborne to airborne. With the advent of this approach, advancements in sensor technology and associated research have introduced systems like AVIRIS (Vane, 1987; Chrien et al., 1990), the compact airborne spectrographic imager (Babey and Anger, 1993), the digital airborne imaging spectrometer (Chang et al., 1993), and the airborne PRISM experiment (Itten et al., 2008; Jehle et al., 2010).

10.3 Hyperspectral remote sensing

Remote sensing satellites provide a variety of data with unique and different characteristics in panchromatic, multispectral, and hyperspectral optical regions. Hyperspectral sensors acquire more than 100 contiguous spectral bands with a narrow bandwidth (5–10 nm) in a wavelength region between 500 and 2500 nm. Hyperspectral sensors are now being widely used for mapping land resources (Harsanyi and Chang, 1994) as multispectral datasets have simple spectral information, which makes the precise identification of similar spectral features difficult.

Unfortunately, due to internal relations between the spatial and spectral sensitivities of sensors, conventional hyperspectral images generally suffer from low spatial resolution as compared to multispectral or panchromatic images. In particular, when they are acquired at a spatial resolution that is higher than the resolution of the sensor (i.e., scanner, microscope), these images have blurring effects related to the optical response of the instrument (point spread function; PSF). To deal with this super-resolution problem, deconvolution techniques can be used to restore the observed data. Such methods require a preprocessing step for PSF estimation (Dobigeon et al., 2014). When a complementary image of higher spatial resolution but lower spectral resolution is acquired simultaneously, it generates a new image with a high spatial and high spectral resolution, which can overcome these limitations.

Various multiband reconstruction techniques, known as hyperspectral pansharpening or multiband image fusion, have shown promising results. However, when such data of good spatial resolution are not available, alternative processing approaches can be considered. Besides, even if these blurring effects are neglected, several macroscopic materials (e.g., vegetation, minerals, and manmade features) generally contribute to the spectrum measured at each single pixel of a hyperspectral image. Specifically, spectral unmixing (SU) or spectral mixture analysis provides comprehensive descriptions of hyperspectral measurements. It consists of extracting the spectral signatures of the main materials present in the scene and quantifying their respective spatial distribution (abundance) over the image (Parente et al., 2010).

Hyperspectral imaging is far more beneficial than multispectral imaging because of the several reasons (Zhu et al., 2011; Goetz, 2009) including:

1. Hyperspectral remote sensing data have high spatial resolution.
2. Hyperspectral data are frequently collected in a distinct spectral range.
3. Bands of hyperspectral data are contiguous and overlapping, making them useful to detect all necessary information.
4. The contiguous spectrum obtained assists atmospheric windows to be recognized for removal from the radiance signal, which is not applicable for multispectral sensors.
5. The signal to noise ratio of the data can be enhanced by comparing pixel spectra, but in multispectral data, it's not possible because of the number of discontiguous bands.
6. The problem of mixed spectra can be solved by directly deriving the relative abundance of materials.
7. The objects or classes of a hyperspectral image can be derived from various spaces such as the spectral space, image space, and character space.

Hyperspectral imagery has been used to detect and map a wide variety of materials for mineral mapping, and to detect soil properties including moisture, organic content, and salinity. Scientists have successfully used hyperspectral imagery to identify snow grain size.

These images have also been used for more than a decade for the detection and identification of divergent surface targets and topographical and geological features (Burke et al., 2001). A review of hyperspectral remote sensing is given by Govender et al. (2007) with a focus on the use of hyperspectral imagery in water resources, flood detection and monitoring, the detection of water quality, and vegetation applications. A comparative study of AVIRIS and Hyperion imagery is presented by Kruse et al. (2003) for mineral mapping. As hyperspectral images have high spectral resolution and low spectral resolution, a single pixel contains spectra in its spectral library from more than one material. Separating each spectrum from a mixed pixel becomes difficult. Chen et al. (2009) described a general technique of spectral unmixing. To solve this problem, an endmember extraction algorithm (EEA) can be used to separate each single spectrum from mixed pixels (Mozaffar et al., 2008).

10.4 The experimental sites

The work was carried out in two study areas, namely (1) Manali, Solang, Dhundi (HP), and nearby areas for snow studies, and (2) Patsio Glacier for glacier studies. The study areas are shown in Fig. 10–1A and B.

10.5 Data used

Four Hyperion scenes of January 12, 15 and 23, 2016 and January 11, 2017 are available for the study area. The Hyperion images covering the study area have been downloaded, and their details are given in Table 10–1. These images are from January as the study area receives the highest level of snowfall in this month. Hyperion images of January 12 and 23, 2016 are shown in Fig. 10–2A and B.

Spectral libraries were created using an SVC GER 1500 Spectroradiometer (350–1050 nm), a snow grid (snow grain size measurement), optic cable with stand, reference panel, global positioning system, snow fork, snow shovel and cutting plate, cold laboratory simulation for contamination study, and a mountain trekking vehicle to collect the samples for contaminant and terrain (adjacency) effects. In the present study, field experiments on the hyperspectral reflectance of snow were conducted using an analytical spectral devices Field Pro Full Range (FR) Spectroradiometer as shown in Fig. 10–3. The instrument operates in a spectral wavelength range of 350–2500 nm. The instrument collects radiance, reflectance, and irradiance measurements.

10.6 Methodology used

A flowchart of the methodology used for developing the spectral signature is given in Figs. 10–4 and 10–5 presents the methodology proposed for contamination study.

(A)

(B)

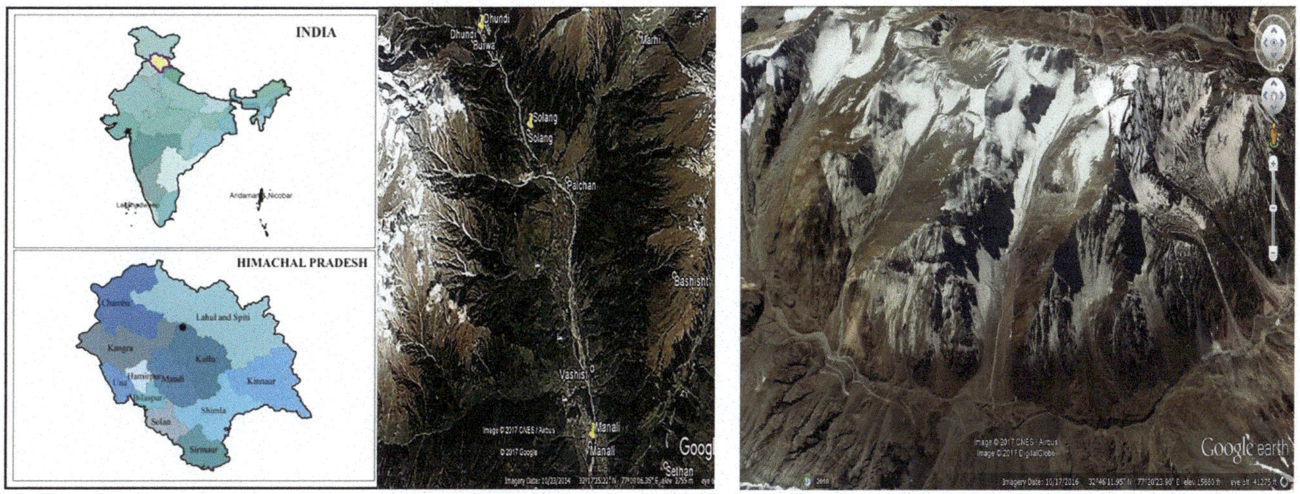

FIGURE 10–1 Study areas (A) Manali, Solang, and Dhundi areas (HP), and (B) Patsio Glacier.

Table 10–1 Suitability of Hyperion bands data.

Date	Suitable bands	Stripping error in bands	Scaling factor
January 12, 2016	8–57 and 77–166	54, 92, 137, and 239	As per the specifications given in the Hyperion manual for scaling factors for fast line-of-sight atmospheric analysis of spectral hypercubes, a scaling factor of 400 is given to all bands up to 56, and 800 is assigned to further bands
January 15, 2016	8–57, 77–97, 99–101, 103, 106–108, 110, 112, 116, 122, 129, 130, 132–134, and 166	6, 54, and 91	
January 23, 2016	8–57, 77–85, 87–94, 97–99, 101, 106, 109, 112, 115, 116, 119, 121, 123, 124, 132–134, 137, 142, and 151	6, 91, 137, and 145	
January 11, 2017	8–57, 77–108, 110, 111, 116, 118, 119, 122, 130–150, 152, 158, 159, 160, 162, and 165–167	8, 92, 143, 137, and 147	

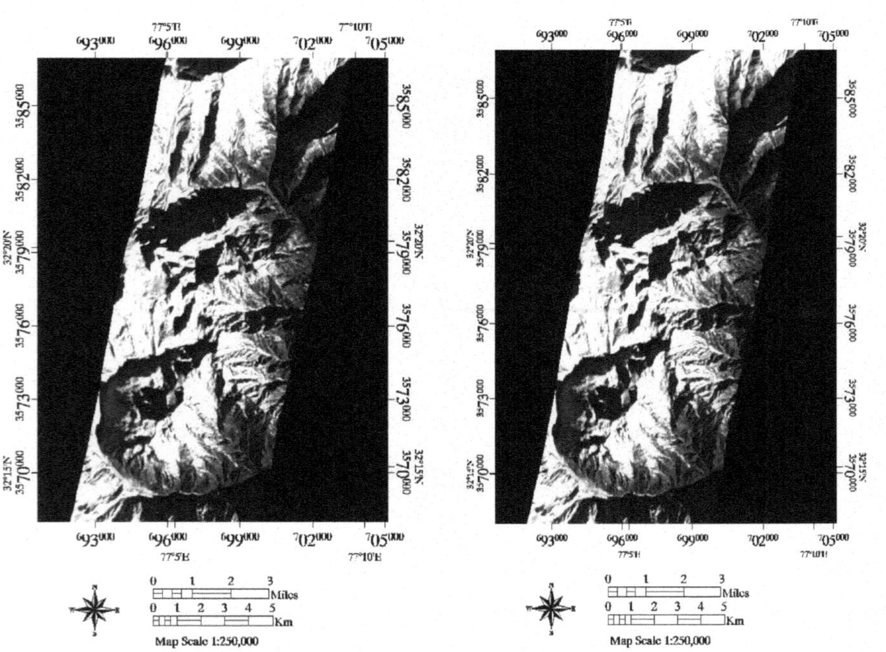

FIGURE 10–2 Hyperion imagery of (A) January 12, 2016, and (B) January 23, 2016.

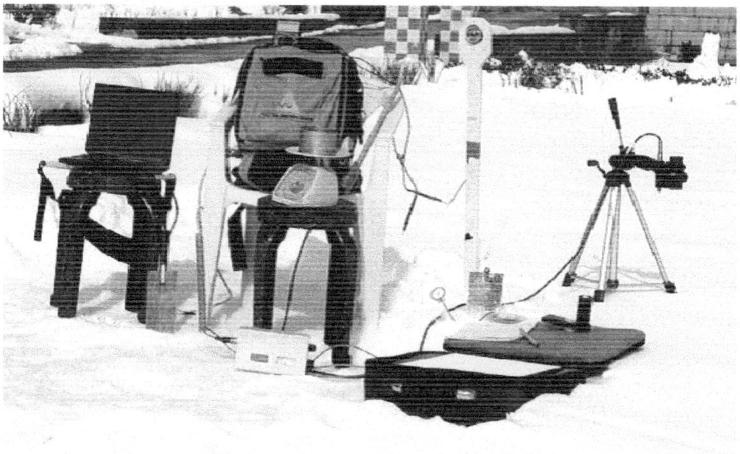

FIGURE 10–3 Field set up of spectroradiometer.

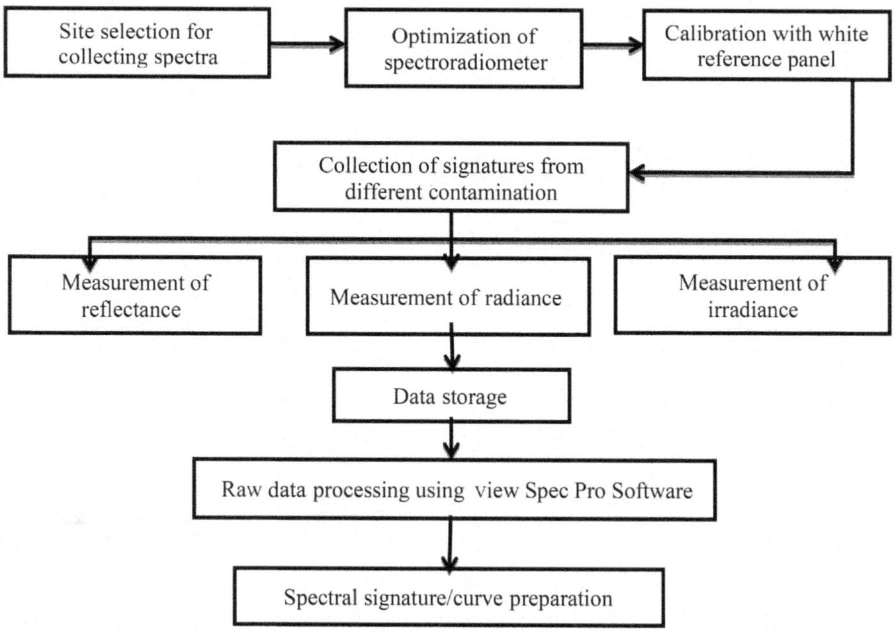

FIGURE 10–4 Methodology used for the generation of spectral signatures.

10.6.1 Preprocessing of hyperion data

The Hyperion has 220 unique bands with a spectral range of 357–2576 nm at a 10 nm bandwidth. The Level 1 radiometric product has a total of 242 bands, but only 198 bands are calibrated [bands 8–57 for visible and near-infrared analysis (VNIR) region and bands 77–224 in shortwave

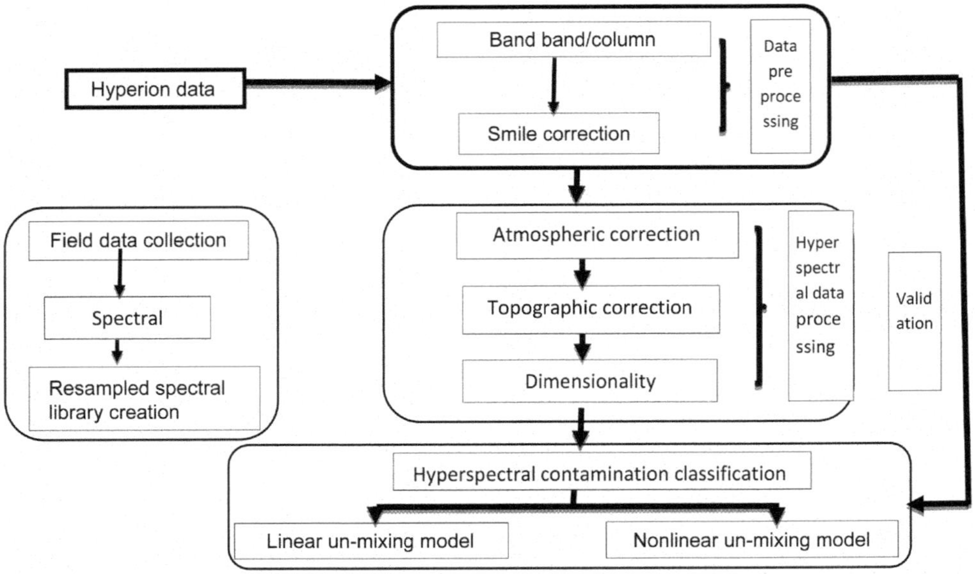

FIGURE 10–5 Flowchart of methodology used for satellite image classification.

infrared region (SWIR) region]. The basic processing of hyperspectral images was done, which includes the removal of bad bands, destripping, and atmospheric correction using Environment for Visualizing Images (ENVI's) fast line-of-sight atmospheric analysis of spectral hypercubes (FLAASH). In each image, band numbers found suitable and bands having stripping errors were identified as given in Table 10–1. Stripping error is removed with the average method.

Atmospheric correction was applied to these hyperspectral data. To process hyperspectral remote sensing datasets, statistical-based relative atmospheric correction methods and physics-based absolute correction models are available. For Hyperion, these values are available for users with the dataset. For performing atmospheric correction on the dataset, ENVI's FLAASH was used. In FLAASH, the mid-latitude winter atmospheric model was used, and in the aerosol model no aerosol was found to be suitable. As per the specifications given in the Hyperion manual, a scaling factor of 400 is given to all bands up to 56, while 800 is assigned to other bands using FLAASH (Table 10–2).

Dimensionality reduction techniques, that is, principal component analysis (PCA), and minimum noise function (MNF) have been applied on all images. For PCA, the eigenvalues for the initial few bands (up to principal component 10) were observed to be high. High eigenvalues indicate a high degree of variance, and, therefore, a high information content. In MNF techniques, bands with large eigenvalues contain data and bands with eigenvalues near 1 contain noise. Hence, MNF bands providing up to 10 eigenvalues were onserved which contain data, while the remaining bands contained the noise.

Spectral profiles of snow and nonsnow areas were compared between images before and after applying atmospheric correction. It was observed that the reflectance of every feature in

Table 10–2 Grain size classification results by spectral angle mapper (SAM) using datasets from January 12, and January 23, 2016.

	Grain index method (% area)		SAM classification (% area)	
Snow grain class	**January 12, 2016**	**January 23, 2016**	**January 12, 2016**	**January 23, 2016**
Fine (up to 0.25 mm)	7.89	15.44	6.38	13.64
Medium (0.25–0.50 mm)	22.64	17.68	10.43	7.21
Coarse (0.50–0.83 mm)	11.96	8.48	7.44	3.56
Dry snow			15.92	7.33
Wet snow			6.82	4.54
Unclassified	57.11	54.40	53.01	63.72

the corrected images increased significantly as expected. An adjacency correction was performed on images while applying the atmospheric correction. There are other methods for adjacency removal like the NIR spectrum similarity method and adjacency correction by the spectral unmixing method. According to the literature, the NIR spectrum similarity method is not so effective in snow clad areas as compared to the spectral unmixing method. For topographic error removal, Atmospheric and Topographic Corrections (ATCOR 3) software has been found to be suitable. It removes various topographic errors like shadow effect and hill shade from images. It generally uses the bidirectional reflectance distribution function model.

Based on simulated atmospheric deposition experiments, spectroscopy methods were applied to analyze the effect of different contamination concentrations on snow reflectance spectra. Then an evaluation of a snow contamination concentration (SCC) retrieval method was employed using silt density index, PCA, back propagation (BP) neural network, and radial basis function (RBF) neural network methods, and their results were compared. The results showed that the neural network model combined with hyperspectral remote sensing data could estimate the SCC well.

10.6.2 Snow grain size measurement

Snow looks bright and white if it is seen by the naked eye, irrespective of its grain size. Clean and deep snow looks the same in the visible part of the spectrum. Snow can be observed in multiple bands (NIR and SWIR), and hence, can be studied as a colorful object in color images. Fresh snow has an extremely fine grain size and good reflectance, but metamorphism and sintering in winter increase the grain size and reduce the reflectance in wavelengths beyond 8 μm. Spectroradiometry focuses on the identification of surface materials by measuring the absorption wavelength and comparing it to the known wavelength of the absorption features.

Granular objects can scatter incoming radiations generally in two ways. First, due to the complex index of refraction of constituent materials, and second, the distribution of grain sizes and shapes of scattering elements. The shape of grains is generally considered to be spherical, and, thus, a uniform surface to volume ratio is assumed while considering them for scattering. Increases in grain size as well as the aging of snow decreases the reflectance of snow in all wavelengths (Dozier et al., 1987). Mainly, snow grain size, that is, dry snow, small grain size snow, medium grain size snow, large grain size snow, and wet snow, is estimated using the snow grain index (SGI), NDSI, and spectral angle mapper (SAM) classification methods. The SAM classification technique is used for snow grain size measurement using the spectral reflectance value of snow, and these spectral signatures are stored in a spectral library using a maximum likelihood classification technique. A spectral library of different snow-cover characteristics from both laboratory-based and in situ experiments is essential for use as reference spectra to select endmembers (Acito et al., 2011).

Negi et al. (2006) measured the snow reflectance in accordance with snow grain size and proposed the SGI as:

$$SGI = \frac{(\text{Reflectance } @ \text{ } 590\text{nm}) - (\text{Reflectance } @ \text{ } 1040-1050\text{nm})}{(\text{Reflectance } @ \text{ } 590\text{nm}) - (\text{Reflectance } @ \text{ } 1040-1050\text{nm})}$$

They observed that with an increasing size of snow grain, SGI increases. The SGI as proposed by Negi et al. (2010), based on the wavelength of satellite sensors, uses green band (0.53–0.60 μm) and NIR band (0.85–0.89 μm) reflectance values for Landsat 8 data:

$$SGI = \frac{\text{Green} - \text{NIR}}{\text{Green} + \text{NIR}}$$

Snow grain size mapping is done using the SGI method based on field-collected hyperspectral reflectance data. Multitemporal and multispectral satellite data have been successfully used in the past to monitor snow cover in a vast and rugged Himalayan terrain. This is possibly due to the unique reflectance characteristics of seasonal snow. Fresh snow has a high reflectance in the red region of the visible spectrum, but this decreases in the SWIR to 1700 nm with minor peaks at approximately 1090–1100, 1830, and 2240 nm with a strong depression around reflectance 1950 and 2050 nm (Negi et al., 2009a,b,c). The estimation of snow cover in the Himalayas is generally complicated due to patchy snow, contamination, vegetation cover, and topography. Thus the satellite sensor used receives reflectance from a mixture of snow and other ambient objects (Negi et al., 2009a,b,c). The NSDI is generally used to monitor snow cover:

$$\frac{NDSI = \text{Reflectance } (\text{Green}) - \text{Reflectance } (\text{SWIR})}{\text{Reflectance } (\text{Green}) + \text{Reflectance } (\text{SWIR})}$$

The grain index (GI) is generally used for snow grain size measurement:

$$\frac{GI = \text{Reflectance } (590\text{nm}) - \text{Reflectance } (1050\text{nm})}{\text{Reflectance } (590\text{nm}) + \text{Reflectance } (1050\text{nm})}$$

For Hyperion imagery, the most suitable bands for the measurement of GI and NDSI are:

$$NDSI = \frac{Ref\ (15) - Ref\ (146)}{Ref\ (15) + Ref\ (146)}$$

$$GI = \frac{Ref\ (24) - Ref\ (90)}{Ref\ (24) + Ref\ (90)}$$

In the present study, snow grain size mapping was carried out using the GI method based on the collected hyperspectral reflectance data. Hyperion band number 24 (central wavelength at 589.62 nm) and band number 90 (central wavelength at 1043.59 nm) were used for both datasets. Snow maps were generated for dry snow, small grain size snow, medium grain size snow, large grain size snow, and wet snow classes. Figs. 10–6 and 10–7 show the spectral distribution of the snow grain sizes for two datasets, namely the January 12, and January 23, 2016 datasets, calculated by GI and NDSI, respectively.

The SAM classification was done using the spectral library approach. The different spectra were collected from the reflectance image of the Hyperion, and these were saved in a spectral library for SAM classification. Image spectra were generated for dry snow, small grain

FIGURE 10–6 Snow grain size maps using grain index on the January 12 (left), and January 23 (right), 2016 datasets.

FIGURE 10–7 Normalized differential snow index derived using Hyperion data of January 12 (left), and January 23 (right), 2016.

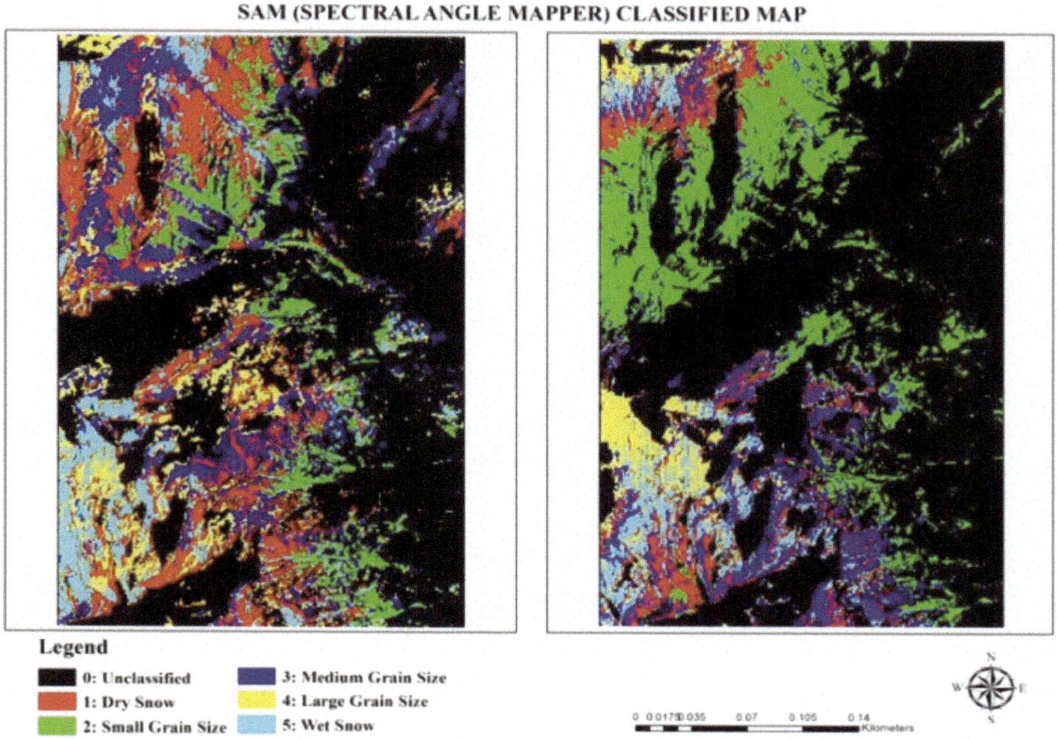

FIGURE 10–8 Spectral angle mapper classified map.

size snow, medium grain size snow, large grain size snow, and wet snow using simple z-profiling in ENVI software.

The spectra were used to classify both the images (Fig. 10–8). In the SAM classification method, some areas were not classified because of the low spectral signature, therefore, such areas were labeled as unclassified class in the output image. This happened in a high altitude area and a shadowed hilly region. It was observed that the SAM is well suited for snow grain size mapping in the Himalayan region of varying altitudes from 4000 m and above. The matrix, comparing different snow grain size classes using the results classification of the SAM methods, showed the overall classified area to be approximately 47% and the unclassified area to be 53% in the January 12, 2016 dataset. The other dataset showed 36.28% to be classified and 63.72% to be unclassified in the January 23, 2016 dataset (Table 10–2).

10.7 Contamination in snow

Contamination in snow is the presence of unwanted constituents or impurities. Beyond the visible wavelength in the NIR and SWIR regions, freshly fallen snow usually has a fine grain size, but the physical properties of snow and any contamination present affect the spectral albedo and reduce the spectral reflectance in wavelengths beyond 0.8 μm. Small amounts of impurities in snow can significantly reduce the snow albedo, and affect the incoming energy balance. Sometimes snow reflectance varies because snowpack contains liquid water content and impurities such as dust, soot, algae, vegetation, coal, and ash. Most observations of visible snow albedo are lower than those calculated for pure snow albedo. Snow has a carbon "soot" concentration naturally or artificially by weight or surface mixture (Warren and Wiscombe, 1980). Contaminations that are likely to have widespread effects on snow albedo are carbon soot, continental dust, fire ash, etc. Among all these contaminants, soot is 50-times more effective than dust, and 250-times more effective than ash in reducing the snow albedo.

Contamination affects the reflectance of snow of different grain sizes to the same extent; although research had been done on the albedo of snow of different grain sizes. Contaminations in snow can be more effective at reducing the albedo if they are located inside the ice grains. Small amounts of contamination in snow cover distributed linearly through larger grain sizes affect snow albedo significantly. Snowmelt is known to carry higher concentrations of contaminants, which adversely affect water quality. Sustainable management and the types of contaminations in snow are, therefore, important for the investigation and simulation of organic contaminants in snow cover.

Minute concentrations of small highly absorbing particles can lower snow albedo in the visible wavelengths by 5%–15%. In undulating terrain like the Himalayas, the atmospheric conditions can change extremely fast, especially in the lower Himalayan region, and the temperature difference is significant between day and night, which generates melt–freeze cycles. Contaminated snow shows a decrease in reflectance in the visible region, whereas grain size within snowpack has more influence in the NIR region. For contaminated snow, reflectance reduces the maximum in the visible region (Warren and Wiscombe, 1980; Warren, 1982) and a lesser effect has been observed in longer wavelengths (Singh et al., 2010; Negi et al., 2006). This spectral property of snow has been used for snow cover monitoring on a regular basis.

Spectral libraries of snow with contamination materials can be generated during field campaigns as these spectral libraries are needed for applying SU models. It is important to get the spectra of snow with different grain sizes during the field data collection. General contamination classes that may be present in snow are soil, coal, carbon soot, dust, algae, ash, and vegetation.

Using spectroradiometer data, Negi et al. (2006) proposed a new contamination index called the snow contamination index (SCI):

$$\frac{\mathrm{SCI} = (\text{Reflectance @ 470 nm}) - (\text{Reflectance @ 590 nm})}{(\text{Reflectance @ 470 nm}) + (\text{Reflectance @ 590 nm})}$$

The SCI as proposed by Negi et al. (2010) uses the true surface reflectance of band 2 (0.45–0.52 μm) and band 3 (0.53–0.60 μm) of Landsat 8 data, and is represented by the equation provided here. They further analyzed that for contaminated snow the SCI remains negative, and in the case of clean snow, a positive value of index or a value close to zero is observed.

$$\mathrm{SCI} = \frac{\mathrm{Blue} - \mathrm{Green}}{\mathrm{Blue} + \mathrm{Green}}$$

For soil and ash contaminations, as the contamination increases, the SCI decreases; however, in the case of coal contamination, a similar trend has not been observed, but overall, the value of SCI remains negative for all types of contamination (Singh et al., 2011).

In the present study, it was planned to measure the spectra of fresh and old snow. To understand the effect of grain size on spectral reflectance, the field experiments were performed in two ways: (1) collecting the snow reflectance and grain-size observations in the top 10 cm of the snowpack every day and then appropriately selecting reflectance plots for different ranges of grain size, and (2) after fresh snowfall (fine-grain snow), the snow reflectance and grain-size observations were taken at 30 minutes intervals. The spectra of snow mixed with different contamination classes with different grain sizes were also measured in the field using a spectroradiometer. The experiments were conducted with a constant quantity of contaminant and a varying quantity of contaminant on various grain sizes of snow to understand the effect of varying amounts of contamination on snow reflectance. These contaminants were used in two categories quantitatively-adding contamination in (1) small quantities, and (2) large quantities. Snow reflectance data were collected between 350 and 2500 nm spectral range and binned at 10 nm intervals by an averaging method. A total of 120 samples were collected during field visits between March 1 and 15, 2018 to Patsio Glacier.

For the contaminated snow reflectance, measurements were carried out for soil, ash, and coal contaminations. Measurements were collected at 1.0 m above the target with a field of view of 25°; making a circle with a diameter of 20 cm. Contamination of 314 mg was spread onto a 20 cm diameter circle, the snow contamination of 1 mg/cm^2. By adding 314 mg of contamination successively, the concentration of contamination was increased. The time period for one complete set of observations was approximately 10–15 minutes, considering the changes in the atmosphere were negligible. To understand the effect of moisture on reflectance, a dielectric moisture meter with a flat capacitive sensor was used. To understand the influence of aging on reflectance, measurements were carried out daily on a plain ground. To study the effect of large amounts contaminants on snow reflectance, first, a small concentration of contamination was added to snow, and second, a larger quantity of contamination was added. These experiments were performed at the same place and time to minimize the effects of other parameters. Each field experiment was repeated four to five times. Details of the instruments used in the field are given in Table 10–3.

During the soil/coal contamination field experiment, the soil was oven dried, ground, and filtered, restricting the size of contaminated particles to below 0.5 mm. For the first part, small packets of 0.1 g each were prepared to observe the subtle variations in snow reflectance due to the addition of these contaminations in snow. Such packets were prepared both

Table 10–3 Instruments used in the field.

S. No	Instrument	Parameter	Remark
1	Spectroradiometer	Spectral reflectance	Analytical spectral devices field Spec Pro FR (350–2500 nm)
2	Moisture meter	Moisture content in snow grain size	Dielectric moisture meter
3	Crystal gauge with magnifying hand lens	Grain size	Equivalent spherical grain diameter
4	Pyrometer	Albedo	Measures shortwave radiation (350–2800 nm)
5	Density meter with weighing machine	Snow density	Density of top 10–15 cm snow surface
6	Thermometers	Snow surface temperature and ambient temperature	
7	Snow stake	Snow thickness	
8	Weighing machine and sieve	Contamination study	Fine grain soil was sprayed on a fixed field of view
9	Manual observations	Cloud cover	Reflectance data of only cloud-free days were selected for analysis

for soil and coal contaminations. While spraying the contaminants over the snow cover area, precautions were taken to distribute this evenly in the instrument's view area. Proper care was taken for other atmospheric parameters like wind, ambient disturbance, etc., for them to a have negligible effect. The snow was moist in nature with a density of 0.24 g/cc and a grain size of 0.1–0.6 mm. The instrument sensor was kept at such a height that a view area of 20 cm in diameter of the snow surface was obtained. The sensor was targeted to collect the observations for clean snow, and after the spraying of soil and coal contaminants equally over the surface. Different wavelength regions were chosen based on the influence of physical change on the snowpack characteristics. The position, strength, and shape of a spectral curve can provide information on smoothly varying spectral properties. The experiment was carried out within 2 minutes to minimize the impact of other factors. To understand the effect of snow depth on spectral reflectance, the field setup used a 50 × 50 cm black sheet. The black sheet was inserted horizontally into the snowpack at different depths from the bottom in the upward direction. Reflectance observations were carried out as shown in the field setup. Then, the soil was weighed in 1 g (low), 5 g (medium), 20 g (high), and 30 g (very high) packets, which were equally distributed on the surface of the snow-covered area. Similarly, the coal was ground and weighed in 0.1 g (low), 0.6 g (medium), and 1.1 g (high) packets, which were sprayed evenly on the surface of snow cover area.

10.7.1 Soil contamination in snow

The spectrum of snow is affected by soil contamination and shows a large drop in the visible region, but reflectance increases beyond the SWIR region, that is, 1500 nm (Negi et al., 2009a,b,c).

Hence changes in reflectance in this region can be used for distinguishing between contaminated snow and patchy snow.

This experiment was planned to study the spectral reflectance of snow and soil simultaneously with reference to the contribution of soil to the snow cover region. Reflectance observations were taken of a 100% snow-covered region, 50% dry soil and 50% snow, 50% moist soil and 50% snow, and a 100% soil covered region. A maximum drop in reflectance takes place in the visible region as the proportion of snow is reduced from 100% to 50%. After a 50% increase of soil in snow, the area has a marginal influence on reflectance. A shift of the peak in the reflectance was observed in the visible region, whereas an increase in reflectance was observed in the VNIR to SWIR region (Fig. 10–9A).

10.7.2 Coal contamination in snow

The study of coal contamination and its effect on snow cover reflectance is a challenging task. Coal has a much larger effect on the reflectance of snow than soil as coal reduces the albedo of snow significantly. Coal contamination absorbs shortwave radiation and increases the melt rate of snow. Increases in contamination shift the peak of reflectance in the visible range at high wavelengths.

The amount of coal powder was evenly distributed over the snow area, and the spectra of snow mixed with coal was taken. This procedure is repeated to collect the spectra of snow mixed with coal at different locations. The experiment was carried out to quantify the effect of coal contamination on snow reflectance. The coal was added in fractions of 20%, 40%, and 50% to specified target areas of 1 g/m^2. There was a drastic fall of reflectance in the visible region as the coal contamination increased, and this remained constant to the short wavelength of the green band. As the amount of contamination increased in different fractions, the reflectance remained constant in the VNIR band (Fig. 10–9B).

10.7.3 Carbon soot contamination in snow

Carbon soot is produced by incomplete combustion caused by humans as one of the major sources of soot from burning coal, wood, and oil for industries, and brush fires for agriculture. Natural forest fires can also contribute soot to atmospheric aerosol, and deposits on snow cover. Soot is of great concern for its climatic effects as absorptive components of haze. The experiment was conducted to quantify the effect of carbon contamination on snow reflectance. Reflectance observations were taken from a 100% snow-covered region, a 50% carbon and 50% snow region, a 33% carbon, 33% coal, and 33% snow region, and a 100% soil covered region. The drop in spectral reflectance is more prominent (90%–10%) in the visible region, and shows more reduction of reflectance in comparison to the same amount of carbon and coal contamination. The rate of the drop of reflectance in the visible channel is lesser as it moves toward added coal contaminated snow (Fig. 10–10A).

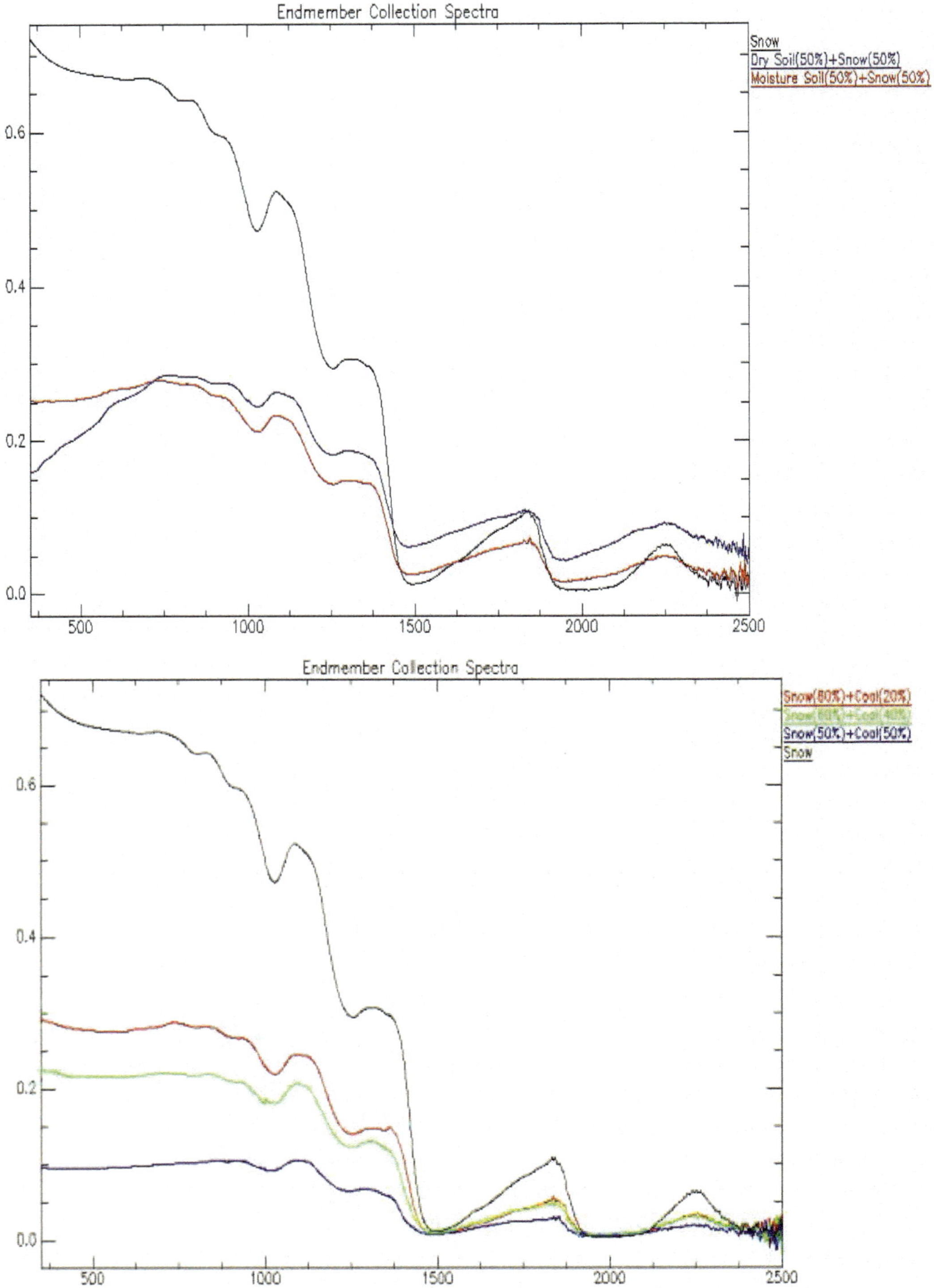

FIGURE 10–9 Reflectance of (A) soil-contaminated and (B) coal-contaminated snow in Patsio Glacier.

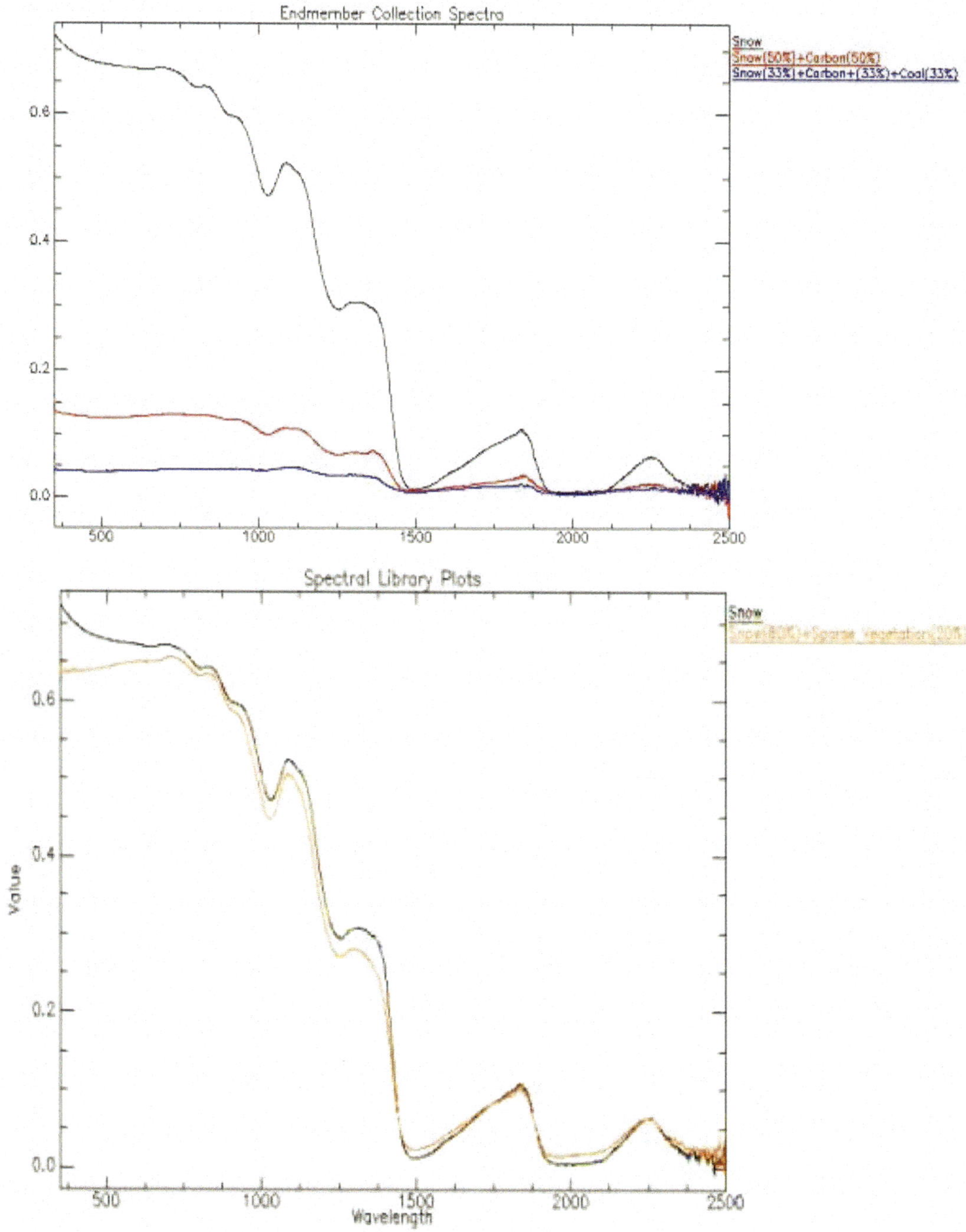

FIGURE 10–10 (A) Carbon contamination and (B) sparsely mixed vegetation in Patsio Glacier.

10.7.4 Ash contamination in snow

Ash is visible in the subsurface of snow cover, which could reduce the snow albedo. Due to the episodic nature of fire, particularly in the Himalayan region, ash is mixed with new snowfall. Since the snow does not melt over immediately after the fall, the ash remains hidden from sunlight once it is covered by fresh snow. The effect of ash on snow albedo is, thus, unlikely to have long term climatic significance.

10.7.5 Sparse vegetation in snow

This experiment was carried out on a mixed area of snow and sparse vegetation to determine its effect on the reflectance value. The 80% snow and 20% sparse vegetation covered target area showed a reflectance pattern of snow as there is a low quantity of vegetation. The reflectance pattern remains the same from the VNIR band to the SWIR band with a drop in reflectance as compared to the reflectance of snow.

10.7.6 Dust contamination in snow

Large amounts of dust may be present in snow melts in summer. This impurity frequently tends to gather at the subsurface rather than washing away with the melt water. During the decay phase, ice sheets are largely covered with thick debris. The insulation of the ice becomes important in the reduction of snow albedo and ablation can be reduced. The delay of ice decay due to protection by dust would cause a noticeable effect in the oxygen isotope record in river sediments.

There are several mountain ranges where dust flies and settles over the snow. In this case, the reflectance in the visible spectrum is much lower than that for clean snow. In many mountain ranges, dust storms often deposit radiatively absorbing dust on the snowpack. Winter and spring storms entrain absorbing dust from desert regions and redistribute optically thick layers to the snow cover as both wet and dry deposition (Dozier et al., 1987). At longer wavelengths, dirty snow has a low reflectance. The grain size of snow also increases due to the absorption of dust particles, and, hence, the absorption increases and the reflectance decreases.

10.7.7 Contamination of algae in snow

Remote sensing through imaging spectroscopy offers the capability to analyze spectral reflectance features that are related to snow algal concentration at a spatial resolution commensurate with the spatial variability of surface cover in alpine basins. *Chlamydomonas nivalis* is the most prevalent alga found in snow fields. Snow algae are found in old, wet snowfields at elevations of over 3000 m (Painter et al., 2001).

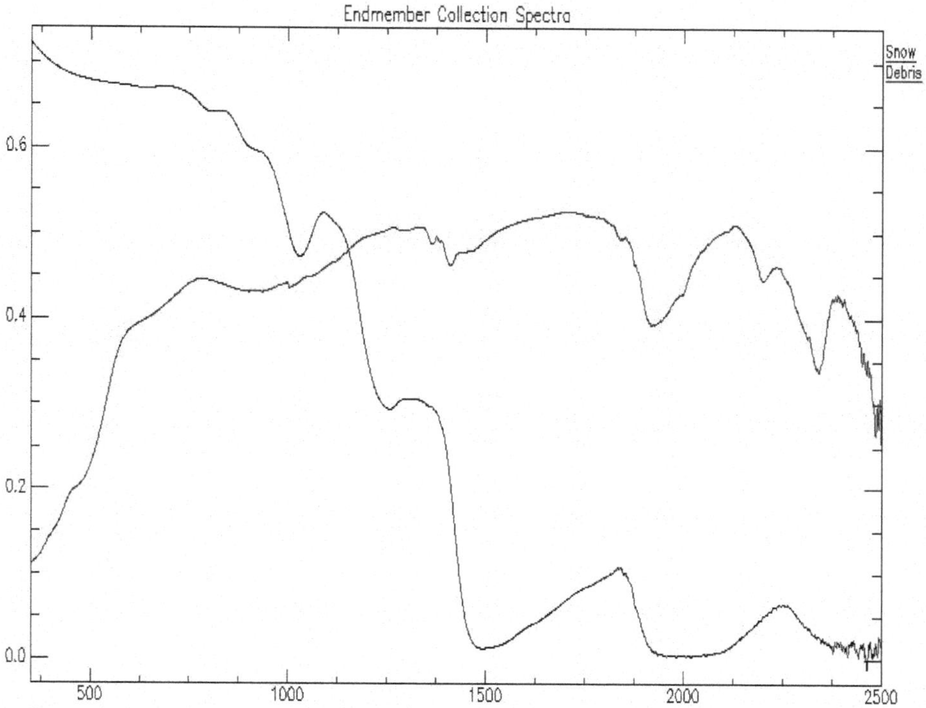

FIGURE 10–11 Reflectance of dparse debris in Patsio Glacier.

10.7.8 Contamination of sparse debris

This experiment was conceded to quantify the effect of debris on snow reflectance. A gradual increase of reflectance in the visible to SWIR region was observed and then a little drop beyond the SWIR region was found. It shows high absorption in the blue band and a low absorption from the green to SWIR region at different wavelengths. The reflectance of snow decreases in the SWIR region as compared to the reflectance of debris in the same wavelength (Fig. 10–11).

In Table 10–4, the minimum and maximum reflectances in the VNIR and SWIR regions are particularly shown. The spectral reflectance of snow with other contaminants (gradually increasing amount) is shown as minimum and maximum reflectances in both regions. For the adjacency effect, the minimum and maximum reflectances of debris at the VNIR and SWIR regions are clearly shown.

10.7.9 Influence of mixed and contaminated snow on normalized differential snow index

The NDSI is found to be helpful in snow cover monitoring. Here, the green band was selected as Hyperion's B15 (498 nm) and the SWIR band as B146 (1608 nm). The NDSI

Table 10–4 Minimum and maximum reflectance of snow with other contaminants including normalized differential snow index (NSDI) and contamination index values.

Class	Visible and near-infrared analysis region (350–850 nm)		Shortwave infrared region (851–2500 nm)			
	Min. reflectance	Max. reflectance	Min. reflectance	Max. reflectance	NDSI	Contamination index
Snow	0.638	0.7231	0.0088	0.6375	0.893	0.0077
Snow (80%) and coal (20%)	0.2752	0.2926	− 0.0051	0.2805	0.864	0.0036
Snow (60%) and coal (40%)	0.2151	0.2261	− 0.0185	0.2186	0.809	− 0.0018
Snow (50%) and coal (50%)	0.0951	0.1054	− 0.0144	0.1058	0.714	− 0.0119
Snow (33%), carbon (33%), and coal (33%)	0.0408	0.044	− 0.0072	0.0464	0.525	− 0.0086
Snow (50%) and carbon (50%)	0.1267	0.1358	− 0.01	0.129	0.781	− 0.0118
Snow (50%) and dry soil (50%)	0.1587	0.2859	0.0398	0.2833	0.443	− 0.1031
Snow (50%) and moist soil (50%)	0.2511	0.2787	0.003	0.2717	0.730	− 0.0244
Snow (80%) and sparse vegetation (20%)	0.6286	0.6567	− 0.0047	0.628	0.867	− 0.0074
Debris	0.1119	0.4425	0.2479	0.5257	− 0.391	− 0.3033

values for different contaminants were estimated, that is, coal, carbon, soil, and sparse vegetation contamination, and these are given in Table 10–4. In this experiment, a number of contaminations were added to pure snow cover areas of 1 m in radius. It was observed that:

1. The NDSI values for coal contamination decreased in comparison to coal contamination in amounts of 20%, 40%, and 50%. This indicates that as the area of vegetation mix increases, the NDSI value decreases.
2. The NDSI values increased by 50% in moist soil contamination of snow in comparison to dry soil contamination of snow.
3. The NDSI values for the snow, coal, and carbon mix (when mixed uniformly in the area) remained positive, but were significantly less than the NDSI value of snow alone.
4. The NDSI values of 20% sparse vegetation contamination in snow were almost same as those of snow because of the low amount of vegetation.
5. The NDSI values for carbon contamination increased in comparison to the same amount of coal contamination.

10.7.10 Contamination index for different levels of contamination using spectroradiometer

The SCI is helpful in snow cover monitoring. Here, the Hyperion bands used include B12 (467 nm) and B24 (589 nm). From SCI, it was observed that (Table 10–4):

1. The SCI for contaminated snow remains negative, while it remains positive or close to zero for clean snow and sparse vegetation contaminated snow.
2. For coal and carbon, as the contamination increases, the SCI decreases.
3. The SCI values increase by 50% in moist soil contamination of snow in comparison to dry soil contamination of snow, which remains negative. The value of SCI remains negative for all types of contaminations.

10.8 Adjacent objects and their effects on snow reflectance

10.8.1 Adjacency effects

The adjacency effect is a well-known phenomenon involving the presence of path interferences between the reflectances from different land cover materials (Burazerović et al., 2012). It refers to the situation where the spectra of multiple land cover materials contribute to the spectrum that is observed for a single pixel. One reason of this interference is related to the interaction of incident and reflected solar radiation at the surface level. The other has to do with backscattering of this radiation into the atmosphere.

In the early 1980s, an increase in Landsat's radiance values from pixels close to the seashore (Kaufman and Joseph, 1982) or from inland water areas (Tanre et al., 1987) was detected and interpreted as the adjacency effect. Hyperspectral satellite imagery of an inland waterbody showed water-leaving reflectance with abnormally high NIR reflectance, which was ascribed to contamination by adjacency effects (Van Mol et al., 2004). Extensive modeling studies have shown the dependence of the adjacency effect on mainly the aerosol optical thickness (AOT) and the aerosol vertical distribution (Santer and Schmechtig, 2000). As the AOT increases, the contribution of the neighboring pixels becomes larger. On the other hand, under clearer atmospheric conditions (i.e., lower AOT), the adjacency effect from more distant pixels becomes more significant due to the dominant contribution of molecular scattering (Minumora et al., 2001).

Furthermore, the higher the aerosol scale height, the stronger the adjacency effects. Approximations have been developed to take into account adjacency effects in atmospheric correction using conventional radiative transfer codes such as second simulation of the satellite signal in the solar spectrum (6S) or moderate resolution transmission (Vermote et al., 1997; Berk et al., 1998). Within an atmospheric correction code, ATCOR for instance, (Richter, 1996, 2008) an adjacency correction is performed in an iterative way by taking into account the ratio of diffuse to direct transmittance and the difference in reflectance between the target pixel and its neighborhood, defined by the range (R) parameter. This horizontal range is highly affected by the aerosol profile. However, this parameter is often unknown as it requires information from light detection and ranging measurements. Therefore the determination of the appropriate range is more or less a trial and error procedure. Although some theoretical modeling and simulations have been performed, examples of operational adjacency corrections for real airborne or satellite data are rarely published (Liang et al., 2001).

An investigation was carried out to understand changes in snow reflectance due to the presence of other surrounding objects. For this purpose, common objects in snow region such as water, sparse vegetation, dry and moist soil, coal, and carbon were considered. Observations were taken for the adjacency factors and their effects on snow albedo were recorded. Suitable locations like debris with snow cover, a waterbody surrounded by snow, snow cover adjacent to the road side, etc., were selected for taking observations. A simple approach was followed to measure adjacency effects so as to create a buffer zone of a snow adjacent feature by distance, and the spectra were measured by distance.

10.8.2 Liquid water content

Wet snow and dry snow have different reflectances in the infrared region. Generally, wet snow is darker than dry snow because liquid water causes grains to form clusters, which act like larger grains so that snow albedo decreases, hence, wet snow appears darker.

10.8.3 Clear waterbody

Waterbodies can be included as adjacent features of snow. The effect of an adjacent feature on the spectral reflectance of snow could be crucial. Singh et al. (2005) found that the reflectance curve of water is reduced to well below 10% and water shows no reflectance as it crosses the 750 nm wavelength.

10.8.4 Vegetation

Dry and green vegetation show similar spectral signatures, except there is a difference in the VNIR band. This is due to the chlorophyll content present in green vegetation and changes in their cellular void spaces due to lack of moisture. Locations having vegetation around snow cover could be utilized to identify and analyze the adjacency effects of vegetation on snow cover.

10.9 Spectral unmixing methods for satellite image classification

10.9.1 Linear unmixing model

Linear spectral unmixing is a standard technique for spectral mixture analysis that infers a set of pure spectral signatures, called endmembers, and fractions of these endmembers, called abundances. The linear unmixing model (LUM) has received considerable attention since it generally consists of an acceptable first-order approximation of the physical processes involved in most scenes of interest. Consequently, it has been used in several research works that aim at developing efficient EEAs that are able to recover pure component signatures in images, and inversion techniques to estimate the abundance coefficients for a given (estimated or a priori known) set of endmembers. Earlier studies dedicated to linear SU were found on geometrical interpretation of the problem.

One of the most popular EEAs is the pixel purity index (PPI), designed to search for a convex geometry in a given dataset that is supposed to represent pure signatures present in those data (Boardman et al., 1995). PPI is the most commonly used method to find extremely pure pixels in multispectral and hyperspectral images. The purest pixel in a given image is computed by repeatedly projecting n-D scatter plots on a random unit vector. A pixel purity image is created where each pixel value corresponds to the number of times that pixel was recorded at endpoints. The PPI function creates an output band or continues its iteration and adds results to an existing output band. Fast PPI methods place the image data into memory and perform the computation in memory, which is faster than the disk-based PPI, but requires adequate memory.

Ideally, ground-based spectra would be required to produce accurate endmembers since endmembers taken from even extremely high spatial resolution imagery may contain multiple surface components. There exists a linear relationship between the fractional abundance of substances comprising the area being imaged in the reflected image. The pixel value of an image indicates the fraction of the pixel that contains the endmember material corresponding to that image. A pixel from this abundance image with a value of 0.92 indicates that 92% of the pixel contains snow and 0.07 indicates that 7% of the pixel contains vegetation. If many pixels have values above 1.0 or below 0.0, it indicates that one or more endmember chosen for analysis is probably not well characterized.

The ENVI tool for the linear mixing model (LMM) has been used in the present study on all images for pixel classification into two classes. The images are classified into snow and nonsnow areas with some errors as the spectra of shadowed areas of snow areas and nonsnow areas interfere with each other. The LMM was attempted using existing inbuilt spectral libraries of snow using ENVI software. The results obtained are to be validated in the field. Endmembers were extracted using the PPI method. Forward MNF was applied on AVIRIS subset data. A total of the first 10 bands were selected having high eigenvalues for further processing. Fast PPI was applied to MNF images with an iteration of 10,000 with a threshold factor of 5. The minimum and maximum threshold values fall in a specified range; all the pointa within this range were selected and exported to n-D visualizer to collect endmembers. As the AVIRIS data cover an area fully covered with snow, a limited number of endmembers were selected for unmixing. The first 10 MNF bands were used to collect the endmembers from the 3D scatter plots through n-D visualization. Linear spectral unmixing was done from selected endmembers, and fraction maps of endmembers were obtained from the MNF images. Fig. 10–12A shows the PPI plot of AVIRIS data, while Fig. 10–12B presents the pure pixels in AVIRIS data.

10.9.2 Nonlinear unmixing model

According to Dias et al., the LUM uses two assumptions. First, the mixing process should occur at a macroscopic scale (Singer et al., 1979). Second, the photons that reach the

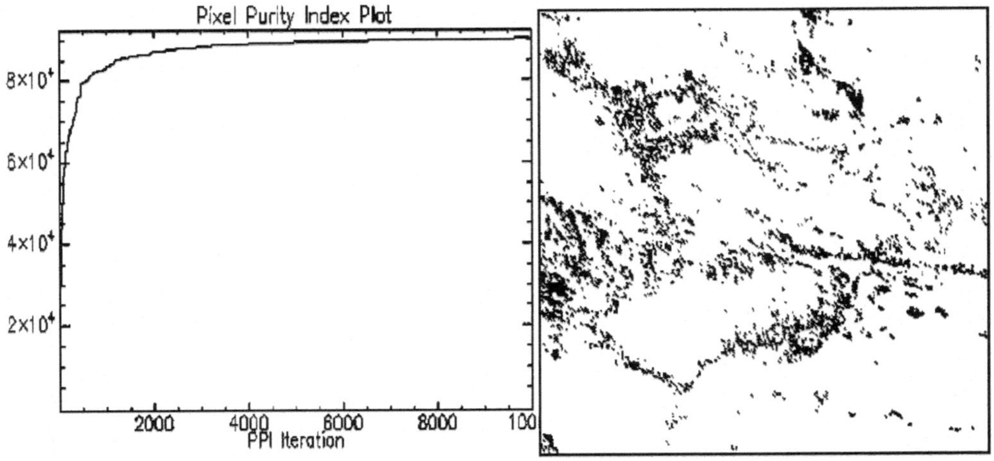

FIGURE 10–12 (A) Pixel purity index plot and (B) pure pixels in airborne visible/infrared imaging spectrometer data.

hyperspectral sensor must interact with only one material as is the case in checkerboard type scenes (Clark et al., 1993). If one of these two assumptions is not satisfied, the LUM may not be valid since various nonlinear effects may occur. To overcome with these nonlinear interactions, more complex mixing models have been proposed, mainly in spectroscopy and remote sensing. These nonlinear models have shown interesting properties to unveil meaningful information that would have been unattainable for the standard LUM (Combe et al., 2008).

10.10 Conclusion

As the grain size of snow increases, a decreasing trend can be seen in the spectral reflectance in the NIR region and higher wavelengths, except in the visible region. In the visible region, contamination decreases with decreasing reflectances. Due to increases in the amount of contamination, a shift in peaks is observed toward the higher wavelengths. There is a significant change in reflectance, which can be observed in the visible region (590–650 nm) and a minor decrease in reflectance can be observed beyond the NIR region. The SWIR region shows no change in reflectance due to the variation of snow depth. As per the literature, it can be concluded that the reflectance of snow always remains higher in the visible region, and remains low in the SWIR region, except for near waterbodies (Negi et al., 2009a,b,c).

In the snow and sparse vegetation mixed area, the reflectance pattern remains the same from the VNIR band to the SWIR band with a drop in reflectance as compared to the reflectance of snow. In the case of coal mixing, as the proportion of coal increased, the reflectance of the visible region was reduced and reflectance decreased in the SWIR region. In case of snow-coal-carbon mix, the rate of drop of reflectance in visible channel is low as compared to coal contaminated snow. This suggests, it may be possible to detect the type and amount of mixing in snow pixels, if hyperspectral data are available. It will be helpful in SU problems in snowbound areas to improve the accuracy of snow cover monitoring.

This study has shown the possibility of the detection of the type of metamorphism using absorption peak behavior at 1025 nm and percentage change in reflectance. Dust and coal contamination are major sources of contamination that influence snow and glaciated regions through avalanches and atmospheric phenomenon. The dust concentration in the snow cover revealed the sporadic high concentrations frequently in spring and large year-to-year variations in the amount deposited from winter to spring.

The modeling of impurities in snow is difficult, but it can be predicted with the help of time series data from several seasons, and registering the frequency and level of local phenomena (Swamy and Brivo, 1996). Thus impurities cannot be modeled, but can be taken into account by checking for increased grain size in visible wavelengths. One of the steepest gradients in the snow spectrum is around 1400 nm. The reflectance is reduced from about 70% to 10% for a grain size of 50 μm, and from about 20% to 1% for a grain size of 1000 μm. This spectral feature is called the NIR edge.

Spectal mixture analysis technique using remotely sensed data was developed more than two decades back but this technique for effective spectral unmixing still remains unexploited. Regardless of the available spatial resolution, the spectral signals collected in natural environments are invariably a mixture of the signatures of the various materials found within the spatial extent of the ground instantaneous field view of the remote sensing imaging instrument.

Further research is required to analyze hyperspectral images along with the ground spectra for analyses of moisture content variation and its effect on reflectance, estimation of snow depth, snow density, and standard deviation of different spectral signatures for sensitive bands of snow cover. The estimation of snow properties such as fractional snow-covered area, snow albedo (affected by grain size), snow wetness, and absorbing impurities (dust and red algae) is also required to be studied from imaging spectrometry. Analyses of the spectral signatures of ambient objects like deep clear water, soil, coniferous trees, wet and dry grass, manmade features, spectral mixing of different objects, and the radiances of different objects with respect to snow may also be carried out in future.

List of abbreviations

AIS airborne imaging spectrometer
AOT aerosol optical thickness
ATCOR Atmospheric and Topographic Corrections
EEA endmember extraction algorithm
FLAASH fast line-of-sight atmospheric analysis of hypercubes
GI grain index
LMM linear mixing model
LUM linear unmixing model
MNF minimum noise function

NDSI normalized differential snow index
NIR near-infrared
PCA principal component analysis
PPI pixel purity index
PSF point spread function
SAM spectral angle mapper
SCC snow contamination concentration
SCI snow contamination index
SGI snow grain index
SU spectral unmixing
SWIR shortwave infrared region
VNIR visible and near-infrared

List of symbols

nm nanometer (10^{-9} m)
μm micrometer (10^{-6} m)
@ at the rate

References

Acito, N., Diani, M., Corsini, G., 2011. Subspace-based striping noise reduction in hyperspectral images. IEEE Trans. Geosci. Remote Sens. 49 (4), 1325–1342.

Akyurek, Z., Sorman, A.Ü., 2002. Monitoring snow-covered areas using NOAAAVHRR data in the eastern part of Turkey. Hydrol. Sci. 47, 243–252.

Babey S.K., Anger, C.D., 1993. Compact airborne spectrographic imager (CASI): a progress review. In: Proceedings of SPIE, Imaging Spectrometry of the Terrestrial Environment, Orlando, FL, vol. 1937, pp. 152–163.

Berk, A., Bernstein, L.S., Anderson, G.P., Acharya, P.K., Robertson, D.C., Chetwynd, J.H., et al., 1998. MODTRAN cloud and multiple scattering upgrades with applications to AVIRIS. Remote Sens. Environ 65, 367–375.

Boardman, J.W., Kruse, F.A., Green, R.O., 1995. Mapping target signatures via partial unmixing of AVIRIS data, Summaries of the Fifth JPL Airborne Earth Science Workshop.

Burazerović, D., Geens, B., Heylen, R., Sterckx, S., Scheunders, P., 2012. Unmixing for detection and quantification of adjacency effects. IEEE International Geoscience and Remote Sensing Symposium pp. 3090–3093.

Burke, H.-H.K., Griffin, M.K., Snow, J.W., Uphan, C.A., Richard, C.M., 2001. Examples of atmospheric characterization using hyperspectral data in the VNIR, SWIR and MWIR, In: Proceedings Volume 4381, Algorithms for Multispectral, Hyperspectral, and Ultraspectral Imagery, VII; Available from: https://doi.org/10.1117/12.437023.

Chang, S.H., Westfield, M.J., Lehmann, F., Oertel, D., Richter, R., 1993. 79-channels airborne imaging spectrometer. In: Proceedings of SPIE, Imaging Spectrometry of the Terrestrial Environment, Orlando, FL, vol. 1937, pp. 164–172.

Chen, J., Jia, X., Yang, W., Matsushita, B., 2009. Generalization of subpixel analysis for hyperspectral data with flexibility in spectral similarity measures. IEEE Trans Geosci Remote Sens 47, 2165–2171. Available from: https://doi.org/10.1109/TGRS.2008.2011432.

Chrien, T.G., Green, R.O., Eastwood, M.L., 1990. Accuracy of the spectral and radiometric laboratory calibration of the Airborne Visible/Infrared Imaging Spectrometer. In: Proceedings of SPIE, Imaging Spectroscopy of the Terrestrial Environment, Orlando, FL, vol. 1298, pp. 37–49.

Clark, R.N., Swayze, G., Gallagher, A., King, T.V., Calvin, W.M., 1993. The U.S. geological survey digital spectral library: version 1: 0.2–3.0 μm, U.S. Geological Survey, Washington, DC, Open File Rep. pp. 93–592.

Combe, J.P., Le Mouélic, S., Sotin, C., Gendrin, A., Mustard, J.F., Le Deit, L., 2008. Analysis of OMEGA/Mars express data hyperspectral data using a multiple endmember linear spectral unmixing model (MELSUM): methodology and first results. Planet Space Sci. 56 (7), 951–975. Available from: https://doi.org/10.1016/j.pss.2007.12.007.

Dobigeon, N., Tits, L., Somers, B., Altmann, Y., Coppin, P., 2014. A comparison of nonlinear mixing models for vegetated areas using simulated and real hyperspectral data. IEEE J. SelTopAppl Earth Obs. Remote. Sens. 7 (6), 1869–1878.

Dozier, J., 1989. Spectral signature of alpine snow cover from the Landsat Thematic Mapper. Remote Sens. Environ. 28, 9–22.

Dozier, J., Davis, R.E., Perla, R., 1987. On the objective analysis of snow microstructure. In: Salm, B., Gubler, H. (Eds.), Avalanche Formation, Movement and Effects. IAHS Publication No. 162, Wallingford, UK, pp. 49–59.

Goetz, A.F.H., 2009. Remote sensing of environmental three decades of hyperspectral remote sensing of the earth: a personal view. Remote Sens. Environ. 113, S5–S16. Available from: https://doi.org/10.1016/j.rse.2007.12.014.

Govender, M., Chetty, K., Bulcock, H., 2007. A review of hyperspectral remote sensing and its application in vegetation and water resource studies. Water SA 33 (2).

Harsanyi, J.C., Chang, C.I., 1994. Hyperspectral image classification and dimensionality reduction: an orthogonal subspace projection approach. IEEE Trans. Geosci. Remote Sens. 32 (4), 779–785.

Hall, D.K., Riggs, G.A., Salomonsan, V.V., 1995. Development of methods for mapping global snow cover using moderate resolution imaging spectroradiometer data. Remote Sens. Environ. 54, 127–140.

Itten, K.I., Dell Endice, F., Hueni, A., Kneubühler, M., Schläpfer, D., Odermatt, D., et al., 2008. APEX—the hyperspectral ESA airborne prism experiment. Sensors 8 (10), 6235–6259.

Jain, S.K., Goswami, A., Saraf, A.K., 2008. Accuracy assessment of MODIS, NOAA and IRS data in snow cover mapping under Himalayan conditions. Int. J. Remote Sens. 29, 5863–5878.

Jehle, M., Hueni, A., Damm, A., Odorico, P.D., Weyermann, J., Kneubiihler, M., et al., 2010. APEX-current status, performance and validation concept. Sensors 2010, 533–537.

Kaufman, Y.J., Joseph, J.H., 1982. Determination of surface albedo and aerosol extinction characteristics from satellite imagery. J. Geophys. Res. 87, 1287–1299.

Kruse, F.A., Boardman, J.W., Huntington, J.F., 2003. Comparison of airborne hyperspectral data and EO-1 hyperion for mineral mapping. IEEE Trans. Geosci. Remote Sens. 41 (6), 1388–1400.

Liang, S., Fang, H., Chen, M., 2001. Atmospheric correction of Landsat ETM + land surfaceimagery—Part 1: Methods. IEEE Trans. Geosci. Remote Sens. 39, 2490–2498.

Minumora, M., Kuze, H., Takeuch, N., 2001. Adjacency effect in the atmospheric correction of satellite remote sensing data: evaluation of the influence of aerosol extinction profiles. Opt. Rev. 8, 133–141.

Mozaffar, M.H., Zoej, M.J.V., Sahebi, M.R., Rezaei, Y., 2008. Vegetation endmember extraction in Hyperion images. Int. Arch. Photogramm. Remote Sens. Spat. Sci. 37, 409–412.

Negi, H.S., Kulkarni, A.V., Singh, S.K., 2006. Effect of contamination and mixed objects on snow reflectance using spectroradiometer in Beas Basin, India. In: Asia-Pacific Remote Sensing, Symposium. International Society for Optics and Photonics.

Negi, H.S., Kulkarni, A.V., Semwal, B.S., 2009a. Estimation of snow cover distribution in Beas basin, Indian Himalaya using satellite data and ground measurements. J. Earth Syst. Sci. 118 (5), 525.

Negi, H.S., Kulkarni, A.V., Semwal, B.S., 2009b. Study of contaminated and mixed objects snow reflectance in Indian Himalaya using spectroradiometer. Int. J. Remote Sens. 30 (2), 315–325.

Negi, H.S., Singh, S.K., Kulkarni, A.V., 2009c. Spectral reflectance measurements in snowbound areas of Indian Himalayas. Hyperspectral Remote Sens. Spectr. Signat. Appl. NIPA 109–120.

Negi, H.S., Singh, S.K., Kulkarni, A.V., Semwal, B.S., 2010. Field-based spectral reflectance measurements of seasonal snow cover in the Indian Himalaya. Int J Remote Sens 31, 2393–2417.

Painter, T.H., Duval, B., Thomas, W.H., Mendez, M., Heintzelman, S., Dozier, J., 2001. Detection and quantification of snow algae with an airborne imaging spectrometer. Appl. Environ. Microbiol. 67 (11), 5267–5272. Available from: https://doi.org/10.1128/AEM.67.11.5267-5272.2001.

Parente, M., Clark, J.T., Brown, A.J., Bishop, J.L., 2010. End-to-end simulation and analytical model of remote-sensing systems: application to CRISM. IEEE Trans. Geosci. Remote Sens. 48 (11), 3877–3888.

Pepe, M., Brivio, P.A., Rampini, A., Rota Nodari, F., Boschetti, M., 2005. Snow cover monitoring in Alpine regions using ENVISAT optical data. Int. J. Remote Sens. 26, 4661–4667.

Santer, R., Schmechtig, C., 2000. Adjacency effects on water surfaces: primary scattering approximation and sensitivity study. Appl. Opt. 39, 361–375.

Singer, R.B., Mccord, T.B., Clark, R.N., 1979. Mars surface composition from reflectance spectroscopy: a summary. J. Geophys. Res 84 (B14).

Singh, S.K., Kulkarni, A.V., Chaudhary, B.S., 2011. Spectral characterization of soil and coal contamination on snow reflectance using hyperspectral analysis. J. Earth Syst. Sci. 120 (2), 321–328.

Singh, S.K., Chaudhary, B.S., Kulkarni, A.V., 2010. Hyperspectral analysis of snow reflectance to understand the effects of contamination and grain size. Ann. Glaciol.

Singh, D., Sharma, V., Juyal, V., 2015. Observed linear trend in few surface weather elements over the northwest Himalayas (NWH) during winter season. J. Earth Syst. Sci. 124, 553–565.

Swamy, A.N., Brivo, P.A., 1996. Hydrological modelling of snowmelt in the Italian Alps using visible and infrared remote sensing, Intern. J. Remote Sens 16, 3169–3188.

Tanre, D., Deschamps, P.Y., Duhaut, P., Herman, M., 1987. Adjacency effect producedby the atmospheric scattering in thematic mapper data. J. Geophys. Res. 92, 12000–12003.

Van Mol, B., Park, Y., Ruddick, K., Nechad, B., 2004. Mapping of chlorophyll and suspended particulate matter maps from CHRIS imagery of the Oostende coresite. In: Lacoste, H. (Ed.), Proceedings of the 2nd CHRIS/Proba Workshop, 28–30 April 2004. ESRIN, Frascati, Italy, pp. 1–8. (Noordwijk, The Netherlands: ESA Special Publications).

Vane, G., 1987. First results from the airborne visible/infrared imaging spectrometer (AVIRIS). In: Proceedings of SPIE, Imaging Spectroscopy II, San Diego, CA, vol. 0834, pp. 166–175.

Vane, G., Goetz, A.F.H., 1988. Terrestrial imaging spectroscopy. Remote Sens. Environ. 24 (1), 1–29.

Vermote, E.F., Tanré, D., Deuzé, J.L., Herman, M., Morcette, J.J., 1997. Second simulation of the satellite signal in the solar spectrum, 6s: an overview. IEEE Trans Geosci Remote Sens 35, 675–686. Available from: https://doi.org/10.1109/36.581987.

Warren, S.G., 1982. Optical properties of snow. Rev Geophys Space Phys 20 (1), 67–89. Available from: https://doi.org/10.1029/RG020i001p00067.

Warren, S.G., Wiscombe, W.J., 1980. A model for the spectral albedo of snow-II, snow containing atmospheric aerosols. J. Atmos. Sci. 37, 2734–2745.

Zhu, J., Zhou, L., Zhang, D., 2011. Identification for building surface material based on hyperspectral remote sensing. In: 19th International Conference on Geoinformatics 2011. IEEE, Shanghai, pp. 1–5, Available from: https://doi.org/10.1109/Geoinformatics.2011.5980687.

11

Remote sensing of inland water quality: a hyperspectral perspective

Shard Chander, Ashwin Gujrati, Aswathy V. Krishna, Arvind Sahay, R.P. Singh

SPACE APPLICATIONS CENTRE, AHMEDABAD, INDIA

11.1 Introduction

Coastal and inland waters are among the most productive natural systems on Earth. Rivers are the primary link between land and ocean systems and serve as the primary channel for delivering significant amounts of dissolved and particulate materials from terrestrial environments to the ocean. Freshwater ecosystems like lakes, streams, and reservoirs, which are comparatively small but spatially complex, are some of the most endangered ecosystems in the world because they are particularly vulnerable to land management changes and climate variability and also pose a major challenge to satellite remote sensing (Hestir et al., 2015). The majority of the human population lives within 60 km of the coast, which poses a threat to the quality of coastal waters and estuaries. The maintenance of the quality of coastal and inland waters is essential for any country as they support the day-to-day activities of people and economic growth activities such as fishing, transport, agriculture, industry, and recreation. Industrial development and the increase in the human population have further affected inland ecosystems and water quality. The monitoring of water quality is, thus, essential in order to characterize waters and identify changes or trends in water quality over time or to be able to respond to emerging water quality problems such as the identification of sediment plumes, harmful algae blooms, and red tides. One of the worst scenarios faced by these ecosystems is eutrophication. Eutrophication is the state of having a high nutrient content and high organic production (Wetzel, 1983). It diminishes water quality by promoting the excessive growth of algae, cyanobacteria (blue-green algae), and macrophytes, resulting in conditions called blooms. The monitoring of algal blooms in inland and oceanic waterbodies plays a significant role, especially when bloom-forming species have an associated toxin production leaving a given ecosystem unsuitable for other organisms to thrive there. These toxins can also affect humans through the consumption of fish and shellfish from toxigenic bloom zones and also by making the water unfit for recreational activities.

Remote sensing is the science of measuring the properties of objects by measuring the amount of radiation they absorb, emit, or reflect at various wavelengths along the

Hyperspectral Remote Sensing. DOI: https://doi.org/10.1016/B978-0-08-102894-0.00017-6

electromagnetic spectrum (Campbell, 2006). Earth observation using remote sensing, particularly for the determination of the quality, quantity, and geographic distribution of resources such as water, is considered as a time and cost-effective way to undertake large-scale monitoring (Okin et al., 2001). Using near real-time satellite data in the operational mode, it is equally possible to derive the current environmental situation at both the local and regional scales. The ability to accurately monitor and map nutrient loading over large expanses of surface water and identify early warning signs of algal blooms could provide a predictive planning and impact analysis tool for water resource managers (Karaska et al., 2004). Remote sensing data over water measures the radiance that originates from sunlight and that passes through the atmosphere, is reflected, absorbed, and scattered by constituents in waterbodies, and is then transmitted back through the atmosphere to the sensor. The outgoing radiance from the water surface is the contribution of both radiances from the water column and from the bottom substrate (in the case of shallow waters).

The optically active constituents (OACs) in a water column include chlorophyll-a (Chl-a), a variety of planktonic species, dissolved organics, suspended sediments with local and distant sources, bottom composition and substrates, and submerged aquatic vegetation along with water depth. These organic and inorganic water constituents are mainly responsible for water color and are also called color producing agents. The assessment of water color is utilized to estimate water quality parameters through the application of aquatic optics (Mishra et al. (2017)). Anclré and Prieur (1977) suggested a bipartite division to all natural waters as either case I or case II. For instance, in open ocean waters (termed case I waters), the primary OACs are pure water itself and phytoplankton (and associated byproducts). The light attenuation is almost entirely related to variations in the size, abundance, pigmentation, and composition of phytoplankton (IOCCG, 2000). It is, thus, relatively straightforward to develop remote sensing models for case 1, that is, open ocean waters.

However, inland and coastal waters belong to the optically complex case II waters, wherein the optical properties depend not only on the phytoplankton, but also on sediments (organic or mineral) and dissolved organic matter (humic, yellow, or gelbstoff substances) (IOCCG, 2000), making it challenging to monitor water quality through remote sensing. Inland waters are also characterized by relatively high phytoplankton, suspended sediment concentrations, and yellow substances as compared to open oceanic environments. Due to their more complex nature, inland waters have more demanding requirements for water constituent retrieval algorithms (IOCCG, 2000). The spectral characteristics of waterbodies and their color are determined by a combination of factors including the radiation incident on the water surface, the radiation scattered and reflected within the water itself, water surface roughness, air—water interface, in-water optical properties, and the angle of observation and illumination, with bottom reflectance an additional consideration in shallow waters.

This chapter is conceived as a way of sharing the actual optical dataset over Indian inland waters either collected through field radiometer or acquired by airborne visible/infrared imaging spectrometer new generation (AVIRIS-NG) instruments during the first phase of a campaign over India (December 2015—March 2016), and interpreting the spectral signature of different types of water. Details of hyperspectral remote sensors and their datasets are provided in Section 11.2. The methods of in situ radiometry data collection, water sample collection, and their analysis in the laboratory, along with satellite/airborne data processing, that is, atmospheric correction and empirical/semianalytical methods, are discussed in Section 11.3. The Section 11.4 of this chapter is dedicated to our experience in understanding the spectralsignatures in terms of various water quality parameters. Finally, brief conclusions along with future direction are given in Section 11.5.

11.2 Hyperspectral remote sensing

Since the early 1960s, multispectral imagery has been used as a data source for water and land observational remote sensing from airborne and satellite systems (Landgrebe, 1999). The primary limiting factor of multispectral sensor systems is that they commonly collect data in three to six spectral bands in the visible and near-infrared regions of the electromagnetic spectrum. This limitation of Earth observation systems has been overcome with the development of hyperspectral sensor technologies. Hyperspectral sensors have made it possible to collect several hundred spectral bands in a single acquisition, producing much more detailed spectral information and can be used in complement to in situ freshwater ecosystem sampling (Hestir et al., 2015). Simultaneous and accurate detection of water quality conditions and parameters, which includes the determination of the trophic status of lakes, characterizing algal blooms, the assessment of ammonia dynamics for wetland treatments, and the estimation and mapping of the concentrations of dissolved organic matter, chlorophyll, or total suspended matter (TSM), are the major advantages of hyperspectral remote sensing technologies (Gower et al., 2007).

Hyperspectral remote sensing imagers acquire many, narrow, contiguous spectral bands throughout the visible, near-infrared, and thermal infrared portions of the electromagnetic spectrum. These sensors typically collect 200 or more bands, which enables the construction of an almost contiguous reflectance spectrum for every pixel in a given scene. This allows for in depth examination of Earth's surface features, which would otherwise be lost within the relatively coarse bandwidths acquired with multispectral sensors (Gower et al., 2007). Hyperspectral images are generated from airborne sensors like NASA's AVIRIS-NG, which delivers 224/425 contiguous channels with wavelengths from 380 to 2500 nm with a pixel size of 4–8 m depending upon the altitude of the sensor (4–8 km). NASA's EO-1 satellite has a hyperspectral instrument named Hyperion that produces 30 m resolution images in 220 spectral bands with wavelengths from 400 to 2500 nm (Schurmer, 2003). The Hyperspectral Imager for the Coastal Ocean (HICO) on the International Space Station was the first space-based, maritime hyperspectral imaging instrument designed specifically to measure water quality parameters over coastal regions. The 90 m spatial resolution with 5 nm spectral bandwidth of the HICO instrument offers a unique capability for characterizing a wide range of water color parameters that can provide a detailed understanding of estuarine and near-coastal environmental conditions (Lucke et al., 2011). Future hyperspectral sensors that are planned to be launched in the near future include PRISMA, EnMAP

HyperSpectral Imager, HISUI, Spaceborne Hyperspectral Applicative Land and Ocean Mission, Hyperspectral Infrared Imager, and Hyperspectral X IMagery (Transon et al., 2018).

The three main levels of processing of any hyperspectral spaceborne/airborne instrument are level 0 (raw data), level 1 (calibrated radiances at-sensor in μW/nm/sr/cm^2), and level 2 (atmospherically-corrected surface reflectances). Although there are a number of software packages available to read and understand hyperspectral data, the ENVI software is the most convenient tool to view and analyze these datasets. For time series analysis or to test analytical codes, researchers should utilize programming languages such as the Matlab ENVI Toolbox or Spectral Python with its libraries.

11.3 Methodology: field and satellite measurements

11.3.1 In situ hyperspectral radiometry

Various researchers have investigated spectral characteristics of inland water quality parameters like Chl-a, cyanobacterial pigment C-phycocyanin, coloured dissolved organic matter (CDOM), suspended particulate matter (SPM) using hyperspectral instruments (Analytical Spectral Devices (ASD) Field Spec FR spectrometer (ASD Inc.) and SVC HR-1024 spectrometer). The ASD spectroradiometer collects radiance (L, μW/nm/sr/cm^2) in the spectral range of 350–2500 nm (Field Spec 3 User Manual, 2006). The acquisition geometry followed by Mobley (1999) is generally used to avoid shadow, sunlight, and sky glint contamination. The protocol followed for measuring above-water radiance and remote sensing reflectance is grouped into two steps. First step is measurement of downward incoming solar radiance on white panel, outgoing surface radiance above water and diffused sky radiance measured pointing opposite to sun direction. Second step is to measure incoming solar irradiance from cosine reflector. To generate above-water remote sensing reflectance, 2% of the diffused sky radiance is subtracted from the outgoing surface radiance above water and its ratio from incoming solar irradiance is taken.

11.3.2 Water sample laboratory analysis

11.3.2.1 Chlorophyll-a

The traditional method for determining Chl-a concentration includes field sample collection and on-spot filtration through 0.7 μm Whatman glass fiber filter (GF/F) paper. The filter paper is preserved and transported carefully in liquid nitrogen to the laboratory for further analysis. In vitro laboratory analyses involve the extraction of chlorophyll pigments from the phytoplankton samples in the filter paper into a solvent, mostly 90% acetone. The optical signals (absorbance) of the Chl-a in the water samples are then measured at four wavelengths, namely 750, 663, 645, and 630 nm using a spectrophotometer (Arar, 1997a), a fluorometer (Arar and Collins, 1997), or high-performance liquid chromatography (HPLC) (Arar, 1997b). Chl-a concentration is then estimated using the formula provided by Jeffrey and Humphrey

(1975). Among all the methods for Chl-a determination, HPLC serves as the most accurate, yet the most time-consuming method (Pinckney et al., 1994).

11.3.2.2 Colored dissolved organic matter

CDOM measurement involves the collection of water samples into acid-washed, precombusted amber colored borosilicate glass bottles using Niskin water samplers (Mannino et al. (2008)). The bottles are rinsed thrice with the sample before the final collection and stored at 4°C. Samples are then transferred to the laboratory and filtered through 47 mm cellulose acetate filters of pore size 0.2 μm under low vacuum to remove particulate and suspended sediment materials from the water samples. Before filtration, the filtration unit should be thoroughly cleaned with distilled water and further rinsed with 10% HCl followed by numerous rinses with distilled water. The filtered samples are allowed to reach the ambient room temperature to avoid temperature difference between samples and the blank. *The absorbance of the filtrate is measured in 10 cm cuvettes on a UV–vis spectrophotometer (400–700) against blank Milli-Q water.*

11.3.2.3 Total suspended matter

TSM can be measured following the protocol described by Tilstone et al. (2002) based on the work of Van der Linde (1998), which suggests filtering a known volume of water through 47 mm Whatman GF/F with a nominal pore size of 0.7 μm. The filters are to be preashed at 450°C for 1 hour, then gently washed in 0.5 L of Milli-Q water to remove those fractions that can be dislodged during filtration, dried at 75°C for 1 hour, preweighed on an analytical balance with an accuracy of 0.1 mg, stored in a desiccator for use within 2 weeks, and finally transferred to clean 50 mm diameter petriplates for transport. Water samples ought to be immediately filtered after collection using a vacuum of 300–400 mmHg, and stored at −20°C until further analysis. For analysis, filters are dried for 24 hours at 75°C and reweighed on the same analytical balance. The gain in weight is a dry weight measure of the particulates present in the water sample expressed in units of milligrams per liter, that is, the ratio of gain in weight to the volume of water filtered.

11.3.3 Atmospheric correction of inland waters

Atmospheric correction of satellite measurements over inland water is challenging due to many factors. Some of the major challenges include terrestrial atmospheric pollution, high turbidity, floating objects, and the adjacency effect of neighboring land pixels, which significantly raise the reflectance of water. Several approaches have been developed to atmospherically correct remotely sensed data from turbid productive waters. The shortwave infrared (SWIR) black-pixel assumption is the most commonly used algorithm due to the advantage that no other assumptions have to be made regarding the optical properties of water. Aquatic atmospheric correction algorithms assume any SWIR reflectance to be from atmospheric aerosols and subtract it from the total radiance spectra (on top of atmospheric radiance) obtained by satellites during the aerosol removal step. This assumption is based on the fact that even shallow water absorbs longer wavelength light and, thus, should be dark in the SWIR bands (Wang and Shi, 2007). The basic approach of SWIR-based methods is described here. First, the spectral ratio of Rayleigh-corrected reflectance at two SWIR bands is used to derive the aerosol model or aerosol spectral shape. Once the aerosol model is determined, the aerosol optical depth (AOD) is derived from the aerosol reflectance at a single SWIR band using a look up table containing aerosol reflectance as a function of AOD. Finally, the atmospheric correction parameters for all visible and near-infrared (NIR) bands are derived from the retrieved AOD and aerosol model through a simple exponential extrapolation. Some of the most commonly used atmospheric correction algorithms include fast line of site atmospheric analysis of spectral hypercubes (Anderson et al., 2002), and 6SV (a version of second simulation of the satellite signal in the solar spectrum) (Vermote et al., 2006). Vanhellemont (2019) has published a new atmospheric correction method dedicated for aquatic applications.

11.3.4 Retrieval of water quality parameters

Ever since the launch of the coastal zone color scanner, the mapping and quantitative estimation of surface biogeophysical constituents such as Chl-a, SPM, and CDOM through the remote sensing of ocean color has gone through a quiet revolution. Extensive studies have been carried out to relate explicitly radiance/reflectance with in-water constituents over the past three decades (Anclré and Prieur, 1977; Gordon et al., 1983; Sathyendranath et al., 1989; Lee et al., 1999). For case II waters, the use of ratio/bio-optical algorithms to estimate water constituents from remotely sensed data is often difficult owing to its optically complex nature. In addition, the bottom reflectance often contributes significantly to observed remote sensing reflectance in shallow bathymetric regions (IOCCG, 2011).

11.3.4.1 Empirical relations

Empirical and semiempirical models are usually based on statistical relationships between in situ measurements of water constituents and radiometric data from satellite sensors. *The formulae used in empirical models are based on a combination of remote sensing reflectance (R_{rs}) at different wavelengths, which will provide the best correlation between reflectance data and the concentration of optically active water constituents. Sem*iempirical models, on the contrary, are based on specific spectral features of absorption and scattering of the constituents governing reflectance. These types of models typically use statistical techniques such as neural networks, least squares, and stepwise regressions to extract the best relationship between R_{rs} and constituent concentrations. *For example, in case I open ocean waters, the OC3 algorithm relates ratios of the maximum of the two blue bands (443 or 490 nm) and green bands (560 nm) to Chl-a with a fourth-order polynomial relationship (O'Reilly et al., 1998).* The red and NIR bands are used for estimating Chl-a over turbid inland waters (Le et al., 2011). Two-band (Gitelson, 1992) and three-band (Gitelson et al., 2007) algorithms have been widely applied to predict Chl-a concentrations in inland waters.

11.3.4.2 Semianalytical solutions

Semianalytical models can be used for describing inland water quality parameters, that is, the optical properties of a given waterbody as a function of all the constituents present within the waterbody. These models use apparent optical properties (AOP) such as reflectance above or below a water surface (R_{rs} or r_{rs}, respectively) to derive inherent optical properties (IOP), namely the absorption and backscattering properties of water. Actually, semianalytical models are a simplified solution to the radiative transfer (RT) equations computed through several analytical and empirical steps. RT theory describes the wavelength-dependent changes in the direction and intensity of electromagnetic radiation. Gordon et al. (1983) provided an early discussion of the analytical approaches for measuring concentrations of water constituents in ocean and coastal waters using remotely sensed data. Semianalytical models for shallow waters have been developed by many researchers (Clark et al., 1987; Lee et al., 1998, 2002) that show that water column properties can be analytically derived from hyperspectral data using an optimization technique. For shallow inland water, with the assumption of vertical homogeneity, the remote sensing reflectance (sr^{-1}) can be written as:

$$R_{rs} = f(a(\lambda), b_b(\lambda), \rho, H, T) \tag{11.1}$$

where a is the absorption coefficient (m^{-1}), b_b is the backscattering coefficient (m^{-1}), ρ is the bottom albedo, H is the bottom depth, and T is a factor that is dependent on the geometry of the incoming/outgoing radiance, that is, subsurface solar zenith angle (θ_w) and subsurface viewing angle (θ_v).

This model divides the total R_{rs} into two contributions, one due to optically deep waters and the other contribution from a finite bottom depth. In place of assuming that the diffuse attenuation coefficient is the same for upward and downward directions, here the different optical path elongation factor is considered for photons from the water column and photons from bottom. *This above water R_{rs} can be converted into subsurface remote sensing reflectance (r_{rs}) using water to air divergence factor and the factor that account for the internal reflection of the water–air interface (Lee et al., 1998).*

$$r_{rs} = r_{rs}^{dp}\left[1 - \exp\left(-\left(\frac{1}{\cos\theta_w} + \frac{1.03(1+2.4u)^{0.5}}{\cos\theta_v}\right)\gamma H\right)\right] \tag{11.2}$$

$$+\frac{1}{\pi}\rho\exp\left(-\left(\frac{1}{\cos\theta_w} + \frac{1.04(1+5.4u)^{0.5}}{\cos\theta_v}\right)\gamma H\right) \tag{11.3}$$

where u, γ, and r_{rs}^{dp} (deep-water remote sensing reflectance) are defined as:

$$u = \frac{b_b}{a + b_b} \tag{11.4}$$

$$\gamma = a + b_b \tag{11.5}$$

$$r_{rs}^{dp} = (0.084 + 0.171u)u \tag{11.6}$$

The total absorption of water constituents was considered as the sum of the components' absorption coefficients, that is, natural water, phytoplankton, and gelbstoff/CDOM. Absorption due to pure water was assumed as constant (Pope and Fry, 1997). The spectral shapes of the CDOM absorption were modeled using exponential law (Bricaud et al., 1981). Backscattering due to pure water was taken from the work of Anclré and Prieur (1977). The backscattering due to particles was estimated using the power law given by Lee et al. (1998). Three types of bottom albedo spectral shapes (normalized at 550 nm) were taken as inputs in the forward simulations of r_{rs}, that is, sand, small vegetation, and heavy vegetation. Fig. 11–1, shows one such example of (A) absorption, (B) backscattering, (C) bottom albedo, and (D) simulated r_{rs} for varying chlorophyll concentrations (1–5 mg/m^3). In this case, all other water optical constituents were assumed to vary with chlorophyll concentration.

For inverting this semianalytical model, any nonlinear least square error minimization scheme can be utilized that starts with the initial guess values. This method uses Levenberg–Marquardt optimization for finding the best possible match. The lower and upper bounds of the unknowns, that is, absorption due to phytoplankton, absorption due to CDOM, and backscattering due to particles were kept in the range of 0–2 m^{-1} based on their range in the natural waters. Maximum iteration along with the tolerance limit was quantified to avoid endless looping. It is an error minimization technique that keeps predicting new values unless the result converges within the given uncertainty.

11.3.4.3 Software

Commercial/freeware software packages are also available to do the forward simulations of remote sensing reflectance data for given water optical constituents. HYDROLIGHT (Mobley,

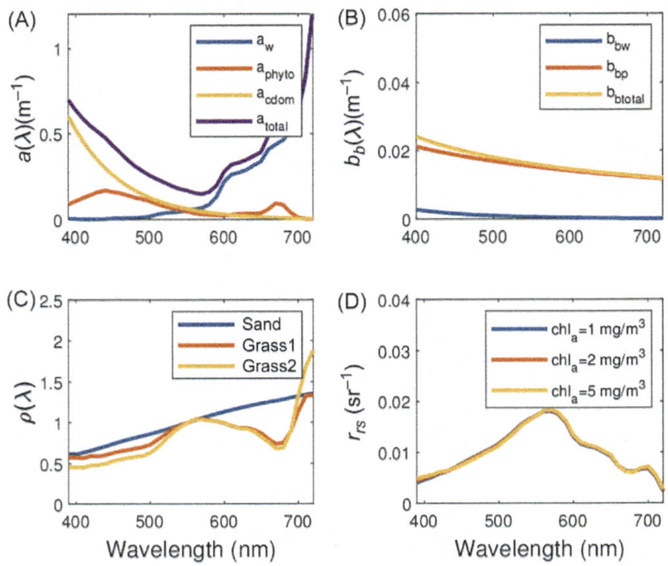

FIGURE 11–1 (A) Absorption, (B) backscattering, (C) bottom albedo, and (D) simulated remote sensing reflectance assuming all other optically active constituents vary as a function of chlorophyll concentration (1–5 mg/m^3).

1998) is one of the most commonly used RT numerical models that computes radiance distribution and the derived quantities of natural waterbodies. It is designed to compute spectral radiance distribution as a function of depth, direction, and wavelength within water using the absorption and scattering properties of a given waterbody. It includes the effects of inelastic scattering by chlorophyll fluorescence, by CDOM fluorescence, and by Raman scattering by the water itself.

Peter (2014) developed an analytic model (WASI 2D) for the direct and diffuse components of the downwelling irradiance in a given water column. *It is a Microsoft Windows-based software tool that has been developed for forward and inverse calculations of major types, namely upwelling radiance above the surface, downwelling irradiance above the surface, remote sensing reflectance above the surface, specular reflectance at the surface, irradiance reflectance below the surface, and bulk absorption of waterbodies.*

Giardino et al. (2012) developed a software package incorporating their bio-optical model-based tool for estimating water quality and bottom properties from remote sensing images (BOMBER). BOMBER is a software package for the simultaneous retrieval of the optical properties of a given water column and bottom from remotely sensed imagery, which makes use of bio-optical models for optically deep and optically shallow waters. This optimization technique allows the user to retrieve maps of Chl-a concentration, SPM concentration, CDOM absorption and, in the case of shallow waters, bottom depth and substrate distributions.

Semianalytical model for bathymetry, unmixing, and concentration assessment is an inversion/optimization approach proposed by Lee et al. (1998, 1999) enhanced to retrieve the concentrations of OACs in the water column (Chl-a, CDOM, and NAP), to account for more than one substratum cover type and to estimate the contribution of substratum to the remote sensing signal (Brando et al., 2009).

11.4 Interpretation of the spectral signatures

Understanding and interpretation of a remotely sensed image requires an understanding of spectral signatures. Spectral signatures are images of reflected solar energy that help in distinguishing materials based on their wavelength-dependent absorptions and scattering. The graph of the spectral reflectance of an object as a function of wavelength is known as the spectral reflectance curve (Lillesand and Kiefer, 1999). Contiguous spectral signatures obtained using hyperspectral imaging allow for detailed analysis through the detection of surface materials and their abundances as well as the inferences of biological and chemical processes (Gower et al., 2007). Chl-a, CDOM, TSM, cyanobacterial bloom, and aquatic macrophytes are primary factors that describe inland water quality and that can be detected through remote sensing. The following five subsections describe the determination of these parameters in various inland waterbodies of India through hyperspectral remote sensing (an ISRO/NASA joint AVIRIS-NG campaign during December 2015–March 2016).

11.4.1 Chlorophyll-a: the fundamental measure of phytoplankton biomass

To understand the eutrophic properties of a waterbody, an important parameter is the phytoplankton biomass and a good indicator of the phytoplankton biomass is Chl-a concentration (Moses et al., 2009). Phytoplankton absorbs light energy from the Sun and produce chemical energy in the form of rich organic materials through photosynthesis. Chlorophyll present in the phytoplankton cell causes two dominant peaks in absorption spectra of hyperspectral images in the visible spectral region, primarily at 440 nm (blue) and secondarily at 675 nm (red). The additional pigments causes broadening of the blue peak and the appearance of additional absorption maxima. The scattering properties of phytoplankton have a direct relationship with remote sensing reflectance. The coefficients of scattering and backscattering are derived from theoretical models or through direct measurements (Bricaud et al., 1983). Differences between waterbodies exist due to variations in the Chl-a contents of the phytoplankton assemblages. This is influenced by both external and internal factors. Factors such as light, nutrients, cell volume and cell size, stratification, and surface temperature influence the Chl- a content in algal cells. The accurate determination of its concentration has great significance regarding the assessment and prediction of water eutrophication conditions.

One such example of Chl- a variation over a freshwater lake (Udaisagar, Udaipur, India) is shown in Fig. 11–2. This image was acquired by AVIRIS-NG in February 2016 and has a spatial resolution of 4 m. Three representative spectra are shown for interpreting the spectral shapes. Udaisagar Lake is different from other lakes due to the fact that it exhibits three varying R_{rs} spectra (marked as S1, S2, and S3 in Fig. 11–2B) from its different zones, which represent low, moderate, and high reflectances.

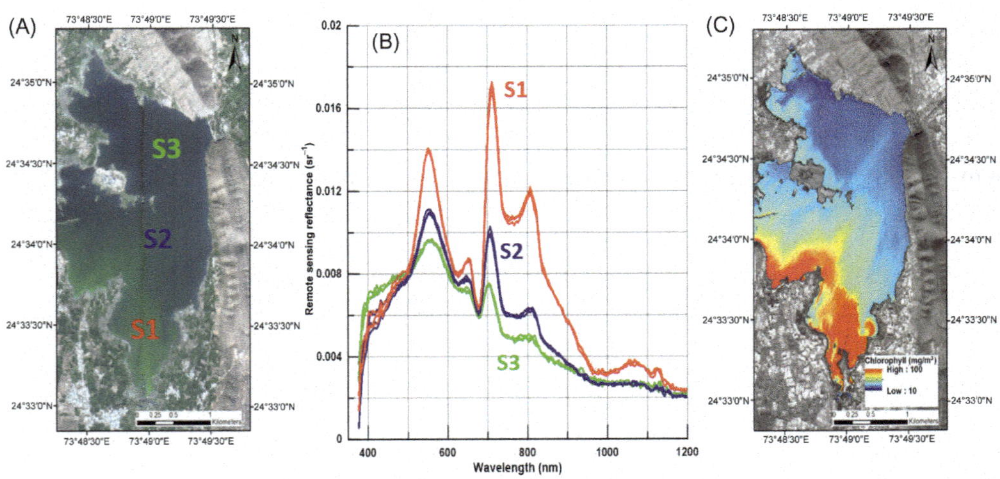

FIGURE 11–2 Airborne visible/infrared imaging spectrometer new generation (A) true color composite image over Udaisagar Lake, (B) three distinct reference spectra at different stations (S1, S2, and S3), and (C) chlorophyll variability map over the region.

Both Chl-a and turbidity were observed to be high in the southern regions (S3) of the lake, which gradually decreased toward the northern region (S1). This trend is exactly represented by the spectra, where the maximum R_{rs} was observed at all wavelengths in the whole spectra along the southern regions of the lake and minimum R_{rs} values at the northern regions. *Throughout the lake, the spectral shape was observed to be similar a major peak at 550 nm, followed by a minor peak at 650 nm and a second major peak at 705 nm and a second minor peak at 810 nm.* The first major and minor peaks are of Chl-a absorption and the second major and minor peaks represent chlorophyll fluorescence. The second set of peaks are generally found in inland waters, where high chlorophyll produces a lot of fluorescence signals. The second major peak shows a gradual shift toward higher wavelengths with increases in peak height. Studies on chlorophyll fluorescence and scattering have proved that an increase in the backscatter ratio increases the amplitude of the chlorophyll fluorescence peak and shifts it slightly toward a longer wavelength (Gower et al., 1999).

11.4.2 Colored dissolved organic matter

Most of the carbon in inland waters is present in dissolved form or dissolved organic carbon (DOC) (Wetzel, 2001). CDOM is the optically active part of DOC and is used as a proxy. Many studies have shown significant correlation between the total amount of DOC and CDOM (Tranvik, 1990). CDOM, also called gelbstoff or yellow substance, is a mixture of organic molecules resulting from the breakdown of higher plants, algae, and bacteria. It absorbs light in the UV and visible parts of the spectrum (exponential decay) and makes water yellow or brown. CDOM alone is difficult to determine by remote sensing since detrital matters present in water also have absorption in the UV and visible regions. Hence it is difficult to separate the signals by remote sensing. CDM represents the combined absorption by CDOM and detrital matter. CDM is an important indicator in water quality parameter estimation. In the following section, CDM variability has been discussed in relation to Chilika Lagoon, Odisha, India using hyperspectral data.

Remote sensing reflectance variation over Chilika Lagoon is shown in Fig. 11–3B. This image was acquired by AVIRIS-NG during December 26–28, 2015 with a spatial resolution of 8 m. Over Chilika Lagoon, a variety of spectra were observed ranging from clear productive water to extreme turbid water marked as S4, S5, and S6 in Fig. 11–3B. In the north sector the spectra, which covered typical turbid water, a peak at 570 nm was observed, while in the south and central sectors, a reflectance peak was observed at 555 nm.

As part of the concurrent ground truth collection with AVIRIS-NG hyperspectral campaign, water samples were collected synchronous with the time of flight. Simultaneous spectroradiometer measurements were also taken over Chilika Lagoon during December 24–28, 2015, and 48 stations were covered. A map of the CDM absorption is shown in Fig. 11–4A and its contiguous spectra is shown in Fig. 11–4B. The variability in CDM absorption spectra at 412 nm shows that in the north sector of Chilika Lagoon, CDM absorption is quite high compared to in the other sectors (5.5 m^{-1} with a standard deviation of 0.06 m^{-1}). In the southern sector and at the outer channel, it is 1.8 m^{-1} with a standard deviation of 0.02 m^{-1}, while in

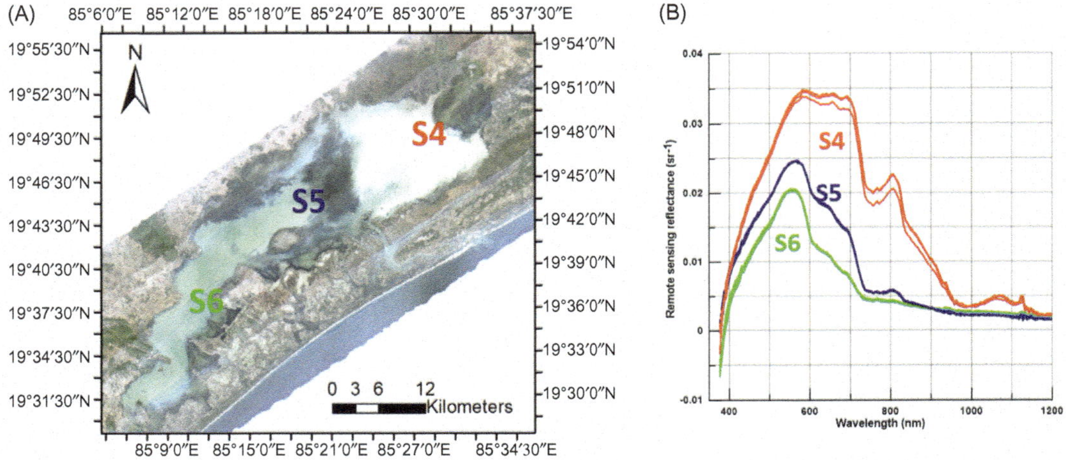

FIGURE 11–3 Airborne visible/infrared imaging spectrometer new generation (A) true color image over Chilika Lagoon and (B) three distinct reference spectra at different stations (S4, S5, and S6) showing the variability of signature over the north, center, and south sections of Chilika Lagoon, respectively.

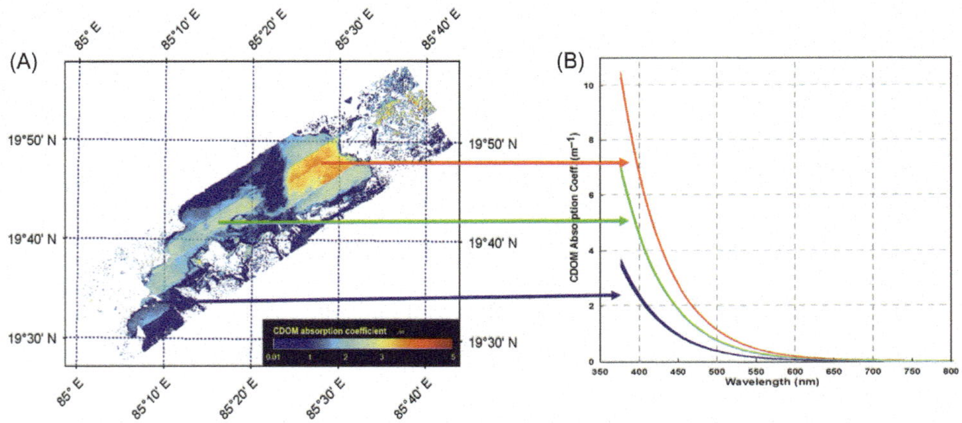

FIGURE 11–4 (A) Map of CDOM absorption coefficient and (B) variability in the slope of CDOM from airborne visible/infrared imaging spectrometer new generation over Chilika Lagoon.

the middle sector it is 3.76 m^{-1} with a standard deviation of 0.22 m^{-1}. At the northern end, tributaries of the Mahanadi River such as Daya, Nuna, and Bhargavi join the lagoon and are responsible for the large freshwater and terrestrial flux into the lagoon. These terrestrial inputs bring a lot of degraded and suspended organic matter into the northern part of the lagoon, which in turn increases the CDM in this region. More detail is provided by Sahay et al. (2019). The advantage of hyperspectral data is that it gives CDM absorption contiguous in the range of 375–425 nm, where the absorption by CDM is strong and separate from Chl-a absorption.

11.4.3 Sediment laden and clear river water

TSM and turbidity being the most visible indicators of water quality, come from soil erosion, run-off, discharges, stirred bottom sediments, or algal blooms. TSM refers to the total quantity measurement of solid material per volume of water (Tassan, 1993). TSM is a specific measurement of all suspended solids, organic or inorganic, by mass. In terms of water quality, high levels of TSM can increase water temperature and decrease dissolved oxygen by absorbing more heat from solar radiation. This additionally causes stratification of a waterbody, which can turn lower layer waters hypoxic due to respiratory and decomposition activities. During the AVIRIS-NG campaign over India, datasets were acquired over the Ganga River at Buxar (February 23, 2016) and Howrah (March 3, 2016) with spatial resolutions of 4–6 m respectively. The representative spectra are marked as S7 and S8 in Fig. 11–5C and D along with the AVIRIS-NG acquired natural color composite image as shown in Fig. 11–5A and B.

Over the Buxar region, the reflectance spectra (S7) show a clear first reflectance peak around 570 nm and another peak around 706 nm. The curves show less diversity when the wavelength is larger than 730 nm. The reflectance peak near 570 nm may be caused by the low absorption of algal pigments and/or the scattering of inorganic suspended materials and phytoplankton cells. The other reflectance peak near 706 nm may be due to the fluorescence

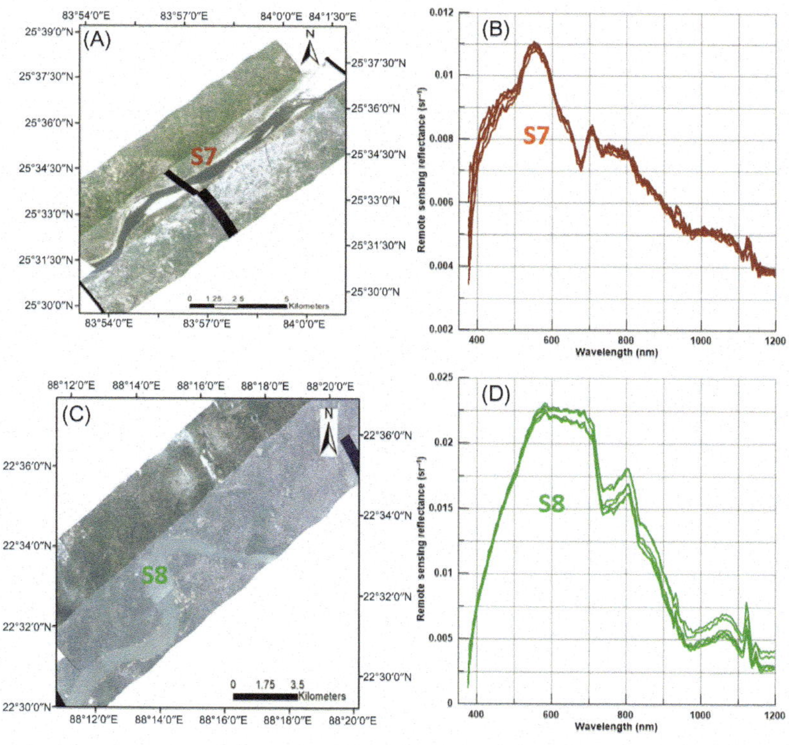

FIGURE 11–5 Airborne visible/infrared imaging spectrometer new generation true color composite image over Ganga River at Buxar (A) and Howrah (B), and their respective reference spectra marked with stations (C) S7 and (D) S8.

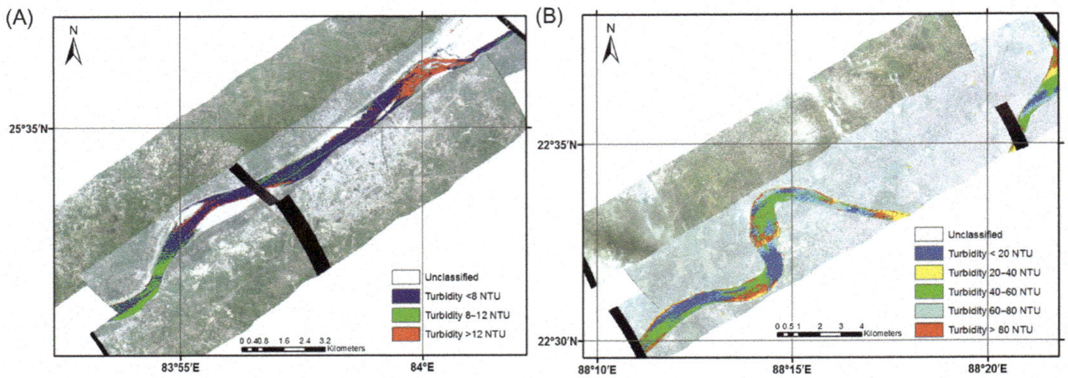

FIGURE 11–6 Turbidity variations over Ganga River at (A) Buxar and (B) Howrah.

of Chl-a. On the other hand, at the Howrah site (Hooghly River, distributary of Ganges River in West Bengal) the reflectance spectra (S8) were found to be highly turbid (beige water) compared to the Ganga River flowing through Buxar. The R_{rs} between 600 and 700 nm was quite high or was nearly double the value obtained for Buxar. The maximum R_{rs} was measured at 580 nm and a second peak was observed at 800 nm, that might be representative of a higher concentration of suspended sediments.

Field measurements were also carried out simultaneously to the AVIRIS-NG flight on February 23, 2016 and February 24–28, 2016 over Buxar and Howrah respectively. However due to bad weather conditions, the AVIRIS-NG imaging was carried out over the Howrah region on March 3, 2016. Over the Ganga river, the water was relatively clear in Buxar (6.87–20 NTU), while it was extremely turbid in Howrah (50–175 NTU). The total suspended sediments were found to be ranging between 42–154 mg/L over Buxar and 75–450 mg/L in Howrah. Based on the spectral matching between in situ and AVIRIS-NG measurements, turbidity maps were prepared for Buxar and Howah (Fig. 11–6A and B, respectively). In Buxar, the spectra were divided into three classes of <8, 8–12, and >12 NTU based on the frequency distribution. Similarly, in Howrah, the spectra were divided into five classes, namely <20, 20–40, 40–60, 60–80, and >80 NTU. More detailed analysis is presented by Chander et al. (2019). Between the two sites over the Ganga River, the Buxar site showed lesser turbidity in comparison to the Howrah site. A possible reason could be the tidal influence of the Bay of Bengal. The passage of ships and ferryboats also keep the sediments in suspension, which further increases the turbidity in this stretch.

11.4.4 Cyanobacterial bloom

Cyanobacteria constitute a microscopic (1–80 μm), highly diverse prokaryotic group that evolved approximately 3.5 billion years ago (Schopf, 2000). They inhabit a wide variety of environments occupying varying climatic zones, which include the most extreme ones like hot springs and deserts as well as cold areas in the Antarctic. Also known as blue-green algae, cyanobacteria are one of the most common types of harmful algal blooms (HABs) in freshwater, saltwater, and brackish environments. Eutrophication in waterbodies and global warming are causes of increased cyanobacterial bloom potentials. Cyanobacteria can produce cyanotoxins (e.g., microcystins and saxitoxins), which pose a threat to human and animal health and interfere with water treatments. Freshwater algal blooms are strongly correlated with phosphorus loading as compared to marine blooms with nitrogen loading. Traditional water sampling techniques are limited for quantifying cyanobacterial blooms due to the nature of cyanobacteria (Kutser, 2004). In contrast to the conventional sampling methods that assume algal cells to be uniformly distributed in a water column, cyanobacterial blooms are often concentrated at specific depths in a water column and tend to form surface scums. This natural structure of cyanobacteria is easily destroyed by research vessels and, thus, does not provide accurate measurements (Kutser, 2004).

Remote sensing techniques can help to overcome these limitations of traditional approaches. These blooms are relatively easy to detect optically because of the surface concentration of cells, the presence of phycocyanin pigments, and the elevated backscatter associated with cell size and the presence of gas vacuoles. Phycocyanin is a pigment–protein complex present in freshwater cyanobacteria with a broad absorption feature at 620 nm and can be detected using a wavelength range of 615–630 nm (Ogashawara et al., 2013). Remote sensing detection methods for the identification of cyanobacterial blooms, thus, rely on algorithms targeting phycocyanin (Kutser, 2009). Another approach that uses the spectral shape, rather than the identification of specific absorption features, has the advantage of less uncertainty in terms of accuracy of atmospheric correction. Wynne et al. (2008) developed a cyanobacterial index (CI) that relies on changes in the spectral shape between 665, 681, and 709 nm caused by the strong scattering by cyanobacteria at around 709 nm.

During the AVIRIS-NG campaign over India, a dataset was acquired over Mahi River, Rajasthan, with a spatial resolution of 4 m that also covered some nearby freshwater lakes (Fig. 11–7A). The reference reflectance spectra over the waterbody marked as station S9, S10, and S11 are shown in Fig. 11–7B. Station S10 and S11 are the typical reflectance spectra of a waterbody with nearby aquatic vegetation. While station S9 shows three reflectance peaks; the first reflectance peak was observed at 555 nm, the second peak was observed at 655 nm, and the third peak was observed at 709 nm. It has been reported by Kudela (2015) that blooms dominated by Microcystis (a cyanobacterial genera) have a strong phycocyanin absorption feature, which when coupled with the Chl-a absorption feature at 680 nm, results in a more pronounced peak in reflectance at approximately 655 nm. The third peak at 709 nm results from chlorophyll fluorescence and scattering. The reflectance spectra peaks should increase in amplitude and shift slightly to longer wavelengths as the concentration increases (Gower et al., 1999). However, in the AVIRIS-NG measured spectra, a small variation in the amplitude of the fluorescence peak was observed, but the reflectance at the peak's red shoulder did not change appreciably.

The CI index algorithm was applied (Wynne et al., 2008, 2010) for the detection of bloom infected lakes using AVIRIS-NG datasets. Fig. 11–7C shows the cyanobacteria index map generated using the AVIRIS-NG dataset, that clearly shows the bloom infected waterbody in

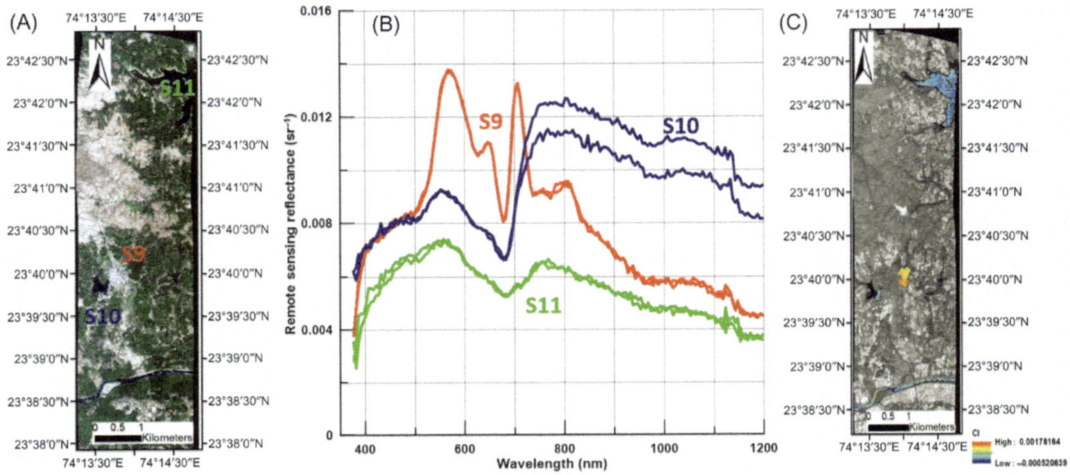

FIGURE 11–7 Airborne visible/infrared imaging spectrometer new generation (A) true color composite image over Mahi River, Rajasthan, (B) three distinct reference spectra of nearby lakes at different stations (S9, S10, and S11), and (C) cyanobacterial index map over the region.

comparison to other freshwater lakes. Similar spectral shape-based approaches are also developed by other authors as well, that is, maximum peak height (Matthews et al., 2012), adaptive reflectance peak height (Ryan et al., 2014), scattering line height, and Aphanizomenon–Microcystis index (Kudela, 2015). With hyperspectral data, it is possible to separate toxic and nontoxic cyanobacterial genera. *At some of the places where nontoxic blooms prior to toxin-producing cyanobacterial genera, can be used as an early warning system for the detection of potentially harmful blooms.*

11.4.5 Aquatic macrophytes

Aquatic macrophytes are one of the most important components of freshwater ecosystems and play a crucial role in providing food and shelter for animals as well as regulating the water chemistry (McLachlan, 1974). They control erosion, improve water quality, and provide an early warning signal for HABs (Everitt and Elder, 2010). These plants may be free-floating or rooted in the bottom sediment and submersed. However, on the other side, aquatic weed infestation is a major environmental challenge globally, threatening the integrity and functioning of most hydrological ecosystems (Everitt and Elder, 2014). This invasion of foreign species leads to a number of ecological impacts, including reduced light and oxygen levels in water columns, which in turn reduces native macrophyte diversity.

Water hyacinth and hydrilla are two exotic, aquatic macrophytes that can cause significant ecological alterations in the invaded community by modifying the natural habitat. Water hyacinth was identified as being amongst the 100 most aggressive invasive species and recognized as one of the top 10 worst weeds in the world by the International Union for Conservation of Nature (Tllez et al., 2008). Although it is native to South America, it has become an environmental and social challenge throughout tropical and subtropical regions

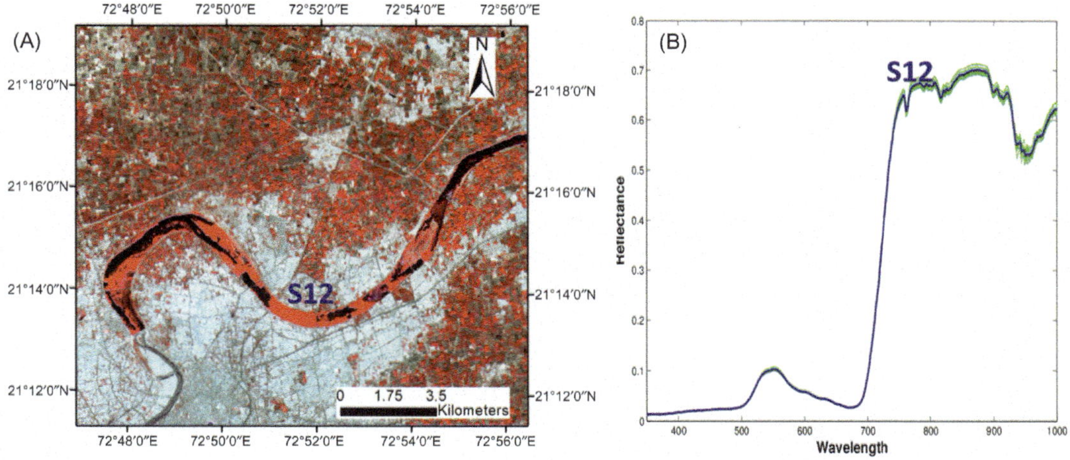

FIGURE 11–8 (A) LANDSAT false color composite image from March 22, 2018 over Tapi River, Surat and (B) the reflectance spectra at station S12 collected for water hyacinth during a field trip.

around the world. Tllez et al. (2008) mapped the worldwide distribution of water hyacinth. Its populations may double every 6–18 days. Its optimum growth occurs between 28°C and 30°C, and it requires abundant nitrogen, phosphorus, and potassium. However, these invasive species can survive in many different temperature, light, water, and soil conditions.

The Tapi River in Surat is facing issues due to the presence of water hyacinth close to a water treatment plant. They can clog reservoirs and reduce water availability for human needs. Fig. 11–8A shows the Landsat-8 false color composite image of the Tapi River, Surat on 22nd March 2018. Later during a field trip over the Tapi River on June 22, 2018, in situ spectral response datasets were acquired for water hyacinth (Fig. 11–8B). The spectral signature of hyacinth had typical characteristics with low reflectance in the visible region of the spectrum from 400 to 700 nm, but high reflectance in the NIR region from 700 to 950 nm. The internal spongy leaf structure reflects high in the NIR region from a wavelength of 700–1000 nm (Lillesand et al., 2008). The low reflectance observed for the visible region is due to the presence of high concentrations of Chl-a, which is an indicator of healthy conditions for the aquatic vegetation.

Water samples were also collected over three locations, one before entering the city (Kamrej), a second one at the Sarthana water treatment plant, and a third one at the outer end (causeway). The nutrient concentration was less before entering the city (ammoniacal nitrogen 0.056 mg/L and phosphate 0.0145 mg/L), while a higher concentration (ammoniacal nitrogen 0.448 mg/L and phosphate 0.05 mg/L) was observed in close proximity to hyacinth. A time series analysis showed that the maximum extent of these macrophytes occurs every February–March period. Less availability of stream flow along with the same concentration of nutrient load could be possible reasons of the high growth. More information on this is provided by Chander et al. (2018). Once established, these aquatic macrophytes are difficult to control and are nearly impossible to eradicate.

11.5 Conclusion

In this chapter an overview is provided for estimating the water quality parameters from remotely sensed measurements of the reflectance over inland waterbodies. This should give a better understanding of the principles involved in the estimation of water quality parameters as well as an overview of the existing methods. Although both multispectral and hyperspectral types of data can be used for the estimation of water quality parameters in waterbodies, hyperspectal measurements provide more information. Hyperspectral channels from visible to IR channels provide better estimation of atmospheric aerosols in turbid water conditions due to the presence of SWIR channels. Apart from this advantage, more spectral information on covarying and other dynamic constituents can be obtained due to the greater number of channels. This type of information is important in describing semianalytical models for RT solutions in inland waterbodies. Due to the high complexity of inland/coastal water types, remote sensing methods are not operational so far in these types of water, except for some regions with prior site-specific knowledge. Water constituent types and their concentrations, bottom type and its albedo, and further air–water interactions make it difficult to develop a generic model that can be applied to all such types of optically complex waters. A dedicated inland water library should be developed for varying concentrations of water quality parameters, their IOP ranges, and their specific inherent optical properties for different geographic regions and seasons. From the perspective of inland waters, where waterbodies are smaller in size, quantitative water quality estimation is likely to remain challenging for the future as well. There are several areas for future research that will reduce uncertainty in water quality parameter estimates in complex waters. Further, different combinations of water constituents can lead to indistinguishable reflectance spectra. These spectra require more detailed analysis by considering covarying relationship between two water quality parameters in terms of their concentration and probability of occurrence together. *AVIRIS-NG class of instrument, due to its higher signal to noise ratio, with better temporal resolution and spatial resolution close to Landsat-8 type instrument (30 m) from spaceborne platform can help in better understanding of freshwater ecosystem.* Systematic and timely information of water quality can help in a wide range of practical applications, for example, impact of eutrophication, sediment suspension/deposition, sediment transport, threats from cyanobacteria blooms for recreational uses, impact of invasive species on local ecosystems, global assessments of biogeochemistry, and the long-term impacts on climate change.

Acknowledgments

The authors would like to express their sincerest gratitude to Shri D.K. Das, Director of SAC for providing overall support. The guidance and encouragement received from Dr. Raj Kumar (Deputy Director EPSA), Dr. A.S. Rajawat (Group Director, GHCAG/EPSA), and Dr. Bimal Bhattacharya (Science Team Leader, AVIRIS-NG Airborne campaign) are greatly acknowledged. The authors also thank the officials of ISRO and NASA for the AVIRIS-NG program whose hyperspectral data was utilized. Thanks are also due to all those who were directly or indirectly associated with this work.

List of abbreviations

ACOLITE new atmospheric correction method dedicated for aquatic applications
AMI Aphanizomenon–Microcystis index
AOD aerosol optical depth
AOPs apparent optical properties
ARPH adaptive reactance peak height
ASD analytical spectral devices
AVIRIS-NG airborne visible/infrared imaging spectrometer new generation
BOMBER bottom properties from Remote sensing images
CDOM colored dissolved organic matter
Chl-a chlorophyll-a
C-PC cyanobacterial pigment C-phycocyanin
CZCS coastal zone color scanner
DOC dissolved organic carbon
FLAASH fast line of site atmospheric analysis of spectral hypercubes
GF/F Whatman glass fiber
HICO The Hyperspectral Imager for the Coastal Ocean
HPLC high-performance liquid chromatography
HypXIM Hyperspectral X IMagery
HyspIRI Hyperspectral Infrared Imager
IOPs inherent optical properties
ISS The International Space Station
IUCN The International Union for Conservation of Nature
LUT look up table
MPH maximum peak height
NIR near infrared
NTU Nephelometric Turbidity Unit (Turbidity)
RTE radiative transfer equations
SAMBUCA semianalytical model for bathymetry, un-mixing, and concentration assessment
SHALOM Spaceborne Hyperspectral Applicative Land and Ocean Mission
SIOPs specific inherent optical properties
SLH scattering line height
SPM suspended particulate matter
SSC suspended sediment concentrations
SVC Spectra Vista Corporation
SWIR shortwave infrared
TSM total suspended matter

List of symbols

6SV a version of second simulation of the satellite signal in the solar spectrum
a the absorption coefficient
b_b is the backscattering coefficient
H is the bottom depth
R_{rs} remote sensing reflectance
θv viewing angle
θw solar zenith angle
ρ is the bottom albedo

References

Anclré, M., Prieur, L., 1977. Analysis of variations in ocean color 1. Limnol. Oceanogr. 22 (4), 709–722.

Anderson, G.P., Felde, G.W., Hoke, M.L., Ratkowski, A.J., Cooley, T.W., James, H., et al., 2002. Modtran4-based atmospheric correction algorithm: FLAASH (fast line-of-sight atmospheric analysis of spectral hypercubes). Proc. SPIE Int. Soc. Opt. Eng. 4725, 65–71.

Arar, E.J., 1997a. Epa method 446.0 in vitro determination of chlorophylls a, b, c + c and pheopigments in marine and freshwater algae by visible spectrophotometry. Nat. Expos. Res. Lab. Off. Res. Dev. U.S. Environ. Prot. Agency 1.2, 1–26.

Arar, E.J., 1997b. Epa method 447.0 determination of chlorophylls a and b and identification of other pigments of interest in marine and freshwater algae using high performance liquid chromatography with visible wavelength detection. Nat. Expos. Res. Lab. Off. Res. Dev. U.S. Environ. Prot. Agency 1.2, 1–26.

Arar, E.J., Collins, G.B., 1997. Epa method 445.0 in vitro determination of chlorophyll a and pheophytin a in marine and freshwater algae by fluorescence. Nat. Expos. Res. Lab. Off. Res. Dev. U.S. Environ. Prot. Agency 1.0, 1–22.

Brando, V.E., Anstee, J.M., Wettle, M., Dekker, A.G., Phinn, S.R., Roelfsema, C., 2009. A physics based retrieval and quality assessment of bathymetry from suboptimal hyperspectral data. Remote Sens. Environ. 113, 755–770.

Bricaud, A., Morel, A., Prieur, L., 1981. Absorption by dissolved organic matter of the sea (yellow substance) in the UV and visible domains1. Limnol. Oceanogr. 26 (1), 43–53.

Bricaud, A., Morel, A., Prieur, L., 1983. Optical-efficiency factors of some phytoplankters. Limnol. Oceanogr. 28, 816–832.

Campbell, J.B., 2006. Introduction to Remote Sensing. *Taylor and Francis, London*, p. 626.

Chander, S., Pompapathi, V., Gujrati, A., et al., 2018. U.D. Growth of invasive aquatic macrophytes over Tapi river. Int. Arch. Photogramm. Remote Sens. Spat. Inf. Sci. XLII-5, 829–833.

Chander, S., Gujrati, A., Hakeem, K.A., Garg, V., Issac, A.M., Dhote, P.R., et al., 2019. Water quality of assessment of river Ganga and Chilika lagoon using AVIRIS-NG hyperspectral data. Curr. Sci. 116 (7), 1172–1181.

Clark, R.N., King, T.V.V., Gorelick, N.S., 1987. Automatic continuum analysis of reflectance spectra. In: Proceedings of Third AIS Workshop (24 June 1987), JPL Publication 8730, Jet Propulsion Laboratory, Pasadena, California. pp. 138–142.

Everitt, J.H., Elder, S.H., 2010. Evaluating airborne multispectral digital video to differentiate giant salvinia from other features in northeast texas. Remote Sens. 2 (10), 2413–2423.

Everitt, J.H., Elder, S.H., 2014. Evaluating MERIS-based aquatic vegetation mapping in lake victoria. Remote Sens. 6 (8), 7762–7782.

Giardino, C., Candiani, G., Bresciani, M., Lee, Z., Gagliano, S., Bomber, M.P., 2012. A tool for estimating water quality and bottom properties from remote sensing images. Comput. Geosci. 45, 313–318.

Gitelson, A., 1992. The peak near 700 nm on radiance spectra of algae and water: relationships of its magnitude and position with chlorophyll concentration. Int. J. Remote Sens. 13 (17), 3367–3373.

Gitelson, A.A., Schalles, J.F., Hladik, C.M., 2007. Remote chlorophyll-a retrieval in turbid, productive estuaries: Chesapeake bay case study. Remote Sens. Environ. 109 (4), 464–472.

Gordon, H., Morel, A., Barber, R., Mooers, N., Bowman, M., Zeitzschel, B., 1983. A review lecture notes on coastal and estuarine studies. Remote Assessment of Ocean Color for Interpretation of Satellite Visible Imagery. Springer-Verlag, NY.

Gower, J.F., Doerffer, R.R., Borstad, G.A., 1999. Interpretation of the 685 nm peak in water-leaving radiance spectra in terms of fluorescence, absorption and scattering, and its observation by MERIS. Int. J. Remote Sens. 20, 1771–1786.

Gower, J.F., Doerffer, R.R., Borstad, G.A., 2007. A review of hyperspectral remote sensing and its application in vegetation and water resource studies. Water 33 (2), 145–151.

Hestir, E.L., Brando, V.E., Bresciani, M., Giardino, C., Matta, E., Villa, P., et al., 2015. Measuring freshwater aquatic ecosystems: the need for hyperspectral global mapping satellite mission. Remote Sens. Environ. 167, 181–195.

IOCCG, 2000. Remote sensing of ocean colour in coastal, and other optically-complex waters. In: Sathyendranath, S. (Ed.), Reports of the International Ocean Colour Coordinating Group, No. 3, IOCCG, Dartmouth, Canada.

IOCCG, 2011. Bio-optical sensors on argo floats. In: Reports of the International Ocean-Colour Coordinating Group. p. 89.

Jeffrey, S.W., Humphrey, G.F., 1975. New spectrophotometric equations for determining chlorophylls a, b, c and c2 in higher plants, algae and natural phytoplankton. Biochemie und Physiologie der Pflanz. 167, 191–194.

Karaska, M.A., Huguenin, R.L., Beacham, J.L., Wang, M.-H., Jensen, J.R., Kaufmann, R.S., 2004. Aviris measurements of chlorophyll, suspended minerals, dissolved organic carbon, and turbidity in the neuse river, North Carolina. Photogramm. Eng. Remote Sens. 70 (1), 125–133.

Kudela, R.M., 2015. Application of hyperspectral remote sensing to cyanobacterial blooms in inland waters. Remote Sens. Environ. 167, 196–205.

Kutser, T., 2004. Quantitative detection of chlorophyll in cyanobacterial blooms by satellite remote sensing. Limnol. Oceanogr. 49, 2179–2189.

Kutser, T., 2009. Passive optical remote sensing of cyanobacteria and other intense phytoplankton blooms in coastal and inland waters. Int. J. Remote Sens. 30 (17), 4401–4425.

Landgrebe, D., 1999. On information extraction principles for hyperspectral data. Cybernetics 1, 1–17.

Le, C.F., Li, Y.M., Zha, Y., Sun, D.Y., Huang, C.C., Zhang, H., 2011. Remote estimation of chlorophyll-a in optically complex waters based on optical classification. Remote Sens. Environ. 115 (2), 725737.

Lee, Z., Carder, K.L., Mobley, C.D., Steward, R.G., Patch, J.S., 1998. Hyperspectral remote sensing for shallow waters: I. A semianalytical model. Appl. Opt. 37 (27), 6329.

Lee, Z., Carder, K.L., Mobley, C.D., Steward, R.G., Patch, J.S., 1999. Hyperspectral remote sensing for shallow waters: 2. deriving bottom depths and water properties by optimization. Appl. Opt. 38 (18), 3831–3843.

Lee, Z.P., Carder, K.L., Arnone, R.A., 2002. Deriving inherent optical properties from water color: a multiband quasi-analytical algorithm for optically deep waters. Appl. Opt. 41 (27), 5755–5772.

Lillesand, T.M., Kiefer, R.W., 1999. Remote Sensing and Image Interpretation. *Wiley & Sons, Inc, New Jersey.*

Lillesand, T.M., Kiefer, R.W., Chipman, J.W., 2008. Remote Sensing and Image Interpretation, sixth ed. *John Wiley and Sons, New York, NY.*

Lucke, R.L., Corson, M., McGlothlin, N.R., Butcher, S.D., Wood, D.L., Korwan, D.R., et al., 2011. Hyperspectral imager for the coastal ocean: instrument description and first images. Appl. Opt. 50, 1501–1516.

Mannino, A., Russ, M.E., Hooker, S.B., 2008. Algorithm development and validation for satellite-derived distributions of DOC and CDOM in the U.S. middle Atlantic bight. J. Geophys. Res., 113(C07051), 1–19.

Matthews, M.W., Bernard, S., Robertson, L., 2012. A new algorithm for detecting trophic status (chlorophyll-a), cyanobacterial-dominanance, surface scums and floating vegetation in coastal and inland waters from meris. Remote Sens. Environ. 124, 637652.

McLachlan, A.J., 1974. Recovery of mud substrate and its associated fauna following a dry phase in a tropical lake. Limnol. Oceanogr. Banner 19 (1), 74–83.

Mobley, C.D., 1998. Hydrolight 4.0 Users Guide. Technical Report. Sequoia Scientific Inc., Mercer Island, WA.

Mobley, C.D., 1999. Estimation of the remote-sensing reflectance from above-surface measurements. Appl. Opt. 38, 7442–7455.

Moses, W.J., Gitelson, A.A., Berdnikov, S., Povazhnyy, V., 2009. Estimation of chlorophyll-a concentration in case ii waters using MODIS and MERIS data-successes and challenges. Environ. Res. Lett. 4 (4), 045005.

Ogashawara, I., Misra, D.R., Mishra, S., Curtarelli, M.P., Stech, J.L., 2013. A performance review of reflectance based algorithms for predicting phycocyanin concentrations in inland waters. Remote Sens. 5, 4774–4798.

Okin, G.S., Roberts, D.A., Murray, B., Okin, W.J., 2001. Practial limits on hyperspectral vegetation discrimination in arid and semi-arid environments. Remote Sens. Environ. 77, 212–225.

O'Reilly, J.E., Maritorena, S., Mitchell, B.G., Siegel, D.A., Carder, K.L., et al., 1998. Ocean color chlorophyll algorithms for seawifs. J. Geophys. Res.: Ocean. 103 (C11), 24937–24953.

Pinckney, J., Papa, R., Zingmark, R., 1994. Comparison of high-performance liquid chromatographic, spectrophotometric, and fluorometric methods for determining chlorophyll a concentrations in estuarine sediments. J. Microbiol. Methods 19 (1), 59–66.

Peter, G., 2014. Wasi-2d: a software tool for regionally optimized analysis of imaging spectrometer data from deep and shallow waters. Comput. Geosci. 62, 208–215.

Pope, R.M., Fry, E.S., 1997. Absorption spectrum (380–700 nm) of pure water. II. integrating cavity measurements. Appl. Opt. 36, 8710–8723.

Ryan, J.P., Davis, C.O., Tufillaro, N.B., Kudela, R.M., Gao, B.C., 2014. Application of the hyperspectral imager for the coastal ocean to phytoplankton ecology studies in Monterey Bay, CA, USA. Remote Sens. 6, 10071025.

Sahay, A., Gupta, A., Motwani, G., Raman, M., Ali, S.M., Shah, M., et al., 2019. Distribution of coloured dissolved and detrital organic matter in optically complex waters of Chilika Lagoon, Odhisha, India, using hyperspectral data of AVIRIS-NG. Curr. Sci. 116 (7), 1166–1171.

Sathyendranath, S., Prieur, L., Morel, A., 1989. A three-component model of ocean colour and its application to remote sensing of phytoplankton pigments in coastal waters. Int. J. Remote Sens. 10 (8), 1373–1394.

Schopf, J.W., 2000. The fossil records; tracing the roots of the cyanobacterial lineage. In: Whitton, B.A., Potts, M. (Eds.), Ecology of Cyanobacteria. pp. 13–35.

Tassan, S., 1993. An improved in-water algorithm for the determination of chlorophyll and suspended sediment concentration from thematic mapper data in coastal waters. Int. J. Remote Sens. 14, 12211229.

Tilstone, G.H., Moore, G.F., Srensen, K., Rttgers, R., Jrgensen, P.V., Vicente, V.M., et al. Revamp regional validation of MERIS chlorophyll products in north sea coastal waters: protocols document. In: REVAMP Methodologies EVG1CT2001-00049, 22, 2002.

Tllez, T., Lopez, E., Granado, G., Prez, E., Lopez, R., Guzmon, J., 2008. The water hyacinth, *Eichhornia crassipes*: an invasive plant in the guadiana river basin spain. Aquat. Invasions 3 (1), 42–53.

Transon, J., Andrimont, R., Maugnard, A., Defourny, P., 2018. Survey of hyperspectral earth observation applications from space in the sentinel-2 context. Remote Sens. 10, 157.

Tranvik, L.J., 1990. Bacterioplankton growth on fractions of dissolved organic carbon of different molecular weights from humic and clear waters. Appl. Environ. Microb. 56, 1672–1677.

Van der Linde, D.W., 1998. Protocol for the determination of total suspended matter in oceans and coastal zones. Jt. Res. Centre, Ispra, Tech. Note I.98, 182.

Vanhellemont, Q., 2019. Adaptation of the dark spectrum fitting atmospheric correction for aquatic applications of the Landsat and Sentinel-2 archives quinten. Remote Sens. Environ. 225, 175–192.

Vermote, E.F., Tanr, D., Deuz, J.L., Herman, M., Morcrette, J.J., Kotchenova, S.Y., 2006. Second Simulation of a Satellite Signal in the Solar Spectrum-Vector (6SV). University of Maryland, College Park, MD.

Wang, M., Shi, W., 2007. The NIR–SWIR combined atmospheric correction approach for modis ocean color data processing. Opt. Express 15 (15), 722–733.

Wetzel, R.G., 1983. Limnology. Saunders College Publishing, Philadelphia, PA.

Wetzel, R.G., 2001. Limnology, Lake and River Ecosystems. third ed.

Wynne, T.T., Stumpf, R.P., Tomlinson, M.C., Warner, R.A., Tester, P.A., Dyble, J., et al., 2008. Relating spectral shape to cyanobacterial blooms in the laurentian great lakes. Int. J. Remote Sens. 29 (12), 3665–3672.

Wynne, T.T., Stumpf, R.P., Tomlinson, M.C., Dyble, J., 2010. Characterizing a cyanobacterial bloom in western lake erie using satellite imagery and meteorological data. Limnol. Oceanogr. 55, 20252036.

12

Efficacy of hyperspectral data for monitoring and assessment of wetland ecosystem

L.K. Sharma[1], Rajashree Naik[1], Prem Chandra Pandey[2]

[1]DEPARTMENT OF ENVIRONMENTAL SCIENCE, SCHOOL OF EARTH SCIENCES, CENTRAL UNIVERSITY OF RAJASTHAN, AJMER, INDIA

[2]CENTER FOR ENVIRONMENTAL SCIENCES & ENGINEERING, SCHOOL OF NATURAL SCIENCES, SHIV NADAR UNIVERSITY, GREATER NOIDA, INDIA

12.1 Introduction

Remote sensing uses the concept of observing the Earth through a bird's eye view, vertically from the top, which started with aerial photography in 1858 using air balloons (Lillesand et al., 1980). This technology was taken forward in the 1970s with the use of aircrafts and then satellites, which have the capacity of scaling down the globe into a synoptic domain, and this unfolded new opportunities of research (Jong, 2007). The Landsat series are pioneer satellite programs launched by the National Aeronautics and Space Administration (NASA) in 1972. These programs opened-up the scope for spatiotemporal analysis. Subsequently, many countries today have their own space agencies, for instance, India has the Indian Space Research Organisation, Japan has the Japan Aerospace Exploration Agency, and the United Kingdom has the United Kingdom Space Agency. Today there are satellites that have high spatial resolution (e.g., GeoEye, SPOT), which are used for urban planning and disaster mitigation, and high temporal resolution (e.g., the moderate resolution imaging spectroradiometer), which are used for local weather and climate studies. Many projects have been conducted worldwide like global land ice measurements from space (Raup et al., 2007); global lake database (Birket and Mason, 1995); global land cover mapping (Congalton et al., 2014); and global rainforest mapping (Rosenqvist et al., 2010). However, there were more questions to be answered, more problems to be solved, and more advancements required in space technology. Understanding Earth through high spatial and temporal datasets requires additional dimensions to be incorporated. This led toward high spectral information. The quest to understand nature in depth spectrally led to hyperspectral remote sensing (HRS). This technology is also known as imaging spectroscopy, which provides imaging spectral information. It is being widely exploited for identification, assessment, mapping, and classification in several research domains such as geology,

Hyperspectral Remote Sensing. DOI: https://doi.org/10.1016/B978-0-08-102894-0.00007-3

limnology, hydrology, pedology, ecology, and wetland ecosystems. The use of hyperspectral datasets is facilitating a deeper understanding of Earth ecosystems, but it is also facing numerous challenges, which are discussed in this chapter.

12.1.1 Wetland ecosystem and its services

An ecosystem is composed of both living organisms known as biotic components and nonliving constituents known as abiotic components. Living organisms consist of flora and fauna interacting among themselves and abiotic components like energy, gaseous and mineral elements, and water. Each type of ecosystem has its own unique structure and function that is related to its floral and faunal composition, flow of energy, and cycling of nutrients. There are different types of ecosystems existing in the nature. Ecosystems are classified broadly into two classes, namely natural and artificial ecosystems. Natural ecosystems are further divided into terrestrial and aquatic ecosystems. Aquatic ecosystems are further classified based on salinity (fresh and saline) and the movement of water (lentic and lotic). Wetlands are like an umbrella that could consist of any type of aquatic ecosystem.

Wetlands are among the most highly productive ecosystems. They exist in wide range of biomes. Wetlands exhibit diversity depending on their process of formation; geographical location; altitude, biological, physical, and chemical constituents and their nature of solubility; climatic factors; nature of hydrological cycle and prevailing water level; sources of water and discharge channels (Lin et al., 2019). These factors lead to diversity in the composition of flora and fauna, which ultimately influences the nutrient cycle and flow of energy. Under the text of the Ramsar Convention (Article 1.1), wetlands are defined as “areas of marsh, fen, peatland or water, whether natural or artificial, permanent or temporary, with water that is static or flowing, fresh, brackish or salt, including areas of marine water the depth of which at low tide does not exceed six meters (RSIS, 2019).”

Wetlands are a giant house of resources. Some of them have been explored and exploited by humans and some are still to be discovered. The services provided through the functioning of diverse ecosystems are called ecosystem services. This has been acknowledged (Lin et al., 2019). The details of ecosystem services are listed in Table 12−1 indicating our dependence for direct use (salt extraction, hydropower generation, rain water collection, bird watching, scientific research, etc.) as well as indirect use (groundwater recharge, wildlife support, external ecosystem support, microclimatic stabilization, nutrient recycle) of ecosystem services.

12.1.2 Role of Ramsar Convention in wetland ecosystems

Observing the destruction of wetlands and the habitat of birds, ornithologists stepped forward toward the conservation of wetlands and, hence, the Ramsar Convention was established in 1971. 18 countries, namely Belgium, Denmark, Finland, France, the Federal Republic of Germany, India, Iran, Ireland, Jordan, the Netherlands, Pakistan, South Africa, Spain, Sweden, Switzerland, Turkey, the Union of Soviet Socialist Republics, and the United Kingdom joined together with intergovernmental agencies like the United Nations Educational, Scientific and Cultural Organization, the Food

Table 12–1 Ecosystem services of wetlands.

S. no.	Ecosystem service	Examples
1	Provisioning	Food, fresh water, fiber, fuel, biochemical and medicines, genetic fuels, salt, energy, ornamental resources
2	Regulating	Climatic regulation, water regulation, water purification, erosion regulation, natural hazard regulation, groundwater recharge, waste water treatment, preventing desertification due to vegetation availability, pollination, maintenance of soil fertility
3	Supporting	Soil formation, nutrient cycling, habitats for species, genetic reservoirs, nutrient dispersal seed dispersal, primary production
4	Cultural	Spiritual, inspirational, recreational, esthetic, educational, and scientific research

and Agricultural Organization, and The United Nations and nongovernmental organizations like World Wildlife Fund and International Union for Conservation of Nature and Natural Resources to establish this convention.

Through these efforts, wetlands have been categorized and classified according to various classification systems. One of the major types of classification is according to habitat type, identifying their importance according to Ramsar International Wetland Classification System given in 1993. 3 major classes were formed, namely marine and coastal wetlands with 12 subclasses, inland wetlands with 14 subclasses, and manmade wetlands with 9 subclasses. To designate these wetlands as Ramsar sites, two broad groups of Criteria A and Criteria B were charted, which were further subdivided into nine criteria based on the uniqueness of the geographical unit, birds, fish, and habitats of endangered flora or fauna.

12.1.3 Global status of wetland ecosystem

Currently there are 2370 Ramsar sites in the world. The Cobourg Peninsula of Australia was the first designated Ramsar site, declared on May 8, 1974. A decadal analysis from 1974 to 2019 states that 291 sites were declared between 1974 and 1984, 477 sites between 1985 and 1995, 874 sites between 1996 and 2006 interval, and 700 between 2007 and 2019. Sundarbans in India is the most newly declared site (January 30, 2019). The geographical distribution of wetlands is found in diverse conditions. Europe ranks first for having the highest number of Ramsar sites (1102), Africa ranks second (396), Asia ranks third (339), North America ranks fourth (218), Latin America and the Caribbean rank fifth (204), and Oceania with the least number of sites (around 82) ranks sixth (RSIS, 2019).

12.1.4 Indian status of wetland ecosystems

Wetlands are classified into various groups based on origin, location, flooding pattern, geographic location, and the ecosystem services that they provide. A widely accepted classification system entitled "Classification of wetlands and deep water habitats" classifies wetlands into five major groups, namely marine, estuarine, riverine, lacustrine, and palustrine for wetland inventory (Cowardin et al., 1979). As the name suggests, marine wetlands are wetlands located near the ocean, estuarine wetlands are placed near estuaries, riverine wetlands originate near rivers, lacustrine wetlands are associated with lakes, and palustrine wetlands belong to any inland wetland that lacks flowing water. Estuarine and marine wetlands have two subclasses, namely intertidal and subtidal. Riverine wetlands are classified into four subclasses, namely tidal, lower perennial, upper perennial, and intermittent. Lacustrine wetlands are classified into two subclasses, namely littoral and limnetic. Palustrine wetlands have no subclasses.

India has a total of 27 (10,568.71 km^2) Ramsar sites including inland, coastal, and man-made wetlands as shown in Fig. 12−1 with their site numbers. These wetlands are of three types, which are mainly inland (3407.19 km^2), marine or coastal (9029.90 km^2), and manmade (2645.74 km^2) (RSIS, 2019). Chilika Lake of Odisha and Keoladeo National Park of Rajasthan is the first declared Ramsar site of India designated on October 1, 1981. Sundarbans wetland was declared on January 30, 2019.

India has 10 biogeographic zones, which are Trans Himalayan zone, Himalayan zone, Desert zone, Semiarid zone, Western Ghats zone, Deccan plateau zone, Gangetic plain zone, Northeast zone, and Coastal zone. The states in these regions overlap, that is, partly occupy multiple zones at the same time. Like Gujarat consists of a coastal zone and an arid zone, and Western Ghats, Maharastra, belongs to the Western Ghats zone, the Deccan plateau zone, and also has coastal zones. The Indian Himalayan region covers 10 Indian states, which are Jammu and Kashmir, Himachal Pradesh, Uttarakhand, Sikkim, Arunachal Pradesh, Meghalaya, Nagaland, Manipur, Mizoram, Tripura, Assam, and West Bengal (ENVIS, assessed on September 27, 2019). Among these, six states have Ramsar sites. Jammu and Kashmir has four sites (Hokera, Surinsar-Mansar, Tsomoriri, and Wular), Himachal Pradesh has three sites (Chandertal, Pong, and Renuka), Manipur has one site (Loktak), Tripura has one site (Rudrasagar), Assam has one site (Deepor beel), and West Bengal has two sites (East Calcutta wetlands and Sundarbans). In short, there are 12 Ramsar sites in the Indian Himalayan region. Collectively, in the desert and semiarid zones of India, there are six Ramsar sites. There are Harike, Kanjli, and Ropar of Punjab, Keoladeo Ghana and Sambhar Salt Lake of Rajasthan, and Nalsarovar of Gujarat. In the states of Western Ghats, Kerala has two sites, namely Asthamudi and Sasthankotta, and Tamil Nadu has the Point Calimere wetland. In the Deccan plateau region, Andhra Pradesh has one site known as Kolleru and Madhya Pradesh has one site known as the Bhoj wetland. In the coastal zones, Odisha has two Ramsar sites, namely Bhitarkanika Mangroves and Chilika.

It is clear that India is blessed with wetlands in all the biogeographic zones ranging from high-altitude regions, fresh water and saline wetlands, coastal-like backwater, lagoons, estuaries, coral reefs, mangroves, swamps, mud flats, and inland wetlands like lakes, ponds, creeks along with wetlands in the flood plain areas (Prasad et al., 2002). All these wetlands exhibit diverse ranges of ecological processes and functions and provide a wide range of services.

India derives many economic benefits from these sites like salt mining, industrial raw materials, commercial products, pharmaceutical products, cosmetic products, biofuels, oil

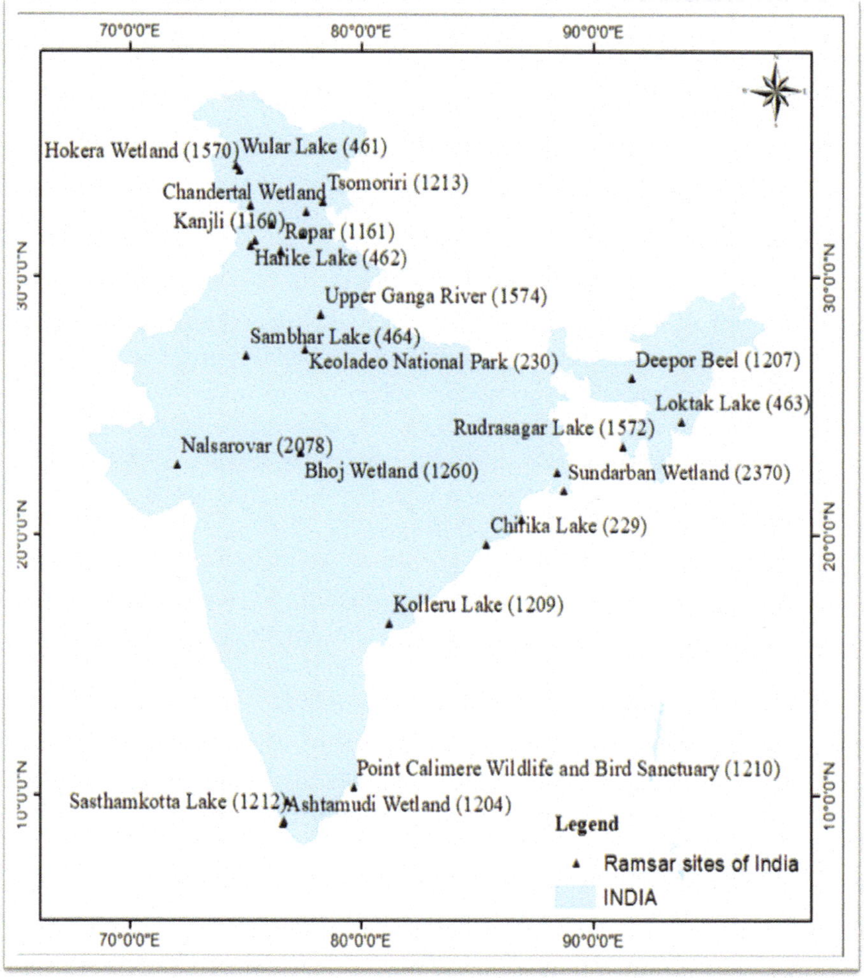

FIGURE 12–1 Map of Indian Ramsar sites.

and energy, renewable energy like solar and wind energy, and livelihoods for local peoples. Tourism sites and bird watching are the most popular use of wetlands. These benefits may not be widespread, but they play a vital role in the Indian economy. Ecologically, these wetlands are critical habitats for winter migratory birds, they recharge groundwater sources, control floods, and regulate microclimates. The states devoid of Ramsar sites include Maharastra, Karnataka, Goa, Bihar, Jharkhand, Telengana, Uttarakhand, Sikkim, Arunachal Pradesh, Meghalaya, Nagaland, and Mizoram. These states may have potential Ramsar sites that have yet to be unveiled.

12.2 Monitoring and assessment of wetlands with multispectral remote sensing

The journey of observing the Earth from above started with the use of aerial photography. With the application of aerial imagery, wetlands could be delineated (Stewart et al., 1980), mapped (Lovvorn and Kirkpatrick, 1982), and the assessment of vegetation growth and water quality (Martyn et al., 1986) became possible, but these data were site specific and also lacked spectral resolutions for characterizing complex ecosystems like wetlands (Mishra, 2014).

The availability of multispectral satellites enabled these shortcomings to be overcome by providing datasets with spatial, spectral, radiometric, and temporal resolutions with larger swath areas that allowed the mapping and monitoring wetland ecosystems from landscape level to regional scales. With the regular advancement of different satellite programs it is possible to get multispectral data in several bands of electromagnetic spectrum, namely 0.45–0.52 μm representing the blue range, 0.52–0.60 μm representing the green range, 0.63–0.69 μm representing the red range, 0.76–0.90 μm representing the near infrared (NIR) range, 1.55–1.75 μm representing the shortwave wave infrared (SWIR) range, 2.08–2.35 μm representing the mid wave infrared range, and 10.4–12.5 μm representing the long wave infrared range.

The Landsat series represents the pioneering mission for Earth observation. It has been providing data through Landsat 1, with a payload multispectral scanner with four bands of green, red, and two NIR bands from 0.5 to 1.1 μm at 60 m spatial resolution since 1972. With the gradual progress currently taking place, it is possible to obtain data from the Landsat 8 that has 11 bands, out of which band 1 to band 7, which represent ultra-blue to SWIR, are available at a 30 m resolution, as well as a panchromatic band at 15 m resolution along with two thermal bands at 100 m resolution. These missions have enabled wetlands to be studied along with their various aspects like identification and delineation (band 1), bathymetric mapping (band 2), wetland vegetation vigor assessment (band 3), shoreline mapping (band 4), hydrophytic vegetation moisture content (band 5), soil moisture estimation of wetland ecosystems (band 10). Gradually with the availability of high temporal resolution like in spectroradiometers, advanced very high resolution radiometers, and in moderate resolution imaging temporal dynamic studies, the succession and phenology of wetland vegetation were obtainable (Yan et al., 2010), but could not be completely depended upon due to the coarser spatial resolution (Ozesmi and Bauer, 2002).

The advancement of multispectral sensors with higher spatial resolution like IKONOS, and Quickbird from DigitalGlobe (GeoEye merged with digitalGlobe) at 5 m or less facilitated the identification of different communities of wetland vegetation that had spectral overlapping, and high spatial resolution data could be differentiated due to the differences in pigments and biomass of vegetation at different seasons.

Since the satellite data available could have radiometric and geometric errors, strip line errors, and other types of noise, these images are subjected to image processing techniques like image preprocessing, image enhancement (contrast stretching, histogram modification, noise filtering), and image segmentation prior to image analysis. To extract information, spatial data mining, thematic map creation and gathering other information about a given wetland, and satellite image classification are done, which are broadly categorized as manual, automated, and hybrid classification methods. Manual visual interpretation is done prior to field knowledge of the analyst and is classified based on the interpretation keys like shape, size, color, tone, texture, patter, shadow, and association. An automated method is done either through supervised or unsupervised classification. Supervised classification is followed by techniques like the use of artificial neural network, maximum likelihood, minimum distance to mean, parallelepiped, Mahalanobis distance, *K* nearest neighbor, decision tree, semantic-based and object-based image segmentation, whereas unsupervised classification is conducted through either Iterative self-organizing data analysis technique or *K* means classification, and hybrid classification techniques combine automated and manual methods for classification (Sunitha and Suresh, 2015) and better understanding of wetland ecosystems.

In summary wetland delineation and wetland mapping (Rapinel et al., 2015; Jahncke et al., 2018), the identification and classification of wetland vegetation (Ludwig et al., 2019), seasonal time series monitoring (Davranche et al., 2013), stages of river inundation and discharge (Cienciala and Pasternack, 2017), soil moisture and salinity mapping (Herrero and Castañeda, 2015; Wigmore et al., 2019), flood regime mapping (Dhiman et al., 2019), water budgeting (Sahoo et al., 2011), ecosystem services assessment at landscape level (Fengqin and Shuwen, 2019), wetland health and risk assessment (Sun et al., 2016), and spatial modeling (Fu et al., 2018) are major aspects in which multispectral remote sensing has been extensively explored.

But this technology is restricted in terms of discriminating aspects of wetlands like soil, water, algae, and vegetation up to the spectral level (Wu, 2017). Their applicability cannot be stretched for segregating fine ecological divisions among wetland vegetation species like microphytes, macrophytes, emergent and submerged (Samiappan et al., 2017) identification of plant species underneath the canopy of dense vegetated wetlands and the estimation of soil moisture content, oil contamination level, metal toxicity, water turbidity quantification, and algal community beneath water. It is the need of the hour to study each component of wetlands in detail, which can be aided by the application of HRS technology as this extends its capability for surface material identification and species level discrimination.

12.3 Monitoring and assessment of wetlands with hyperspectral remote sensing

The quest to understand nature in depth spectrally led to HRS technology. These sensors provide data in a combination of two sensing techniques, namely imaging and spectrometry. These sensors provide spatial information of reflected or emitted electromagnetic radiation in accordance with variations in signal strength (Eismann, 2012), which ultimately give contiguous spectral bands, more than hundred data in cube with two spatial dimensions (x, y) and a spectral dimension (z) Various multispectral missions provide courser to high spatial resolution datasets, but with broad spectral bands. This limits their efficiency to discriminate two objects on the ground at a spectral level (Wu, 2017).

Hyperspectral remote sensors collect spatial as well as spectral information for hundreds of contiguous bands within the visible to SWIR region through the use of portable spectroradiometers and airborne and spaceborne sensors within the range of 0.3–2.5 μm along with bandwidth stretching between 2 and 10 nm. But the challenges of hyperspectral data include image calibration, data redundancy, and dimensionality issues. Systematic and basic approaches are followed for processing, including analyzing a large amount of information.

Since hyperspectral platforms collect radiance information, these are affected by atmospheric variations, as a result of which, authentic information about a given wetland is not obtained. So these data are subjected to geometric, radiometric, or sensor corrections, and are then atmospherically corrected to get an exact surface reflectance. Atmospheric correction is done using either a scene-based empirical approach like internal average reflectance, flat field, empirical line, quick atmospheric correction or using a radiative transfer model-based approach like ATmospheric CORrection, ATmospheric REMoval Program, Atmospheric CORrection Now, fast line-of-sight atmospheric analysis of spectral hypercube, high accuracy atmospheric correction for hyperspectral data (Kale et al., 2017). Thereafter data dimensionality is reduced for minimum storage space requirement through techniques like principal components analysis, minimum noise fraction, or independent component analysis. These processed data can now be used directly for unsupervised classification or to collect endmembers for supervised classification through algorithms like automated morphological endmember extraction, convex cone analysis, spatial–spectral endmember extraction, iterative error analysis, N-FINDR, pixel purity index, simplex growing algorithm, and vertex component analysis (Kale et al., 2017).

Image classification is required for in depth understanding of wetland ecosystems and their prevailing structures and composition. Unlike multispectral image classification, these data need different sorts of classifiers to avail maximum information. Classifications are done on the basis of training sample (supervised, unsupervised, and semisupervised), input distribution (parametric and nonparametric), pixel information (subpixel and per pixel), information based (spectral, contextual, spectral–spatial), and a number of classifier-based (single and ensemble). The most common and accepted classifiers for hyperspectral datasets include the support vector machine (SVM), random forest, spectral angle mapper, binary encoding, and deep learning classifiers.

Hyperspectral classifiers identify objects with the use of spectral endmembers in spectral libraries. Spectral libraries of riparian wetland soil (hydric), agricultural soil, hydrophytes, vegetation of marshy and swampy areas, etc., are being developed to use the full mapping potential of wetland ecosystems (Dudley et al., 2015) for sustainable management. This data are being widely exploited for finding, identifying, and mapping in geology, limnology, hydrology, pedology, ecology, phenology, and wetland ecosystems. There is a high prospect of extracting an enormous amount of information in complex wetland ecosystems like interdunal lakes (Timm and McGarigal, 2012); reefs (Bajjouk et al., 2019); in saline soil mapping (Khadim et al., 2019); coastal areas (Pengra et al., 2007); littoral areas (Karpouzli et al., 2004); tidal marshes (Turpie et al., 2015); salt marshes (Judd et al., 2007); estuarine areas (Lunetta et al., 2009), etc.

Water quality parameters like turbidity, total dissolved solids, and dissolved organic matter (Matthews et al., 2010) are assessed with the help of these data. The highest peak of reflectance for water is at 0.3 μm in the blue spectral range. At the NIR region and further, there is complete absorption. The study of water and its properties presents a challenge when using multispectral data (Sharma et al., 2015), but with the availability of hyperspectral data, the scope of in depth research of water bodies and aquatic ecosystems have widened up. This includes the classification of the trophic level of lakes and the determination of total dissolved solids (Thiemann and Kaufmann, 2002; Koponen et al., 2002).

Hyperspectral data present high efficiency in the identification and prediction of soil physicochemical parameters like soil texture, organic carbon, total nitrogen, phosphorous, and potassium (Yu et al., 2017). Hyperspectral data studies are applied for soil erosion (Chabrillat et al., 2014); soil moisture content (Yanmin et al., 2010); identification, estimation, and mapping of heavy metal contamination (Lamine et al., 2019); land use/cover changes (Pandey et al 2018); soil available nutrients (Qi et al., 2018); and soil types (Vibhute et al., 2015). The assessment of colored dissolved organic matter, a good indicator of water quality, could be discriminated at 0.565 and 0.645 μm and total dissolved solids at 0.62–0.67 μm (Lunetta et al., 2009).

These datasets are highly capable of discriminating between different algae at generic and species levels compared to at multispectral level (Kudela et al., 2015). This is useful for accounting food availability for birds, amphibians, fish, and fauna in a wetland. Identification along with discrimination of algal blooms and survival competition (Bi et al., 2019), anthocyanin content and antioxidant activity analysis (Huang et al., 2017), and ecophysiology (Stratoulias et al., 2015) are also possible.

Coral reefs, an integral part of marine ecosystems, are extremely hostile to climate change. Habitat mapping of these diverse and dynamic organisms has been inhibited by the availability of spatial and spectral resolution datasets, which are not custom-made to coral reefs. These have led to ambiguous benthic organism identification, but with the help of in situ hyperspectral data (Garcia et al., 2018) could discriminate benthic corals reefs, algae, and sand types.

Wetland plants play a role in ecological, economic, and hydrological processes. Wetland vegetation varies with the availability of water, their physiological structures, biochemical constituents like chlorophyll, carotene, xanthophyll, cellulose, lignin, and different molecular make ups (Chen et al., 2017). It is important to know the type of prevailing diverse macrophytic vegetation at every season. With the help of these data, wetland floral species discrimination is done best at 0.745, 0.746, 0.892, 0.932, 0.934, 0.958, 0.961, and 0.989 μm after the reduction of considerable data dimensionality (Anne et al., 2014). At the one end, these data enable the detection of species, genotypes, or individuals of submerged or floating vegetation, and at the other end, they also provide the location and identification of individual flora in a diverse community of similar species in a wetland ecosystem. This dataset enables chlorophyll estimation for biomass assessment as well as productivity of a wetland (Igamberdiev et al., 2011). The identification of wetland species under different stress conditions due to differences in biochemical constituents plays a vital role in determining ecological indicators

(Siles et al., 2019). Examples of the widespread use of this in different wetland ecosystems can be found in Section 12.6.

12.4 Details of hyperspectral images for wetland monitoring

Hyperspectral datasets are available through portable field spectroradiometers, airborne sensors, and spaceborne sensors and will also be available through future missions also. Each type of sensor has its own advantages and limitations. Existing airborne sensors are capable of providing high spectral resolution (0.005–0.010 μm), spatial resolution (0.002–0.020 μm), and high signal to noise ratio (500:1). Spaceborne hyperspectral sensors are as capable as airborne sensors as these could be used for inaccessible areas. They cover a high spectral resolution of about 10 nm and have comparatively low signal to noise ratios. With the advancement and commercialization of this technology, datasets are becoming more easily available. There are various companies like ITRES Research Ltd. and space agencies like NASA, and the European Space Agency (ESA), etc., that operate hyperspectral scanners. This technology is applicable to all the components of wetlands including solids, liquids, and gases at both the microscopic and macroscopic levels. Since airborne hyperspectral data acquisitions have advantages of use at acceptable weather conditions, flight line arrangement, time schedules, time calibration measurement spectral and spatial resolution intense wetland ecosystem assessment is possible at higher accuracy. But the high cost involvement limits its widespread usage for large study areas. This ultimately gives the widest scope to spaceborne hyperspectral sensors.

12.4.1 Airborne hyperspectral sensors

Any object on the surface of Earth is identifiable either directly or indirectly based on its spectral signatures (Jasmine and Pattabiraman, 2018). Airborne sensors provide a vantage point to locate, identify, and discriminate Earth surface materials in hundreds of square kilometers at a single time (Birk, 1992), which also enables the creation of precise maps of submerged and emergent plant species, benthic organisms, minerals, and the quantification of soil moisture content. For wetland studies, airborne hyperspectral sensors are generally flown at altitudes stretching from 1500 to 3000 m with spatial resolutions ranging from 1 to 4 m (Judd et al., 2007). Datasets of these sensors are also capable of providing underwater biogeochemical maps at different scales extending even to submeter scales, addressing various scientific and research questions of wetland ecosystems. There are a number of airborne sensors in use. These datasets are made available to the public by companies like ITRES Research Ltd., GER, Earth Search Science Inc. and also by space agencies like NASA and ESA. Examples of such sensors are the compact airborne spectral imager, airborne visible/infrared imaging spectrometer (AVIRIS), hyperspectral mapping, airborne imaging spectroradiometer for applications, hyperspectral digital imagery collection experiment, Probe-1. Details regarding different remote sensors can be found in the article by Pandey et al. (2019b).

12.4.2 Spaceborne hyperspectral sensors

Spaceborne sensors are able to collect data of Earth surfaces without terrain or weather constraints. Once they are on-board, they provide data regularly till they are decommissioned. A large number of research and scientific groups have easy access to data even at comparatively cheap rates. Spaceborne sensors provide a track of historical data due to the presence of temporal resolution. These sensors are devoid of frequent changes in imaging geometry. Because of these advantages, the HRS technology has been taken forward toward spaceborne sensors from airborne sensors. There are a number of spaceborne sensors from space agencies of China, Europe, and the United States. Hyperion, a sensor of Earth Observation 1 (EO-1), was on-board in 2000, and provided the first spaceborne data that became available to the public. This is enabled its widespread use in wetland ecosystems and dynamics study. Examples of spaceborne sensors are Hyperion, medium resolution imaging spectrometer, compact high-resolution imaging spectrometer, etc. Extensive information of ongoing and future missions can be consulted in the literature by Pandey et al. (2019b).

12.5 Application of hyperspectral images for wetland ecosystems

HRS is being continuously experimented with and explored in a wide range of wetland ecosystems like mangroves, rivers, reservoirs, peatlands, swamps, marshes, fresh water and saline lakes, and bogs, etc. Hyperspectral datasets having high spatial and spectral resolutions are used for multiple uses for wetlands like mapping (Kamal and Phinn, 2011), monitoring, soil and water quality assessment (Zhang et al., 2011), species discrimination (Panigrahy et al., 2012), metal toxicity analysis (Salmelin et al., 2018), identification of invasive species (Bustamante et al., 2016), (Song et al., 2012; El-Magd and El-Zeiny, 2014), chlorophyll estimation and biomass assessment (Kalacska et al., 2015; Turpie et al., 2015), mineral mapping (Anne et al., 2014), benthic and submerged flora identification (Hill et al., 2014), and organic carbon estimation (Shao et al., 2016). An intensive review of the literature has been done from 2011 to 2019 and presented in Table 12–2.

From Table 12–2, it can be inferred that all the types of wetland ecosystems have been explored with the use of this advanced technology. Out of the five wetland types, marine wetlands have received more attention than estuarine, lacustrine, riverine, and palustrine wetlands. This can be understood better with the illustration provided in Fig. 12–2. The researches conducted on marine ecosystems make up about 50% of all marine ecosystem research and the rest of the researches aggregate to 50%. Out of the other classes, lacustrine wetlands contribute to about 18.42% of researches, while riverine and palustrine ecosystems both contribute 13.15%. Estuarine wetlands are the least researched with about 5.26% of research cover.

Table 12–2 Application of hyperspectral images for wetlands.

S. no.	Type of wetland	Application	Platform	References
1	Marine	Mangrove mapping	Compact airborne spectral imager	Kamal and Phinn (2011)
2	Riverine	Species discrimination	Portable spectrometer	Lee et al. (2011)
3	Riverine	Saline soil monitoring using vegetation indices as proxies	Portable spectrometer	Zhang et al. (2011)
4	Marine	Mangrove species discrimination	Portable spectrometer	Panigrahy et al. (2012)
5	Palustrine	Total phosphorous of reservoir	Airborne imaging spectroradiometer for applications	Song et al. (2012)
6	Palustrine	Soil moisture and fertility gradient of peatland	Hyperspectral mapping	Middleton et al. (2012)
7	Estuarine	Species discrimination	Hyperspectral mapping	Adam et al. (2012)
8	Marine	Mapping	Portable spectrometer	Schweitzer et al. (2012)
9	Marine	Mangrove species discrimination	Portable spectrometer	Manjunath et al. (2013)
10	Marine	Estimation of foliar pigment concentration	Portable spectrometer	Proctor and He (2013)
11	Marine	Mangrove species chlorophyll a content	Portable spectrometer	Flores-de-Santiago et al. (2013)
12	Marine	Mangrove floristic composition	Hyperion	Kumar et al. (2013)
13	Marine	Benthic, submerged, and floating aquatic vegetation, benthic red algae, bare sand	Spectroscopic aerial mapping system with on-board navigation	Hill et al. (2014)
14	Marine	Turbidity and chlorophyll content of water	Portable spectrometer	El-Magd and El-Zeiny (2014)
15	Marine	Mangrove mapping	Hyperion	Jia et al. (2014)
16	Marine	Soil minerals and organic carbon	Hyperion	Anne et al. (2014)
17	Lacustrine	Water constituents, water column heights, and benthic substrate	Multispectral infrared and visible imaging spectrometer	Giardino et al. (2015)
18	Lacustrine	Chlorophyll estimation	Airborne imaging spectroradiometer for applications-Eagle	Stratoulias et al. (2015)
19	Palustrine	Model total chlorophyll and nitrogen concentrations of a bog	Compact airborne spectral imager-2	Kalacska et al. (2015)
20	Estuarine	Ecological properties such as species composition, biomass, hydrology, and evapotranspiration	Hyperspectral infrared imager	Turpie et al. (2015)
21	Lacustrine	Concentration of chlorophyll-a and the total amount of suspended solids	Airborne imaging spectroradiometer for applications	Vinciková et al. (2015)
22	Riverine	Invasive species detection	Compact airborne spectral imager	Bustamante et al. (2016)
23	Marine	Invasive species detection	HySpex	Le Bris et al. (2016)
24	Lacustrine	Leaf chlorophyll and nitrogen content	Portable spectrometer	Guo and Guo (2016)
25	Marine	Mangrove species discrimination	Portable spectrometer	Prasad and Gnanappazham (2016)

26	Riverine	Chromophoric dissolved organic matter, dissolved organic carbon, phytoplankton, and total suspended matter	Portable spectrometer	Shao et al. (2016)
27	Marine	Mangrove species mapping	Hyperion	Salghuna and Pillutla (2017)
28	Palustrine	Above ground biomass	Light detection and range and compact airborne spectral imager	Luo et al. (2017)
29	Marine	Salt monitoring	Hyperion	Riaza et al. (2017)
30	Lacustrine	Mapping	1A series satellite (HJ-1A) CCD/HSI camera	Chen et al. (2017)
31	Lacustrine	Cadmium toxicity	Portable spectrometer	Salmelin et al. (2018)
32	Lacustrine	Wetland demarcation	Portable spectrometer	Saluja et al. (2018)
33	Marine	Mapping	Hyperspectral imager HSI2, Pleiades, and Unmanned aerial vehicle data	Lamine et al. (2018)
34	Riverine	Harmful algal bloom mapping	Airborne HIS by NASA Glenn Research Center	Sawtell et al. (2019)
35	Marine	Mapping	Hyperion	Kumar et al. (2019)
36	Marine	Organic matter	Airborne visible/infrared imaging spectrometer-next generation	Sahay et al. (2019)
37	Marine	Mangrove species and biomass assessment	Hyperion	Pandey et al. (2019a)
38	Palustrine	Above ground biomass	Portable spectrometer	Li et al. (2019)

Note: CCD, charged coupled device; *HySpex*, Hyperspectral.

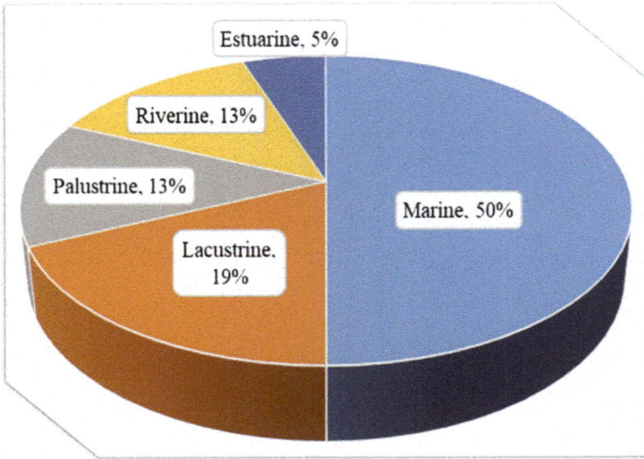

FIGURE 12–2 Applications for different types of wetlands.

Marine wetlands play a critical role in stabilizing shorelines, acting as barriers to debris accumulation, and preventing erosion that save lives and millions of properties (Armitage et al., 2019). Currently out of 2341 Ramsar sites, 41.64% (975) are marine wetlands (RSIS, 2019). These have attracted major attention from the scientific community for the application of hyperspectral datasets. But along with marine wetlands, other types do play a vital role in balancing the global ecosystem. Functions like providing feeding, breeding, and nesting grounds for migratory birds (Wang et al., 2016), acting as nitrogen sinks (Jordan et al., 2011), carbon dioxide sinks (Kayranli et al., 2010), providing livelihoods (Van Dam et al., 2013), and proving sources of minerals and biomass (Vijay and Pinto, 2016), etc., as valuable services are also provided by other wetlands.

From the application point of view, a major part of researches have been conducted toward mapping and monitoring (26.31%) and species discrimination (18.42%), which together constitute 44.73% of studies conducted, whereas all other applications together constitute about 55.26%, which is illustrated in Fig. 12–3. Applications have to be focused on all the domains of wetland ecosystems from soil, water, and vegetation, which include soil moisture and soil fertility gradient, soil contamination and metal toxicity, organic matter estimation and other soil parameters, water turbidity, total suspended particles and other water parameters, assessment of benthic substrates, chlorophyll estimation and above ground biomass of wetland vegetation, and evapotranspiration.

Focusing on the type of platforms used out of the three (airborne, spaceborne, and portable spectroradiometer), airborne sensors and portable spectroradiometers have been comparatively applied at large scale with 44.73% and 36.84% together constituting 81.57%, whereas spaceborne platforms are applied only 18.42% as illustrated in Fig. 12–4. This gives a wide scope for the application of future spaceborne missions in a wide range of wetland ecosystems.

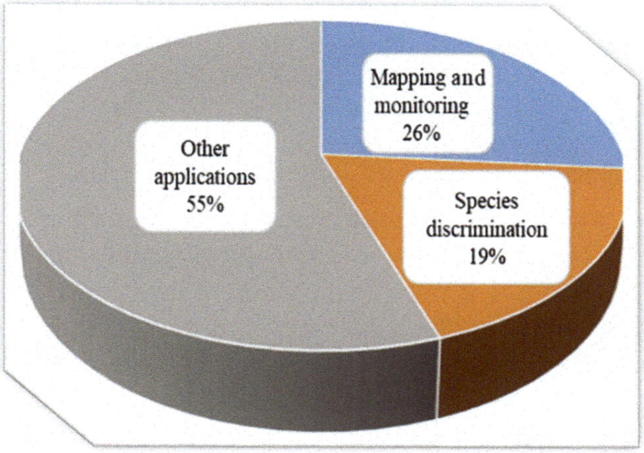

FIGURE 12–3 Application for different purposes.

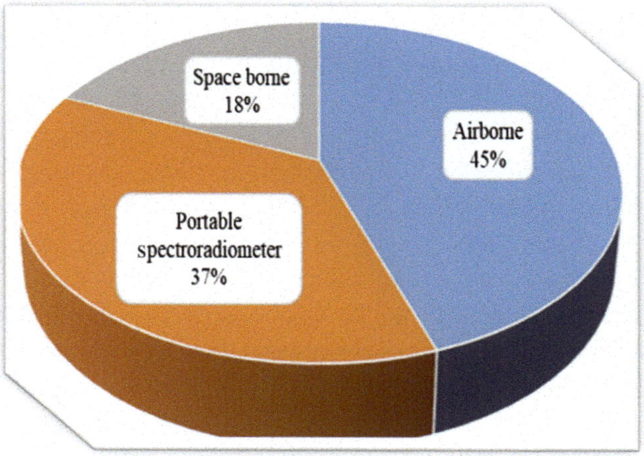

FIGURE 12–4 Application with different platforms.

12.6 Future scope and challenges of hyperspectral remote sensing for wetland ecosystem

12.6.1 Future scopes

Presently, the most explored and applied hyperspectral imaging spectroradiometers are airborne sensors. These studies are conducted at the regional level. A large number of researches have been done and validated for airborne sensorshowever there are still many fields to unveil. The ongoing and forthcoming spaceborne missions have enormous capability for the monitoring and assessment of various wetland ecosystems at the global level. There is need to have a spectral library of different wetland organisms like submerged periphyton algae, fungi, bacteria, and detritus microphytes, rooted and nonrooted floating vegetation, sprawling and tall emergent macrophytes, shoreline vegetation, sedges, grasses, aquatic weeds, and microbes in hydric soils, which are the primary consumers of wetland food webs. Since wetlands differ with their nature of origin, geographical location, and climatic conditions, their soil and water qualities also change accordingly over the seasons. The monitoring and quantification of their physicochemical parameters will provide further information about their trophic status, turbidity level, pollution and contamination agents along with invasive species. This, in turn, will be used to predict soil, water, and biomass productivity levels and to estimate carbon and nitrogen pools. More results that are accurate can be achieved with the development of data processing techniques, enhanced image fusion methods, and classifiers for benthic and submerged vegetation.

12.6.2 Challenges

In spite of all the benefits of hyperspectral data, there are certain challenges that need to be solved. The major shortcomings are low spatial resolution, mixed pixel, high dimensionality, limited availability of training samples, unavailability of time series data, expensive airborne data, data geometry, and the need for efficient hardware and software (Chen et al., 2017). HRS for wetland ecosystems is far more challenging than terrestrial ecosystem studies. Wetland ecosystems have greater vertical and horizontal strata intermixing with hydrological regime. Low spatial resolution data sets are intervened by chlorophyll content, lignin, cellulose, water content, pigment concentration which make it difficult to obtain correct signature profiles. The mixture of the signatures for floating, emergent, submerged, and benthic vegetation due to the dominance of the surrounding marsh, swamp, peat, bog, sands make it difficult for analysis (Igamberdiev et al., 2011).

Above all, there are more issues that need to be resolved before this technology is widely applied. Currently data are available only in small amounts, which needs to improve for large coverage, high spatial resolutions, the incorporation of fine sensors, and the availability of discrimination models. Such improvement would encourage more applicability.

12.7 Applicability of hyperion image for Sambhar Salt Lake, a saline wetland

12.7.1 Study area

Sambhar Salt Lake is a Ramsar site as declared on March 23, 1990. The lake comes under the jurisdiction of Salt department, Ministry of Commerce and Industry, Department of Industrial Policy and Promotion. The lake is elliptical in shape and shallow in nature with a 225 km^2

surface area. It has a 22.5 km maximum length and a 3.2–11.2 km maximum width. It is located at 27°00′N latitude and 75°00′E longitude with a 552,000 ha catchment area. It has no outlets, but receives water from four major rivers, namely the Roopnagar, Mendha, Kharian, and Khandel rivers. The study area experiences a dry season from March to June, with temperatures ranging from 24.4°C to 36.7°C with winter temperatures ranging from 11.7°C to 31.7°C. Sambhar Lake is a critical wintering ground for migratory birds like flamingos, spoonbills, and stints due to the presence of halophiles that survive in harsh saline conditions

12.7.2 Methodology

The Hyperion data was acquired from the official site of the united state geological survey Earth Explorer. The data acquired was from October 12, 2012. The acquired data was processed for subsequent steps before obtaining the spectral profiles of different land use land cover (LULC) classes. The Hyperion data was subject to the removal of bad lines and bands, and then was atmospherically and geometrically corrected. Then the spectral profiles were generated.

12.7.3 Results

The spectral profile curves show the graph between wavelength and reflectance values. The spectral profiles of terrestrial vegetation (vegetation of Aravalli region), saline vegetation, saline soil, salt crust, saline water in the lake, saline water in the saltpans, red algae, and green algae are illustrated in this section. These figures show the reflectance curve in the range between 500 and 2250 nm on all the classes.

In Fig. 12–5, it is clear that there are distinct differences in the spectral properties of the two algal communities. There are 11 peaks at 536, 749, 753, 938, 1125, 1255, 1625, 1649, 1750, 2000, 2125 nm and troughs at 688, 750, 813, 997, 1188, 1475, and 1640 nm for the red algae; whereas only 7 crests at 623, 625, 688, 755, 880, 1110, 1125 nm and troughs at 624, 630, 750, 780, 1000, 1120, and 1188 nm for the green algae. The highest crest for the red algae is at 938 nm, whereas for the green algae, it is at 623 nm.

In Fig. 12–6, it is clear that there are distinct differences in the spectral properties of water in different classes. There are 4 peaks at 437, 600, 753, 875, 1125, 1125 nm and troughs at 753 and 999 nm, which becomes flat after 1132 nm, for the water in the lake; whereas crests appeared at 668, 813, 1125, 1620, 2063 nm and troughs at 750, 1000, 1500, 1999, and 2250 nm for the water in the saltpans. The highest crest for the water in the lake area is at 600 nm, whereas in the saltpans, it is at 668 nm.

In Fig. 12–7, it is observed that there are peaks at 1500, 1750, 1998, 2150, and 2375 nm, whereas troughs are at 2125 and 2188 nm. The peaks of saline vegetation are at 900, 1125, 1755, 2015, and 2249 nm, whereas troughs are at 1000, 1500, 2000, and 2125 nm. The highest peak is observed at 1125 nm.

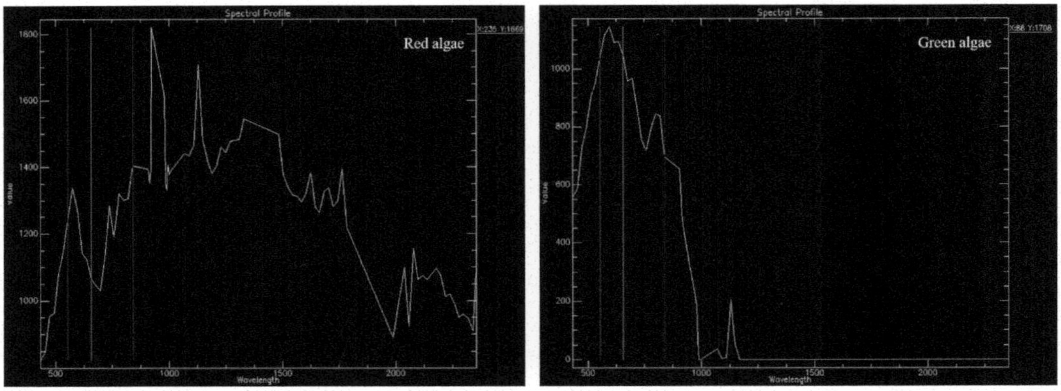

FIGURE 12–5 Spectral profiles of red and green algal communities.

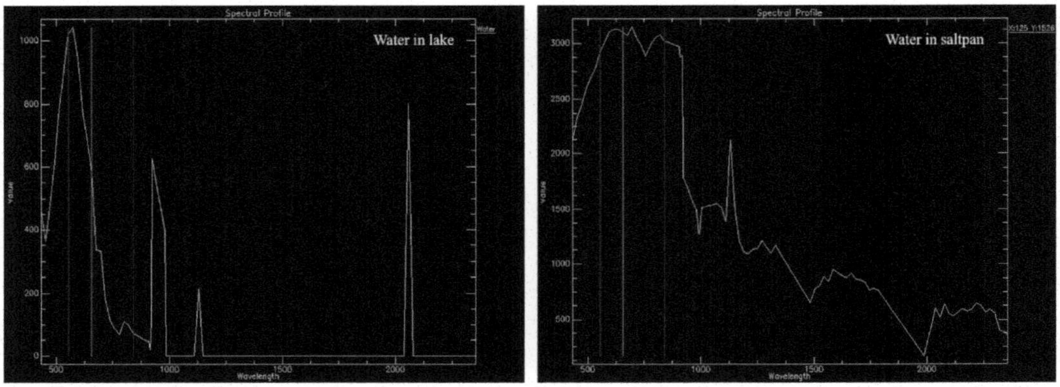

FIGURE 12–6 Spectral profile of water in lake and saltpans.

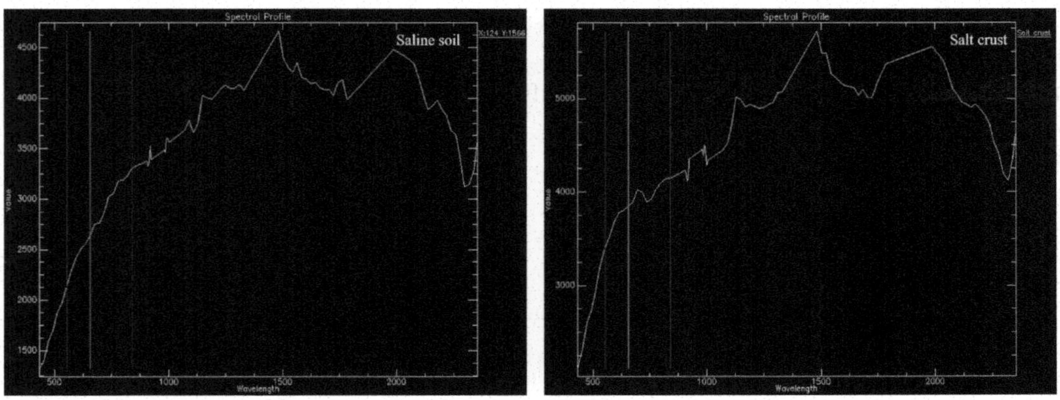

FIGURE 12–7 Spectral profiles of saline soil and vegetation.

12.7.4 Discussion

The green algae (Chlorophyceae) found to be present in the Sambhar Lake are *Chlamydomonas* sp., *Dunelialla salina, Mycanthococcus, Oedogonium* sp., *Oocystis* sp., *Rhizoclonium* sp., and the blue green algae (Cyanophyceae) present are *Anabaena fertilissima, Rao Anabaena laxa, Anabaenopi circularis, Anabaenopsis arnoldii, Aptekarj Aphanocapsa* sp., *Aphanothece halophytica rerny, Arthrospira platensis (Nordst), Chroococcus minor (Kutz.), Gloeothece* sp., *Lyngbya* sp., *Merismopedia* sp., *Microcoleus* sp., *Myxosarcina concinna, Nostoc sphaericum, Oscillatoria subbrevis, Scmid Oscillatoria. p., Oscillatoria jasorvensis, Vouk Oscillatoria salina, Oscillatoria slmplicisslma, Gomont Phormdium p., Rameria* sp., *Scytonema* sp., *Spirulina subslalsa, Corst Synechococcu elongatu aeg, Synechocystis* sp., while Bacilariophyceae is represented by Closterium sp., *Cosmarium* sp., *Cymbella* sp., *Melosira* sp., *Navicula* sp., *Nitzschia* sp., *Synedra* sp., and *Euglena* sp. (Upasani, 2008). The difference in the biochemical constituents in the saline water as well as the age, number, and composition of these algae could be the probable cause of the difference in spectral profiles. The signatures of soil differ because of moisture content, organic matter content, particle size distribution, presence and absence of particle size, soil mineralogy, and soil structure. Due to the presence of salt, physicochemical properties differ. The presence of cations like Na^{+}, Ca^{2+}, K^{+}, Mg^{2+}, Fe^{2+}, and Mn^{2+} and anions like CO^{3-}, HCO^{3-}, Cl^{-}, NO^{3-}, SO_4^{2-}, and PO_4^{3-} (Cherekar and Pathak, 2016) cause the spectral properties to differ from normal soil spectral profiles. The signatures of saline vegetation are dominated by saline soil properties in the study area. Since the vegetation is sparse, it was difficult to find pure pixels for saline vegetation that are represented by halophytes like *Suaeda fruticosa, Salsola baryosma,* and *Cressa cretica,* tree species like *Salvadora oleoides, Capparis decidua,* and *Prosopis cineraria,* and shrubs like *Tamarix dioica* and *Mimosa hamata.*

Due to the availability of hyperspectral data for the study area, spectral signatures of different LULCs were obtained. However, the biggest limitation is that the dataset is from 2012. Since, this lake is a temporary and seasonal lake, it has undergone different climatic changes and subsequently changes in biogeochemical properties. Currently, physicochemical and biological properties of soil, water, and algae are being conducted as a result properties are only estimated but not quantified with the available technology.

Acknowledgments

Authors are thankful to Central University Rajasthan for support. Dr. Pandey is thankful to Shiv Nadar University for help and support.

List of abbreviations

AIS airborne imaging spectrometer
AVIRIS airborne visible/infrared imaging spectrometer

CHRIS compact high-resolution imaging spectrometer
ESA European Space Agency
GER Geophysical and Environmental Research Corporation
HRS hyperspectral remote sensing
MIVIS multispectral infrared and visible imaging spectrometer
NASA The National Aeronautics and Space Administration
NIR near infrared
SVM support vector machines
SWIR short wave infrared
UNESCO United Nations Educational, Scientific and Cultural Organization

List of symbols

μm micrometer
% percentage
°C degree Celsius

References

Adam, E.M., Mutanga, O., Rugege, D., Ismail, R., Bag, P., Africa, S., 2012. Discriminating the papyrus vegetation (*Cyperus papyrus* L.) and its co-existent species using random forest and hyperspectral data resampled to. Int. J. Remote Sens. 33 (20), 552–569. Available from: https://doi.org/10.1080/01431161.2010.543182.

Anne, N.J.P., Abd-elrahman, A.H., Lewis, D.B., Hewitt, N.A., 2014. Modeling soil parameters using hyperspectral image reflectance in subtropical coastal wetlands. Int. J. Appl. Earth Obs. Geoinf. 33, 47–56. Available from: https://doi.org/10.1016/j.jag.2014.04.007.

Armitage, A.R., Weaver, C.A., Kominoski, J.S., Pennings, S.C., 2019. Resistance to hurricane effects varies among wetland vegetation types in the marsh–mangrove ecotone. Estuaries Coasts. 1–11.

Bajjouk, T., Mouquet, P., Ropert, M., Quod, J.P., Hoarau, L., Bigot, L., et al., 2019. Detection of changes in shallow coral reefs status: towards a spatial approach using hyperspectral and multispectral data. Ecol. Indic. 96 (June 2018), 174–191. Available from: https://doi.org/10.1016/j.ecolind.2018.08.052.

Bi, X., Lin, S., Zhu, S., Yin, H., Li, Z., Chen, Z., 2019. Species identification and survival competition analysis of microalgae via hyperspectral microscopic images. Optik 176, 191–197. Available from: https://doi.org/10.1016/j.ijleo.2018.09.077.

Birk, R.J., (1992). Airborne hyperspectral sensor profiles. In: Workshop at the International Space Year International Symposium on Spectral Sensing Research.

Birkett, C.M., Mason, I.M., 1995. A new global lakes database for a remote sensing program studying climatically sensitive large lakes. J. Gt. Lakes Res. 21 (3), 307–318. Available from: https://doi.org/10.1016/S0380-1330(95)71041-3.

Bustamante, J., Aragonés, D., Afán, I., Luque, C.J., Pérez-Vázquez, A., Castellanos, E.M., et al., 2016. Hyperspectral sensors as a management tool to prevent the invasion of the exotic cordgrass *Spartina densiflora* in the Doñana wetlands. Remote. Sens. 8 (12). Available from: https://doi.org/10.3390/rs8121001.

Chabrillat, S., Milewski, R., Schmid, T., Rodriguez, M., Escribano, P., Palacios-orueta, A., et al., 2014. Potential of hyperspectral imagery for the spatial assessment of soil erosion stages in agricultural semi-arid Spain at different German Research Centre for Geosciences, Telegrafenberg, 14473 Potsdam, Germany EEZA-CSIC, Departmento de Desertificación y Geoecología, 04120 Almería, Spain, pp. 2918–2921.

Chen, B., Chen, L., Lu, M., Xu, B., 2017. Wetland mapping by fusing fine spatial and hyperspectral resolution images. Ecol. Model. 1–12. Available from: https://doi.org/10.1016/j.ecolmodel.2017.01.004.

Cherekar, M.N., Pathak, A.P., 2016. Chemical Assessment of Sambhar Soda Lake, a Ramsar Site in India. J. Water Chem. Technol. 38 (4), 244–247. Available from: https://doi.org/10.3103/S1063455X1604010X.

Cienciala, P., Pasternack, G.B., 2017. Floodplain inundation response to climate, valley form, and flow regulation on a gravel-bed river in a Mediterranean-climate region. Geomorphology 282, 1–17. Available from: https://doi.org/10.1016/j.geomorph.2017.01.006.

Congalton, R.G., Gu, J., Yadav, K., Thenkabail, P., Ozdogan, M., 2014. Global land cover mapping: a review and uncertainty analysis. Remote Sens. 6 (12), 12070–12093. Available from: https://doi.org/10.3390/rs61212070.

Cowardin, L.M., Carter, V., Golet, F.C., LaRoe, E.T., 1979. Classification of Wetlands and Deepwater Habitats of the United States. U.S. Department of the Interior, Fish and Wildlife Service, Washington, DC.

Davranche, A., Poulin, B., Lefebvre, G., 2013. Mapping flooding regimes in Camargue wetlands using seasonal multispectral data. Remote Sens. Environ. 138, 165–171. Available from: https://doi.org/10.1016/j.rse.2013.07.015.

Dhiman, R., Vishnuradhan, R., Arun, T.I.E., 2019. Flood risk and adaptation in Indian coastal cities: recent scenarios. Appl. Water Sci. 1–16. Available from: https://doi.org/10.1007/s13201-018-0881-9.

Dudley, K.L., Dennison, P.E., Roth, K.L., Roberts, D.A., Coates, A.R., 2015. A multi-temporal spectral library approach for mapping vegetation species across spatial and temporal phenological gradients. Remote Sens. Environ. 167, 121–134. Available from: https://doi.org/10.1016/j.rse.2015.05.004.

Eismann, M.T., 2012. Hyperspectral Remote Sensing. SPIE Press. ISBN: 9780819487872.

El-Magd, I.A., El-Zeiny, A., 2014. Quantitative hyperspectral analysis for characterization of the coastal water from Damietta to Port Said, Egypt. Egypt J. Remote Sens. Space Sci. 17 (1), 61–76. Available from: https://doi.org/10.1016/j.ejrs.2014.02.001.

Fengqin, Y., Shuwen, Z., 2019. Ecosystem service decline in response to wetland loss in the Sanjiang Plain, Northeast China. Ecol. Eng. 130 (January), 117–121. Available from: https://doi.org/10.1016/j.ecoleng.2019.02.009.

Flores-de-Santiago, F., Kovacs, J.M., Flores-Verdugo, F., 2013. The influence of seasonality in estimating mangrove leaf chlorophyll-a content from hyperspectral data. Wetl. Ecol. Manag. 21 (3), 193–207. Available from: https://doi.org/10.1007/s11273-013-9290-x.

Fu, Y., Zhao, J., Peng, W., Zhu, G., Quan, Z., Li, C., 2018. Spatial modelling of the regulating function of the Huangqihai Lake wetland ecosystem. J. Hydrol. 564, 283–293. Available from: https://doi.org/10.1016/j.jhydrol.2018.07.017.

Garcia, R.A., Lee, Z., Hochberg, E.J., 2018. Hyperspectral shallow-water remote sensing with an enhanced benthic classifier. Remote Sens. 10 (147), 1–25.

Giardino, C., Bresciani, M., Valentini, E., Gasperini, L., Bolpagni, R., Brando, V.E., 2015. Airborne hyperspectral data to assess suspended particulate matter and aquatic vegetation in a shallow and turbid lake. Remote Sens. Environ. 157, 48–57. Available from: https://doi.org/10.1016/j.rse.2014.04.034.

Guo, C., Guo, X., 2016. Estimating leaf chlorophyll and nitrogen content of wetland emergent plants using hyperspectral data in the visible domain. Spectrosc. Lett. 49 (3), 180–187. Available from: https://doi.org/10.1080/00387010.2015.1118126.

Herrero, J., Castañeda, C., 2015. Temporal changes in soil salinity at four saline wetlands in NE Spain. Catena 133, 145–156. Available from: https://doi.org/10.1016/j.catena.2015.04.017.

Hill, V.J., Zimmerman, R.C., Bissett, W.P., Dierssen, H., Kohler, D.D.R., 2014. Evaluating light availability, seagrass biomass, and productivity using hyperspectral airborne remote sensing in Saint Joseph's Bay, Florida. Estuaries Coasts 37 (6), 1467–1489. Available from: https://doi.org/10.1007/s12237-013-9764-3.

Huang, L., Zhou, Y., Meng, L., Wu, D., He, Y., 2017. Comparison of different CCD detectors and chemometrics for predicting total anthocyanin content and antioxidant activity of mulberry fruit using visible and near infrared hyperspectral imaging technique. Food Chem. 224, 1–10. Available from: https://doi.org/10.1016/j.foodchem.2016.12.037.

Igamberdiev, R.M., Grenzdoerffer, G., Bill, R., Schubert, H., Bachmann, M., Lennartz, B., 2011. Determination of chlorophyll content of small water bodies (kettle holes) using hyperspectral airborne data. Int. J. Appl. Earth Obs. Geoinf. 13 (6), 912–921. Available from: https://doi.org/10.1016/j.jag.2011.04.001.

Jahncke, R., Leblon, B., Bush, P., Larocque, A., 2018. Mapping wetlands in Nova Scotia with multi-beam RADARSAT-2 Polarimetric SAR, optical satellite imagery, and Lidar data. Int. J. Appl. Earth Obs. Geoinf. 68 (September 2017), 139–156. Available from: https://doi.org/10.1016/j.jag.2018.01.012.

Jasmine, S.G., Pattabiraman, V., 2018. Improved pure pixel identification algorithms to determine the endmembers in hyperspectral images. Comput. Electr. Eng. 71 (July), 515–532. Available from: https://doi.org/10.1016/j.compeleceng.2018.07.023.

Jia, M., Zhang, Y., Wang, Z., Song, K., Ren, C., 2014. Mapping the distribution of mangrove species in the Core Zone of Mai Po Marshes Nature Reserve, Hong Kong, using hyperspectral data andhigh-resolution data. Int. J. Appl. Earth Obs. Geoinf. 33 (1), 226–231. Available from: https://doi.org/10.1016/j.jag.2014.06.006.

De Jong, S.M., 2007. Basics of Remote Sensing. pp. 1–15 (Chapter 1). https://doi.org/10.1007/978-1-4020-2560-0.

Jordan, S.J., Stoffer, J., Nestlerode, J.A., 2011. Wetlands as sinks for reactive nitrogen at continental and global scales: a meta-analysis. Ecosystems 14 (1), 144–155. Available from: https://doi.org/10.1007/s10021-010-9400-z.

Judd, C., Steinberg, S., Shaughnessy, F., Crawford, G., 2007. Mapping salt marsh vegetation using aerial hyperspectral imagery and linear unmixing in Humboldt Bay, California mapping salt marsh vegetation using aerial hyperspectral. BioOne 27 (4), 1144–1152.

Kalacska, M., Lalonde, M., Moore, T.R., 2015. Estimation of foliar chlorophyll and nitrogen content in an ombrotrophic bog from hyperspectral data: scaling from leaf to image. Remote Sens. Environ. 169, 270–279. Available from: https://doi.org/10.1016/j.rse.2015.08.012.

Kale, K.V., Solankar, M.M., Nalawade, D.B., Dhumal, R.K., Gite, H.R., 2017. A research review on hyperspectral data processing and analysis algorithms a research review on hyperspectral data processing and analysis algorithms. Proc. Nat. Acad. Sci. India A: Phys. Sci. 87 (4), 541–555. Available from: https://doi.org/10.1007/s40010-017-0433-y.

Kamal, M., Phinn, S., 2011. Hyperspectral data for mangrove species mapping. Remote Sens. 3, 2222–2242. Available from: https://doi.org/10.3390/rs3102222.

Karpouzli, E., Malthus, E.T.J., Place, E.C.J., 2004. Hyperspectral discrimination of coral reef benthic communities in the western Caribbean. Coral Reefs 23, 141–151.

Kayranli, B., Scholz, M., Mustafa, A., Hedmark, Å., 2010. Carbon storage and fluxes within freshwater wetlands: a critical review. Wetlands 30 (1), 111–124. Available from: https://doi.org/10.1007/s13157-009-0003-4.

Khadim, F.K., Su, H., Xu, L., Tian, J., 2019. Soil salinity mapping in Everglades National Park using remote sensing techniques and vegetation salt tolerance. Phys. Chem. Earth. Available from: https://doi.org/10.1016/j.pce.2019.01.004.

Koponen, S., Pulliainen, J., Kallio, K., Hallikainen, M., 2002. Lake water quality classification with airborne hyperspectral spectrometer and simulated MERIS data. Remote Sens. Environ. 79, 51–59.

Kudela, R.M., Palacios, S.L., Austerberry, D.C., Accorsi, E.K., Guild, L.S., Torres-perez, J., 2015. Remote sensing of environment application of hyperspectral remote sensing to cyanobacterial blooms in inland waters. Remote Sens. Environ. 167, 196–205.

Kumar, T., Panigrahy, S., Kumar, P., Parihar, J.S., 2013. Classification of floristic composition of mangrove forests using hyperspectral data: case study of Bhitarkanika National Park, India. J. Coast. Conserv. 17 (1), 121–132. Available from: https://doi.org/10.1007/s11852-012-0223-2.

Kumar, T., Mandal, A., Dutta, D., Nagaraja, R., Dadhwal, V.K., 2019. Discrimination and classification of mangrove forests using EO-1 Hyperion data: a case study of Indian Sundarbans. Geocarto Int. 34 (4), 415–442. Available from: https://doi.org/10.1080/10106049.2017.1408699.

Lamine, S., Petropoulos, G.P., Singh, S.K., Szabó, S., Bachari, N.E.I., Srivastava, P.K., et al., 2018. Quantifying land use/land cover spatio-temporal landscape pattern dynamics from Hyperion using SVMs classifier and FRAGSTATS. Geocarto Int. 33 (8), 862–878. Available from: https://doi.org/10.1080/10106049.2017.1307460.

Lamine, S., Petropoulos, G.P., Brewer, P.A., Bachari, N., Srivastava, P.K., Manevski, K., et al., 2019. Combined geochemistry and field spectroradiometry in the United Kingdom. Sensors 762, 2–16.

Le Bris, A., Rosa, P., Lerouxel, A., Cognie, B., Gernez, P., Launeau, P., et al., 2016. Hyperspectral remote sensing of wild oyster reefs. Estuarine Coast Shelf Sci. 172, 1–12. Available from: https://doi.org/10.1016/j.ecss.2016.01.039.

Lee, B.S., Mcgwire, K.C., Fritsen, C.H., 2011. Identification and quantification of aquatic vegetation with hyperspectral remote sensing in western Nevada. Int. J. Remote Sens. 32 (24), 9093–9117. Available from: https://doi.org/10.1080/01431161.2010.549850.

Li, W., Dou, Z., Wang, Y., Wu, G., Zhang, M., Lei, Y., et al., 2019. Estimation of above-ground biomass of reed (*Phragmites communis*) based on in situ hyperspectral data in Beijing Hanshiqiao Wetland, China. Wetl. Ecol. Manag. 27 (1), 87–102. Available from: https://doi.org/10.1007/s11273-018-9644-5.

Lillesand, T.M., Kiefer, R.W., Chipman, J.W., 1980. Remote sensing and image interpretation. Geograph. J. 146. Available from: https://doi.org/10.2307/634969.

Lin, W., Xu, D., Guo, P., Wang, D., Li, L., Gao, J., 2019. Exploring variations of ecosystem service value in Hangzhou Bay Wetland, Eastern China. Ecosyst. Serv. 37 (August 2018), 100944. Available from: https://doi.org/10.1016/j.ecoser.2019.100944.

Lovvorn, J.R., Kirkpatrick, C.M., 1982. Analysis of freshwater wetland vegetation with large-scale color infrared aerial photography. J. Wildl. Manag. 46 (1), 61–70.

Ludwig, C., Walli, A., Schleicher, C., Weichselbaum, J., Riffler, M., 2019. A highly automated algorithm for wetland detection using multi-temporal optical satellite data. Remote Sens. Environ. 224 (October 2017), 333–351. Available from: https://doi.org/10.1016/j.rse.2019.01.017.

Lunetta, R.S., Knight, J.F., Paerl, H.W., Streicher, J., Peierls, B.L., Gallo, T.O.M., et al., 2009. Measurement of water colour using AVIRIS imagery to assess the potential for an operational monitoring capability in the Pamlico Sound Estuary, USA. Int. J. Remote 30 (13), 3291–3314.

Luo, S., Wang, C., Xi, X., Pan, F., Qian, M., Peng, D., et al., 2017. Retrieving aboveground biomass of wetland *Phragmites australis* (common reed) using a combination of airborne discrete-return LiDAR and hyperspectral data. Int. J. Appl. Earth Obs. Geoinf. 58, 107–117. Available from: https://doi.org/10.1016/j.jag.2017.01.016.

Manjunath, K.R., Kumar, T., Kundu, N., Panigrahy, S., 2013. Discrimination of mangrove species and mudflat classes using in situ hyperspectral data: a case study of Indian Sundarbans. GIScience Remote Sens. 50 (4), 400–417. Available from: https://doi.org/10.1080/15481603.2013.814275.

Martyn, R.D., Noble, R.L., Bettoli, P.W., Maggio, R.C., 1986. Mapping aquatic weeds with aerial color infrared photography and evaluating their control by grass carp. J. Aquat. Plant. Manag. 24, 46–56.

Matthews, M.W., Bernard, S., Winter, K., 2010. Remote sensing of cyanobacteria-dominant algal blooms and water quality parameters in Zeekoevlei, a small hypertrophic lake, using MERIS. Remote Sens. Environ. 114 (9), 2070–2087.

Middleton, M., Närhi, P., Arkimaa, H., Hyvönen, E., Kuosmanen, V., Treitz, P., et al., 2012. Ordination and hyperspectral remote sensing approach to classify peatland biotopes along soil moisture and fertility gradients. Remote Sens. Environ. 124, 596–609. Available from: https://doi.org/10.1016/j.rse.2012.06.010.

Mishra, N.B., 2014. Wetlands: remote sensing. Encycl. Nat. Resourc. Land. (October), 566–574. Available from: https://doi.org/10.1081/E-ENRL-120049156.

Ozesmi, S.L., Bauer, M.E., 2002. Satellite remote sensing of wetlands. Wetl. Ecol. Manag. 10 (5), 381–402.

Pandey, P.C., Manevski, K., Srivastava, P., Petropoulos, G., 2018. The use of hyperspectral earth observation data for land use/cover classification: present status, challenges and future outlook. In: Thenkabail, P. (Ed.), Hyperspectral Remote Sensing of Vegetation, vol. 4. second ed., pp. 147–173 (Chapter 8).

Pandey, P.C., Anand, A., Srivastava, P.K., 2019a. Spatial distribution of mangrove forest species and biomass assessment using field inventory and earth observation hyperspectral data. Biodivers. Conserv. Available from: https://doi.org/10.1007/s10531-019-01698-8.

Pandey, P.C., Koutsia, N., Petropoulos, G.P., Srivastava, P.K., Ben Dor, E., 2019b. Land use/land cover in view of earth observation: data sources, input dimensions, and classifiers—a review of the state of the art. Geocarto Int. 1-32.

Panigrahy, S., Kumar, T., Manjunath, K.R., 2012. Hyperspectral leaf signature as an added dimension for species discrimination: case study of four tropical mangroves. Wetl. Ecol. Manag. 20 (2), 101–110. Available from: https://doi.org/10.1007/s11273-011-9245-z.

Pengra, B.W., Johnston, C.A., Loveland, T.R., 2007. Mapping an invasive plant, *Phragmites australis*, in coastal wetlands using the EO-1 Hyperion hyperspectral sensor. Remote Sens. Environ. 108, 74–81.

Prasad, K.A., Gnanappazham, L., 2016. Multiple statistical approaches for the discrimination of mangrove species of Rhizophoraceae using transformed field and laboratory hyperspectral data. Geocarto Int. 31 (8), 891–912. Available from: https://doi.org/10.1080/10106049.2015.1094521.

Prasad, S.N., Ramachandra, T.V., Ahalya, N., Sengupta, T., Kumar, A., Tiwari, A.K., 2002. Conservation of wetlands of India—a review. Trop. Ecol. 43 (1), 173–186.

Proctor, C., He, Y., 2013. Estimation of foliar pigment concentration in floating macrophytes using hyperspectral vegetation indices. Int. J. Remote Sens. (September), 8012–8027. Available from: https://doi.org/10.1080/01431161.2013.828183.

Qi, H., Paz-Kagan, T., Karnieli, A., Jin, X., Li, S., 2018. Evaluating calibration methods for predicting soil available nutrients using hyperspectral VNIR data. Soil Tillage Res. 175 (August 2017), 267–275. Available from: https://doi.org/10.1016/j.still.2017.09.006.

Rapinel, S., Hubert-moy, L., Clément, B., 2015. Combined use of LiDAR data and multispectral earth observation imagery for wetland habitat mapping. Int. J. Appl. Earth Obs. Geoinf. 37, 56–64. Available from: https://doi.org/10.1016/j.jag.2014.09.002.

Raup, B., Ka, A., Kargel, J.S., Bishop, M.P., Hamilton, G., Lee, E., et al., 2007. Remote sensing and GIS technology in the Global Land Ice Measurements from Space (GLIMS) project. Comput. Geosci. 33 (1), 104–125. Available from: https://doi.org/10.1016/j.cageo.2006.05.015.

Riaza, A., Buzzi, J., Moral, B., Carrère, V., Richter, R., Buzzi, J., et al., 2017. Monitoring salt crusts on an AMD contaminated coastal wetland using hyperspectral Hyperion data (Estuary of the River Odiel, SW Spain). Int. J. Remote Sens. 38 (12), 3735–3762. Available from: https://doi.org/10.1080/01431161.2017.1302621.

Rosenqvist, A., Shimada, M., Chapman, B., 2000. The global rain forest mapping project—a review. Int. J. Remote Sens. 21 (6-7), 1375–1387.

RSIS, 2019. Ramsar Sites in World. Available from: <https://rsis.ramsar.org/rissearch/?pagetab = 2&f%5B0%5D = wetlandTypes_en_ss%3AMarine%20or%20coastal%20wetlands> (accessed 15.06.19).

Sahay, A., Gupta, A., Motwani, G., Raman, M., Ali, S.M., Shah, M., et al., 2019. Distribution of coloured dissolved and detrital organic matter in optically complex waters of Chilika lagoon, Odisha, India, using hyperspectral data of AVIRIS-NG. Curr. Sci. 116 (7), 1166–1171.

Sahoo, A.K., Pan, M., Troy, T.J., Vinukollu, R.K., Sheffield, J., Wood, E.F., 2011. Reconciling the global terrestrial water budget using satellite remote sensing. Remote Sens. Environ. 115 (8), 1850–1865. Available from: https://doi.org/10.1016/j.rse.2011.03.009.

Salghuna, N.N., Pillutla, R.C.P., 2017. Mapping mangrove species using hyperspectral data: a case study of Pichavaram Mangrove Ecosystem, Tamil Nadu. Earth Syst. Environ. 1 (2), 1–12. Available from: https://doi.org/10.1007/s41748-017-0024-8.

Salmelin, J., Pölönen, I., Puupponen, H.H., Hämäläinen, H., Karjalainen, A.K., Väisänen, A., et al., 2018. Hyperspectral imaging of macroinvertebrates—a pilot study for detecting metal contamination in aquatic ecosystems. Water Air Soil Pollut. 229 (9). Available from: https://doi.org/10.1007/s11270-018-3963-2.

Saluja, R., Prasad, S., Garg, J.K., 2018. Field spectroradiometry for discrimination of wetland components: a case study of a tropical inland wetland in India. Wetl. Ecol. Manag. 26 (5), 915–930. Available from: https://doi.org/10.1007/s11273-018-9620-0.

Samiappan, S., Turnage, G., Hathcock, L., Yao, H., Kincaid, R., Moorhead, R., et al., 2017. Classifying common wetland plants using hyperspectral data to identify optimal spectral bands for species mapping using a small unmanned aerial system—a case study. IEEE 5924–5927.

Sawtell, R.W., Anderson, R., Tokars, R., Lekki, J.D., Shuchman, R.A., Bosse, K.R., et al., 2019. Real time HABs mapping using NASA Glenn hyperspectral imager. J. Gt. Lakes Res. Available from: https://doi.org/10.1016/j.jglr.2019.02.007.

Schweitzer, A., Bordeaux, U., De, 2012. Hyperspectral field database in support to coastal wetland mapping. IEEE (June 2014), . Available from: https://doi.org/10.1109/IGARSS.2012.6350384.

Shao, T., Song, K., Du, J., Zhao, Y., Liu, Z., Zhang, B., 2016. Retrieval of CDOM and DOC using in situ hyperspectral data: a case study for potable waters in Northeast China. J. Indian Soc. Remote. Sens. 44 (1), 77–89. Available from: https://doi.org/10.1007/s12524-015-0464-2.

Sharma, L.K., Kumari, B., Gupta, S., 2015. Remote sensing strategy for the study of algal monitoring. Algae and Environmental Sustainability. Springer, pp. 155–164.

Siles, G., Charland, A., Voirin, Y., Bénié, G.B., 2019. Integration of landscape and structure indicators into a web-based geoinformation system for assessing wetlands status. Ecol. Inform. Available from: https://doi.org/10.1016/j.ecoinf.2019.05.011.

Song, K., Li, L., Li, S., Tedesco, L., Hall, B., Li, L., 2012. Hyperspectral remote sensing of total phosphorus (TP) in three central Indiana water supply reservoirs. Water Air Soil Pollut. 223 (4), 1481–1502. Available from: https://doi.org/10.1007/s11270-011-0959-6.

Stewart, W.R., Carter, V., Brooks, P.D., 1980. Inland (non-tidal) wetland mapping. Photogram. Eng. Remote Sens. 46 (5), 617–628.

Stratoulias, D., Balzter, H., Zlinszky, A., Tóth, V.R., 2015. Assessment of ecophysiology of lake shore reed vegetation based on chlorophyll fluorescence, field spectroscopy and hyperspectral airborne imagery. Remote Sens. Environ. 157, 72–84. Available from: https://doi.org/10.1016/j.rse.2014.05.021.

Sun, T., Lin, W., Chen, G., Guo, P., Zeng, Y., 2016. Wetland ecosystem health assessment through integrating remote sensing and inventory data with an assessment model for the Hangzhou Bay, China. Sci. Tot. Environ. 566–567, 627–640. Available from: https://doi.org/10.1016/j.scitotenv.2016.05.028.

Sunitha, A., Suresh, B.G., 2015. Satellite image classification methods and techniques: a review. Int. J. Comput. Appl. 119 (8), 20–25. Available from: https://doi.org/10.1021/acs.orglett.7b02266.

Thiemann, S., Kaufmann, H., 2002. Lake water quality monitoring using hyperspectral airborne data—a semi empirical multisensory and multi-temporal approach for the Ecklenburg Lake District, Germany. Remote Sens. Environ. 81, 228–237.

Timm, B.C., McGarigal, K., 2012. Fine-scale remotely-sensed cover mapping of coastal dune and salt marsh ecosystems at Cape Cod National Seashore using Random Forests. Remote. Sens. Environ. 127, 106–117. Available from: https://doi.org/10.1016/j.rse.2012.08.033.

Turpie, K.R., Klemas, V.V., Byrd, K., Kelly, M., Jo, Y.H., 2015. Prospective HyspIRI global observations of tidal wetlands. Remote Sens. Environ. 167, 206–217. Available from: https://doi.org/10.1016/j.rse.2015.05.008.

Upasani, V.N., 2008. Microbiological Studies on Sambhar Lake (Salt of Earth) Rajasthan, India. In: Proceedings of TAAL 2007: The 12th World Lake Conference, pp. 448–450.

Van Dam, A.A., Kipkemboi, J., Rahman, M.M., Gettel, G.M., 2013. Linking hydrology, ecosystem function, and livelihood outcomes in African papyrus wetlands using a bayesian network model. Wetlands 33 (3), 381–397. Available from: https://doi.org/10.1007/s13157-013-0395-z.

Vibhute, A.D., Kale, K.V., Dhumal, R.K., Mehrotra, S., 2015. Soil type classification and mapping using hyper-spectral remote sensing data. In: International Conference on Man and Machine Interfacing (MAMI), 1 (978-1-5090), 0–3.

Vijay, R., Pinto, S.M., 2016. A multi-temporal analysis for change assessment and estimation of algal bloom in Sambhar Lake, Rajasthan. Environ. Monit. Assess. 188 (9).

Vinciková, H., Hanuš, J., Pechar, L., 2015. Spectral reflectance is a reliable water-quality estimator for small, highly turbid wetlands. Wetl. Ecol. Manag. 23 (5), 933–946. Available from: https://doi.org/10.1007/s11273-015-9431-5.

Wang, R., Wu, F., Chang, Y., Yang, X., 2016. Waterbirds and their habitat utilization of artificial wetlands at Dianchi Lake: implication for waterbird conservation in Yunnan–Guizhou plateau lakes. Wetlands 36 (6), 1087–1095. Available from: https://doi.org/10.1007/s13157-016-0823-y.

Wigmore, O., Mark, B., Mckenzie, J., Baraer, M., Lautz, L., Blanca, C., 2019. Sub-metre mapping of surface soil moisture in proglacial valleys of the tropical Andes using a multispectral unmanned aerial vehicle. Remote Sens. Environ. 222 (December 2018), 104–118. Available from: https://doi.org/10.1016/j.rse.2018.12.024.

Wu, Q., 2017. GIS and remote sensing applications in wetland mapping and monitoring. Compr. Geogr. Inf. Syst. (September 2017), 140–157. Available from: https://doi.org/10.1016/b978-0-12-409548-9.10460-9.

Yan, Y.E., Ouyang, Z.T., Guo, H.Q., Jin, S.S., Zhao, B., 2010. Detecting the spatiotemporal changes of tidal fl ood in the estuarine wetland by using MODIS time series data. J. Hydrol. 384 (1), 156–163.

Yanmin, Y., Youqi, C., Yingbin, H., Tang, P., Na, W., 2010. Soil moisture monitoring using hyper-spectral remote sensing technology. In: International Conference on Geoscience and Remote Sensing, pp. 373–376.

Yu, H., Kong, B., Wang, G., Du, R., Qie, G., 2018. 2018. Prediction of soil properties using a hyperspectral remote sensing method. Arch. Agron. Soil Sci. 64 (4), 546–559.

Zhang, T.T., Zeng, S.L., Gao, Y., Ouyang, Z.T., Li, B., Fang, C.M., et al., 2011. Using hyperspectral vegetation indices as a proxy to monitor soil salinity. Ecol. Indic. 11 (6), 1552–1562. Available from: https://doi.org/10.1016/j.ecolind.2011.03.025.

SECTION IV

Hyperspectral Remote Sensing Application in Soil and Mineral Exploration

13

Spectroradiometry as a tool for monitoring soil contamination by heavy metals in a floodplain site

Salim Lamine[1,2], Manish Kumar Pandey[3], George P. Petropoulos[4], Paul A. Brewer[2], Prashant K. Srivastava[5,6], Kiril Manevski[7,8], Leonidas Toulios[9], Nour-El-Islam Bachari[10], Mark G. Macklin[11]

[1]FACULTY OF NATURAL SCIENCES AND LIFE AND EARTH SCIENCES, UNIVERSITY AKLI MOHAND OULHADJ OF BOUIRA, BOUIRA, ALGERIA
[2]DEPARTMENT OF GEOGRAPHY AND EARTH SCIENCES, UNIVERSITY OF ABERYSTWYTH, CEREDIGION, UNITED KINGDOM
[3]INSTITUTE OF ENVIRONMENT AND SUSTAINABLE DEVELOPMENT, BANARAS HINDU UNIVERSITY, VARANASI, INDIA
[4]DEPARTMENT OF GEOGRAPHY, HAROKOPIO UNIVERSITY OF ATHENS, KALLITHEA, ATHENS, GREECE
[5]REMOTE SENSING LABORATORY, INSTITUTE OF ENVIRONMENT AND SUSTAINABLE DEVELOPMENT, BANARAS HINDU UNIVERSITY, VARANASI, INDIA
[6]DST-MAHAMANACENTRE FOR EXCELLENCE IN CLIMATE CHANGE RESEARCH, BANARAS HINDU UNIVERSITY, VARANASI, INDIA
[7]DEPARTMENT OF AGROECOLOGY, AARHUS UNIVERSITY, TJELE, DENMARK
[8]SINO-DANISH CENTER FOR EDUCATION AND RESEARCH, EASTERN YANQIHU CAMPUS, BEIJING, P.R. CHINA
[9]DEPARTMENT OF SOIL WATER RESOURCES, INSTITUTE OF INDUSTRIAL & FORAGE CROPS, HELLENIC AGRICULTURAL ORGANIZATION (HAO) "DEMETER" (FORMER NAGREF), DIRECTORATE GENERAL OF AGRICULTURAL RESEARCH, LARISSA, GREECE
[10]FACULTY OF BIOLOGICAL SCIENCES, UNIVERSITY OF SCIENCES AND TECHNOLOGY HOUARI BOUMEDIENE, BAB EZZOUAR, ALGERIA
[11]SCHOOL OF GEOGRAPHY, COLLEGE OF SCIENCE, UNIVERSITY OF LINCOLN, LINCOLN, UNITED KINGDOM

Hyperspectral Remote Sensing. DOI: https://doi.org/10.1016/B978-0-08-102894-0.00002-4

13.1 Introduction

With the growing population and increasing demands for food, identifying contaminated soil sites and their management has turned into a consequential research area (Henke and Petropoulos, 2013; Liu et al., 2013; Luo et al., 2012). Thus an effective and efficient solution is required in identifying and resolving the risk due to these contaminations (Choe et al., 2008a; Cai et al., 2012; Al Maliki et al., 2014; Pandit et al., 2010). Traditional techniques for the assessment of heavy metal soil contamination (HMSC) are time-consuming as well as expensive (Foulds et al., 2014; Choe et al., 2008b; Djokić et al., 2011; Zhang et al., 2011). These techniques generally involve field-based soil/sediment sampling, wet chemical digestion followed by laboratory analysis, and interpolating the results to generate spatial risk maps (Srivastava et al., 2011, 2012; Sharma et al., 2012). With technological advancement, hyperspectral remote sensing has become a preferential solution for detecting HMSC in an effective and efficient manner (Farrand and Harsanyi, 1997; Ferrier, 1999; Lamine et al., 2017a; El Islam et al., 2017; Meharrar and Bachari, 2014).

River systems usually get contaminated by metals such as lead (Pb), zinc (Zn), cadmium (Cd), and copper (Cu) through mineralization from drain catchments. In the 18th and the 19th centuries, base metal mining was at its peak due to a lack of effective regulations for averting the release of these contaminants into waterbodies. Floods have been the main force behind these contaminant dispersions, which result in land segmentation that remains for 10–100 years until their remobilization through surface or riverbank loss. The resultant contaminations in floodplain soils and sediments are hazardous not only to human health, but also to agricultural products and the environment (Johnston, 2002; Foulds et al., 2014; Gozzard et al., 2011; Mayes et al., 2013).

Technological advances in hyperspectral remote sensing have been widely applied in HMSC studies as they are able to provide assessments in a rapid and cost-effective way. Hyperspectral imaging has been explored in the past for mapping the distribution of heavy metals in affected areas (You et al., 2011; Srivastava et al., 2013; Lamine et al., 2014; Pandey et al., 2018; Rosero-Vlasova et al., 2016). Hyperspectral imaging is an efficient method for finding the properties of soil along with its mineral concentration due to the output produced, which is spectrally rich as well as spatially continuous. This output can be utilized for mapping as well as soil contamination monitoring. A relatively cost-effective method is reflectance spectroradiometry, which is based on chemistry (Choe et al., 2008b; Summers, 2009; Ben-Dor et al., 2002; Wu et al., 2005; Ren et al., 2009). The resultant spectral signatures from the soil are characterized into three types as illustrated in Fig. 13–1,

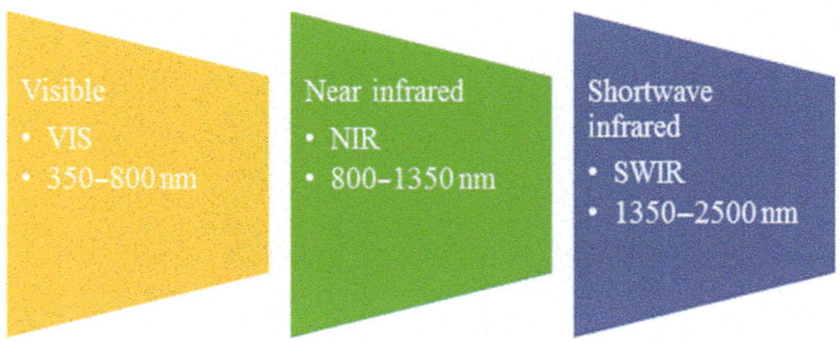

FIGURE 13–1 The portrayal of the spectral signatures obtained from the soil based on their reflectance values in the specific bands of the electromagnetic spectrum.

based on their reflectance in the corresponding electromagnetic spectrum (Horta et al., 2015; Nocita et al., 2015; Song et al., 2015).

Spectroradiometers are handy, which is the reason for them being popular among soil science researchers. A number of researchers have utilized their functionality after proper calibration in estimating soil properties such as carbon and nitrogen, sand and clay, cation exchange capacity and pH (Soriano-Disla et al., 2013; Stenberg et al., 2010), etc. The use of visible and near-infrared (VNIR) reflectance and soil spectra is explored for estimating soil (Schwartz et al., 2011; Shi et al., 2014).

The current work explored the use of field spectroradiometry in combination with geochemical analyses of Pb, Zn, Cu, and Cd to enable the quantification and modeling of HMSC. The objectives of the current work are:

1. Collection of field- as well as lab-based spectra from contaminated soils and the formation of the associated spectral libraries.
2. Identification of the specific spectral interval associated with the modeling of the HMSC through statistical discrimination analyses.
3. Collection and geochemical analysis of the soil samples.
4. Development and validation of the heavy metal prediction model (HMPM) using soil metal concentration and spectral reflectance data.

The current work explores, for the first time, the prospective of contaminant metals that are discriminated based on their spectra in a floodplain site. The main research hypotheses are:

1. The spectra of the soil display variances in specific wavelengths that endorse for their spectral discrimination.
2. Heavy metal concentrations can be obtained from spectra with improved accuracy.
3. The sample with the highest heavy metal concentration (color of the soil will be darker) would carry the lowest reflectance/highest absorbance and the reflectance would increase in the ratio of decreasing heavy metal concentration.

13.2 Distribution and vulnerabilities of heavy metals in the United Kingdom

Regardless that metal mining was stopped long back, high concentrations of heavy metals were observed in the west draining rivers after the flooding of June 2012. The concentrations in the flood sediments were surprisingly far above the permitted level as per the national and European standards (Foulds et al., 2014). It was observed (Macklin et al., 1997, 2006; Dennis et al., 2003; Brewer et al., 2005) that massive floods could've caused the distribution and overbank sedimentation of these highly contaminated constituents from the catchments of ceased metal mining sites. Even minute contaminants could lead to a catastrophe in the flexibility, dynamism, and arrangement of the ecosystem. It was observed (Kooistra et al., 2003; Smith et al., 2009; Ning et al., 2012) that sheep can ingest high concentrations of heavy metals per day (1685 mg of Pb, 486 mg of Zn, and 60 mg Cu) during winter in the Ystwyth valley.

The extraction of Pb, Zn, and Cu from West Wales was found to have been reported a long time ago, as far back as roman times or the bronze ages in some civilizations. Mining of Pb and Zn was at a peak in the middle of the 19th century and ceased early in the 20th century (Lamine et al., 2017a,b; Lamine and Petropoulos, 2013; Evans et al., 2018; Petropoulos et al., 2016).

The vulnerability of heavy metals was observed in offal by many researchers. For example, Rodríguez-Estival et al. (2012) observed that 91.4% of cattle and 13.5% of sheep were carrying high Pb blood concentrations. Later it was found that two cattle had Pb concentrations equivalent to clinical poisoning. Bovine species are also found to be susceptible to Pb poisoning, especially young ones (Neathery and Miller, 1975; Ward et al., 1978) Human health was at risk being a part of this food chain. The threat was substantial taking into consideration the events of floods occurring during the past century, which definitely would've contaminated the soil with heavy metals in West Wales and associated areas in the United Kingdom. This calls for state of the art techniques such as hyperspectral remote sensing to portray qualitatively and quantitatively the heavy metal contamination and offer a solution to counter the vulnerabilities associated with this (Foulds et al., 2014; Lamine et al., 2014).

13.3 Materials and methods

13.3.1 Study area and soil sampling

Bow Street in West Wales, the United Kingdom was identified as the study area. A triangular plot of 40 Ha of land was taken and divided into a sequence of experimental plots to be managed by Aberystwyth University's Institute of Biological, Environmental and Rural Sciences. The foremost land cover of the study area was a forage crop used for sheep and cattle grazing (Fig. 13–2).

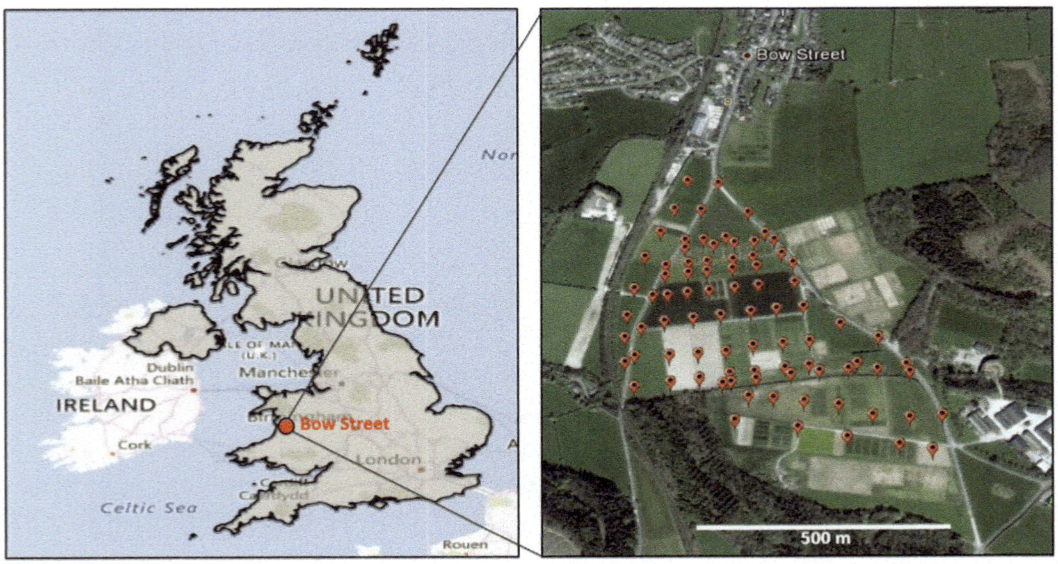

FIGURE 13–2 The topographical location of the study area and the identified 85 sampling points.

Overall, 85 surface soil samples (0–5 cm) were identified, which are depicted in Fig. 13–2. The characteristics of the soil samples were:

1. They were collected after removing the vegetation from the soil surface.
2. They were taken in a ratio of 500 g per 5 spot samples in one go from an area of 1 m^2.
3. Initially, samples were kept in wet strength soil bags followed by their treatment in an oven for 48 hours at 40°C.

The output of Foulds et al. (2014) was used as a benchmark for the current work. The benchmark proved that the flood sediments were polluted toward the higher end of the contamination scale. It was found that crop fodder harvested in flood-affected regions carried up to 1900 mg of Pb per kilogram of sediment, which were prone to causing cattle contamination and death.

13.3.2 Field and laboratory spectral measurements

An ASD (analytical spectral devices) FieldSpec3 portable spectroradiometer in hand-held mode was procured in August 2014. It has a spectral resolution of 3 nm in the 350–1000 nm range and 10 nm in the 1001–2500 nm range. The ranges interpolated to 1 nm during the measurements. The spectral measurement in the field took place before the soil samples were taken from the 85 locations. The instrument was properly warmed up for 30 minutes to avoid errors caused by spectroradiometer array. After the vegetation was removed, five spectral measurements were carried out from each of the sample locations and these were averaged to a single spectrum. After this, soil samples were taken from the exact spots that the spectra were measured. This whole procedure was implemented in clear sky condition from 10 a.m. to 2 p.m. with the sun as the only source of illumination. Prior to this, white panel reference data was recorded for each of the soil measurements. The readings were taken with the help of a pistol grip pointed toward each soil sample at a distance of 50 cm. The radius of the field of view (FOV) was set at 3.5 cm. The radius was calculated using Eq. (13.1), where R is the radius of the FOV, H is the height from the soil to the sensor in the pistol grip, and angle of view (AOV) is the angle of view of the sensor (8°).

$$R = tg\left(\frac{\mathrm{AOV}}{2}\right) \times H \times 100\ (\mathrm{cm}) \tag{13.1}$$

The same procedure was carried out in the laboratory, the only difference was the source of illumination, which was a 100 W reflectorized halogen lamp aligned at 12° to the probe body, and the sensed spot had a diameter of dprobe = 1.1 cm with a FOV = 1.33 cm^2. Here the spectrum collection was carried out using a high-intensity contact probe (CP; direct contact with the soil). Here, the soil with a particle size <2 mm was placed in a black plastic dish equivalent to the size of a Petri dish and the CP was aligned in direct contact with the soil and then the spectra were registered. The measurements were repeated three times followed by it being averaged to a single spectrum (Figs. 13–3 and 13–4).

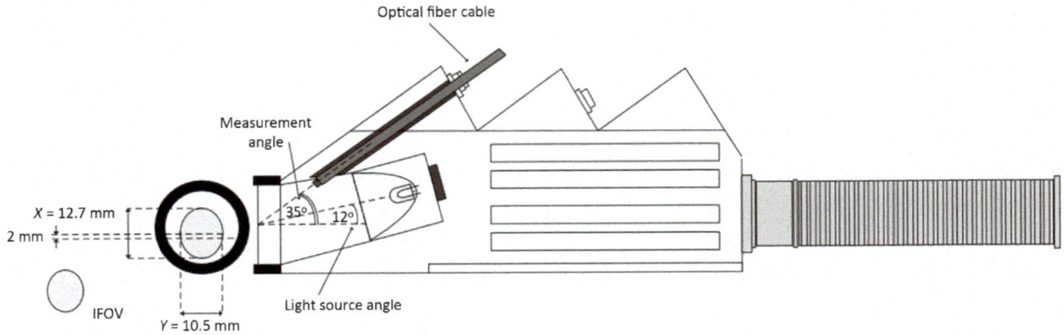

FIGURE 13–3 Illustration of the analytical spectral devices high-intensity contact probe according to ASD Inc. (ASD, 2008). *X* and *Y* are the height and width respectively of the field of view (FoV).

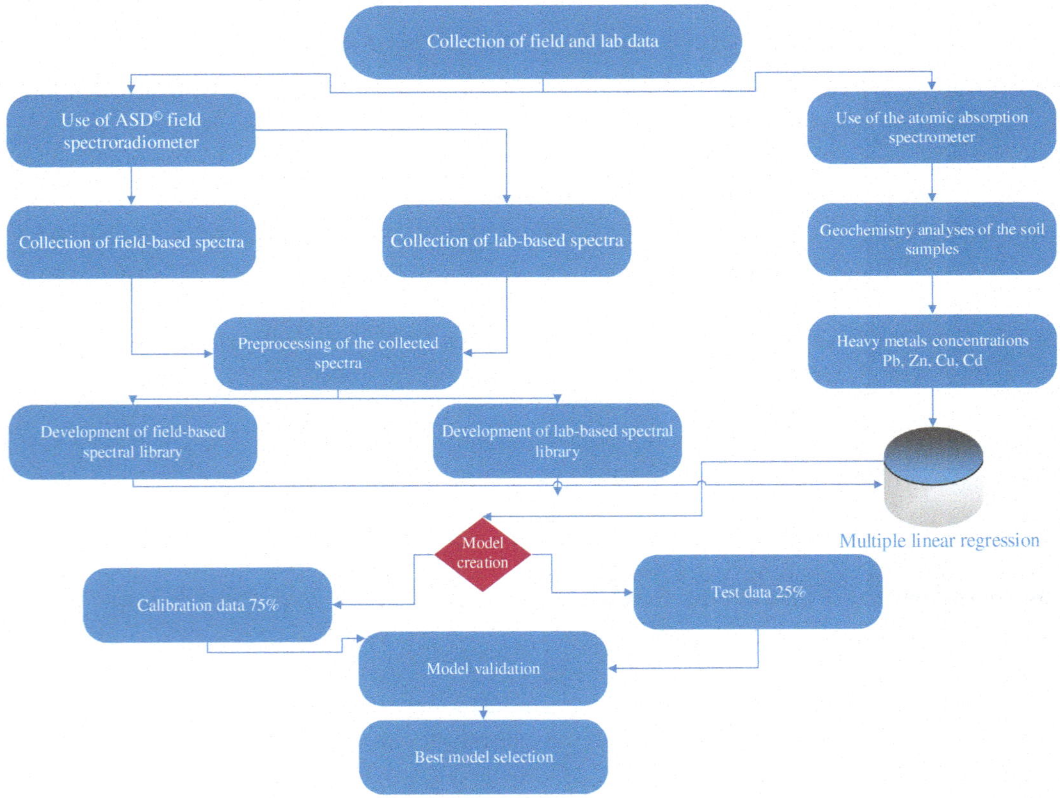

FIGURE 13–4 Flowchart of the steps used in the current study.

13.3.3 Geochemical analysis of the soil samples

A number of acids and their combinations are found to be effective in the decomposition of soil or rock samples like hydrofluoric–perchloric–nitric or perchloric–nitric acid. Generally iron

(Fe(III)) oxide minerals are not affected by single acids. The steps involved in the geochemical analysis of the 85 soil samples were:

1. Nitric acid was used for the extraction of heavy metals (Cd, Cu, Pb, and Zn).
2. The soil samples were put in an oven for drying at 40°C.
3. Soil samples with particle size <63 μm were taken as the main source of contamination with the highest concentration (Wolfenden and Lewin, 1977) of heavy metals for the examination.
4. Soil samples were weighed to 0.5 ± 0.005 g using a weighing boat followed by their allocation in clearly labeled boiling tubes.
5. An automatic dispenser was used to add 2 mL of concentrated nitric acid to the samples.
6. The tubes were placed into a digestion block and exposed to 100°C for 1 hours.
7. The boiling tubes were taken out of the block and allowed to cool.
8. An automatic dispenser was again used to add 18 mL of distilled water to the contents of the tubes. A whirlimixer (Fisher Scientific Ltd., United Kingdom) was used for a thorough mixing of the contents.
9. The mixes were covered in cling film and left overnight for the suspended particles to settle down.
10. Without blocking the capillary tube, the samples were sprayed into the flame of an atomic absorption spectrometer (PerkinElmer Inc., United States).
11. Though the dilution factor used was 40, this could be varied using an automatic dilutor by preparing serial dilutions of $10\times$. These were applicable in cases where the concentration was more than the calibration range of the spectrometer.
12. To control the analytical activities, certified reference material (GBW 07307 stream sediment) was made available.

13.3.4 Data processing and statistics

The resultant field and lab spectra were corrected by removing their continuum as suggested by Clark and Roush (1984) followed by their normalization to achieve maximum spectral absorption features. Continuum removal analysis is the standard transformation used in land cover spectral discrimination (Manevski et al., 2011, 2012, 2017). Continuum is a convex hull with straight-line segments that are fitted over the spectrum and need to be removed through division or rationing relative to the spectrum (Meer, 2006).

To find the broader spectral bands sensitive to heavy metal concentrations, an analysis of variance (ANOVA) is used. The analysis was carried at each wavelength from 350 to 2500 nm for each of the fields as well as the lab-based spectral libraries at a 95% confidence level. After the ANOVA, a correlation analysis was carried out between the spectral features and the heavy metal concentrations followed by calculating the Pearson's correlation coefficient. This coefficient is an excellent indicator of dependence between two quantities and is obtained by dividing the covariance of the two variables by the product of their standard deviations. This is estimated using Eq. (13.2), where E is the expected value, μ is the mean,

Cov is the covariance between *X* and *Y*, and *Corr* is the correlation coefficient (Song et al., 2015).

$$Corr(X, Y) = \frac{Cov(X, Y)}{\sigma X \sigma Y} = \frac{E[(X - \mu_X)(Y - \mu_Y)]}{\sigma X \sigma Y} \tag{13.2}$$

The higher the coefficient, the stronger the linear correlation is. After this, stepwise multiple linear regression (SMLR) was performed to create the HMSC models. In this, at each step, the independent variable (wavelength) and the smallest probability of *F* was entered, if the probability was sufficiently small. In the case of large probability, the variables in the regression equations were removed. The terminating condition (Zhao et al., 2013) is when no more variables are eligible for inclusion or removal. The generated regression model is represented in Eq. (13.3).

$$\text{HMSC (mg/kg)} = [A_n \, R_{350-2500} + B] \times 1000 \tag{13.3}$$

where HMSC is the heavy metal soil concentration (mg/kg), A_n is the slope of the regression (n coefficients of the regression), $R_{350-2500}$ is the reflectance wavelength varying from 350 to 2500 nm, B is the regression constant, and the result was multiplied by 1000 in order to obtain the concentration of heavy metal in milligrams per kilogram (mg/kg). The framework can be represented in the form of a flowchart as seen in Fig. 13–4.

13.4 Results and discussion

13.4.1 Soil descriptive statistics

The current study focuses on four major heavy metals, namely Pb, Zn, Cu, and Cd (Foulds et al., 2014) as the main contaminants. The descriptive statistics of the geochemical analyses are summarized in Table 13–1. The high standard deviation of Pb and Zn (1037.96 and 59.85, respectively) implies large spatial variability and the possible existence of "hot spots," which are represented in Figs. 13–5 and 13–6. The concentrations of Cu and Cd showed lower

Table 13–1 Descriptive statistics of heavy metal concentrations in the soil of the contaminated area in Bow Street, United Kingdom, based on 85 samples.

mg/kg	Pb	Zn	Cu	Cd
Max	4600	361	249	2
Min	220	82	13	1
Median	670	140	32	1
Mean	1100	156	47	1
Stdev	1037.959	59.850	42.869	0.204

Max, Min, Median, Mean, and Stdev are maximum, minimum, median, mean, and standard deviation, respectively. The minimum detection limits of the atomic absorption spectrometer were 0.8, 1.5, 1.5, and 15 mg/kg for Cd, Cu, Zn, and Pb, respectively.

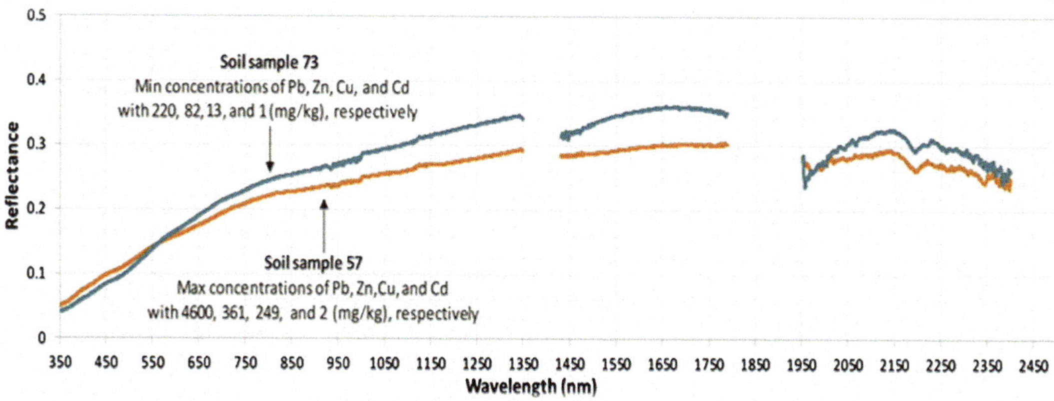

FIGURE 13–5 Mean spectra for soil samples characterized by low (sample 57) and high (sample 73) concentrations of heavy metals in the study site.

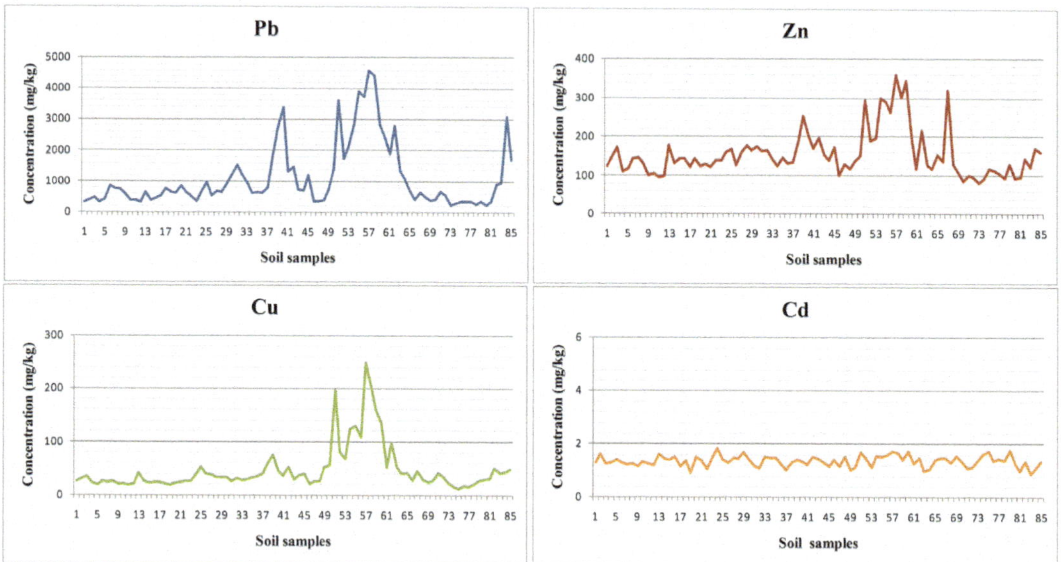

FIGURE 13–6 Mean ($n = 85$) variation in concentrations of the four heavy metals found in the study site.

magnitudes and variations. Pb was found in the highest concentrations among the metals, its concentrations could be attributed to the variation of the soil reflectance.

The third hypothesis of this study is supported by Fig. 13–5, which depicts a lower reflectance for the highest contaminated soil sample (sample 57 in this case) as compared to the reflectance of the least contaminated soil sample (sample 73 in this case). More exploration is required to support this claim. The figure clearly depicts that the reflectance is governed by the concentration of heavy metals in each of the samples. From Fig. 13–6, it is clear that sample 57 had the maximum concentrations of the four heavy metals, whereas sample 73 had the minimum concentrations pertaining to the lower reflectance value.

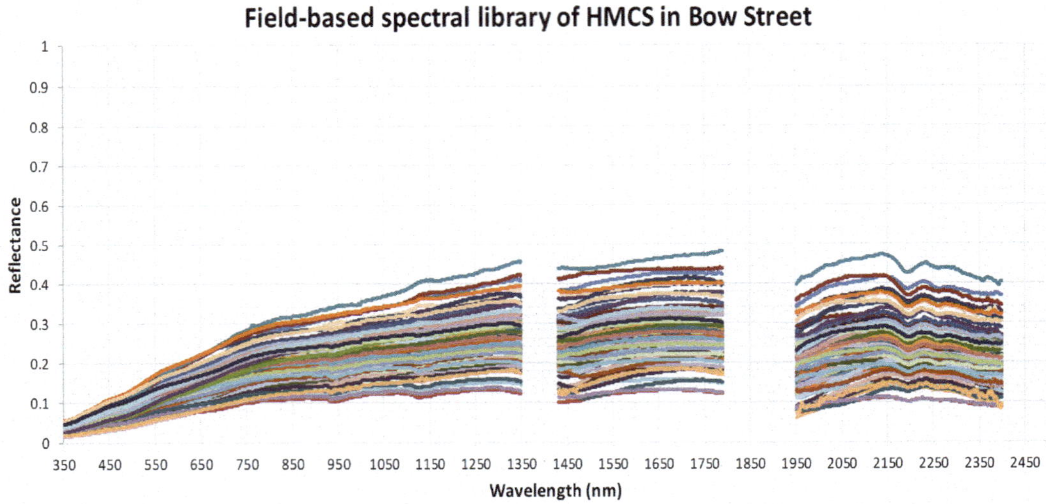

FIGURE 13–7 Field-based spectral library of heavy metal soil contamination (HMSC) at the Bow Street site. Spectral regions related to water vapor absorption (1350–1430, 1790–1950, and 2400–2500 nm) have been removed.

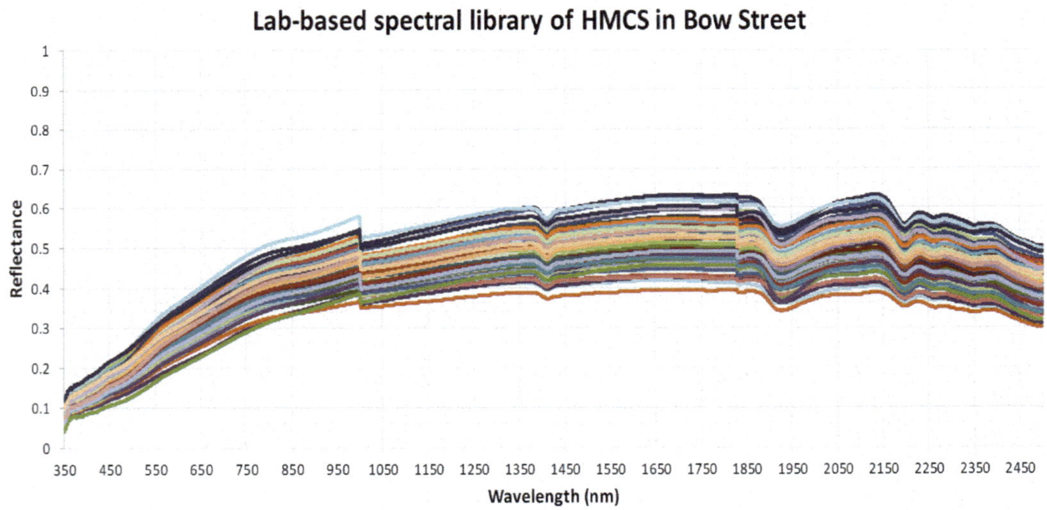

FIGURE 13–8 Lab-based spectral library of the heavy metal soil contamination (HMSC) at the Bow Street site.

13.4.2 Creation of field- and lab-based spectral libraries

The spectral libraries obtained from the field and the lab are composed of 85 spectra each and are depicted in Figs. 13–7 and 13–8, respectively. The soil spectra showed a continued increase in the VIS and relatively consistent in the NIR and SWIR. There are some lags in 1400, 1900, and 2200, which can be attributed to water and clay absorption. There are some visible variations for both the libraries in the NIR and majorly in SWIR, which again could be attributed to the disparity in the soil properties like moisture, clay, and organic matter content as well as the content of heavy metals. The resulting spectral libraries enrich the spectral database and may serve as reference spectra for HMSC in the United Kingdom. This also enables the validation of the reflectance information to be extracted from the radiance data from various remote sensing platforms and, thus, can play a crucial role in tracking temporal changes of the soil spectra over sampling locations.

13.4.3 Statistical discrimination analysis

The results of the reflectance after continuum removal at each wavelength through the process of ANOVA are depicted in Figs. 13–9 and 13–10 for the field- and lab-based spectral libraries, respectively. The red dashed line in the figures denotes the critical *P* value (.05). Below this value, significant results are achieved (shaded gray, which represents the spectral regions that contain at least one significantly different soil spectral from the others).

The output clearly demonstrates the significant differences in the mean continuum removed field-based soil spectra in the VIS spectrum range of 350–800 nm and narrower windows in the NIR and SWIR. For lab-based soil spectra, the most significant result was obtained in the VNIR range of 360–1270 nm and fewer windows at the end of the NIR. Though there was no significant achievement in reducing the wavelengths with the help of ANOVA, it helped in the statistical modeling of HMSC by serving as an input.

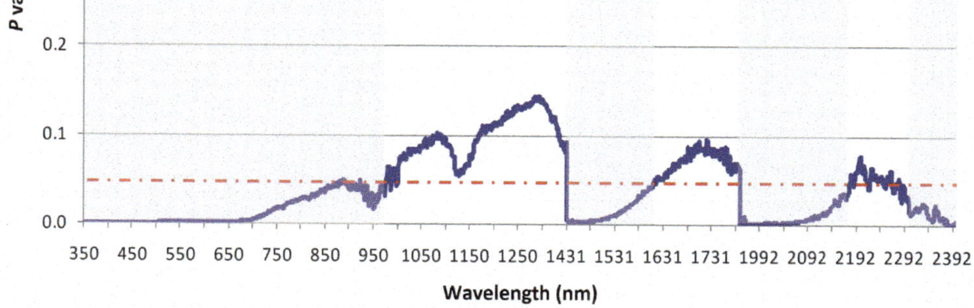

FIGURE 13–9 Wavelength-intervals shaded gray depict statistically significant differences between the field-based spectra. The red dashed line denotes the limit for statistical significance (95% confidence level).

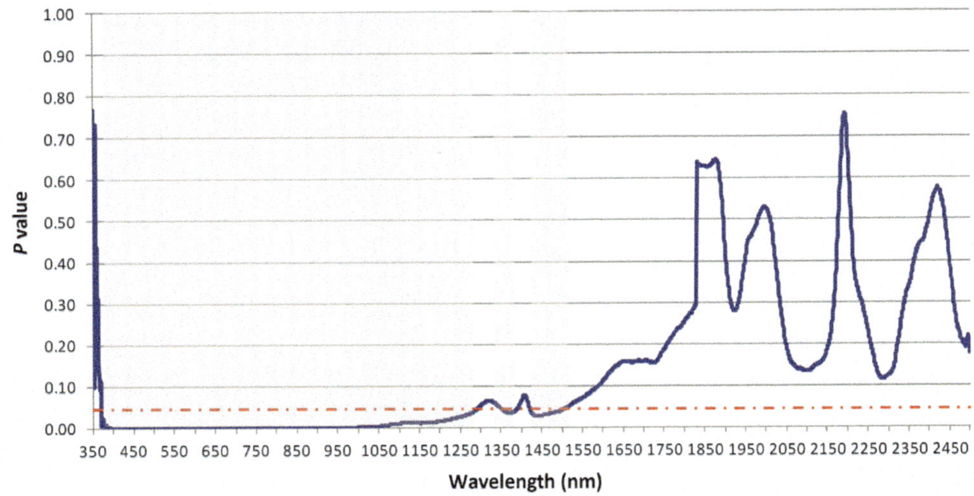

FIGURE 13–10 Wavelength-intervals shaded gray depict statistically significant differences between the lab-based spectra. The red dashed line denotes the limit for statistical significance (95% confidence level).

13.4.4 Model development and validation

To find the most useful spectral bands that contribute in the most efficient way to building the SMLR model, the evaluation parameter known as the coefficient of determination (R^2, the square of the Pearson correlation coefficient) was used. These values for the field- and lab-based prediction models are listed in Tables 13–2 and 13–3, respectively. In the regression analysis, the independent variable used was the spectral bands, whereas the dependent variable used was the concentration of heavy metals. For each of the steps in the SMLR modeling algorithm, the independent variables wih the lowest probability are progressively integrated till the terminating conditions are met. The performance quality for each of the calibrations was evaluated using R^2. A model was proposed by Song et al. (2015) to assess aluminum, copper, and chrome contamination both in the soil and water taken from a mining area in China. They derived spectral features of these metals from the measured spectra. They established a linear correlation between spectral wavebands and heavy metal concentrations. Cd and Pb contaminations were explored in Chinese soils (Liu et al., 2016) using a spectroradiometer and R^2 was obtained based on SMLR modeling of −0.65 to 0.82 for Cd and 0.78−0.88 for Pb. The result obtained in the current work is slightly similar and occasionally better than previous studies. The model could be tested and improved upon using intact or transformed spectra [e.g., the logarithm of reciprocal spectra (Liu et al., 2016); derivative spectra (Stenberg et al., 2010)] and other statistical models [e.g., generalized regression neural network (Dong et al., 2016)].

Based on Table 13−2, the four developed field-based HMPMs were:

$$Pb_{FSpec} = [-320.758R_{354} + 456.742R_{389} - 94.144R_{582} + 92.316R_{1719} - 82.081R_{1775} - 0.172] \times 1000$$

Table 13–2 Summary of the selected spectral bands and regression coefficients for the field-based spectral library using stepwise multiple linear regression.

	Model coefficients for the studied heavy metals			
Spectral bands (nm)	**Pb**	**Zn**	**Cu**	**Cd**
354	− 320.758	–	–	–
366	–	− 64.043	–	–
367	–		− 81.125	–
368	–	–	42.275	–
374	–	71.865	–	–
386	–	− 57.897	–	–
388	–	90.868	–	–
389	456.742	–	64.551	–
393	–	66.374	–	–
394	–	− 96.782	–	–
434	–	–	− 23.652	–
582	− 94.144	–	–	–
586	–	− 6.142	–	–
1348	–	0.965	–	–
1719	92.316	–	–	–
1775	− 82.081	–	–	–
1951	–	–	–	0.008
1978	–	–	–	− 0.007
Constant	− 0.172	0.139	0.026	0.001
R^2	0.671	0.697	0.561	0.123

A dash denotes that the spectral band was not included in the model equation for the relevant heavy metal.

$$Zn_{FSpec} = [-64.043R_{366} + 71.865R_{374} - 57.897R_{386} + 90.868R_{388} + 66.374R_{393} - 96.782R_{394} - 6.142R_{586} + 0.965R_{1348} + 0.139] \times 1000$$
$$Cu_{FSpec} = [-81.125R_{367} + 42.275R_{368} + 64.551R_{389} - 23.652R_{434} + 0.026] \times 1000$$
$$Cd_{FSpec} = [0.008R_{1951} - 0.007R_{1978} + 0.001] \times 1000$$

Based on Table 13–3, the four developed lab-based HMPMs were:

$$Pb_{LSpec} = (90.729R_{356} - 25.105R_{618} - 0.057) \times 1000$$
$$Zn_{LSpec} = (-4.369R_{358} + 5.055R_{368} + 9.101R_{376} - 78.747R_{470} + 127.870R_{475} - 53.910R_{484} - 0.048) \times 1000$$
$$Cu_{LSpec} = (2.502R_{359} - 0.628R_{651} - 0.016) \times 1000$$
$$Cd_{LSpec} = (-0.001R_{1465} + 0.002) \times 1000$$

During analysis, it was observed that a number of wavelengths that correlated with heavy metal soil concentrations were in VIS bands. This could be attributed to:

1. Soil properties such as molecular structure, organic matter content, etc.
2. Chemical responses to the contaminants.
3. The physical environment.

Table 13–3 Summary of the selected spectral bands and regression coefficients for the lab-based spectral library using stepwise multiple linear regression.

	Coefficients of the studied heavy metals			
Spectral bands (nm)	**Pb**	**Zn**	**Cu**	**Cd**
356	90.729	–	–	–
358	–	−4.369	–	–
359	–	–	2.502	–
368	–	5.055	–	–
376	–	9.101	–	–
470	–	−78.747	–	–
475	–	127.870	–	–
484	–	− 53.910	–	–
618	−25.105	–	–	–
651	–	–	−0.628	–
1465	–	–	–	−0.001
Constant	−0.057	−0.048	−0.016	0.002
R^2	0.641	0.642	0.428	0.048

A dash denotes that the spectral band was not included in the model equation for the relevant heavy metal.

Most of the heavy metals do not offer any spectral features in NIR and SWIR, which demands further exploration of the physical relationship between the spectral data and the heavy metals and could be affected by the presence or absence of other inorganic components such as iron cations, carbonate cations, and phosphate, etc. (Siebielec et al., 2004). Thus the mathematical relations such as HMPM can be used for prediction, testing, calibration, and validation (Siebielec et al., 2004; Mohamed et al., 2018).

The efficient behavior of the ASD field spectroradiometer makes it a wonderful instrument for the precise collection of spectral data. The establishment of a correlation between the estimated heavy metal concentrations and the predicted heavy metal contents guide us in the direction of a feasible solution using SMLR models to build efficient and effective predictive models based on spectral measurements and geochemical variables from lab-based analyses (Liu et al., 2016; Leone, 2000; Bachari et al., 2004). Thus the current research highlights and contributes to the growing field of hyperspectral imaging through the amalgamation of the state of the art features of field spectroradiometry data and lab-based geochemical analyses to predict heavy metal contamination.

13.5 Conclusion

In order to quantify and model heavy metal contamination, the current work puts emphasis on the latent role of field- and laboratory-based hyperspectral data along with geochemical data of Pb, Zn, Cu, and Cd. This is carried out in a highly contaminated floodplain site in

Wales in the west of the United Kingdom. The results of the work endorse the hypothesis of predefined study. First, they support the spectral discrimination of the soil in a study which confirms the fact that soil spectral signatures exhibit differences in specific wavelengths of electromagnetic spectrum. Second, with the help of a field spectroradiometer, heavy metal concentrations can be retrieved from the spectral reflectance data in the range of 350–2500 nm with rational accuracy. Third, the samples with the highest heavy metal concentrations possess the lowest reflectance and the reflectance increases as the concentration of heavy metals decreases.

A total of 85 soil samples from the contaminated area were used for deriving the field- and lab-based spectral features that were used to develop two spectral libraries. These libraries in combination were used to build eight heavy metal prediction models using SMLR. The output establishes great prospect for the prediction of HMSC in a highly contaminated floodplain through the combination of soil geochemical analyses and spectroradiometry.

The main challenge of heavy metal contamination in West Wales and the surrounding areas in the United Kingdom can be encountered effectively using hyperspectral spectroradiometry, which offers a state of the art, fast, and efficient facility to map HMSC. Though spectral features of soil in the spectral range of 350–2500 nm are complex, identifying the exact spectral wavebands corresponding to Pb, Cu, Zn, and Cd concentrations unaffected by the chemical composition and surviving in the physical conditions of the soil surface is a noteworthy challenge. The offered prediction model in the current study can provide an unconventional solution to predict heavy metal soil contamination using field and laboratory hyperspectral measurements. This can be used as a model in the future over a hefty area using space-borne hyperspectral sensors such as Hyperion, AVIRIS, EnMAP, and CHIRS Proba, etc.

Acknowledgments

The authors would like to thank Prof William Haresign (Emeritus Professor at the Institute of Biological, Environmental and Rural Sciences, Animal Systems Research Group, Aberystwyth University) for facilitating the access to the study site and his advice on-field protocol. The authors appreciate the effort of the laboratory engineer Wynne Ebenezer in assisting the geochemistry analyses. The authors also thank the anonymous reviewers for their constructive and useful comments.

List of abbreviations

ANOVA analysis of variance
HMPM heavy metal prediction models
HMSC heavy metal soil contamination

List of symbols

A_n the slope of the regression
Cd Cadmium
Cu Copper

n	coefficients of the regression
Pb	Lead
$\boldsymbol{R_{350\text{-}2500}}$	the reflectance wavelength varying from 350 to 2500 nm
Zn	Zinc

References

Al Maliki, A., Bruce, D., Owens, G., 2014. Prediction of lead concentration in soil using reflectance spectroscopy. Environ. Technol. Innov. 1, 8–15. Available from: https://doi.org/10.1016/j.eti.2014.08.002.

ASD, 2008. Integrating Sphere User Manual, ASD Document 600660. ASD Inc., Boulder, CO.

Bachari, N.E.I., Khodja, S., Belbachir, A., 2004. Multispectral analysis of satellite images. In: Proceedings of the XXth International Society for Photogrammetry and Remote Sensing. Congress Geo-Imagery Bridging Continents, Istanbul, Turkey, 12–23 July 2004, pp. 1071–1073.

Ben-Dor, E., Patkin, K., Banin, A., Karnieli, A., 2002. Mapping of several soil properties using dais-7915 hyperspectral scanner data—a case study over clayey soils in Israel. Int. J. Remote Sens. 23, 1043–1062. Available from: https://doi.org/10.1080/01431160010006962.

Brewer, P.A., Dennis, I.A., Macklin, M.G., 2005. The use of geomorphological mapping and modelling for identifying land affected by metal contamination on river floodplains. DEFRA Res. Dev. Rep. SP 0525, 58.

Cai, Q.-Y., Mo, C.-H., Li, H.-Q., Lü, H., Zeng, Q.-Y., Li, Y.-W., et al., 2012. Heavy metal contamination of urban soils and dust in Guangzhou, south China. Environ. Monit. Assess. 185, 1095–1106. Available from: https://doi.org/10.1007/s10661-012-2617-x.

Choe, E., Kim, K.-W., Bang, S., Yoon, I.-H., Lee, K.-Y., 2008a. Qualitative analysis and mapping of heavy metals in an abandoned Au–Ag mine area using NIR spectroscopy. Environ. Geol. 58, 477–482. Available from: https://doi.org/10.1007/s00254-008-1520-9.

Choe, E., van der Meer, F., van Ruitenbeek, F., van der Werff, H., de Smeth, B., Kim, K.-W., 2008b. Mapping of heavy metal pollution in stream sediments using combined geochemistry, field spectroscopy, and hyperspectral remote sensing: a case study of the rodalquilar mining area, SE Spain. Remote Sens. Environ. 112, 3222–3233. Available from: https://doi.org/10.1016/j.rse.2008.03.017.

Clark, R.N., Roush, T.L., 1984. Reflectance spectroscopy: quantitative analysis techniques for remote sensing applications. J. Geophys. Res. Solid Earth 89, 6329–6340. Available from: https://doi.org/10.1029/jb089ib07p06329.

Dennis, I.A., Macklin, M.G., Coulthard, T.J., Brewer, P.A., 2003. The impact of the October–November 2000 floods on contaminant metal dispersal in the river swale catchment, North Yorkshire, UK. Hydrol. Process. 17, 1641–1657. Available from: https://doi.org/10.1002/hyp.1206.

Djokić, B.V., Jović, V., Jovanović, M., Ćirić, A., Jovanović, D., 2011. Geochemical behaviour of some heavy metals of the grot flotation tailing, southeast Serbia. Environ. Earth Sci. 66, 933–939. Available from: https://doi.org/10.1007/s12665-011-1303-6.

Dong, J., Dai, W., Xu, J., Li, S., 2016. Spectral estimation model construction of heavy metals in mining reclamation areas. Int. J. Environ. Res. Publ. Health 13, 640. Available from: https://doi.org/10.3390/ijerph13070640.

El Islam, B.N., Fouzia, H., Khalid, A., 2017. Combination of satellite images and numerical model for the state followed the coast of the bay of Bejaia–Jijel. Int. J. Environ. Geoinf. 4, 1–7. Available from: https://doi.org/10.30897/ijegeo.306486.

Evans, A., Lamine, S., Kalivas, D.P., Petropoulos, G.P., 2018. Exploring the potential of EO data and GIS for ecosystem health modeling in response to wildfire: a case study in central Greece. Environ. Eng. Manage. J. 17, 9.

Farrand, W.H., Harsanyi, J.C., 1997. Mapping the distribution of mine tailings in the coeur d'alene river valley, idaho, through the use of a constrained energy minimization technique. Remote Sens. Environ. 59, 64–76. Available from: https://doi.org/10.1016/S0034-4257(96)00080-6.

Ferrier, G., 1999. Application of imaging spectrometer data in identifying environmental pollution caused by mining at rodaquilar, Spain. Remote Sens. Environ. 68, 125–137. Available from: https://doi.org/10.1016/s0034-4257(98)00105-9.

Foulds, S.A., Brewer, P.A., Macklin, M.G., Haresign, W., Betson, R.E., Rassner, S.M.E., 2014. Flood-related contamination in catchments affected by historical metal mining: an unexpected and emerging hazard of climate change. Sci. Total Environ. Available from: https://doi.org/10.1016/j.scitotenv.2013.12.079.

Gozzard, E., Mayes, W.M., Potter, H.A.B., Jarvis, A.P., 2011. Seasonal and spatial variation of diffuse (non-point) source zinc pollution in a historically metal mined river catchment, UK. Environ. Pollut. 159, 3113–3122. Available from: https://doi.org/10.1016/j.envpol.2011.02.010.

Henke, J.M., Petropoulos, G.P., 2013. A GIS-based exploration of the relationships between human health, social deprivation and ecosystem services: the case of wales, UK. Appl. Geogr. 45, 77–88. Available from: https://doi.org/10.1016/j.apgeog.2013.07.022.

Horta, A., Malone, B., Stockmann, U., Minasny, B., Bishop, T.F.A., McBratney, A.B., et al., 2015. Potential of integrated field spectroscopy and spatial analysis for enhanced assessment of soil contamination: a prospective review. Geoderma 241, 180–209. Available from: https://doi.org/10.1016/j.geoderma.2014.11.024.

Johnston, D., 2002. A Metal Mine Strategy for Wales. Environment Agency Wales, pp. 1–11.

Kooistra, L., Wanders, J., Epema, G.F., Leuven, R.S.E.W., Wehrens, R., Buydens, L.M.C., 2003. The potential of field spectroscopy for the assessment of sediment properties in river floodplains. Anal. Chim. Acta 484, 189–200. Available from: https://doi.org/10.1016/s0003-2670(03)00331-3.

Lamine, S., Petropoulos, G.P., 2013. Evaluation of the Spectral Angle Mapper "SAM" Classification Technique using Hyperion Imagery. In: Proceedings of the European Space Agency Living Planet Symposium, Edinburgh, UK, 9–13 September 2013.

Lamine, S., Brewer, P.A., Petropoulos, G.P., Kalaitzidis, K., Manevski, K., Macklin, M.G., et al., 2014. Investigating the potential of hyperspectral imaging (HSI) for the quantitative estimation of lead contamination in soil (LCS). In: Hyperspectral Imaging and Applications Conference, Ricoh Arena, Coventry, UK, 2014.

Lamine, S., Petropoulos, G.P., Singh, S.K., Szabó, S., Bachari, N.E.I., Srivastava, P.K., et al., 2017a. Quantifying land use/land cover spatio-temporal landscape pattern dynamics from Hyperion using SVMs classifier and FRAGSTATS®. Geocarto Int. 1–47. Available from: https://doi.org/10.1080/10106049.2017.1307460.

Lamine, S., Saunders, I., Boukhalfa, S., Petropoulos, G., Bachari, N.E.I., Brewer, P., et al., **2017**. Phytoremediation of heavy metals–contaminated soils by two willow species Salix viminalis and Salix dasyclados. In: International Conference Environnement, Agriculture et Biotechnologie, Bouira, Algeria, 27–28th November 2017b.

Leone, A., 2000. Multivariate analysis of laboratory spectra for the assessment of soil development and soil degradation in the southern apennines (Italy). Remote Sens. Environ. 72, 346–359. Available from: https://doi.org/10.1016/s0034-4257(99)00110-8.

Liu, Y., Wen, C., Liu, X., 2013. China's food security soiled by contamination. Science 339, 1382. Available from: https://doi.org/10.1126/science.339.6126.1382-b.

Liu, K., Zhao, D., Fang, J.-Y., Zhang, X., Zhang, Q.-Y., Li, X.-K., 2016. Estimation of heavy-metal contamination in soil using remote sensing spectroscopy and a statistical approach. J. Indian Soc. Remote Sens. 45, 805–813. Available from: https://doi.org/10.1007/s12524-016-0648-4.

Luo, X.-S., Yu, S., Zhu, Y.-G., Li, X.-D., 2012. Trace metal contamination in urban soils of China. Sci. Total Environ. 421, 17–30. Available from: https://doi.org/10.1016/j.scitotenv.2011.04.020.

Macklin, M.G., Hudson-Edwards, K.A., Dawson, E.J., 1997. The significance of pollution from historic metal mining in the pennine orefields on river sediment contaminant fluxes to the north sea. Sci. Total Environ. Available from: https://doi.org/10.1016/s0048-9697(96)05378-8.

Macklin, M.G., Brewer, P.A., Hudson-Edwards, K.A., Bird, G., Coulthard, T.J., Dennis, I.A., et al., 2006. A geomorphological approach to the management of rivers contaminated by metal mining. Geomorphology 79, 423–447. Available from: https://doi.org/10.1016/j.geomorph.2006.06.024.

Manevski, K., Manakos, I., Petropoulos, G.P., Kalaitzidis, C., 2011. Discrimination of common mediterranean plant species using field spectroradiometry. Int. J. Appl. Earth Obs. Geoinf. 13, 922–933. Available from: https://doi.org/10.1016/j.jag.2011.07.001.

Manevski, K., Manakos, I., Petropoulos, G.P., Kalaitzidis, C., 2012. Spectral discrimination of mediterranean maquis and phrygana vegetation: results from a case study in greece. IEEE J. Sel. Top. Appl. Earth Obs. Remote. Sens. 5, 604–616. Available from: https://doi.org/10.1109/jstars.2012.2190044.

Manevski, K., Jabloun, M., Gupta, M., Kalaitzidis, C., 2017. Field-scale sensitivity of vegetation discrimination to hyperspectral reflectance and coupled statistics. In: Sensitivity Analysis in Earth Observation Modelling. Elsevier B.V., Amsterdam, Netherlands, pp. 103–121. Available from: http://dx.doi.org/10.1016/b978-0-12-803011-0.00006-9.

Mayes, W.M., Potter, H.A.B., Jarvis, A.P., 2013. Riverine flux of metals from historically mined orefields in england and wales. Water Air Soil. Pollut. 224. Available from: https://doi.org/10.1007/s11270-012-1425-9.

Meer, F.V.D., 2006. Indicator kriging applied to absorption band analysis in hyperspectral imagery: a case study from the rodalquilar epithermal gold mining area, SE Spain. Int. J. Appl. Earth Obs. Geoinf. 8, 61–72. Available from: https://doi.org/10.1016/j.jag.2005.07.001.

Meharrar, K., Bachari, N.E.I., 2014. Modelling of radiative transfer of natural surfaces in the solar radiation spectrum: development of a satellite data simulator (SDDS). Int. J. Remote Sens. 35, 1199–1216. Available from: https://doi.org/10.1080/01431161.2013.876116.

Mohamed, E.S., Saleh, A.M., Belal, A.B., Gad, A.A., 2018. Application of near-infrared reflectance for quantitative assessment of soil properties. Egypt J. Remote. Sens. Space. Sci. 21, 1–14. Available from: https://doi.org/10.1016/j.ejrs.2017.02.001.

Neathery, M.W., Miller, W.J., 1975. Metabolism and toxicity of cadmium, mercury, and lead in animals: a review. J. Dairy. Sci. 58, 1767–1781. Available from: https://doi.org/10.3168/jds.s0022-0302(75)84785-0.

Ning, Y., Li, J., Cai, W., Shao, X., 2012. Simultaneous determination of heavy metal ions in water using near-infrared spectroscopy with preconcentration by nano-hydroxyapatite. Spectrochim. Acta A 96, 289–294. Available from: https://doi.org/10.1016/j.saa.2012.05.034.

Nocita, M., Stevens, A., van Wesemael, B., Aitkenhead, M., Bachmann, M., Barthès, B., et al., 2015. Soil spectroscopy: an alternative to wet chemistry for soil monitoring. In: Advances in Agronomy. Elsevier B.V., Amsterdam, pp. 139–159. Available from: http://dx.doi.org/10.1016/bs.agron.2015.02.002.

Pandey, P.C., Manevski, K., Srivastava, P.K., Petropoulos, G.P., 2018. The use of hyperspectral earth observation data for land use/cover classification: present status, challenges, and future outlook. In: Thenkabail, P. (Ed.), Hyperspectral Remote Sensing of Vegetation, first ed. Taylor & Francis CRC Press, London, pp. 147–173.

Pandit, C.M., Filippelli, G.M., Li, L., 2010. Estimation of heavy-metal contamination in soil using reflectance spectroscopy and partial least-squares regression. Int. J. Remote. Sens. 31, 4111–4123. Available from: https://doi.org/10.1080/01431160903229200.

Petropoulos, G.P., Ireland, G., Lamine, S., Griffiths, H.M., Ghilain, N., Anagnostopoulos, V., et al., 2016. Operational evapotranspiration estimates from Seviri in support of sustainable water management. Int. J. Appl. Earth Obs. Geoinf. 49, 175–187. Available from: https://doi.org/10.1016/j.jag.2016.02.006.

Ren, H.-Y., Zhuang, D.-F., Singh, A.N., Pan, J.-J., Qiu, D.-S., Shi, R.-H., 2009. Estimation of as and cu contamination in agricultural soils around a mining area by reflectance spectroscopy: a case study. Pedosphere 19, 719–726. Available from: https://doi.org/10.1016/s1002-0160(09)60167-3.

Rodríguez-Estival, J., Barasona, J.A., Mateo, R., 2012. Blood the and δ-alad inhibition in cattle and sheep from a pb-polluted mining area. Environ. Pollut. 160, 118–124. Available from: https://doi.org/10.1016/j.envpol.2011.09.031.

Rosero-Vlasova, O.A., Pérez-Cabello, F., Montorio Llovería, R., Vlassova, L., 2016. Assessment of laboratory VIS–NIR–SWIR setups with different spectroscopy accessories for characterisation of soils from wildfire burns. Biosyst. Eng. 152, 51–67. Available from: https://doi.org/10.1016/j.biosystemseng.2016.06.011.

Schwartz, G., Eshel, G., Ben-Dor, E., 2011. Reflectance spectroscopy as a tool for monitoring contaminated soils. In: Soil Contamination. InTech, London. Available from: http://dx.doi.org/10.5772/23661.

Sharma, N.K., Bhardwaj, S., Srivastava, P.K., Thanki, Y.J., Gadhia, P.K., Gadhia, M., 2012. Soil chemical changes resulting from irrigating with petrochemical effluents. Int. J. Environ. Sci. Technol. 9, 361–370. Available from: https://doi.org/10.1007/s13762-012-0039-5.

Shi, T., Chen, Y., Liu, Y., Wu, G., 2014. Visible and near-infrared reflectance spectroscopy—an alternative for monitoring soil contamination by heavy metals. J. Hazard. Mater. 265, 166–176. Available from: https://doi.org/10.1016/j.jhazmat.2013.11.059.

Siebielec, G., McCarty, G.W., Stuczynski, T.I., Reeves, J.B., 2004. Near- and mid-infrared diffuse reflectance spectroscopy for measuring soil metal content. J. Environ. Qual. 33, 2056. Available from: https://doi.org/10.2134/jeq2004.2056.

Smith, K.M., Abrahams, P.W., Dagleish, M.P., Steigmajer, J., 2009. The intake of lead and associated metals by sheep grazing mining-contaminated floodplain pastures in mid-Wales, UK: I. Soil ingestion, soil–metal partitioning and potential availability to pasture herbage and livestock. Sci. Total Environ. 407, 3731–3739. Available from: https://doi.org/10.1016/j.scitotenv.2009.02.032.

Song, L., Jian, J., Tan, D.-J., Xie, H.-B., Luo, Z.-F., Gao, B., 2015. Estimate of heavy metals in soil and streams using combined geochemistry and field spectroscopy in the Wan-Sheng mining area, chongqing, china. Int. J. Appl. Earth Obs. Geoinf. 34, 1–9. Available from: https://doi.org/10.1016/j.jag.2014.06.013.

Soriano-Disla, J.M., Janik, L.J., Viscarra Rossel, R.A., Macdonald, L.M., McLaughlin, M.J., 2013. The performance of visible, near- and mid-infrared reflectance spectroscopy for prediction of soil physical, chemical, and biological properties. Appl. Spectrosc. Rev. 49, 139–186. Available from: https://doi.org/10.1080/05704928.2013.811081.

Srivastava, P.K., Gupta, M., Mukherjee, S., 2011. Mapping spatial distribution of pollutants in groundwater of a tropical area of india using remote sensing and gis. Appl. Geomat. 4, 21–32. Available from: https://doi.org/10.1007/s12518-011-0072-y.

Srivastava, P.K., Han, D., Gupta, M., Mukherjee, S., 2012. Integrated framework for monitoring groundwater pollution using a geographical information system and multivariate analysis. Hydrol. Sci. J. 57, 1453–1472. Available from: https://doi.org/10.1080/02626667.2012.716156.

Srivastava, P.K., Singh, S.K., Gupta, M., Thakur, J.K., Mukherjee, S., 2013. Modeling impact of land-use change trajectories on groundwater quality using remote sensing and GIS. Environ. Eng. Manage. J. 12, 2343–2355. Available from: https://doi.org/10.30638/eemj.2013.287.

Stenberg, B., Viscarra Rossel, R.A., Mouazen, A.M., Wetterlind, J., 2010. Visible and near infrared spectroscopy in soil science. Advances in Agronomy. Elsevier B.V., Amsterdam, pp. 163–215. Available from: http://dx.doi.org/10.1016/s0065-2113(10)07005-7.

Summers, D., 2009. Discriminating and Mapping soil Variability With Hyperspectral Reflectance Data (Ph.D. thesis). Adelaide University, Adelaide.

Ward, N.I., Brooks, R.R., Roberts, E., 1978. Lead levels in sheep organs resulting from pollution from automotive exhausts. Environ. Pollut. 17, 7–12. Available from: https://doi.org/10.1016/0013-9327(78)90050-2.

Wolfenden, P.J., Lewin, J., 1977. Distribution of metal pollutants in floodplain sediments. CATENA 4, 309–317. Available from: https://doi.org/10.1016/0341-8162(77)90030-3.

Wu, Y., Chen, J., Wu, X., Tian, Q., Ji, J., Qin, Z., 2005. Possibilities of reflectance spectroscopy for the assessment of contaminant elements in suburban soils. Appl. Geochem. 20, 1051–1059. Available from: https://doi.org/10.1016/j.apgeochem.2005.01.009.

You, D., Zhou, J., Wang, J., Ma, Z., Pan, L., 2011. Analysis of relations of heavy metal accumulation with land utilization using the positive and negative association rule method. Math. Comput. Modell. 54, 1005–1009. Available from: https://doi.org/10.1016/j.mcm.2010.11.028.

Zhang, B., Wu, D., Zhang, L., Jiao, Q., Li, Q., 2011. Application of hyperspectral remote sensing for perspective monitoring in mining areas. Environ. Earth Sci. 65, 649–658. Available from: https://doi.org/10.1007/s12665-011-1112-y.

Zhao, K., Valle, D., Popescu, S., Zhang, X., Mallick, B., 2013. Hyperspectral remote sensing of plant biochemistry using bayesian model averaging with variable and band selection. Remote Sens. Environ. 132, 102–119. Available from: https://doi.org/10.1016/j.rse.2012.12.026.

Further reading

Liu, M., Liu, X., Li, J., Li, T., 2012. Estimating regional heavy metal concentrations in rice by scaling up a field-scale heavy metal assessment model. Int. J. Appl. Earth Obs. Geoinf. 19, 12–23. Available from: https://doi.org/10.1016/j.jag.2012.04.014.

Tan, K., Niu, C., Jia, X., Ou, D., Chen, Y., Lei, S., 2020. Complete and accurate data correction for seamless mosaicking of airborne hyperspectral images: A case study at a mining site in Inner Mongolia, China. *ISPRS* J Photogramm Remote 165, 1–15. Available from: https://doi.org/10.1016/j.isprsjprs.2020.04.022.

Xu, X., Ren, M., Cao, J., Wu, Q., Liu, P., Lv, J., 2020. Spectroscopic diagnosis of zinc contaminated soils based on competitive adaptive reweighted sampling algorithm and an improved support vector machine. Spectrosc. Lett. 53 (2), 86–99. Available from: https://doi.org/10.1080/00387010.2019.1696828.

Wang, M., Yang, K.M., Zhang, W., 2020. Hyperspectral monitoring of maize leaves under copper stress at different growth stages. Remote Sens. Lett. 11(4), 343–352. Available from: https://doi.org/10.1080/2150704X.2020.1716408.

14 Hyperspectral remote sensing applications in soil: a review

Huan Yu[1], Bo Kong[2], Qing Wang[3], Xian Liu[4], Xiangmeng Liu[1]

[1]COLLEGE OF EARTH SCIENCES, CHENGDU UNIVERSITY OF TECHNOLOGY, CHENGDU, P.R. CHINA

[2]INSTITUTE OF MOUNTAIN HAZARDS AND ENVIRONMENT, CHINESE ACADEMY OF SCIENCES, CHENGDU, P.R. CHINA

[3]DEPARTMENT OF GEOGRAPHY AND ENVIRONMENTAL RESOURCES, SOUTHERN ILLINOIS UNIVERSITY, CARBONDALE, IL, UNITED STATES

[4]DEPARTMENT OF PLANT BIOLOGY, SOUTHERN ILLINOIS UNIVERSITY, CARBONDALE, IL, UNITED STATES

14.1 Introduction

Soil is defined as the weathered upper crust of the Earth's solid surface where the lithosphere, biosphere, atmosphere, and hydrosphere interact with each other (Ben-Dor et al., 2009). Soil characteristics including both physical and chemical information, provide important support for understanding transformations that occur in environmental systems (Vicente and Filho, 2011). The rapid and reliable assessment of soil characteristics is an important step in agricultural and natural resource management (Das et al., 2015). The traditional methods for observing soil properties involve field sampling and laboratory analysis. After that, the soil properties are usually mapped using an interpolation technique to transform the point data into a surface. This process is costly, time-consuming, and labor-consuming, and cannot provide accurate information for large spatial areas (Choe et al., 2008).

In the late 1990s, along with the booming remote sensing (RS) technology, predictive soil mapping was introduced. Remotely sensed (airborne or satellite) data generally cover a great spatial region and can be collected on certain schedules, providing a continuous observation of the soil (Silvestri et al., 2003). Compared to direct field measures, RS-based soil studies are timely, nondestructive, and unveil the spatial patterns of soil attributes. However, multispectral broadband-based RS has a limitation in terms of the quantitative estimation of physical and chemical properties primarily because of its low spectral resolution, resulting in the loss of critical information available in specific narrow bands (Sahoo et al., 2015). Therefore

Hyperspectral Remote Sensing. **DOI: https://doi.org/10.1016/B978-0-08-102894-0.00011-5**

hyperspectral remote sensing (HRS) techniques, which enable a precise recording of the spectrum and a detailed analysis of the spectral properties of a soil surface, has attracted growing interest from soil scientists and researchers (Tong et al., 2014).

There has been a wealth of HRS applications for soil studies in the past 40 years, which invokes the necessity to compose a comprehensive overview of these legacies. Such a review would be critical to understand the current opportunities and future challenges of HRS-based soil studies. The objectives of this chapter are to provide an overview of the field in terms of the applications of HRS in the extraction of soil properties including mineral, nutrient, organic carbon, moisture, salinity, and soil texture, and then to identify the opportunities and challenges and provide suggestions for a new direction for advancing its development.

14.2 Hyperspectral remote sensing application in soil mineral identification

Conventional mineral mapping based on geological survey is labor-intense, expensive, and inefficient. Such traditional surveys involve extensive fieldwork, structural mapping, the study of landforms, petrography, and mineralogy and geochemical analyses (Kusuma et al., 2012; Ramakrishnan et al., 2013; Ramakrishnan and Bharti, 2015). RS, on the other hand, can provide a much less expensive solution to easily access and collect mineral information over a large spatial area. This is due to the fact that the spectral information captured by remote sensors can be used to infer hydrothermal alteration minerals or alteration mineral zones, which are further associated with different types of mineralization systems (Carrino et al., 2018). The advances achieved in HRS technology allow clay minerals, sulfate minerals, carbonate minerals, iron oxides, and silica to be mapped to further estimate the spatial variations of alteration facies (Meer et al., 2012). The application of HRS in soil mineral identification has a long history. In 1981, a shuttle multispectral infrared radiometer was tested for the feasibility of using five closely spaced spectral channels in the 2.2–2.5 μm region to directly identify minerals (Goetz et al., 1982), and it was the very beginning of such applications (Goetz, 2009). This specific spectral range has been confirmed to have a good detection ability for many types of minerals (Fig. 14–1).

Over the past three decades, there have been extensive research attempts using ground, airborne, and spaceborne hyperspectral sensors for the determination and mapping of soil minerals. The latest applications of HRS in soil mineral discrimination have been listed and summarized in Table 14–1.

In the past, a number of airborne and spaceborne HRS systems have been designed and developed by various agencies and numerous soil (rock) mineral studies have been accomplished using remotely sensed datasets. Moreover, many studies of soil contamination, in particular, were performed using HRS technology (Ong and Cudahy, 2014; Davies and Calvin, 2017; Zhao et al., 2018). Although most of these studies were preliminary investigations for solving specific problems, it is important to understand that HRS can be successfully implemented in soil studies for determining and mapping soil minerals as well as monitoring soil contamination. Therefore it is expected that HRS will have important application prospects in the exploration of soil minerals.

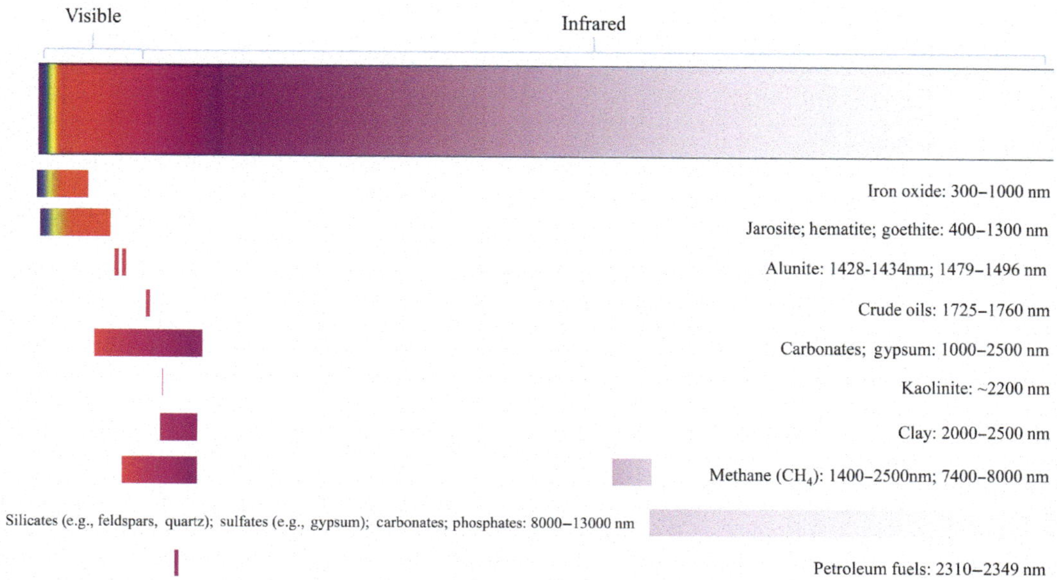

FIGURE 14–1 The specific spectral range for detecting different types of minerals.

Table 14–1 The applications of hyperspectral remote sensing in soil mineral discrimination over the past five years.

Sensor	Method	Target	Research
AHS	Continuum removal analysis	Lignite open-pit mines	Notesco et al. (2014)
AISA	Linear spectral unmixing (LSU)	Lithological mapping	Feng et al. (2018)
AVIRIS	Tetracorder spectral-shape matching system	Advanced argillic alteration	Swayze et al. (2014)
	Diagnostic spectral features (DSF)	Geothermal indicator minerals	Littlefield and Calvin (2014)
CASI/SASI	Mixture tuned matched filtering (MTMF)	Alteration mineral identification	Zhao et al. (2015)
Chinese Tiangong-1 Hyperspectral Imager	Spectral angle mapper (SAM)	Gold bearing veins	Liu et al. (2017)
HyMap	Matched filtering (MF)	Alteration minerals including kaolinite, montmorillonite, sericite (muscovite/illite), calcite, chlorite, epidote, and goethite	Molan et al. (2014)
	DSF	Geothermal indicator minerals	Littlefield and Calvin (2014)
	Support vector machine	Regolith-geology map for Ni exploration	Boissieu et al. (2017)
	MTMF	Mineral assemblies for discriminating potential high sulfidation epithermal targets	Carrino et al. (2018)
Hyperion	MTMF	Porphyry copper	Zadeh et al. (2014)
	LSU	Gold mineral alteration	Pour et al. (2014)
	MF	Mineral resource prospecting	Liu et al. (2016)
	Spectral feature fitting; SAM	Altered minerals	Rani et al. (2017)
	SAM	Hydrothermally altered and weathered minerals	Govil et al. (2018)

(*Continued*)

Table 14–1 (Continued)

Sensor	Method	Target	Research
ProSpecTIR	DSF	Geothermal indicator minerals	Littlefield and Calvin (2014)
	Partial least square regression	Petroleum hydrocarbons	Scafutto et al. (2017)
	Band rationing (BR); relative-absorption band depth (RBD); spectral correlation mapper (SCM); MF	Direct hydrocarbon detection	Asadzadeh and Filho (2016)
SEBASS	MTMF	Rock-forming minerals	Riley et al. (2017)
	LSU	Lithological mapping	Feng et al. (2018)
	MF	Methane plumes	Scafutto et al. (2018)
Specim hyperspectral system	Visual inspection	Ferric iron crystal field	Murphy et al. (2014)
	SAM	Quartz and feldspars	Notesco et al. (2015)
	SAM	Kaolinite, calcite, dolomite, quartz, feldspars, and gypsum	Notesco et al. (2016)
	SAM	Hydrocarbon-induced rock alterations	Sun and Khan (2016)
WorldView-3	BR; RBD; SCM; MF	Direct hydrocarbon detection	Asadzadeh and Filho (2016)

14.3 Hyperspectral remote sensing application in soil nutrient prediction

As definitive indicators of soil fertility, soil nutrients play a pivotal role in agricultural productivity, food security, and agroecological sustainability (Nowak et al., 2015). A timely and accurate mapping of soil nutrients can be particularly helpful for reducing the loss of soil nutrients, therefore, improving agricultural fertilization management. The monitoring of soil nutrients in farmlands traditionally merely relies on field sampling and laboratory analysis, which are inefficient and time-consuming. On the other hand, because HRS is capable of sensing the slightest spectral changes in soil nutrients, the data provided by HRS has become an important information source for modeling the soil nutrients (Song et al., 2018).

According to records available on Web of Science, there are only a few studies that were dedicated to monitoring soil nutrients based on HRS information. Song et al. (2018) collected 1,297 soil samples and measured the content of soil total nitrogen (TN), soil available phosphorus (AP), and soil available potassium (AK) in Zengcheng, north of the Pearl River Delta, China. In their study, HRS images (115 bands) of the Chinese Environmental 1A (HJ1A) satellite were used as auxiliary variables after passing through reduce-dimension process using Pearson correlation analysis and principal component analysis. They compared different prediction models including simple linear regression, support vector machine (SVM), random forest, and back-propagation neural network (BPNN), and the models were all trained on both field samples and preprocessed HRS variables. The model validations were based on 324 independent data points. Their study compared the ability of different linear, nonlinear machine learning, and hybrid kriging models in predicting soil nutrient contents (TN, AP, and AK) using HRS images as auxiliary variables. They concluded that the application of hyperspectral imaging vis–NIR (visible-near-infrared) data with a BPNN model was the most efficient method for mapping and monitoring soil nutrients at the regional scale (Song et al., 2018).

Yu et al. (2018) proposed a new HRS method for soil property modeling, and their study area is located in Shenzha County of the Qiangtang Plateau, northwestern Qinghai–Tibet Plateau where the dominated land cover is alpine grasslands. In their study, hyperspectral data were collected in a total of 67 sample points. At the same time, soil samples were obtained at the locations and soil properties including organic carbon, TN, total potassium, and total phosphorus were measured. Correlations of the soil properties with original bands and enhanced spectral variables derived from both field and satellite hyperspectral data were analyzed. Regression models that explained the relationships were further developed to map the soil properties. The results showed that (1) the soil properties had significant correlations with the vis–NIR bands and the wavelengths of 1720–1738 nm, and (2) the stepwise regression models based on the satellite hyperspectral imaging derived enhanced spectral variables and produced reasonable spatial distributions of the soil properties and relative root mean square error values of 68.9%, 46.3%, 31.4%, and 45.5% for soil organic carbon, TN, total phosphorus, and total potassium, respectively, were obtained. Thus this study implied that the hyperspectral data-based method provided great potential to predict soil properties. Their study explored the relationships between the soil content such as organic carbon, TN, total potassium, and total phosphorus and the hyperspectral reflectance of a *Stipa purpurea* canopy. This established regression models of the soil properties with spectral variables derived from hyperspectral data and developed a new method for mapping the soil properties of the alpine grasslands dominated by *S. purpurea* and further for evaluating the growth conditions of the *S. purpurea* based on HRS (Yu et al., 2018).

The number of papers published on HRS technology for monitoring soil nutrient elements is small and the dates of these publication are recent, which means that this aspect of HRS is just beginning. There are many types of hyperspectral sensors at present, and similar studies still need to be conducted in other areas and for different land cover species.

14.4 Hyperspectral remote sensing application in soil organic carbon estimation

Soil organic carbon (SOC) plays a major role with respect to many chemical and physical processes in soil environments (Gomez et al., 2008a). SOC provides a primary source of nutrients for plants, aids in the aggregation of particles, develops soil structure, increases water storage capacity, and provides a habitat for soil biota (Schoonover and Crim 2015). The spatial distribution of SOC concentration of surface soil is an important soil property in crop management for guiding fertilizer and chemical applications (Chen et al., 2000). Various approaches have been conducted to map the SOC concentration of crop fields in which conventional SOC stock mapping methods involve the collection and analysis of point soil samples, the calibration of a spatial prediction function, and the interpolation of the function over the whole study area (Minasny et al., 2013). These methods are expensive and time-consuming because of the large number of samples required to acquire the high spatial variability of SOC (Stevens et al., 2010). In the past few years, HRS has provided a powerful tool for multiscale and rapid monitoring of SOC and many ground, airborne, and spaceborne hyperspectral sensors have been tested (Table 14–2).

There are fewer applications of spaceborne hyperspectral sensors than airborne hyperspectral sensors in Table 14–2. As remotely sensed satellite hyperspectral data offer a synoptic view and a repetitive coverage, which are two important advantages compared to ground observations and airborne hyperspectral data, the study of the potential of satellite hyperspectral data for SOC prediction becomes a major issue for digital soil mapping development. Furthermore,

Table 14–2 Current studies on the estimation of soil organic carbon using hyperspectral remote sensing.

Sensor	Method	Research
AHS	Partial least square regression (PLSR); penalized-spline signal regression; support vector machine (SVM)	Stevens et al. (2010)
	PLSR	Stevens et al. (2008); Bartholomeus et al. (2011); Stevens et al. (2012); Denis et al. (2014); Steinberg et al. (2016); Fernández et al. (2016)
	Simple linear regression (SLR); stepwise multiple linear regression (SMLR); PLSR	Peón et al. (2017a)
APEX	PLSR	Castaldi et al. (2018)
	PLSR; random forest	Castaldi et al. (2019)
AISA	PLSR	Kanning et al. (2016)
AVNIR	Multiple linear regression (MLR)	DeTar et al. (2008)
CASI	Principal component analysis (PCA); SMLR	Uno et al. (2005)
	PLSR	Stevens et al. (2006)
	SVM	Gholizadeh et al. (2018)
DAIS	Visible and near-infrared analysis	Ben-Dor et al. (2002)
EnMAP	PLSR; step-down variable selection algorithm	Rosero-Vlasova et al. (2018)
HyMap	PLSR; MLR	Selige et al. (2006)
	PLSR	Patzold et al. (2008); Schwanghart and Jarmer (2011); Eisele et al. (2012); Gomez et al. (2012); Hbirkou et al. (2012); Gerighausen et al. (2012); Vohland et al. (2017)
Hyperion	SMLR; artificial neural network	Jaber et al. (2011)
	PLSR	Gomez et al. (2008a); Lu et al. (2013); Zhang et al. (2013); Castaldi et al. (2016); Minu et al. (2017)
	Ordinary least squares using the minimum noise fraction eigenvectors	Castaldi et al. (2014)
	SLR; SMLR; PLSR	Peón et al. (2017b)
	Stepwise regression; minimum regression; PLSR; principle component regression	Nowkandeh et al. (2018)
HyperSpecTIR	PLSR	Hively et al. (2011)
HyspIRI	PLSR	Castaldi et al. (2016)
PRISMA	PLSR	Castaldi et al. (2016)
ProSpecTIR	PLSR	Franceschini et al. (2015)
RDACS/H-3	PLSR	Bajwa and Tian (2005)
TASI	PLSR	Eisele et al. (2012)
	PLSR; cubist regression	Pascucci et al. (2014)

the estimation of SOC using airborne and satellite hyperspectral sensors has mainly been restricted to small agricultural or bare soil areas, and it is still under the testing phases. In addition, the modeling technique used for SOC prediction in most studies was the simple statistical method, which has several drawbacks related to the physical interpretation of results and the complexity of transferring the models from one sensor to another (Gomez et al., 2008a; Peón et al., 2017a, b).

14.5 Hyperspectral remote sensing application in soil moisture retrieval

Soil moisture is a critical process in the water cycle and its assessment is of paramount importance in forecasting changes in the water balance of a region (Salvucci et al., 2002). Information on soil moisture is of importance for irrigation scheduling and for improving crop yield, but also in water management in the context of climate change and for the evaluation of the anthropic ecosystem impact (Krapez and Olioso, 2011). Direct measurements of soil moisture are typically (in particular at large scales) costly, time-consuming, invasive, and user and method dependent (Vereecken et al., 2010). As an alternative to ground point data, RS data have been seen as a promising approach to evaluate soil moisture characteristics in different landscapes and sampling areas by providing a regional description of water redistribution at different temporal and spatial resolutions (McCabe and Wood, 2006). HRS is considered to be a promising tool for the rapid quantification of soil moisture and some applications have been tested (Table 14–3).

In the past, numerous studies have been conducted, mainly in the artificial environment of a laboratory or outdoor field, to characterize the physical processes linking soil moisture to its reflectance. Although new studies have demonstrated that profile soil moisture could be estimated well by the assimilation of HRS data into a hydrological model, these could only serve as case studies from which other HRS users can start in order to create quantitative soil moisture maps. This approach has not yet been fully studied or developed in this innovative direction though it appears to be promising and necessary because many problems such as low signal-to-noise ratios, unreliable spectral band response, atmospheric influence on raw data, the need to position samples on the ground, and the lack of image pixel-based physical or chemical models related to soil moisture content remain unsolved.

14.6 Hyperspectral remote sensing application in soil salinity detection

Soil salinity has remained one of the major and most widespread land degradation problems for a long time and it substantially limits crop productivity (Epstein et al., 1980). The mapping and monitoring of salt-affected soils are needed so that proper and timely decisions can be made to modify management practices or undertake reclamation and rehabilitation

Table 14–3 Current studies on the retrieval of soil moisture using hyperspectral remote sensing.

Research	Sensor	Method	Conclusion
Ben-Dor et al. (2002)	DAIS	Visible and near-infrared analysis	It was possible to obtain reliable prediction equations for soil moisture using the DAIS spectral information.
Haubrock et al. (2008)	HyMap	Normalized soil moisture index (NSMI)	The NSMI was appropriate for modeling surface soil moisture from high spectral-resolution remote sensing data.
Finn et al. (2011)	ITD-SWIR	General linear model	A significant statistical correlation of the hyperspectral instrument data and the soil moisture probe data at 5.08 cm depth was determined, but models for the 20.32 and 30.48 cm depths were not able to estimate soil moisture to the same degree.
Krapez and Olioso (2011)	Hymap	Linear spectral unmixing	A method combining temperature, vegetation indexes, and albedo was proposed to evaluate soil moisture. A better correlation was, thus, expected between the calculated moisture and the measured one.
Sobrino et al. (2012)	AHS	Polynomial formulation with temperature emissivity and NDVI	The soil moisture from AHS data could be obtained with a root mean square error of 0.05 m^3/m^3 compared with ground measurements.
Zhang et al. (2013)	Hyperion	Partial least square regression (PLSR)	High accuracy for all soil properties from laboratory spectra, but low accuracy for soil moisture from Hyperion spectra were obtained.
Maltese et al. (2013)	AHS	Thermal inertia	The thermal inertia method could be applied to sparsely vegetated soil if the solar radiation reaching the ground was accurately estimated. A corrective coefficient taking into account actual sky cloudiness throughout the day allowed better estimates of thermal inertia and, thus, of soil water content.
Song et al. (2014)	Hyperion	Modified water-cloud model	Results revealed an average absolute deviation and average absolute relative deviation of 0.051 cm^3/cm^3 and 19.7%, respectively, between the estimated soil moisture and the field measurements.
Babaeian et al. (2015)	EnMAP	Stepwise multiple linear regression	This spectral approach performed reasonably well in terms of predicting soil water retention characteristics and unsaturated hydraulic conductivity.
Zeng et al. (2017)	Hyperion	PLSR	Results indicated that PLSR is a powerful tool for soil moisture estimation from hyperspectral data.

(Metternicht and Zinck, 2003). Conventional techniques available for identifying and monitoring these salt-affected soils are expensive, time-consuming, and require intensive sampling to characterize spatial variability (Shepherd and Walsh, 2002). By providing fast, timely, relatively cheap, and repetitive data, HRS plays an important role in detecting, mapping, and monitoring salt-affected surface features. The capabilities of HRS for salinity mapping have been tested and a variety of airborne and spaceborne hyperspectral sensors have been used for identifying and monitoring salt-affected areas (Table 14–4).

HRS has been widely used to identify and map salt-affected areas and many studies have shown that hyperspectral data can be used to quantify the characteristics of saline soils at various scales. These studies mainly identified and mapped salt-affected soils, but did not characterize them in terms of their severity. Furthermore, most of these studies stay in the experimental stage, and specific application examples are less apparent. In addition, various linear, nonlinear regression, and RS image classification methods are used to identify and map soil salinity, which creates confusion when choosing models or methods. All these problems indicate that there is still a need to explore universal quantitative models and methods, and extensively apply them to practical work to test their actual effect.

Table 14–4 Current studies on the detection of soil salinity using hyperspectral remote sensing (HRS).

Research	Sensor	Method	Conclusion
Ben-Dor et al. (2002)	DAIS	Visible and near-infrared analysis (VNIRA)	The VNIRA method was a promising strategy for quantitative soil surface mapping; furthermore, the method could even be improved if a better quality of HRS data were used.
Dehaan and Taylor (2003)	HyMap	Convex geometry	Spectral endmembers unmixed from imagery could map both soil and vegetation indicators of salinity.
Shrestha et al. (2005)	HyMap	Linear spectral unmixing (LSU)	LSU provided more realistic results for mapping "desert-like" surface features than the spectral angle matching technique.
Farifteh et al. (2007)	HyMap	Partial least square regression (PLSR); artificial neural network (ANN)	The relation between soil salinity and soil reflectance could be approximated by a linear function.
Weng et al. (2008)	Hyperion	PLSR; stepwise regression (SWR)	The PLSR method was a more suitable technique than stepwise regression for quantitative estimation of soil salt content in a large area.
DeTar et al. (2008)	AVNIR	Multiple linear regression (MLR)	Some soil properties could be accurately detected using airborne remote sensing over nearly bare fields, and it was possible to produce a fine-resolution, farm-size, soil map showing the in-field distribution of these properties.
Dutkiewicz et al. (2009)	HyMap; Hyperion; CASI	Partial spectral unmixing	Hyperspectral imagery could improve discrimination of vegetation and mineral indicators of surface salinity and had the potential to improve traditional soil and salinity mapping based on multispectral satellite imagery and aerial photointerpretation.
Weng et al. (2010)	Hyperion	Univariate linear regression	The satellite hyperspectral data had the potential to predict soil salt content in a large area.
Wu et al. (2010)	Hyperion	SVM	The procedure for estimating the extent of land degradation was feasible and the SVM algorithm was an efficient method for mapping land degradation.
Ghosh et al. (2012)	Hyperion	LSU	Various severity classes of salt-affected soils could be reliably mapped using LSU analysis.
Hamzeh et al. (2013)	Hyperion	Linear regression modeling	Soil salinity could be estimated by satellite-based hyperspectral vegetation indices, but validation of obtained models for independent data was essential for selecting the best model.
Zhang et al. (2014)	Hyperion	PLSR	The variation of Ca^{2+} content was consistent with salt deposition. The correlation between Ca^{2+} content and reflectance was in accordance with salt-lake evolution.
Li et al. (2014)	HJ1A	Decision tree classification	Compared to multispectral data, the hyperspectral imagery could be more accurate and efficient for land salinization information extraction.

(*Continued*)

Table 14–4 (Continued)

Research	Sensor	Method	Conclusion
Lei et al. (2014)	HJ1A	MLR	Both the measured hyperspectral soil salinity monitoring model and HRS image soil salinity inversion model had good accuracy. The correction HRS image soil salinization monitoring model could better improve the model accuracy under the condition of regional scale soil salinization monitoring, and using this method to carry out the soil salinization quantitative remote sensing monitoring was feasible.
Kumar et al. (2015)	Hyperion	Pearson product-moment correlation analysis (PPCA); SWR	The HRS data showed the potential to assess the severity of salt-affected soils for a large area, which may be useful for identifying areas for carrying out reclamation measures and management planning.
Li et al. (2015)	HJ1A	Normal soil salt content response index; soil adjusted salinity index	The new method significantly improved the accuracy of soil salt content mapping, and HRS data could be used to map soil salt content precisely and were suitable for monitoring soil salt content on a large scale.
Moreira et al. (2015)	Hyperion	SVM	Compared with Operational Land Imager (OLI), the narrowband salinity indices of Hyperion produced a lower root mean square error for soil salinity estimates and better discrimination between saline and nonsaline soils using SVM classification.
Hamzeh et al. (2016)	Hyperion	SVM; SAM; minimum distance; maximum likelihood	The study evaluated the feasibility of hyperspectral and multispectral satellite imagery for categorical and quantitative mapping of salinity stress in sugarcane fields. It was concluded that categorical mapping of salinity stress was the best option for monitoring agricultural fields.
Neto et al. (2017)	ProSpecTIR; HyspIRI	Ordinary least squares; PLSR; multilayer perceptron; extreme learning machine	The study verified the applicability of images obtained by the hyperspectral airborne sensor for the estimation of soil salt content on a per-pixel basis; the performance of the narrowband sensors to estimate soil salt content was better than the performance of the broadband instruments.
Liu et al. (2018)	CASI	General linear model	The forced invariance approach was able to improve the retrieval accuracy of soil salinity at a depth of 10 cm. Consequently, the vegetation suppression method had the potential to improve quantitative estimation of soil properties with multivariate statistical methods.

14.7 Hyperspectral remote sensing application in soil texture acquisition

Soil texture, defined by the composition of particle size, namely sand, silt, and clay, is an important land environmental variable because it plays a key role in soil degradation and water transport processes, controlling soil quality and its productivity (Hillel 1980; Blume et al., 2010). Knowledge of soil texture variability is crucial for the implementation of site-specific farming management strategies that allow for more efficient use of resources such as water and fertilizers, therefore, reducing costs and environmental impacts (Castaldi et al., 2015). Methods for the mapping of soil texture at various scales are needed for applications, but, typically, a large number of samples must be collected and analyzed in order to adequately estimate the soil texture spatial variability in traditional ways (Curcio et al., 2013). The expensive conventional methods and field surveys are currently driving users to develop indirect estimation methods based on proximal and remote sensors (ground-based or airborne) including reflectance spectroscopy (Brown et al., 2006; Ben-Dor et al., 2009). The potential for the estimation of soil texture by HRS has been evaluated in many studies (Table 14–5).

Laboratory and airborne imaging spectroscopy of bare soils have been shown to have considerable potential for the estimation of soil texture with promising results. However, only a few studies exist that determine soil texture directly from satellite hyperspectral imagery. Although this approach holds great potential for digital soil mapping with satellite hyperspectral imagery, soil texture assessment from image data acquired by spaceborne systems is a more difficult issue, mainly due to atmospheric distortions and the low spatial and spectral resolution of sensors (Mulder et al., 2011). In order to be able to fully exploit data from forthcoming hyperspectral satellites, information on several issues related to sensor spatial and

Table 14–5 Current studies on the estimation of soil texture using hyperspectral remote sensing (HRS).

Research	Sensor	Method	Conclusion
Chabrillat et al. (2002)	AVIRIS; HyMap	Mixture tuned matched filtering	Such hyperspectral remote sensing methods, along with modest field and laboratory analyses, could facilitate the construction of clay hazard maps, as long as the soils were adequately exposed.
Selige et al. (2006)	HyMap	Partial least square regression (PLSR) and multiple linear regression	The proposed methodology provided a means of simultaneously estimating topsoil organic matter and texture in a rapid and nondestructive manner, whilst avoiding the spatial accuracy problems associated with spatial interpolation.
Gomez et al. (2008b)	HyMap	Continuum removal analysis (CRA); PLSR	VNIR/SWIR airborne hyperspectral data processed by the PLSR technique allow for accurate mapping of clay and $CaCO_3$ contents, which will contribute significantly to the digital mapping of soil properties.
Lagacherie et al. (2008)	HyMap	CRA	The performances of clay and $CaCO_3$ estimations decreased from the laboratory to airborne scales. The main factors inducing uncertainties in the estimates were radiometric and wavelength calibration uncertainties of the airborne sensor as well as possible residual atmospheric effects.
Hively et al. (2011)	HyperSpecTIR	PLSR	13 out of the 19 soil properties were predicted with $R^2 > 0.5$ including sand, silt, and clay.
Gomez et al. (2012)	HyMap	PLSR	Four out of the eight soil properties ($CaCO_3$, iron, clay, and cation-exchange capacity) were suitable for mapping using hyperspectral data, and both accurate local predictions and good representations of spatial structures were observed.

(*Continued*)

Table 14–5 (Continued)

Research	Sensor	Method	Conclusion
Casa et al. (2013)	MIVIS; CHRIS-PROBA	PLSR	Despite the lack of SWIR bands and a lower spatial resolution, CHRIS did show a comparable potential to MIVIS in terms of accuracy and of prediction ability for soil texture (clay, silt, and sand).
Garfagnoli et al. (2013)	Hyper SIM-GA	CRA; PLSR	The comparison between the hyperspectral-derived maps of clay and the correspondent inverse distance weighted (IDW) interpolations of the measured clay content values demonstrated that the results were encouraging and reliable.
Ciampalini et al. (2015)	Hyper SIM-GA	Data fusion	The ability of combining data from relatively inexpensive and efficient sensing methods for producing reliable and high-resolution maps of clay content was showed.
Dutta et al. (2015)	AVIRIS	Visible and near-infrared analysis	The approach was feasible and provided insights into the accuracy and uncertainty of the approach for soil textural properties (sand, silt, and clay).
Castaldi et al. (2015)	PRISMA	PLSR	A priori knowledge of the soil moisture class could reduce the error of clay estimation when using HRS data.
Castaldi et al. (2016)	Hyperion; EnMAP; PRISMA; HyspIRI	PLSR	Hyperspectral data from the forthcoming satellite missions could marginally improve mapping and monitoring soil texture as compared to current imagers.
Kanning et al. (2016)	AISA	PLSR	Providing a cost-effective possibility to spatially predict soil parameters. The adaptation to hyperspectral satellite data will allow for the regionalization of soils for larger areas.

spectral resolution and range as well as on calibration and validation issues, is still required. The development of more physically-based models in this context would offer a real step forward toward the generalization of estimation approaches, but at present, this still seems to be an elusive objective (Casa et al., 2013).

14.8 Opportunities and challenges

This chapter deals with the applications of HRS on soil property extraction including mineral identification, nutrient, organic carbon, moisture, salinity, and soil texture. In order to better demonstrate the application of HRS in soil, a synoptic scheme chart was created to show the different HRS platforms with diverse applications and how they may support soil monitoring (Fig. 14–2). Although HRS provides new insights in terms of theory and methodology for studying soil attributes, there is much work to be done, both experimentally and theoretically, before the physical and chemical processes predicting these soil parameters are full understood.

With the increasing resolution of hyperspectral imaging RS sensors, more and more information dimensions are obtained. At the same time, the amount of RS data acquired is also showing an obvious growth. The "big data" feature has become significant. While obtaining a large amount of hyperspectral image data, people are also facing the problem of how to maximize the use of these massive data. Although some progress has been made in the technology of hyperspectral data

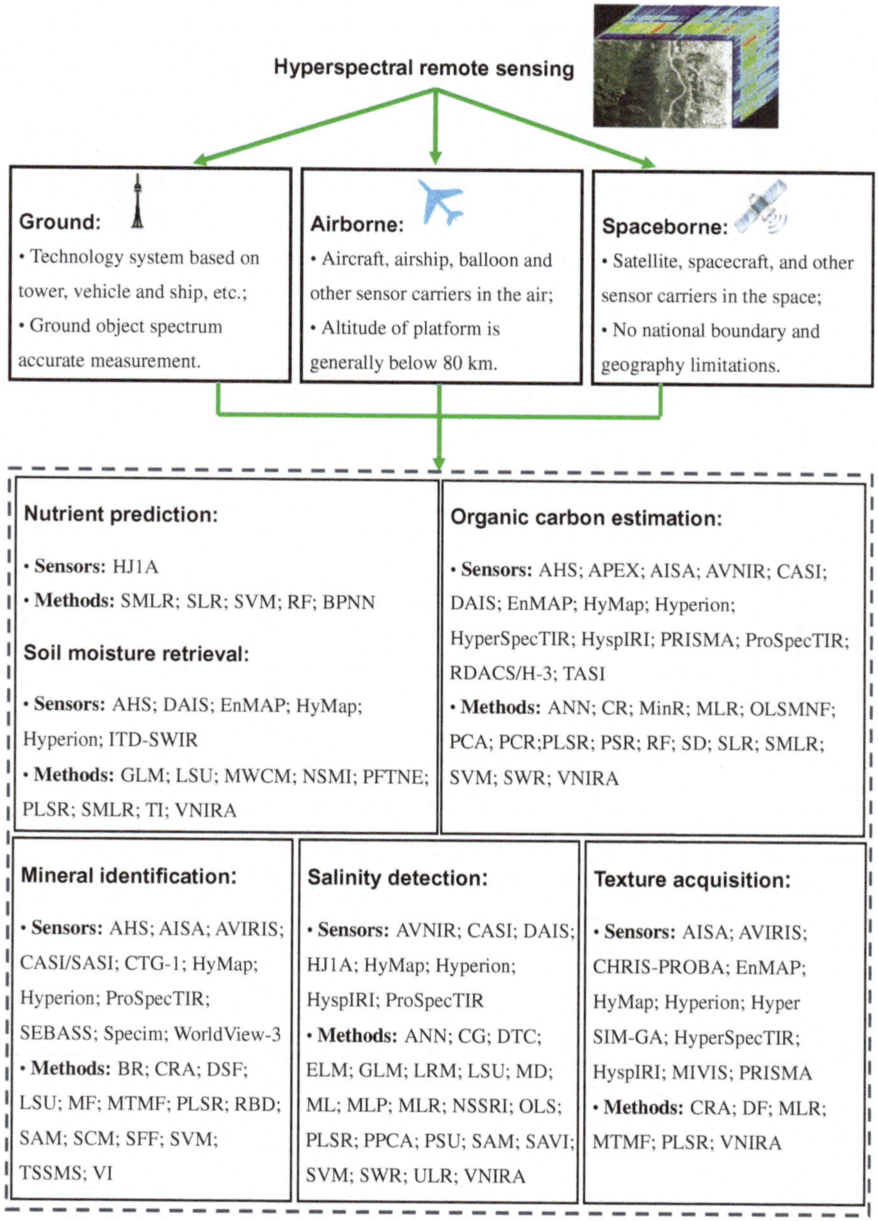

FIGURE 14–2 A synoptic scheme chart showing different hyperspectral remote sensing platforms with diverse applications and how they may support soil monitoring.

classification and information extraction, these still lag behind the development of sensors in general. Therefore there is still a long way to go in terms research on hyperspectral data classification and information extraction. How to effectively achieve HRS data mining, information extraction, high-efficiency data compression, and high-speed data transmission are some of the most important issues to be solved in the future.

On one hand, new data mining technologies such as deep learning provide opportunities for the efficient processing of large amounts of HRS data. Deep learning proposes a method for computers to automatically learn pattern features and incorporate feature learning into the process of model building, thus, reducing the incompleteness caused by artificial design features. At present, some machine learning applications with deep learning as the core method have achieved excellent recognition or classification performance beyond the existing algorithms. On the other hand, in view of the huge amount of HRS data, more effective feature parameters must be extracted for soil attributes monitoring from the original measurement data. There are two ways to realize this process, namely band selection and feature extraction. It is well known that bands with more information, less data correlation, a large spectral difference of soil attributes, and good separability should be chosen as the best working bands, and that feature extraction indexes or methods that are easy to implement with higher extraction accuracy should be developed. However, due to the different research objects and regions, the best spectral parameters or characteristic indices of the same soil attribute are also different. Most of the existing spectral indices are based on limited datasets, which makes monitoring models lack universality in the selection and application of characteristic parameters.

Soil science is facing many technical difficulties and challenges in its development process. These problems cannot be avoided when HRS technology is introduced into soil parameter retrieval. For example, the excavation of the interaction mechanisms between soil parameters has always been a research hotspot in soil science, and it is difficult to establish an HRS monitoring model with a physical and chemical basis. But it also provides a new technical idea for monitoring soil properties by HRS, for example, by aiming at the parameters with weak spectral reflectance characteristics, we try to indirectly detect them by other parameters closely related to their physical and chemical properties in soil. Furthermore, for working areas with vegetation coverage, the monitoring of soil properties is indirectly realized through the spectral characteristics of the vegetation. In this case, the interactional mechanisms of vegetation and soil should also be integrated into the construction of the HRS monitoring model.

Among the specific methods of model construction, linear mathematical analysis methods are widely used including principle component regression, multiple linear regression, partial least square regression, and univariate linear regression. Some studies have explored methods of nonlinear mathematical analysis such as SVM, genetic algorithm, and artificial neural network technology. These nonlinear methods can compensate for the shortcomings of linear methods to some extent and improve the prediction accuracy of models. On the basis of these studies, new artificial intelligence modeling methods should be attempted such as linear and nonlinear coupling, machine learning, and their application in soil attribute monitoring should be explored. Furthermore, the development from an empirical model to a physical model will improve the universality and robustness of a given model.

Soil is an interrelated and interacting entirety of minerals, organic matter, water, air, and other substances. Its spectral reflectance characteristics are also a comprehensive reflection of various physical and chemical properties. Therefore while discussing the method of model building, different soil types must also be taken into account because of the great differences in their composition. Because the compositions of different soils are quite different, their spectral reflectance curves are quite varied even when the soil parameters are the same. When HRS technology is used to monitor soil properties in large-scale regions, different soil types are often involved. Therefore determining methods to reduce the impact of components and establish a relationship between spectral reflectance and monitoring objectives for a variety of soils is a challenging issue. Even for the same type of soil, soil properties such as soil water content and organic matter affect the reflectance together. Only by distinguishing their coupled effects on the reflectance, can their quantitative monitoring be better realized. Therefore in the application of HRS for the monitoring of soil parameters, more attention should be paid to the application of multivariate models involving multiple parameters, rather than to single-factor models.

Soil spectra not only have high similarity and spatial variability, but also strong temporal dynamics. Especially for the monitoring of soil attributes in vegetation-covered areas, the spectral variation of soil attributes will be more distinct with time due to the influence of the seasonal characteristics of vegetation. Therefore by making full use of the advantages of HRS in distinguishing the subtle differences of a surface, and combining this with the temporal and dynamic characteristics of soil attributes, the accuracy of detecting and monitoring soil attributes will be greatly improved. However, at present, due to the redundancy of spectral resolution and the limitation of temporal resolution in HRS data as well as the complexity of the relevant models, further data accumulation and method exploration are still needed to realize the monitoring of soil parameters considering the dynamic process of vegetation growth.

With the successful development and launch of ground-based, airborne, and spaceborne hyperspectral sensors worldwide, it is easier to obtain reliable and time-sensitive hyperspectral data of land surfaces. Moreover, the new features of high spatial resolution, high spectral resolution, and high temporal resolution of HRS technology have become more and more obvious. However, optical HRS means may be affected by many factors such as dust, rust, plowing, particle size distribution, vegetation coverage, littering, and physical and biogenic crusts, which make the use of a single type of optical HRS data rather limited and problematic when striving for quantitatively accurate information. Sensors with different working modes and wavelength ranges can provide a variety of detecting means and methods, which can form complementary sets of information to improve monitoring accuracy. For this reason, multidata fusion and multiscale data assimilation will become another research hotspot in HRS monitoring of soil properties. Fusing hyperspectral data with other types of sensor data such as light detection and ranging (LiDAR), synthetic aperture radar (SAR), and high spatial resolution images will have wide application prospects if it can expand the monitoring range or improve the detection accuracy. In addition, with the continuous improvement of the temporal and spatial resolution of HRS data and the diversity of data acquisition methods (e.g., ground-based, airborne, and spaceborne), HRS data assimilation has great potential for application. Multiscale data assimilation will improve the monitoring accuracy of the land surface process and promote the comprehensive application of multiresolution (temporal, spatial, and spectral) HRS data in soil science.

Acknowledgments

This study was supported by the National Natural Science Funds of China (grant no. 41871357), the Sichuan Basic Science and Technology Project (grant no. 18YYJC1148), the Branch of Mountain Sciences, Kathmandu Center for Research and Education, CAS-TU, Chengdu, China (grant no. Y8R3310310), the Hundred Young Talents Program of the Institute of Mountain Hazards and Environment (grant no. SDSQB-2015-02), the One-Three-Five Project of Chinese Academy of Sciences (grant no. SDS-135−1708) and the Science and Technology Service Network Program of the Chinese Academy of Sciences (grant no. Y8R2020022).

Conflict of interest

The authors have declared no conflict of interest.

List of abbreviations

AK	available potassium
ANN	artificial neural network
AP	available phosphorus
BPNN	back-propagation neural network
HJ1A	Chinese Environmental 1A Satellite
HRS	hyperspectral remote sensing
MLR	multiple linear regression
PLSR	partial least square regression
RS	remote sensing
SLR	linear regression
SOC	soil organic carbon
SVM	support vector machine
TN	total nitrogen
VIS−NIR	visible-near-infrared

References

Asadzadeh, S., Filho, C.R.S., 2016. Investigating the capability of WorldView-3 superspectral data for direct hydrocarbon detection. Remote Sens. Env. 173, 162−173.

Babaeian, E., Homaee, M., Montzka, C., Vereecken, H., Norouzi, A.A., 2015. Towards retrieving soil hydraulic properties by hyperspectral remote sensing. Vadose Zone J. 14 (3), 1−17.

Bajwa, S.G., Tian, L.F., 2005. Soil fertility characterization in agricultural fields using hyperspectral remote sensing. Trans. ASABE 48 (6), 2399−2406.

Bartholomeus, H., Kooistra, L., Stevens, A., van Leeuwen, M., van Wesemael, B., Ben-Dor, E., et al., 2011. Soil organic carbon mapping of partially vegetated agricultural fields with imaging spectroscopy. Int. J. Appl. Earth Obs. 13, 81−88.

Ben-Dor, E., Chabrillat, S., Demattê, J.A.M., Taylor, G.R., Hill, J., Whiting, M.L., et al., 2009. Using imaging spectroscopy to study soil properties. Remote Sens. Environ. 113, 38−55.

Ben-Dor, E., Patkin, K., Banin, A., Karnieli, A., 2002. Mapping of several soil properties using DAIS-7915 hyperspectral scanner data—a case study over clayey soils in Israel. Int. J. Remote Sens. 23, 1043−1062.

Blume, H.P., Brümmer, G., Horn, R., Kandeler, E., Kögel-Knabner, I., Kretzschmar, R., et al., 2010. Scheffer/Schachtschabel Lehrbuch der Bodenkunde, sixteenth ed. Spektrum Akademischer Verlag, Heidelberg.

Boissieu, D.E., Sevin, F., Cudahy, B., Mangeas, T., Chevrel, M., Ong, S., et al., 2017. Regolith-geology mapping with support vector machine: a case study over weathered Ni-bearing peridotites, New Caledonia. Int. J. Appl. Earth Obs. Geoinf. 64, 377–385.

Brown, D.J., Shepherd, K.D., Walsh, M.G., Mays, M.D., Reinsch, T.G., 2006. Global soil characterization with VNIR diffuse reflectance spectroscopy. Geoderma 132, 273–290.

Carrino, T.A., Crósta, A.P., Toledo, C.L.B., Silva, A.M., 2018. Hyperspectral remote sensing applied to mineral exploration in southern Peru: a multiple data integration approach in the Chapi Chiara gold prospect. Int. J. Appl. Earth Obs. Geoinf. 64, 287–300.

Casa, R., Castaldi, F., Pascucci, S., Palombo, A., Pignatti, S., 2013. A comparison of sensor resolution and calibration strategies for soil texture estimation from hyperspectral remote sensing. Geoderma 197–198, 17–26.

Castaldi, F., Casa, R., Castrignanò, A., Pascucci, S., Palombo, A., Pignatti, S., 2014. Estimation of soil properties at the field scale from satellite data: a comparison between spatial and non-spatial techniques. Eur. J. Soil Sci. 65, 842–851.

Castaldi, F., Palombo, A., Pascucci, S., Pignatti, S., Santini, F., Casa, R., 2015. Reducing the influence of soil moisture on the estimation of clay from hyperspectral data: a case study using simulated PRISMA data. Remote Sens. 7, 15561–15582.

Castaldi, F., Palombo, A., Santini, F., Pascucci, S., Pignatti, S., Casa, R., 2016. Evaluation of the potential of the current and forthcoming multispectral and hyperspectral imagers to estimate soil texture and organic carbon. Remote Sens. Environ. 179, 54–65.

Castaldi, F., Chabrillat, S., Jones, A., Vreys, K., Bomans, B., van Wesemael, B., 2018. Soil organic carbon estimation in croplands by hyperspectral remote APEX data using the LUCAS topsoil database. Remote Sens. 10 (2), 153.

Castaldi, F., Hueni, A., Chabrillat, S., Ward, K., Buttafuoco, G., Bomans, B., et al., 2019. Evaluating the capability of the Sentinel 2 data for soil organic carbon prediction in croplands. ISPRS J. Photogramm. Remote Sens. 147, 267–282.

Chabrillat, S., Goetz, A.F.H., Krosley, L., Olsen, H.W., 2002. Use of hyperspectral images in the identification and mapping of expansive clay soils and the role of spatial resolution. Remote Sens. Environ. 82, 431–445.

Chen, F., Kissel, D.E., West, L.T., Adkins, W., 2000. Field-scale mapping of surface soil organic carbon using remotely sensed imagery. Soil Sci. Soc. Am. J. 64, 746–753.

Choe, E., van der Meer, F., van Ruitenbeek, F., van der Werff, H., de Smeth, B., Kim, K.W., 2008. Mapping of heavy metal pollution in stream sediments using combined geochemistry, field spectroscopy, and hyperspectral remote sensing: a case study of the Rodalquilar mining area, SE Spain. Remote Sens. Environ. 112, 3222–3233.

Ciampalini, A., André, F., Garfagnoli, F., Grandjean, G., Lambot, S., Chiarantini, L., et al., 2015. Improved estimation of soil clay content by the fusion of remote hyperspectral and proximal geophysical sensing. J. Appl. Geophys. 116, 135–145.

Curcio, D., Ciraolo, G., Asaro, F.D., Minacapilli, M., 2013. Prediction of soil texture distributions using VNIR–SWIR reflectance spectroscopy. Proc. Environ. Sci. 19, 494–503.

Das, B.S., Sarathjith, M.C., Santra, P., Sahoo, R.N., Srivastava, R., Routray, A., et al., 2015. Hyperspectral remote sensing: opportunities, status and challenges for rapid soil assessment in India. Curr. Sci. 108 (5), 860–868.

Davies, G.E., Calvin, W.M., 2017. Mapping acidic mine waste with seasonal airborne hyperspectral imagery at varying spatial scales. Environ. Earth Sci. 76 (12), 432.

Dehaan, R.L., Taylor, G.R., 2003. Image-derived spectral endmembers as indicators of salinisation. Int. J. Remote Sens. 24 (4), 775–794.

Denis, A., Stevens, A., van Wesemael, B., Udelhoven, T., Tychon, B., 2014. Soil organic carbon assessment by field and airborne spectrometry in bare croplands: accounting for soil surface roughness. Geoderma 226–227, 94–102.

DeTar, W.R., Chesson, J.H., Penner, J.V., Ojala, J.C., 2008. Detection of soil properties with airborne hyperspectral measurements of bare fields. Trans. ASABE 51 (2), 463–470.

Dutkiewicz, A., Lewis, M., Ostendorf, B., 2009. Evaluation and comparison of hyperspectral imagery for mapping surface symptoms of dryland salinity. Int. J. Remote Sens. 30 (3), 693–719.

Dutta, D., Goodwell, A.E., Kumar, P., Garvey, J.E., Darmody, R.G., Berretta, D.P., et al., 2015. On the feasibility of characterizing soil properties from AVIRIS data. IEEE Trans. Geosci. Remote Sens. 53 (9), 5133–5147.

Eisele, A., Lau, I., Hewson, R., Carter, D., Wheaton, B., Ong, C., et al., 2012. Applicability of the thermal infrared spectral region for the prediction of soil properties across semi-arid agricultural landscapes. Remote Sens. 4, 3265–3286.

Epstein, E., Norlyn, J.D., Rush, D.W., Kingsbury, R.W., Kelly, D.B., Cunningham, G.A., et al., 1980. Saline culture of crops: a genetic approach. Science 210, 399.

Farifteh, J., van der Meer, F., Atzberger, C., Carranza, E.J.M., 2007. Quantitative analysis of salt-affected soil reflectance spectra: a comparison of two adaptive methods (PLSR and ANN). Remote Sens. Environ. 110, 59–78.

Feng, J., Rogge, D., Rivard, B., 2018. Comparison of lithological mapping results from airborne hyperspectral VNIR–SWIR, LWIR and combined data. Int. J. Appl. Earth Obs. Geoinf. 64, 340–353.

Fernández, S., Peón, J., Recondo, C., Calleja, J.F., Guerrero, C., 2016. Spatial modelling of organic carbon in burned mountain soils using hyperspectral images, field datasets, and NIR spectroscopy (Cantabrian Range, NW Spain). Land. Degrad. Dev. 27, 1479–1488.

Finn, M.P., Lewis, M., Bosch, D., Giraldo, M., Yamamoto, K., Sullivan, D.G., et al., 2011. Remote sensing of soil moisture using airborne hyperspectral data. GIScience Remote Sens. 48 (4), 522–540.

Franceschini, M.H.D., Demattê, J.A.M., da Silva Terra, F., Vicente, L.E., Bartholomeus, H., de Souza Filho, C.R., 2015. Prediction of soil properties using imaging spectroscopy: considering fractional vegetation cover to improve accuracy. Int. J. Appl. Earth Obs. 38, 358–370.

Garfagnoli, F., Ciampalini, A., Moretti, S., Chiarantini, L., Vettori, S., 2013. Quantitative mapping of clay minerals using airborne imaging spectroscopy: new data on Mugello (Italy) from SIM-GA prototypal sensor. Eur. J. Remote Sens. 46, 1–17.

Gerighausen, H., Menz, G., Kaufmann, H., 2012. Spatially explicit estimation of clay and organic carbon content in agricultural soils using multi-annual imaging spectroscopy data. Appl. Environ. Soil Sci. 2012, 868090.

Gholizadeh, A., Zizala, D., Saberioon, M., Boruvka, L., 2018. Soil organic carbon and texture retrieving and mapping using proximal, airborne and Sentinel-2 spectral imaging. Remote Sens. Environ. 218, 89–103.

Ghosh, G., Kumar, S., Saha, S.K., 2012. Hyperspectral satellite data in mapping salt-affected soils using linear spectral unmixing analysis. J. Indian Soc. Remote Sens. 40 (1), 129–136.

Goetz, A.F.H., 2009. Three decades of hyperspectral remote sensing of the Earth: a personal view. Remote Sens. Environ. 113, S5–S16.

Goetz, A.F.H., Rowan, L.C., Kingston, M.J., 1982. Mineral identification from orbit: initial results from the shuttle multispectral infrared radiometer. Science 218, 1020–1024.

Gomez, C., Rossel, R.A.V., McBratney, A.B., 2008a. Soil organic carbon prediction by hyperspectral remote sensing and field vis–NIR spectroscopy: an Australian case study. Geoderma 146, 403–411.

Gomez, C., Lagacherie, P., Coulouma, G., 2008b. Continuum removal versus PLSR method for clay and calcium carbonate content estimation from laboratory and airborne hyperspectral measurements. Geoderma 148, 141–148.

Gomez, C., Lagacherie, P., Coulouma, G., 2012. Regional predictions of eight common soil properties and their spatial structures from hyperspectral vis–NIR data. Geoderma 189, 176–185.

Govil, H., Gill, N., Rajendran, S., Santosh, M., Kumar, S., 2018. Identification of new base metal mineralization in Kumaon Himalaya, India, using hyperspectral remote sensing and hydrothermal alteration. Ore Geol. Rev. 92, 271–283.

Hamzeh, S., Naseri, A.A., AlaviPanah, S.K., Mojaradi, B., Bartholomeus, H.M., Clevers, J.G.P.W., et al., 2013. Estimating salinity stress in sugarcane fields with spaceborne hyperspectral vegetation indices. Int. J. Appl. Earth Obs. Geoinf. 21, 282–290.

Hamzeh, S., Naseri, A.A., Alavipanah, S.K., Bartholomeus, H., Herold, M., 2016. Assessing the accuracy of hyperspectral and multispectral satellite imagery for categorical and quantitative mapping of salinity stress insugarcane fields. Int. J. Appl. Earth Obs. Geoinf. 52, 412–421.

Haubrock, S.N., Chabrillat, S., Kuhnert, M., Hostert, P., Kaufmann, H., 2008. Surface soil moisture quantification and validation based on hyperspectral data and field measurements. J. Appl. Remote Sens. 2, 023552.

Hbirkou, C., Pätzold, S., Mahlein, A.-K., Welp, G., 2012. Airborne hyperspectral imaging of spatial soil organic carbon heterogeneity at the field-scale. Geoderma 175, 21–28.

Hillel, D., 1980. Applications of Soil Physics. Academic Press Inc, Amsterdam.

Hively, W.D., McCarty, G.W., Reeves, J.B., Lang, M.W., Oesterling, R.A., Delwiche, S.R., 2011. Use of airborne hyperspectral imagery to map soil properties in tilled agricultural fields. Appl. Environ. Soil Sci. 2011, 358193.

Jaber, S.M., Lant, C.L., Al-Qinna, M.I., 2011. Estimating spatial variations in soil organic carbon using satellite hyperspectral data and map algebra. Int. J. Remote Sens. 32, 5077–5103.

Kanning, M., Siegmann, B., Jarmer, T., 2016. Regionalization of uncovered agricultural soils based on organic carbon and soil texture estimations. Remote Sens. 8, 927.

Krapez, J.C., Olioso, A., 2011. A combination of temperature, vegetation indexes and albedo, as obtained by airborne hyperspectral remote sensing, for the evaluation of soil moisture. QIRT J. 8, 187–200.

Kumar, S., Gautam, G., Saha, S.K., 2015. Hyperspectral remote sensing data derived spectral indices in characterizing salt-affected soils: a case study of Indo-Gangetic plains of India. Environ. Earth Sci 73, 3299–3308.

Kusuma, K.N., Ramakrishnan, D., Pandalai, H.S., 2012. Spectral pathways for effective delineation of high-grade bauxites: a case study from the Savitri River Basin, Maharashtra, India, using EO-1 Hyperion data. Int. J. Remote Sens. 33 (22), 7273–7290.

Lagacherie, P., Baret, F., Feret, J.B., Madeira-Netto, J., Robbez-Masson, J.M., 2008. Estimation of soil clay and calcium carbonate using laboratory, field and airborne hyperspectral measurements. Remote Sens. Environ. 112 (3), 825–835.

Lei, L., Tashpolat, T., Ding, J.L., Jiang, H.N., Ardak, K., 2014. Study on the soil salinization monitoring based on measured hyperspectral and HSI data. Spectrosc. Spectr. Anal. 34 (7), 1948–1953.

Littlefield, E.F., Calvin, W.M., 2014. Geothermal exploration using imaging spectrometer data over Fish Lake Valley, Nevada. Remote Sens. Environ. 140, 509–518.

Liu, L., Zhou, J., Jiang, D., Zhuang, D., Mansaray, L.R., Hu, Z., et al., 2016. Mineral resources prospecting by synthetic application of TM/ETM +, Quickbird and Hyperion data in the Hatu area, West Junggar, Xinjiang, China. Sci. Rep. 6, 21851.

Liu, L., Feng, F., Rivard, B., Xu, X., Zhou, J., Han, L., et al., 2017. Mapping alteration using imagery from the Tiangong—1 hyperspectral spaceborne system: example for the Jintanzi gold province. China Int. J. Appl. Earth Obs. Geoinf. 64, 275–286.

Liu, L., Ji, M., Buchroithner, M., 2018. A case study of the forced invariance approach for soil salinity estimation in vegetation-covered terrain using airborne hyperspectral imagery. ISPRS Int. J. Geoinf 7, 48.

Li, J., Zhao, G., Chang, C., Liu, H., 2014. Land salinization information extraction method based on HSI hyperspectral and TM imagery. Spectrosc. Spectr. Anal. 34 (2), 520–525.

Li, J., Pu, L., Zhu, M., Dai, X., Xu, Y., Chen, X., et al., 2015. Monitoring soil salt content using HJ-1A hyperspectral data: a case study of coastal areas in Rudong County, Eastern China. Chin. Geog. Sci. 25 (2), 213–223.

Lu, P., Wang, L., Niu, Z., Li, L., Zhang, W., 2013. Prediction of soil properties using laboratory vis–NIR spectroscopy and Hyperion imagery. J. Geochem. Explor. 132, 26–33.

Maltese, A., Capodici, F., Ciraolo, G., Loggia, G.L., 2013. Map soil water content under sparse vegetation and changeable sky conditions: comparison of two thermal inertia approaches. J. Appl. Remote Sens. 7, 073548.

McCabe, M., Wood, E., 2006. Scale influences on the remote estimation of evapotranspiration using multiple satellite sensors. Remote Sens. Environ. 105 (4), 271–285.

Meer, F.D., Werff, H.M.A., Ruitenbeek, F.J.A., Hecker, C.A., Bakker, W.H., Noomen, M.F., et al., 2012. Multi- and hyperspectral geologic remote sensing: a review. Int. J. Appl. Earth Obs. Geoinf. 14, 112–128.

Metternicht, G.I., Zinck, J.A., 2003. Remote sensing of soil salinity: potentials and constraints. Remote Sens. Environ. 85 (1), 1–20.

Minasny, B., McBratney, A.B., Malone, B.P., Wheeler, I., 2013. Digital mapping of soil carbon. In: Sparks, D.L. (Ed.), Advances in Agronomy, 118. Elsevier Academic Press Inc, San Diego, CA, pp. 1–47.

Minu, S., Shetty, A., Minasny, B., Gomez, C., 2017. The role of atmospheric correction algorithms in the prediction of soil organic carbon from Hyperion data. Int. J. Remote Sens. 38 (23), 6435–6456.

Molan, Y.E., Refahi, D., Tarashti, A.H., 2014. Mineral mapping in the Maherabad area, eastern Iran, using the HyMap remote sensing data. Int. J. Appl. Earth Obs. Geoinf. 27 (4), 117–127.

Moreira, L.C.J., Teixeira, A.S., Galvão, L.S., 2015. Potential of multispectral and hyperspectral data to detect saline-exposed soils in Brazil. GIScience Remote Sens. 52 (4), 416–436.

Mulder, V.L., de Bruin, S., Schaepman, M.E., Mayr, T.R., 2011. The use of remote sensing in soil and terrain mapping—a review. Geoderma 162, 1–19.

Murphy, R.J., Schneider, S., Monteiro, S.T., 2014. Consistency of measurements of wavelength position from hyperspectral imagery: use of the ferric iron crystal field absorption at 900 nm as an indicator of mineralogy. Geosci. Remote Sens. IEEE Trans. 52, 2843–2857.

Neto, O.C.R., Teixeira, A.D.S., Leão, R.A.O., Moreira, L.C.J., Galvão, L.S., 2017. Hyperspectral remote sensing for detecting soil salinization using ProSpecTIR-VS aerial imagery and sensor simulation. Remote Sens. 2017 (9), 42.

Notesco, G., Kopačková, V., Rojík, P., Schwartz, G., Livne, I., Dor, E.B., 2014. Mineral classification of land surface using multispectral LWIR and hyperspectral SWIR remote-sensing data. A case study over the Sokolov Lignite Open-Pit Mines, the Czech Republic. Remote Sens. 6, 7005–7025.

Notesco, G., Ogen, Y., Ben-Dor, E., 2015. Mineral classification of Makhtesh Ramon in Israel using hyperspectral longwave infrared (LWIR) remote-sensing data. Remote Sens. 7, 12282–12296.

Notesco, G., Ogen, Y., Ben-Dor, E., 2016. Integration of hyperspectral shortwave and longwave infrared remote-sensing data for mineral mapping of Makhtesh Ramon in Israel. Remote Sens. 8, 318.

Nowak, B., Nesme, T., David, C., Pellerin, S., 2015. Nutrient recycling in organic farming is related to diversity in farm types at the local level. Agric. Ecosyst. Environ. 204, 17–26.

Nowkandeh, S.M., Noroozi, A.A., Homaee, M., 2018. Estimating soil organic matter content from Hyperion reflectance images using PLSR, PCR, MinR and SWR models in semi-arid regions of Iran. Environ. Dev 25, 23–32.

Ong, C.C.H., Cudahy, T.J., 2014. Mapping contaminated soils: using remotely-sensed hyperspectral data to predict pH. Eur. J. Soil Sci. 65, 897–906.

Pascucci, S., Casa, R., Belviso, C., Palombo, A., Pignatti, S., Castaldi, F., 2014. Estimation of soil organic carbon from airborne hyperspectral thermal infrared data: a case study. Eur. J. Soil Sci. 65, 865–875.

Patzold, S., Mertens, F.M., Bornemann, L., Koleczek, B., Franke, J., Feilhauer, H., et al., 2008. Soil heterogeneity at the field scale: a challenge for precision crop protection. Precis. Agric. 9, 367–390.

Peón, J., Fernández, S., Recondo, C., Calleja, J.F., 2017a. Evaluation of the spectral characteristics of five hyperspectral and multispectral sensors for soil organic carbon estimation in burned areas. Int. J. Wildland Fire 26 (3), 230–239.

Peón, J., Recondo, C., Fernández, S., Calleja, J.F., Miguel, E.D., Carretero, L., 2017b. Prediction of topsoil organic carbon using airborne and satellite hyperspectral imagery. Remote Sens. 9, 1211.

Pour, B.A., Hashim, M., Marghany, M., 2014. Exploration of gold mineralization in a tropical region using Earth Observing-1 (EO1) and JERS-1 SAR data: a case study from Bau gold field, Sarawak, Malaysia. Arab. J. Geosci. 7 (6), 2393–2406.

Ramakrishnan, D., Nithya, M., Singh, K.D., Bharti, R., 2013. A field technique for rapid lithological discrimination and ore mineral identification: results from Mamandur polymetal deposit, India. J. Earth Syst. Sci. 122 (1), 1–14.

Ramakrishnan, D., Bharti, R., 2015. Hyperspectral remote sensing and geological applications. Curr. Sci. 108 (5), 879–891.

Rani, N., Mandla, V.R., Singh, T., 2017. Spatial distribution of altered minerals in the gadag schist belt (GSB) of Karnataka, Southern India using hyperspectral remote sensing data. Geocarto Int. 32 (3), 225–237.

Riley, D.N., Aslett, Z., Taranik, J.V., 2017. Mapping rock-forming minerals at boundary canyon, Death Valley National Park, California, using aerial SEBASS thermal-infrared hyperspectral image data. Int. J. Appl. Earth Obs. Geoinf. 64, 326–339.

Rosero-Vlasova, O.A., Vlassova, L., Perez-Cabello, F., Montorio, R., Nadal-Romeroa, E., 2018. Modeling soil organic matter and texture from satellite data in areas affected by wildfires and cropland abandonment in Aragon, Northern Spain. J. Appl. Remote Sens. 12 (4), 042803.

Sahoo, R.N., Ray, S.S., Manjunath, K.R., 2015. Hyperspectral remote sensing of agriculture. Curr. Sci. 108 (5), 848–859.

Salvucci, G.D., Saleem, J.A., Kaufmann, R., 2002. Investigating soil moisture feedbacks on precipitation with tests of granger causality. Adv. Water Resour. 25, 1305–1312.

Scafutto, R.D.P.M., Filho, C.R.S., Riley, D.N., Oliveira, W.J., 2017. Hyperspectral remote sensing detection of petroleum hydrocarbons in mixtures with mineral substrates: implications for onshore exploration and monitoring. ISPRS J. Photogramm. Remote Sens. 128, 146–157.

Scafutto, R.D.P.M., Filho, C.R.S., Riley, D.N., Oliveira, W.J., 2018. Evaluation of thermal infrared hyperspectral imagery for the detection of onshore methane plumes. Significance for hydrocarbon exploration and monitoring. Int. J. Appl. Earth Obs. Geoinf. 64, 311–325.

Schoonover, J.E., Crim, J.F., 2015. An introduction to soil concepts and the role of soils in watershed management. J. Contemp. Water Res. Educ. 154 (1), 21–47.

Schwanghart, W., Jarmer, T., 2011. Linking spatial patterns of soil organic carbon to topography—a case study from south-eastern Spain. Geomorphology 126, 252–263.

Selige, T., Böhner, J., Schmidhalter, U., 2006. High resolution topsoil mapping using hyperspectral image and field data in multivariate regression modeling procedures. Geoderma 136, 235–244.

Shepherd, K.D., Walsh, M.G., 2002. Development of reflectance spectral libraries for characterization of soil properties. Soil Sci. Soc. Am. J. 66, 988–998.

Shrestha, D.P., Margate, D.E., van der Meer, F., Anh, H.V., 2005. Analysis and classification of hyperspectral data for mapping land degradation: an application in southern Spain. Int. J. Appl. Earth Obs. Geoinf. 7, 85–96.

Silvestri, S., Marani, M., Marani, A., 2003. Hyperspectral remote sensing of salt marsh vegetation, morphology and soil topography. Phys. Chem. Earth 28, 15–25.

Sobrino, J.A., Franch, B., Mattar, C., Jiménez-Muñoz, J.C., Corbari, C., 2012. A method to estimate soil moisture from airborne hyperspectral scanner (AHS) and ASTER data: application to SEN2FLEX and SEN3EXP campaigns. Remote Sens. Environ. 117, 415–428.

Song, X., Ma, J., Li, X., Leng, P., Zhou, F., Li, S., 2014. First results of estimating surface soil moisture in the vegetated areas using ASAR and Hyperion data: the Chinese Heihe River Basin case study. Remote Sens. 6, 12055–12069.

Song, Y.Q., Zhao, X., Su, H.Y., Li, B., Hu, Y.M., Cui, X.S., 2018. Predicting spatial variations in soil nutrients with hyperspectral remote sensing at regional scale. Sensors 18, 3086.

Steinberg, A., Chabrillat, S., Stevens, A., Segl, K., Foerster, S., 2016. Prediction of common surface soil properties based on vis–NIR airborne and simulated EnMAP imaging spectroscopy data: prediction accuracy and influence of spatial resolution. Remote Sens. 8, 613.

Stevens, A., van Wesemael, B., Vandenschrick, G., Toure, S., Tychon, B., 2006. Detection of carbon stock change in agricultural soils using spectroscopic techniques. Soil Sci. Soc. Am. J. 70, 844–850.

Stevens, A., van Wesemael, B., Bartholomeus, H., Rosillon, D., Tychon, B., Ben-Dor, E., 2008. Laboratory, field and airborne spectroscopy for monitoring organic carbon content in agricultural soils. Geoderma 144, 395–404.

Stevens, A., Udelhoven, T., Denis, A., Tychon, B., Lioy, R., Hoffmann, L., et al., 2010. Measuring soil organic carbon in croplands at regional scale using airborne imaging spectroscopy. Geoderma 158, 32–45.

Stevens, A., Miralles, I., van Wesemael, B., 2012. Soil organic carbon predictions by airborne imaging spectroscopy: comparing cross-validation and validation. Soil Sci. Soc. Am. J. 76, 2174–2183.

Sun, L., Khan, S., 2016. Ground-based hyperspectral remote sensing of hydrocarbon-induced rock alterations at cement, Oklahoma. Mar. Pet. Geol. 77, 1243–1253.

Swayze, G.A., Clark, R.N., Goetz, A.F.H., Livo, K.E., Breit, G.N., Kruse, F.A., et al., 2014. Mapping advanced argillic alteration at Cuprite, Nevada, using imaging spectroscopy. Econ. Geol. 109, 1179–1221.

Tong, Q., Xue, Y., Zhang, L., 2014. Progress in hyperspectral remote sensing science and technology in China over the past three decades. IEEE J. Sel. Top. Appl. Earth Obs. Remote Sens 7 (1), 70–91.

Uno, Y., Prasher, S.O., Patel, R.M., Strachan, I.B., Pattey, E., Karimi, Y., 2005. Development of field-scale soil organic matter content estimation models in Eastern Canada using airborne hyperspectral imagery. Can. Biosyst. Eng. 47, 9–14.

Vereecken, H., Weynants, M., Javaux, M., Pachepsky, Y., Schaap, M.G., van Genuchten, M.Th, 2010. Using pedotransfer functions to estimate the van Genuchten–Mualem soil hydraulic properties: a review. Vadose Zone J. 9, 795–820.

Vicente, L.E., Filho, C.R.S., 2011. Identification of mineral components in tropical soils using reflectance spectroscopy and advanced spaceborne thermal emission and reflection radiometer (ASTER) data. Remote Sens. Environ. 115, 1824–1836.

Vohland, M., Ludwig, M., Thiele-Bruhn, S., Ludwig, B., 2017. Quantification of soil properties with hyperspectral data: selecting spectral variables with different methods to improve accuracies and analyze prediction mechanisms. Remote Sens. 9, 1103.

Weng, Y., Gong, P., Zhu, Z., 2008. Soil salt content estimation in the Yellow River delta with satellite hyperspectral data. Can. J. Remote Sens. 34 (3), 259–270.

Weng, Y., Gong, P., Zhu, Z., 2010. A spectral index for estimating soil salinity in the Yellow River Delta region of China using EO-1 Hyperion data. Pedosphere 20 (3), 378–388.

Wu, J., Liu, Y., Wang, J., He, T., 2010. Application of Hyperion data to land degradation mapping in the Hengshan region of China. Int. J. Remote Sens. 31 (19), 5145–5161.

Yu, H., Kong, B., Wang, G.X., Du, R.X., Qie, G.P., 2018. Prediction of soil properties using a hyperspectral remote sensing method. Arch. Agron. Soil Sci. 64 (4), 546–559.

Zadeh, M.H., Tangestani, M.H., Roldan, F.V., Yusta, I., 2014. Sub-pixel mineral mapping of a porphyry copper belt using EO-1 Hyperion data. Adv. Space Res. 53 (3), 440–451.

Zeng, W.Z., Lei, G.Q., Zhang, H.Y., Hong, M.H., Xu, C., Wu, J.W., et al., 2017. Estimating root zone moisture from surface soil using limited data. Ecol. Chem. Eng. 24 (4), 501–516.

Zhang, T., Li, L., Zheng, B., 2013. Estimation of agricultural soil properties with imaging and laboratory spectroscopy. J. Appl. Remote Sens. 7, 73587.

Zhang, T., Shao, Y., Gong, H., Li, L., Wang, L., 2014. Salt content distribution and pale climatic significance of the Lop Nur “Ear” feature: results from analysis of EO-1 Hyperion imagery. Remote Sens. 6, 7783–7799.

Zhao, Y.J., Qin, K., Sun, Y., Liu, D.C., Tian, F., Pei, C.K., et al., 2015. Progress of geological survey using airborne hyperspectral remote sensing data in the Gansu and Qinghai Regions. Acta Geol. Sin. 89 (5), 1783–1784.

Zhao, L., Hu, Y.M., Zhou, W., Liu, Z.H., Pan, Y.C., Shi, Z., et al., 2018. Estimation methods for soil mercury content using hyperspectral remote sensing. Sustainability 10, 2474.

15

Mineral exploration using hyperspectral data

Arindam Guha

GEOSCIENCES GROUP, NATIONAL REMOTE SENSING CENTRE, INDIAN SPACE RESEARCH ORGANIZATION, HYDERABAD, INDIA

15.1 Introduction

Reflectance spectroscopy is a branch of science that studies the reflectance spectra of natural targets and identifies characteristic dips or kinks in their spectral profiles as the key to analyzing their chemistry and atomic structure (Clark, 1999). Spectral profiles or reflectance spectra are plots of reflectance with respect to wavelength of electromagnetic radiation. Reflectance spectroscopy is a primarily laboratory-based analytical technique that is used to investigate "material properties" in different disciplines of science based on the interaction of electromagnetic radiation and matter (Green et al., 1998). Imaging spectroscopy or hyperspectral remote sensing in the solar reflected spectrum [i.e., visible near-infrared (VNIR) and shortwave-infrared (SWIR) electromagnetic domains] has been regarded as an extension of laboratory-based reflectance spectroscopy for delineating surface mineralogy. Imaging spectroscopy works with similar principles to those of its laboratory counterpart, but with global and regional perspectives (Clark et al., 2003). However, all the rocky planetary surfaces in the solar system have been studied using imaging spectrometers as the diagnostic absorption features imprinted on the reflectance spectra of the rocks and minerals of these rocky planets can be identified from any distance provided the spectrometer used has the requisite number of spectral channels and sensitivity for recording such absorption features (Green et al., 1998; Clark et al., 2003). Imaging spectrometers acquire spectral data as an image using different types of detector arrays moving along a specified direction along with a satellite or aircraft and these spectrometers have the requisite number of continuous and contiguous spectral bands. The molecules of land, water, and atmosphere interact with incident solar radiation in the spectral domain of 400–2500 nm (i.e., solar reflected region) through absorption, scattering, and reflectance processes. Imaging spectrometers are generally operative within the solar reflected region to record spectra as images derived from the mentioned interactions. The main objective of geological spectroscopy is to spectrally delineate and discriminate different rocks and minerals using their diagnostic absorption features, which are the result of interactions between land and incident electromagnetic radiation (Green et al., 1998; Clark et al., 2003). One of the primary objectives of geological

Hyperspectral Remote Sensing. DOI: https://doi.org/10.1016/B978-0-08-102894-0.00012-7

spectroscopy has been the discrimination of rocks and minerals in spatial domain by segmenting natural surfaces based on their distinct spectral profiles (Clark, 1999). The mineralogical information retrieved by an imaging spectrometer is used as an input for identifying potential areas of detailed geological exploration.

The absorption features or spectral features imprinted on reflectance spectra are the result of different electronic processes such as electronic transitions, charge transfer processes, and crystal fields (Clark, 1999). In addition to these, molecular vibrational processes (resulting due to the vibration of molecular bonds and legands) also imprint absorption features, which are diagnostic for the identification of minerals (Hunt, 1979; Clark, 1999). The spectral features of rocks and minerals resulting due to electronic processes and vibrational processes are recorded at different wavelengths. Most of the diagnostic absorption features of minerals do not overlap with the absorption features of atmospheric particulate materials. This has made imaging spectroscopy the most widely used compositional or mineral mapping tool for planets (e.g., for the mapping of pyroxenes, ilmenite on the Moon, and mineralogical variations in anorthosite on Mars), for mapping different rock types, and also for mapping the surface signatures of mineralization (alteration zone and "caprock" of minerals) on the Earth's surface.

In this chapter, the atomic processes involved in imprinting the absorption features in the reflectance spectra of rocks and minerals are discussed along with how these absorption signatures can be upscaled to map the surface mineralogy of Earth. Also the significance of these studies in terms of providing key information for detailed geoexploration or mineralogical studies is shown (Johnson et al., 1991; Tompkins and Pieters, 1999; Lucey, 2004; Kruse, 2012; Van der Meer et al., 2012). In this chapter, it was attempted to analyze the reflectance spectra of rocks and minerals within the spectral domain of 350–2500 nm (within the spectral range of VNIR–SWIR).

An overview of different data processing and analytical techniques involved in mapping different rock types and minerals is also provided; which are primarily implemented either using the specific absorption features of minerals and rocks or using the overall spectral contrast of rock recorded in a part or the entire range of the spectral domain of interest. The different potential applications of hyperspectral data processing for targeting different types of mineral deposits is also discussed. These maps are used as supplementary and complimentary information to update the contacts between different lithounits (i.e., rock types) and also help to identify surface evidence of mineral enrichment resulting due to different endogenetic and exogenetic ore-forming processes.

15.2 Spectroscopy of rocks and minerals

The physical basis of spectroscopy lies in the interactions between mineral surfaces and the incident photons (i.e., sunlight for imaging spectroscopy related applications). As photons incident upon a mineral surface, some are reflected from its grain surfaces, some part of the photons passes through the grain, and some are absorbed (Clark et al., 2003). Those photons that are reflected from grain surfaces or refracted through a particle are said to be "scattered photons." Scattered photons may encounter another grain or be scattered away from the surface. These scattered energies are generally detected using a spectrometer. Mineral surfaces having temperatures above absolute "zero" also emit electromagnetic radiation; which is of lower intensity in comparison to the reflected energy recorded by detectors operative in the VNIR–SWIR domain. Emitted energies are recorded in the higher wavelengths [thermal infrared (TIR) region] using different types of spectrometers (e.g., Fourier thermal infrared spectrometer). Emitted electromagnetic energies also experience the same sequence of interaction (reflection, absorption, scattering) as that experienced by reflected energy. Kinks or dips imprinted on the emissivity spectra are also important in identifying minerals and their modal abundance (Buettner and Kern, 1965; Christensen et al., 2000). The utility of emittance spectroscopy in geological exploration is not discussed in this chapter.

In this chapter, the physical basis of reflectance spectroscopy and the major atomic processes involved in imprinting a spectral feature on the spectra of rocks and minerals in the VNIR–SWIR region are discussed along with their utility in the field of mineral exploration (Clark, 1999). The shape and intensity of scattered energy are dependent on complex interactions between light and matter, which are the combined effect of reflection and refraction from the index of refraction boundaries; the process is also known as scattering (Clark et al., 2003). Scattered energies from rocks and minerals are further modified by their absorption by the medium through which reflected light passes. The combined effect of scattering and absorption controls the number of photons received from a surface at a detector placed on a spaceborne and airborne platform. The spectral and spatial resolution of a spectrometer and its radiometry also play a significant role in the details at which the spectral data are collected from a distance. Here, the spectral features of spectrally diagnostic rock-forming minerals (mineral groups) are discussed along with the atomic and molecular processes that are responsible for imprinting these spectral features. A summary of the spectral features of these minerals is also given in Table 15–1.

15.2.1 Olivine

Diagnostic absorption feature of olivine is at 10 μm in the TIR domain. However, olivine has a strong absorption feature at 1 μm; which is stable and can be used for remote detection (Mustard and Sunshine, 1999) (Fig. 15–1A). This feature is primarily attributed to the crystal field effect of iron associated with iron-rich olivine. Iron, like other transition elements, has unfulfilled "d-orbitals" in its atomic structure. These d-orbitals have identical energies in an isolated ion, but the energy levels split when the atom is located in a crystal field (Burns, 1974). This splitting of the energy states of d-orbitals allows an electron of d-orbital to move between the orbitals having different energy levels (from a lower level into a higher level). The movement of these electron would be possible if an electron absorbs the equivalent energy matching with the energy difference between the above-mentioned energy states of

Table 15–1 Spectral features of different rock-forming minerals. (Cloutis, 1996 and Clark, 1999)

Mineral group		Spectral resolution required for spatial mapping
Olivine	Near 1 and 0.65 μm	At least 30 nm is required to permit the lowest level of quantitative compositional information about olivine minerals (in the 1 μm wavelength region)
Pyroxene	0.9,1.05, 2.35(approx) μm	130–500 nm
Amphibole	Absorption features near 1.1, 1.4, and 2.3–2.4 μm	Approximately 60 nm
Feldspar	1.1–1.3 μm	At least 200 nm can be used to constrain major (Ca, Na) and minor (Fe) element abundances and grain sizes
Clay	1.4, 1.9, 2.1, 2.16, 2.2 μm	15–20 nm is required to differentiate clay minerals. Spectral resolution on the order of 100 nm is only useful for identifying the presence or absence of clay minerals
Carbonate	2.2–2.4 μm (prominent feature at 2.3 μm)	10–20 nm
Iron oxides/hydroxides	0.5, 0.85, 0.95 μm	Differences in band positions between different species may be as small as 10 nm

d orbitals. The differences in the energy levels are controlled by the valence state of the atom (e.g., Fe^{2+}, Fe^{3+}), its coordination number, and the symmetry of the site it occupies. Crystal field varies with crystal structure from mineral to mineral. The amount of splitting of energy varies and the same ion (e.g., Fe^{2+}) would produce different absorptions, making specific mineral identification possible with spectroscopy as each of these minerals have a unique crystal structure. The mapping of olivine-bearing rocks using hyperspectral data is important as olivine-bearing host rock contains many economic minerals like chromite and platinum group of elements (PGE), etc.

15.2.2 Pyroxene

The VNIR–SWIR reflectance spectra of orthopyroxene are characterized by two major absorption bands situated near 0.9 and 1.05 μm (Cloutis, 1996; Cloutis et al., 2010). These features, which are of roughly equal intensity (Fig. 15–1), are attributable to crystal field transitions in ferrous iron, which preferentially shift to higher wavelengths with the increment of Fe^{2+} ions. The reflectance spectra of clinopyroxenes are of two types, namely type A and type B. This categorization is based on the presence of their absorption band. In type A, clinopyroxene spectra have two major absorption bands near 1.05 μm (band I) and 2.35 μm (band II). Calcium is strongly partitioned into the M2 crystallographic site with iron, which often makes up any deficiencies in calcium M2 site occupancy. Absorption bands result due to crystal field transitions in ferrous iron situated in the M2 crystallographic sites. In type B, clinopyroxene spectra have two main absorption bands near 0.9 and 1.15 μm, which partially

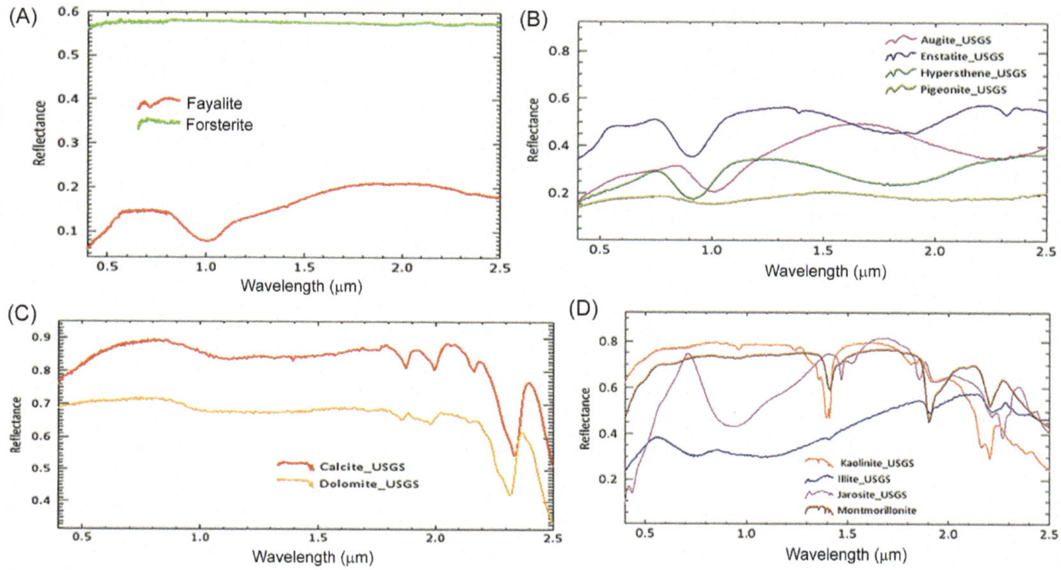

FIGURE 15–1 (A) Spectral profiles of olivine (reflectance spectra of two major endmembers of olivine are plotted). (B) Spectral profiles of a few important pyroxenes. (C) Spectral profiles of calcite and dolomite. (D) Spectral Profiles of a few important clay minerals. *USGS spectral library available with ENVI 5.5. software package. Clark, R.N., Swayze, G.A., Wise, R.A., Livo, K.E., Hoefen, T.M., Kokaly, R.F., et al., 2007. USGS Digital Spectral Library Splib06a, No. 231, US Geological Survey. Available from: https://doi.org/10.3133/ds231.*

often overlap each other. In Fig. 15–1B, enstatite and augite have type A spectra, while hypersthenes and pigeonite have type B spectra. Diagnostic absorption features imprinted on the orthopyroxene and clinopyroxene reflectance spectra will be characterized with larger absorption depth and the consequent decrease in the overall reflectance, absorption depth as the particle size of mineral grains increase. The mapping of different pyroxenes is important as many mineral deposits are associated with pyroxene-bearing rock such as different mineral deposits of magmatic origin (early magmatic deposits, volcanic massive sulfide deposits, etc.).

15.2.3 Carbonate minerals

The spectra of common anhydrous endmember carbonate minerals contain seven strong absorption features having wavelengths greater than 1.6 μm due to vibrations of the carbonate radical. Positions, widths, and spacing between these absorption features are diagnostic of mineralogy (Gaffey, 1987). Most diagnostic absorption features are around 2.3 μm (Fig. 15–1C). Carbonate-bearing rocks are known for hosting strata bound base metal deposits, phosphate, etc. Limestone also has economic value for its usage in the steel industry and also as a building material. Carbonate minerals are common minerals in the alteration mineral assemblage of hydrothermal deposits.

15.2.4 Clay and mica minerals

Clay and mica minerals have absorption features recorded in the SWIR domain due to the vibration of their molecular bonds (Farmer, 1968). The vibration of these bonds may occur due to bending, stretching, and/or rotation, and their different combinations (Clark, 1999). The frequency of vibration depends on the strength of the bond and mass of the atom or element making these bands (Clark, 1999). For a molecule with *N* numbers, there could be $3N-6$ normal modes of vibrations. These vibrations are known as "fundamentals." In addition to fundamentals, molecules can have different frequencies (multiple fundamental frequencies) of vibration. Multiple vibrations of single fundamentals are known as "overtones," while vibrations resulting due to combinations of different fundamental modes are known as "combinations" (Clark, 1999). The combination of bending of metal-OH bond and stretching of O-H bond would imprint absorption features within the spectral domain of 2.1 to 2.3 μm. These absorption features are key for identifying clay minerals (Fig. 15–1D). In addition to the above-mentioned vibrational features of clay minerals, clay minerals also have absorption feature at 1.4 μm, resulted from the overtone of stretching vibration of OH bond. The combinations of two molecular vibration modes (resulted from bending of H-OH bond and stretching of OH bond) result spectral feature at 1.9 μm. The spectral features of different clays within the spectral domain of 2.1 to 2.3 μm are generally used as these absorption features are the key to delineate different types of clay minerals. The vibrational features of different clay minerals are a few nanometers apart and narrow (e.g., kaolinite has an absorption feature with minima at 2.16 μm, while sericite has an absorption feature at 2.2 μm). Distinguishing between different clay minerals is important for characterizing the surface alteration assemblage associated with mineral deposits. For example, sericitisation is often associated with porphyry copper (Cu) deposits.

The spectral features of muscovite are more prominent than those of biotite, and the spectral features of muscovite are important in mapping muscovite mineral using hyperspectral data. Muscovite is a key mineral in the altered mineral assemblage associated with hydrothermal deposits (Kishida and Kerrich, 1987).

15.2.5 Iron oxides and iron hydroxides

The absorption features found in iron minerals are primarily due to two electronic processes, namely charge transfer and crystal field effect. In charge transfer, electrons move from ion to ion or ligand to ligand. Absorption features caused by charge transfers are diagnostic of the mineralogy of these iron oxides and hydroxides. The strengths of their charge transfer features are typically hundreds to thousands of times stronger and these features are recorded in the ultraviolet and visible electromagnetic domains (sharp fall in reflectance from 0.5 μm causing red color in iron oxide), while crystal field features are recorded in the near-infrared spectral domain (around 0.9 μm). The reflectance spectra of iron-bearing samples containing hematite and goethite are characterized with a reflectance minimum or an absorption feature near 0.9 μm. The iron absorption features at 0.9 μm are significantly reduced in terms of

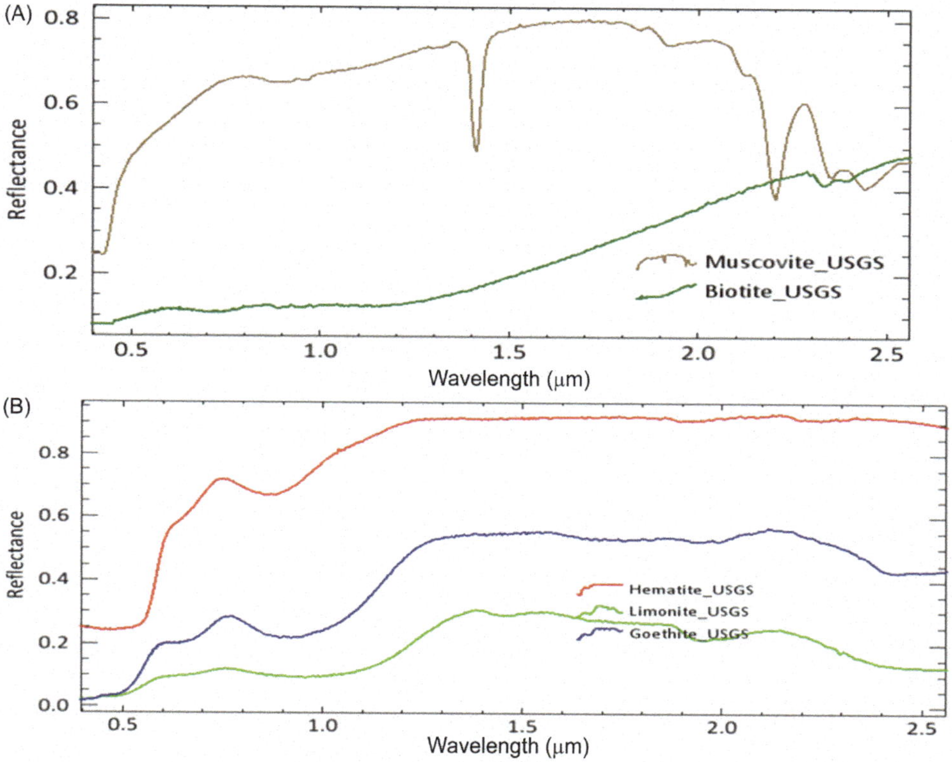

FIGURE 15–2 Spectral profiles of major minerals. (A) Spectral profiles of two major phyllosilicates. (B) Spectral profiles of iron oxides and hydroxides. *USGS spectral library available with ENVI 5.5. software package. Clark, R.N., Swayze, G.A., Wise, R.A., Livo, K.E., Hoefen, T.M., Kokaly, R.F., et al., 2007. USGS Digital Spectral Library Splib06a, No. 231, US Geological Survey. Available from: https://doi.org/10.3133/ds231; Kokaly, R.F., Clark, R.N., Swayze, G.A., Livo, K.E., Hoefen, T.M., Pearson, N.C., et al., 2017. USGS Spectral Library Version 7: U.S. Geological Survey Data Series 1035, 61 pp. Available from: https://doi.org/10.3133/ds1035.*

absorption depth when the grain size of a mineral is reduced. A large surface area to volume ratio of small grain-sized iron minerals results in a greater proportion of grain boundaries where crystal field effects are different from each other. This reduces the absorption strength of spectral features. Iron oxides and iron hydroxides are spectrally conspicuous and wider in terms of wavelength range of the absorption feature and these features are often used the key for delineating these minerals in spatial domain using superior multispectral data (Fig. 15–2).

15.2.6 Sulfide

Some sulfide minerals have characteristic reflectance spectra due to the movement of their electrons from two different energy levels known as the "valence band" and "conduction band" (Clark, 1999). In the valence band or lower energy level, the electron remains attached to the atom, while electrons move freely throughout the lattice in the conduction band or higher energy level. The difference between these energy levels is called the band gap (Clark, 1999). Few semiconductor minerals have considerable band gaps. The movement of electrons within the band gap results (after receiving the energy from the photon) in step-like reflectance spectra of sulfide minerals in the visible domain. Cinnabar (Hgs) has a strong step-like reflectance spectrum, where the reflectance sharply falls at around 0.5 μm and spectral features extend toward lower wavelengths. The spectral features of rocks and minerals are summarized in Table 15–1.

15.3 Hyperspectral sensors suitable for mineral exploration

Hyperspectral data with high spatial and spectral resolution are often required for studies related to advanced stage mineral exploration. Most of surface alterations, caprocks associated with mineralization, are localized features and these features are required to be mapped in detail to detect mineralogical variations. The high spectral resolution of hyperspectral sensors could help to delineate the finer spectral contrast of mineralogically similar rocks. On the other hand, the high signal to noise ratio (SNR) of hyperspectral sensors allow radiometric contrast to be preserved in order to detect the subtle or feeble kinks in spectral profiles. It is also essential that the surface mineralogical proxies should be mapped in high spatial resolution as these surface signatures are required to be synergized/integrated with other sets of geological, geochemical, and geophysical data that are collected with higher spatial resolution. Therefore hyperspectral sensors mounted on airborne platforms are more effective in the advanced stage of exploration where there is need to focus on small areas with spectral data with high spectral, spatial, and radiometric resolutions. On the other hand, hyperspectral sensors mounted on spaceborne platforms that have moderate spatial resolution and relatively poor SNR, are effective in mapping spectrally conspicuous host rocks of mineral deposits, residual enrichment deposits, and surface mineralogical proxies occupying larger spatial domain.

Here, details of the most popularly used hyperspectral sensors suitable for mineral exploration studies from airborne and spaceborne platforms are provided (Table 15–2). The details of a few prospective spaceborne sensors for mineral exploration and geological mapping are also provided.

15.4 Broad overview of hyperspectral data processing steps for geological exploration

Delineation of surface signatures of mineralization from hyperspectral datasets is not an easy task and different sets of data are required for mapping surface mineralogy accurately. Although commercial software packages are available to calibrate, process, and validate the results of mineral mapping from hyperspectral datasets, still, different degrees of customization are required to incorporate new algorithms, especially those algorithms that are efficient in extracting targets within the spatial extent of a pixel based on mathematical modeling of

Table 15–2 Specification of hyperspectral sensors having potential utility for mineral exploration studies (launched/to be launched)

Sensor name	Specification	Year/proposed year of launch	Agency
Spaceborne hyperspectral sensors			
Hyperion	Spatial resolution: 30 m Swath: 7.75 km Spectral range: 220 unique channels; VNIR (70 channels, 356 nm–1058 nm) and SWIR (172 channels, 852 nm–2577 nm) SNR details: 61 (550 nm); 147 (700 nm); 110 (1125 nm); 40 (2125 nm)	2000	NASA
Enmap	Spatial resolution: 30 m Swath: 30 km Spectral range: VNIR (420–1000 nm): 88 spectral bands with bandwidth of 8.1 ± 1.0 nm and SNR >400:1 at 495 nm SWIR (900–2450 nm): 154 bands with spectral bandwidth of 12.5 ± 1.5 nm with SNR 170:1 at 2200 nm (*Source*: www.enmap.org)	2020	Space Administration Division of the German Aerospace Center (Deutsches Zentrum für Luft- und Raumfahrt (DLR))
Spatial resolution: Hyperspectral Imaging suite (HISUI)	Spectral range: 400–2500 nm with 185 spectral bands (10–12.5 nm spectral resolution) with SNR (30% albedo) 450 at 620 nm $\geq$300 at 2100 nm (*Source*: https://hyspiri.jpl.nasa.gov/downloads/2018_Workshop/day3/14_1808_HISUI_Matsunaga_04b.pdf)	2019 Swath: 20 km ; Spatial Resolution: 30 m	Japanese Ministry of Economy, Trade, and Industry (METI)
Airborne hyperspectral sensor			
Advanced visible infrared imaging spectrometer (AVIRIS)-next generation	Spatial resolution: 27 μm × 27 μm Swath: 20 km Spectral range: 5 nm ± 0.5 nm within the spectral domain of 380–2510 nm. Data resolution of 14 bits	Presently operational	National Aeranautics and Space Administration (NASA), United States
HyMap	Spatial resolution: 3–10 m spatial resolution depending upon the height of the platform Spectral range: HyMap data have 126 spectral bands spanning the wavelength interval of 0.45–2.5 μm. The SNR measured outside the aircraft with a sun angle of 30 degree and a 50% reflectance standard is more than 500:1 except near the major atmospheric water absorption bandwidth	Presently operational	HyMap is an airborne hyperspectral imaging sensor that was developed in Australia and is manufactured by Integrated Spectronic
Digital Airborne Imaging Spectrometer (DAIS)	Spectral range: 72 bands from visible to shortwave-infrared wavelengths		DLR, Germany

SNR, Signal to noise ratio; *SWIR*, shortwave-infrared; *VNIR*, visible near-infrared.

spectral behaviour of complex background to enhance their spectral contrast from the target of interest.

It is essential that spectral data are well calibrated to the reflectance and image spectra are synchronized with the laboratory or field spectra of different references used in the field of mineral exploration (spectra of different rock types, alteration rocks, etc.). Both "scene-based" and physics-based atmospheric calibration algorithms are used to calibrate hyperspectral data for mineral mapping. The main requirement for hyperspectral data calibration is to convert the "radiance spectra" to "reflectance spectra" by estimating solar irradiance based on deriving the influences of atmospheric parameters on the incident solar energy and reflected energy of the land surface under study. Before attempting target mapping, it is necessary to ensure that the calibrated reflectance spectra have preserved the shape of the reference or laboratory or field derived reflectance spectra of rocks and minerals. It is also necessary to ascertain whether the diagnostic absorption features of minerals are preserved in the spectra in terms of the accurate wavelength of different spectrometric parameters (absorption minima, shoulders of absorption, depth of absorption, etc.). This is often attempted by comparing the image spectra of known targets with the spectra of the target collected in the field or laboratory.

Once the hyperspectral data are calibrated and the spectral integrity of the targets from ground to image are ensured, several processing/preprocessing steps are generally followed for processing hyperspectral data for mineral mapping. Each of these processing steps and their issues are briefly discussed here.

15.4.1 Signal to noise ratio estimation

The assessment of SNR is an integral part of hyperspectral data preprocessing as it evaluates the quality of the scene to be processed for mineral mapping. There are few established methods to calculate the SNR from data (van der Meer et al., 2002). Generally, a homogeneous (spectrally homogeneous, that is, the spectral signature of the object under study does not change spatially) area is identified in the image as "region of interest (ROI)" to calculate the SNR of the data based on deriving the mean to variance ratio of the reflectance values of the above mentioned ROI. Large silica-rich granitic surface outcrop or fresh surface exposure of quartzite or sandstone or sand sheet can be used as homogeneous areas for SNR calculation as these rocks are spectrally invariant in the VNIR–SWIR domain. There are also other methods available to calculate the SNR of data and researchers should consult the literature before selecting the optimum method for SNR estimation (van der Meer et al., 2002). In general, an SNR of 200:1 or higher (the SWIR domain) is suitable for mineral mapping as some feeble, but diagnostic spectral features of different minerals, which are important for mineral exploration studies, can only be detected in hyperspectral data acquired with an appreciable SNR.

15.4.2 Data dimensionality reduction

Minimum noise fraction (MNF) is a spectral transformation method that is used to determine the inherent dimensionality of image data, to segregate noise in data, and also to allow for the optimal data dimensionality to be selected to reduce the computational requirements of subsequent processing for the efficient mapping of a target (Green et al., 1988). The MNF transform as modified from the work of Green et al. (1988) and implemented in commercial software packages, is a linear transformation that consists of two consecutive spectral transformation or rotation methods. The first rotation uses the principal components of the noise covariance matrix to decorrelate and rescale the noise in the data under study (this process is known as noise whitening). This results separation of noise from hyperspectral data with noise that has unit variance and no band-to-band correlations. The second rotation uses the principal components (PC) that have been derived from the original image data after they have been noise-whitened by the first rotation and rescaled by the noise standard deviation (Green et al., 1988). In this regard, transformation coefficients in principal components (noise removed) are used to determine the percent of total variance explained by each of the principle components. The inherent dimensionality of the data is determined by examining the final eigenvalues of the MNF bands (these bands are derived after implementing second PC transformation). The entire data space can be divided into two parts with one part associated with large eigenvalues and coherent eigenimages, and a complementary part with near-unity eigenvalues and noise-dominated images. The coherent portions are used for further processing. An inverse MNF transform using those MNF bands (large eigenvalues) having significant spectral information is essential for getting the desired results in spectral mapping. This approach is generally followed for identifying bands having considerable spectral information. This method is efficient to identify spectrally pure targets and it also helps in improving the computational efficiency. Some commercial packages have routines to perform and analyze MNF products and researchers may use the same or similar software with the mentioned routine to perform data dimensionality.

15.4.3 Endmember extraction

Endmembers or targets to be mapped have to be spectrally characterized before applying the mapping algorithm to hyperspectral data. These endmembers must be spectrally unique and must have diagnostic spectral features that are well translated from the field to the pixels of hyperspectral data. Endmembers can be identified by following three different approaches (van der Meer et al., 2002). One can collect the spectral profile from an image itself by identifying each rock unit, altered rock assemblage, or mineral of interest using proper geolocation and other geological details of the target exposed on the surface. A spectra collected thus can easily be used to map the spatial distribution of features. Image spectra of target preserve the absorption features of target in addition to spectral artifacts introduced due to sensor or other unknown atmospheric effects. Another method of finding spectral endmembers is to utilize data statistics. There are few established statistical methods that are used to characterize spectral endmembers after identifying the pure pixels in an image. Some of the most popular methods of spectral endmember characterization are the pixel purity index and simplex methods, etc. The mathematical principles of these methods, which identify targets spectrally, are well discussed in the literature (Veganzones and Grana, 2008). These statistical methods are suitable when users have little idea about the mineralogy of different targets present in a scene.

Endmember characterization using statistical methods work well in the prefield data processing stage. In this stage of data processing, geologist try to identify and map major spectral variables for detailed field check. Subsequently, the pre-field map is refined with ground truth data. A classical, widely used approach of endmember selection in mineral exploration related studies is either field-based or laboratory-based characterization of spectral parameters of the target of interest. In mineral exploration, researchers generally background knowledge on the targets to be mapped based on previous geological records. The spectral library of targets developed using above-mentioned methods are needed to be continuum removed (normalization of spectral signature based on average background reflectance measured in the spectra) if the spectral endmembers and the scene are acquired in different solar illumination conditions. Moreover, hyperspectral data in which endmembers to be mapped are also required to be continuum removed. This is essential to bring image spectra and field/laboratory spectra under the same measurement framework or reference to subdue the effects of measurement condition (illumination, measurement geometry, etc.).

15.4.4 Spectral mapping

The ultimate aim of the utilization of hyperspectral data in mineral exploration is to map surface mineralogy proxy of mineral deposits and also to prepare detailed lithological maps delineating host rock and other controls of mineralization. Several mapping algorithms are available to map targets. Mapping algorithms can be subdivided broadly into per pixel and subpixel methods. Per pixel methods are often proved useful in mapping lithology, while subpixel methods are mainly used for mapping alteration zones having smaller size or spatial dimensions in comparison to the pixel size of the scene. Most popular commercial software packages routinely apply few important mapping algorithms. A number of important mapping algorithms and their utility in mineral exploration are briefly summarized in Table 15–3.

15.5 Application of hyperspectral remote sensing in mineral exploration

Potential uses of hyperspectral data in different geological applications are discussed in the literature. Hyperspectral data have been used to map surface alterations associated with different hydrothermal deposits. Different types of alteration minerals are formed with different types of hydrothermal deposits. Hydrothermal process is a form of metasomatic process (i.e., metasomatism allows exchange of chemical components between the fluids and the wall rocks, i.e., the rocks through which hydrothermal solutions propagates). There are different types of hydrothermal reactions. Each reaction forms specific mineral assemblages. Hydrothermal fluids and their geochemistry play a significant role in carrying different metals and contribute to localizing mineral deposits. Magmatic fluids, both vapor and hypersaline liquid, are regarded as the primary source of many components (including metals and their ligands) (Hedenquist and Lowenstern, 1994). These components get further concentrated in magmas in various ways by incorporating metals from various sources. For example, the leaching of metals from

Table 15–3 Spectral mapping algorithms and their utility in mineral exploration.

Name of mapping algorithm	Category	Concept	Utility
Spectral angle mapper	Per pixel	The algorithm determines the similarity between two spectra by calculating the angle (measured in the n dimension) between them. In the algorithm, spectra are treated as vectors in the hyperspectral data space with dimensionality equal to the number of bands of hyperspectral data.	This technique, when used on calibrated reflectance data, is relatively insensitive to illumination and albedo effects. Therefore spectra collected under different illumination condition can be compared. This method is suitable for preparing updated maps of host rock or detailed lithological maps of a study area where rock units have considerable spectral contrast in the spectral dimension of hyperspectral data.
Spectral information divergence (SID)		SID is a spectral classification method that uses a divergence measure to match the image spectra with the reference spectra. The smaller the divergence, the more likely the pixels would be similar and the target is appropriately classified.	The output from SID is a classified image and a set of rule images like spectral angle mapper and, therefore, a single map of all endmembers (suitable for lithology mapping) can be generated as a single output. SID has the capability to subdue intra-rock variation and enhance inter-rock spectral contrast.
Support vector machine (SVM)		SVM is a classification system derived from statistical learning theory. It separates the different classes using a separation surface that maximizes their spectral contrasts. The surface is called the optimal hyperplane, and the data points closest to the hyperplane are called support vectors. These vectors are used to mathematically construct the hyperplane in the n dimension of hyperspectral data. SVM also can perform as a nonlinear classifier through the use of nonlinear kernels.	Useful for lithological mapping. The spectral contrast between different rocks is enhanced. SVM is suitable for deriving automated lithological maps by incorporating intratarget variability.
Band ratios		It is a simple method that can be used to delineate a target based on spectral features using the spectral bands coinciding with the wavelength of absorption minima and highest reflectance maxima defining the edge of the absorption feature.	Generally, it is used for mapping both alteration zone and lithology. This method is effective for multispectral and hyperspectral equally data.

(Continued)

Table 15–3 (Continued)

Name of mapping algorithm	Category	Concept	Utility
Linear spectral unmixing	Subpixel	Linear spectral unmixing determines the relative abundance of different targets that are present in the pixels of a scene. The algorithm is suitable for the spectral characterization of all the endmembers present in the scene and it decomposes each pixel spectra to derive the relative spatial abundance of each target within the pixel.	The linear unmixing method is applicable where targets of interest or other rocks are exposed side by side and the geological complexity of the area is relatively less. This method works well to delineate "residual enrichment deposits" like bauxite, etc., in which the spectral characterization of different spectral endmembers (bauxite, laterite, and associated host rock) is possible. In a nutshell, this method can be used when the spectral details of all the endmembers in the scene are known and the number of endmembers is lesser than the number of spectral bands of hyperspectral data.
Matched filtering (MF)		MF is a partial unmixing or spectral decomposition technique to find the abundances (i.e., fraction of the area of the pixel occupied with the target) of target(s) of interest in each pixel of a hyperspectral image. This technique maximizes the response of the target of interest and suppresses the response of the composite unknown background. Background spectral behavior is derived from the mean of the data covariance matrix.	MF provides a rapid means of detecting specific materials using a library or image endmember spectra as the reference. It does not require knowledge of all the endmembers within an image. This technique may find some false positives if the material has a limited spatial distribution within the pixel and the spectral contrast between the target and background is low.
Mixed tuned matched filtering (MTMF)		MTMF is used to perform MF and also to derive an infeasibility image as the supplement MF map to accurately detect subpixel targets. The infeasibility image is used to reduce the number of false positives that are sometimes found when using MTMF.	This algorithm reduces false positives in detecting specific anomalies like mineral alteration zone or surface caprock as it derives an infeasibility map.
Constrained energy minimization (CEM)		CEM is similar to the MF technique as only the spectral profiles of targets are required. CEM uses a finite impulse response filter based on the endmember spectra and pixels are passed through the filter to enhance the spectral response of the desired target while minimizing its output energy resulting from a background.	The strengths of the CEM technique are its ability to deal with a variety of spectral backgrounds and its efficiency to accommodate nonlinear mixing among background materials. It is suitable for mapping of alteration minerals.

adjacent country rocks under the influence of acid magmatic vapors mixed with deeply circulating meteoric waters contribute significantly in concentrating metals in hydrothermal solutions (Hedenquist and Lowenstern, 1994). The main controlling factors of the alteration processes are: (1) the nature of wall rocks; (2) the composition of the fluids; and (3) the concentration, activity, and chemical potential of the fluid components such as H^+, CO_2, O_2, K^+, S_2, etc. (Piranjo, 2008). The factors of hydrothermal alteration contribute in the formation of different types of altered products depend upon the relative influence of the mentioned factors. For example, potassic alteration with mineral assemblages of K-feldspar and biotite, etc., are associated with porphyry and epithermal mineral systems usually accompanied by sulfides (chalcopyrite, pyrite, and molybdenite). Alterations associated with Cu–porphyry have successfully been targeted based on mapping different alteration assemblages including kaolinite, montmorillonite, sericite (muscovite/illite), calcite, chlorite, epidote, and goethite using an airborne hyperspectral sensor (HyMap sensor) (Molan et al., 2014). On the other hand, propyllite alteration assemblages are associated with altered volcanic rocks. Propyllite alteration is formed by water (H_2O) and carbon dioxide (CO_2)-rich fluid (with the presence of sulfur). Typical minerals formed for this type of alteration are epidote, chlorite, carbonates, albite, K-feldspar, and pyrite. Minerals like sericite, iron (Fe)-oxides, montmorillonite, and zeolite may also be found at places. These alteration assemblages are often associated with epithermal gold deposits (Piranjo, 2008). Quartz–sericite–pyrite or phyllic alteration are associated with wide varieties of mineral deposits from Archean volcanogenic massive sulfides and gold quartz lodes to recent epithermal systems (Piranjo, 2008). Other types of alteration such as tourmalinization (boron metasomatism) are associated with Sn–W deposits with greisen affinity (Fe-rich tourmalines), whereas magnesium (Mg)-rich tourmalines are found with massive sulfide deposits and strata bound W deposits (Piranjo, 2008).

All of these alteration assemblages can be targeted using hyperspectral data as most alteration assemblages have diagnostic absorption features in the SWIR domain. In one study, multiple endmember spectral mixture analysis was applied to analyze SWIR band data of the HyMap imaging spectrometer to map hydrothermal alteration associated with epithermal gold mineralization in the Rodalquilar caldera complex in Spain (Bedini, 2009).

Therefore it has been observed that different hydrothermal alteration processes indicate different types of mineral deposits. Most altered minerals from hydrothermal processes are characterized with absorption features resulting from vibrational and electronic processes (specially, iron hydroxide alterations). The surface signatures of hydrothermal processes are also preserved on rock exposures, which can be effectively delineated using hyperspectral data as altered rocks are rich in different types of clay minerals, iron-hydroxides, carbonates, etc. These minerals have diagnostic absorption features in their reflectance spectra, which can be effectively delineated using hyperspectral data. Hyperspectral data have been extensively used to detect the absorption features of altered rocks associated with gold mineralization in different parts of the world. For example, the presence of ammonium–illite on the Earth's surface has been used as a proxy to detect the possibility of Carlin-type gold deposits. Imaging spectroscopy can detect the presence of ammonium–illite more effectively than conventional methods like X-ray diffraction analysis, etc. (Browning, 2014). Like gold deposits, the surface signatures of Cu deposits were also targeted using hyperspectral data. Spaceborne hyperspectral data from the Hyperion sensor were used to map different alterations associated with porphyry–Cu deposits based on the implementation of mixture tuned matched filtering algorithms (Zadeh et al., 2014).

Surface processes such as oxidation and reduction are known to have their roles in in contributing situ enrichment of Cu as well as other metals such as Zn, Ag, and Au that occur at or near the surface (Robb, 2013). The process is generally known as supergene enrichment and it is the product of the oxidation and hydrolysis of sulfide. In the regolith profile, associated with supergene enrichment of base metal deposit, Fe is often found, as hematite or goethite/limonite (Robb, 2013). Upper clay- and Fe-oxy and hydroxide-rich capping, which also contains the skeletal outlines of the original sulfide minerals, is referred to as a gossan. Gossans are useful indicators of the previous existence of sulfidic ore (Robb, 2013). In the past, hyperspectral data (Hyperion data) were processed to map regolith and gossans using an object-oriented image classification technique (Farooq and Govil, 2014).

Hyperspectral remote sensing data are also used extensively for mapping "skarn" deposits. Skarn assemblages are products of the metasomatic replacement of carbonate rocks (limestone and dolomite) by varieties of calc–silicate mineral assemblages during either contact or regional metamorphic processes. Mineral deposits associated with skarn assemblages are referred to as skarn deposits. These deposits are the result of contact metamorphism and metasomatism associated with the intrusions of granite into carbonate rocks. A wide variety of mineral deposits and metal associations fall into the category of skarn deposits. These metals include W, Sn, Mo, Cu, Fe, Pb–Zn, and Au ores. The VNIR–SWIR spectral bands of HyMap data were used to estimate abundance and the levels of Tschermak substitution in white micas as well as mapping of the Mg-Fe chemistry of chlorite associated Cu–porphyry skarn deposits in Neveda (Cudahy et al., 2001).

Volcanogenic massive sulfide (VMS) ore deposits also known for the alteration assemblages associated with them. VMS deposits are generally metal deposits of Cu–Zn associated with volcanic sources emplaced under submarine conditions during their formation. Their main alteration reactions are chloritisations, sericitization, and silicifications. The alteration assemblages produced from these chemical reaction processes have diagnostic absorption features in the SWIR domain, which can be mapped using hyperspectral data having spectral bands suitable to delineate the spectral features of these minerals. The potential use of ground, laboratory, and airborne optical remote sensing methods for the detection of hydrothermal alteration zones associated with the Izok Lake VMS deposit in Nunavut, Canada, has been well demonstrated. Ground spectroscopy in that area indicated there is a systematic trend in the variation of the wavelength position of the Fe–OH absorption feature of biotite/chlorite in the proximal and distal part of the VMS deposits (Laakso et al., 2015). In another study, variation in white mica mineralogy was detected in Archaean submarine hydrothermal systems associated with VMS deposits using airborne hyperspectral data of the HyMap sensor (van Ruitenbeek et al., 2012).

The spectrometric parameters, especially depth and width, of the absorption features recorded by hyperspectral sensors have also been used as the proxy to estimate the relative grade of ore deposits. However, hyperspectral data need to be precisely calibrated for such studies and the target of interest should occupy few pixels in the image so that the reflectance spectra of the "pure target" are only recorded in the hyperspectral image. Models are often developed to estimate grade using hyperspectral data based on deriving the mathematical relation between the grade of a deposit and the spectrometric parameters of the absorption features such as depth, width, etc., of diagnostic spectral feature of the ore sample (Magendran and Sanjeevi, 2014). However, it is essential to remember that the viewing geometry and grain size also influence the width and depth of absorption features.

Spectral features imprinted on the reflectance spectra of Rare Earth Element (REE) minerals are also proved useful in estimating the grade of certain REE minerals. The diagnostic absorption features of Nd^{3+} at 741 nm and Sm^{3+} at 1250 nm recorded by the spectra of drill cores were used to develop proxies based on relating the absorption depth of these features with the relative concentration of the REE elements (Turner et al., 2014). In another study, Hyperion data were used to map different grade variations in Fe ore in Nomaundi mine in India. It was observed that the spectral parameters (width, area, and wavelength position) of the absorption features at 850–900 nm (NIR feature) and 2150–2250 nm (SWIR feature) of Hyperion image spectra of Fe exposures in Noamundi mine in India correlate well with the concentration of Fe oxide and alumina (gangue), respectively. Area, width, and depth of the NIR spectral feature has a high degree of correlation with Fe oxide concentration, while same parameters of the SWIR spectral feature of sample correlate well with alumina concentration (Magendran and Sanjeevi, 2014).

Bauxite is formed on surface during residual enrichment process, which preferentially remove silica and iron from host rock to enrich them in alumina (Schellmann, 1994). The Al–OH vibrational feature of bauxite was used as a key spectral endmember to process the Hyperion data to delineate bauxite enrichment in the Savitri River Basin, Maharashtra (Kusuma et al., 2012).

Hyperspectral data have also been well used for mapping hydrocarbon seepages. In one study, HyMap airborne hyperspectral data were used to delineate hydrocarbon spectral features. Hydrocarbon-bearing reference objects were characterized by absorption maxima at wavelengths of 1730 and 2310 nm (Hörig et al., 2001). Hydrocarbon seepages were also detected using spectral analysis of reflectance spectra in the Patrick Draw area of Wyoming (Khan and Jacobson, 2008). In this study, the presence of hydrocarbon seepages were characterized with increased presence of higher proportions of clays within anomalous samples compared to non-anomalous or background samples (Khan and Jacobson, 2008).

In the absence of seamless spaceborne hyperspectral data for the entire globe, advanced multispectral spectroradiometers having spectral channels at wavelengths sensitive to mineralogy have been increasingly used to map spectral anomalies associated with mineralization. In this regard, VNIR–SWIR spectral bands of the Advanced Spaceborne Thermal Emission and Reflection (ASTER) radiometer were used extensively in different studies related to mineral exploration. The potential utility of ASTER data to map minerals associated with hydrothermal alteration zones and caprock of supergene enrichment deposits, host rocks of important minerals like chromite, apatite, PGE, etc., and different residual enrichment deposits like bauxite have been extensively discussed in the literature (Guha et al., 2013, 2014, 2015, 2019a,b; Mars and Rowan, 2006; Pour and Hashim, 2011, 2012). In a few studies, airborne and spaceborne hyperspectral data and advanced multispectral data with ground data were combinedly used for mineral mapping. In these studies, it was observed that airborne hyperspectral data were better in mapping alteration minerals with respect to spaceborne hyperspectral data and advanced multispectral data. The conjugate use of ASTER TIR bands with AVIRIS SWIR bands for mapping altered rocks associated with hot spring deposits in the Lower, Midway, and Upper Geyser Basins of Yellowstone National Park highlighted the complementary utility of TIR band spectroscopy and SWIR band spectroscopy for mineral mapping (Hellman and Ramsey, 2004). However, the poor SNR of spaceborne hyperspectral data was regarded as a hindrance in the effective mapping of mineralogy and often derived similar results with respect the mineral maps obtained from processing multispectral data (Hubbard and Crowley, 2005; Kruse et al., 2002; Zhang and Pazner, 2007).

Hyperspectral data from airborne platforms often collect high volumes of hyperspectral data with high spectral, spatial, and radiometric resolutions. These data are capable of recording the wider spectrum of variabilities in the spectra of rocks and minerals from one pixel to the other in the same scene. A specific rock generally has its diagnostic feature; which is consistent in all the pixels (in terms of wavelength range of the spectral feature) occupied with the rock. However, portions of the same rock spectra may vary from place to place (or pixel to pixel) due to localized mineralogical variations or weathering effect. These spectral variations can be recorded (hyperspectral data can record these subtle spectral variations) and used for data processing using machine learning algorithms (MLAs) to prepare updated lithological maps. MLAs are used to automatically extract information from data following statistical approach. The potential utility of MLAs for lithological mapping has been demonstrated to map different rocks of greenstone–granitoid systems associated with gold mineralization in Hutti, India (Kumar et al., 2020).

The potential use of spectral transformation techniques on hyperspectral data and their capability in mapping rocks types and minerals have also been explored. Different principal component (PC) image composites (using VNIR and SWIR bands, respectively) derived from fine resolution spectral bands of advanced visible infrared spectrometer-next generation (AVIRIS-NG) data enhanced the lithology efficiently in the Pur Banera area of Rajasthan (Fig. 15–3). PC image composites often represent the entire spectral variances of hyperspectral datasets using fewer bands (i.e., PC bands) and this could help in delineating different rock types efficiently without carrying out detailed spectral analysis of each rock (Fig. 15–3). Similarly, noise-whitened PC bands, that is, MNF bands, are suitable for delineating different components of igneous layered complex at Sittampundi, India (Fig. 15–4).

It is important to understand the role of geobotany in detecting surface anomalies in any mineral deposits. The spectral characterization of geobotanical anomaly and regolith (elemental/mineralogical variations) may provide valuable input on the extension of concealed or covered mineral deposits. A geobotanical anomaly associated with hydrocarbon seepages was used as a proxy to identify a seepage area in Qingyang Oilfield. It was observed that leaf mesophyll structure was sensitive to oil/gas microseepage (Huang et al., 2019). In this study,

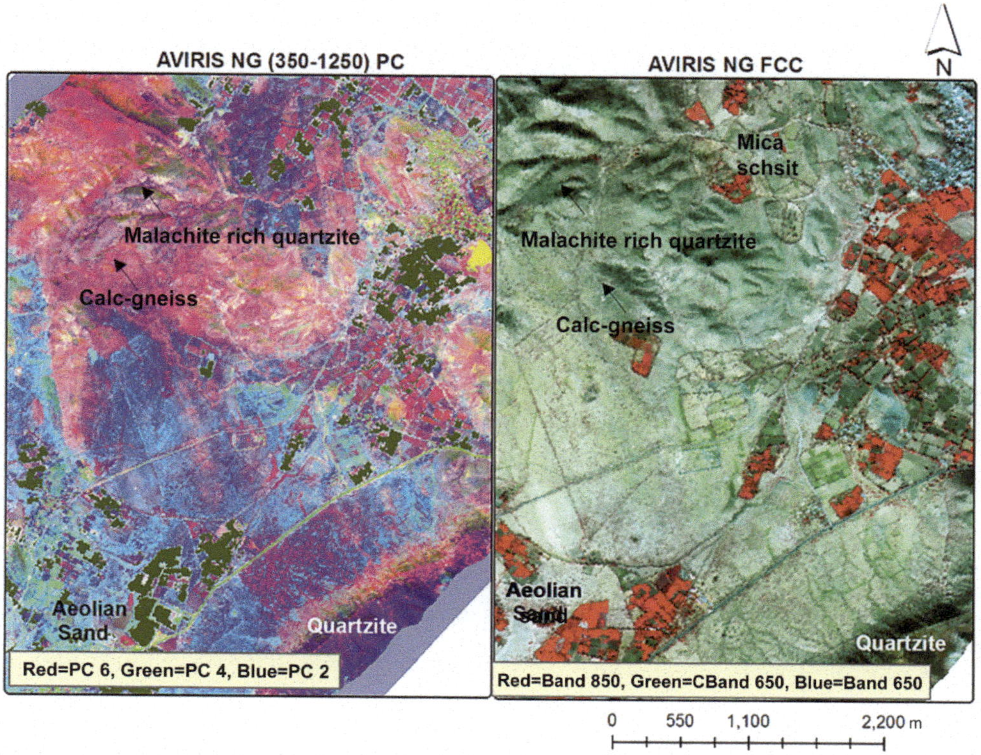

FIGURE 15–3 (A) PC composite prepared from advanced visible infrared spectrometer-next generation (AVIRIS-NG) data (resolution of 8m) and (B) false color composite prepared from AVIRIS-NG–visible near-infrared bands (*red* = spectral bands of 950 nm; *green* = 650 nm, and *blue* = 560 nm).

multivariate regression equations were derived using varying gasoline volumes as the dependent variable and the spectral feature parameters of leaves as the independent variables based on the processing of Compact Airborne Spectrographic Imager (CASI)/Shortwave infrared Airborne Spectrographic *Imager* (SASI) airborne hyperspectral data. A regression equation was derived with the highest correlation coefficient.

15.6 Requirement and future research focus

15.6.1 Data requirement and related approach

In the present scenario, the major issue that has limited the operational use of hyperspectral data for different real-time mineral exploration projects is the limited availability of hyperspectral data with appreciable specifications (i.e., appreciable swath, spatial resolution, spectral resolution, and SNR). Hyperspectral data of different specifications for different stages of mineral exploration are required. A larger swath (10 km or more) is essential for detailed lithological mapping of strata bound deposits like Fe and for delineating their spatial extent.

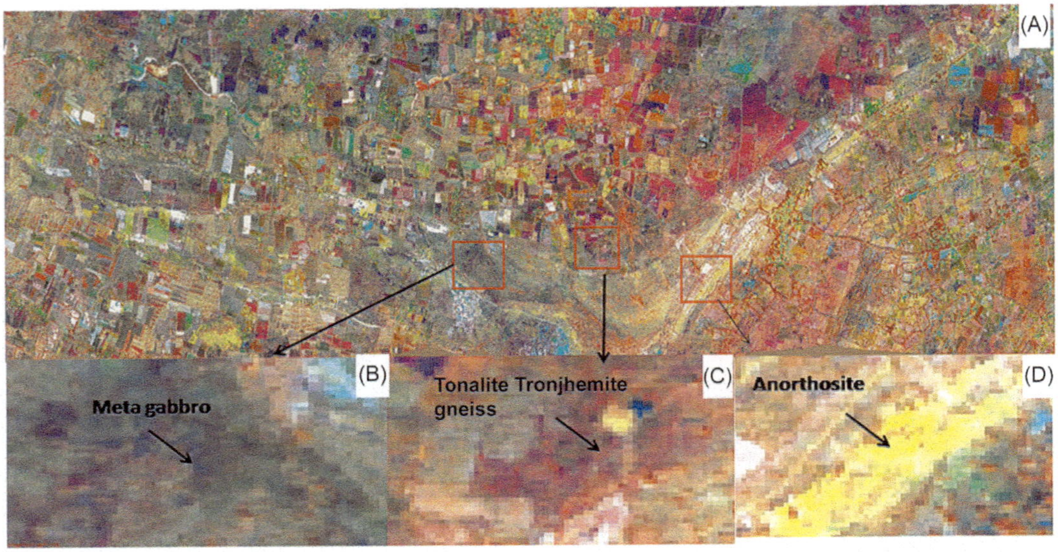

FIGURE 15–4 (A) Advanced visible infrared spectrometer-next generation derived minimum noise fraction (MNF) image composite delineating major components (i.e., intrusive rocks) in Sittampundi layered igneous complex (R = MNF band 20, G = MNF band 14, B = MNF band 10), (B) blue (gabbro), (C) pink (granodiorite–gneiss), and (D) yellow (anorthosite).

It is essential that the spatial extension of host rocks is accurately mapped for exploring these deposits before identifying hotspot areas for detailed exploration. On the other hand, airborne hyperspectral sensors such as the AVIRIS-NG, which has a better SNR, spatial resolution, and spectral resolution with respect to spaceborne hyperspectral data like those from Hyperion are useful for detailed study like the mapping of alteration zones along a specified structural control or delineating different grades of ore within banded hematite quartzite or similar Fe deposits. There will be few spaceborne hyperspectral sensors till 2021 and this limits the utilization of hyperspectral data in earth science applications. Present spaceborne sensors such as the Hyperion, Hysis, etc., are limited by data availability and the data quality required for utilization in different geological applications, which require hyperspectral data with an appreciable swath without compromising the spatial resolution and SNR of spectral bands. This limits the use of hyperspectral remote sensing for identifying new prospect areas of different minerals.

It has been understood that a significant SNR is required in data (at least 100:1 or higher in the SWIR region) to detect the subtle absorption signatures of different minerals. Hyperspectral data from spaceborne sensors can be used to carry out lithological mapping for larger spatial domains and for exploring virgin areas as the data coverage of hyperspectral sensors is large and cheaper than that of airborne hyperspectral survey.

On the other hand, airborne hyperspectral sensors appear to be able to meet the data requirements for different active geoexploration programs, but airborne hyperspectral survey is a costly affair and cannot be used unless a certain amount of background knowledge on the deposit is available. Airborne sensors often have an edge over spaceborne sensors as they can be customized to collect data for a required area and swath for specific geological applications. Spatial resolution can also be manipulated as per the requirements of a given study. However, airborne hyperspectral data often cannot be used for multidisciplinary requirements if a study area covered using an airborne sensor is too small for using the same area for different applications. Therefore the application of specific requirements versus the cost of the data needs to be analyzed before initiating airborne acquisition. Furthermore, airborne hyperspectral data need to be procured within the guidelines of the data acquisition policy of the country and necessary approvals are to be acquired before initializing the hyperspectral data acquisition from an airborne platform. In a nutshell, airborne hyperspectral data have appreciable spectral, spatial resolutions and may play an important role in mineral exploration, but crunch in availability of spaceborne hyperspectral data limits the utilization of hyperspectral data in exploring virgin areas using imaging spectroscopy. In this context, the conjugate use of advanced multispectral data and spaceborne hyperspectral data is necessary to find surface proxies of mineral deposits using airborne hyperspectral data and extending these proxies to map the virgin areas using multispectral data after establishing quantitative relations between mineral enrichment and spectral signatures of surface proxies (alteration zones, caprock of mineral deposits, etc.) recorded by hyperspectral and multispectral spectroradiometers.

15.6.2 Requirement of improving data processing approach

Hyperspectral data and advanced multispectral data often provide valuable information on surface mineralogical signatures by mapping alteration zones, caprocks of supergene enrichment deposits, and host rocks of mineral deposits such as chromite, ilmenite, and REE, etc. The value of this information in exploration model only can be realized if these alterations maps, lithological maps are combined with geophysical data to derive the three-dimensional the perspective of surface alterations for relating these surface indications with the ore-forming processes;which work in four dimensions (three spatial dimensions and time). Synergy between surface and subsurface information is important for deriving better utility from hyperspectral data in mineral exploration. A few attempts have been made to focus on combining hyperspectral/multispectral data-derived mineral maps with ground geophysical data (Mondal et al., 2019; Rani et al., 2018). Another important aspect would be the crosslinking of different disciplines of Earth sciences to derive new insights into the derived results obtained from the processing of hyperspectral data in the context of mineral exploration. For example, surface mineral maps, if linked with surface run off data and hydrological data on aquifers (porosity, type of aquifer, etc.) may derive important information on ground water quality. Further linkages between geology, ecology, and environment would provide better perspectives on the requirement of hyperspectral data for assessing the environmental impact of certain mineral deposits.

Presently, artificial intelligence approaches are increasingly being used by implementing Machine Learning Algorithms (MLAs) on hyperspectral data due to their superior spectral and radiometric resolutions often coupled with moderate to high spatial resolution. The implementation of MLAs is aimed to derive better classification accuracy from hyperspectral data and deriving automated lithological maps using the image spectra of hyperspectral data. Traditional classification algorithms have limitations in exploiting the utility of the high data dimensionality of hyperspectral data. MLAs provide valuable inputs for the selection of optimum spectral endmembers and provide new approaches for the reduction of data dimensionality for feature selection and extraction. Machine learning also contributes immensely to improving classification accuracy via optimized clustering.

In hyperspectral remote sensing, different partial subpixel mapping methods are used to delineate the fractional areal abundance of a given target within a pixel. At present, few algorithms are successfully being used for this purpose. It is essential to derive advanced algorithms to optimally characterize the spectral contrast between a target and background. There are no standardized methods to validate subpixel mineral abundance maps and there is subjectivity in fixing a threshold value to segregate alteration zones or similar targets from these subpixel abundance maps. Therefore attempts are required to derive a standardized approach to fix the threshold (based on deriving certain mathematical criteria) for delineating targets of interest (i.e., alteration minerals, etc.) in subpixel mineral maps and also for validating these mineral maps. This would help in increasing the reproducibility of these surface anomaly maps and this, in turn, would augment the operational use of hyperspectral data in the field of mineral exploration.

List of abbreviations

ASTER	Advanced Spaceborne Thermal Emission and Reflection radiometer
AVIRIS-NG	advanced visible infrared imaging spectrometer-next generation
AVIRIS	Advanced visible infrared imaging spectrometer
DAIS	digital airborne imaging spectrometer
EnMAP	Environmental Mapper and Analysis Programme
HISUI	hyperspectral imaging suite
MF	matched filtering
MLA	machine learning algorithms
MNF	minimum noise fraction
MTMF	mixture tuned matched filtering algorithm
PGE	platinum group of elements
PC	principal components
ROI	Region of interest
SAM	spectral angle mapper
SID	spectral information divergence
SNR	signal to noise ratio
SVM	support vector machine
SWIR	shortwave-infrared
TIR	Thermal Infra red
VMS	volcanogenic massive sulfide
VNIR	visible near-infrared

References

Bedini, E., 2009. Mapping lithology of the Sarfartoq carbonatite complex, southern West Greenland, using HyMap imaging spectrometer data. Remote Sen. Envi 113 (6), 1208–1219.

Browning, D.A., 2014. Hyperspectral remote sensing in mineral exploration: ammonium–illite as a pathfinder for gold. <http://adsabs.harvard.edu/abs/2014PhDT.85B>.

Buettner, K.J., Kern, C.D., 1965. The determination of infrared emissivities of terrestrial surfaces. J. Geophys. Res. 70, 1329–1337.

Burns, R.G., 1974. The polarized spectra of iron in silicates: olivine. A discussion of neglected contributions from Fe^{2+} ions in M (1) sites. Am. Mineral.: J. Earth Planet. Mater. 59 (5–6), 625–629.

Christensen, P.R., Bandfield, J.L., Hamilton, V.E., Howard, D.A., Lane, M.D., Piatek, J.L., et al., 2000. A thermal emission spectral library of rock-forming minerals. J. Geophys. Res. Planets 105, 9735–9739.

Clark, R.N., 1999. Spectroscopy of rocks and minerals, and principles of spectroscopy. Man. Remote Sens. 3 (3–58), 2–2.

Clark, Roger N., Swayze, G.A., Livo, K.E., Kokaly, R.F., Sutley, S.J., Dalton, J.B., et al., 2003. Imaging spectroscopy: Earth and planetary remote sensing with the USGS Tetracorder and expert systems. J. Geophys. Res. Planets 108 (E12).

Clark, R.N., Swayze, G.A., Wise, R.A., Livo, K.E., Hoefen, T.M., Kokaly, R.F., et al., 2007. USGS Digital Spectral Library Splib06a, No. 231, US Geological Survey. Available from: https://doi.org/10.3133/ds231.

Cloutis, E.A., 1996. Review article hyperspectral geological remote sensing: evaluation of analytical techniques. Inter. J. Remote Sens 17 (12), 2215–2242.

Cloutis, E.A., Klima, R.L., Kaletzke, L., Coradini, A., Golubeva, L.F., McFadden, L.A., et al., 2010. The 506 nm absorption feature in pyroxene spectra: nature and implications for spectroscopy-based studies of pyroxene-bearing targets. Icarus 207 (1), 295–313.

Cudahy, T.J., Wilson, J., Hewson, R., Linton, P., Harris, P., Sears, M., et al., 2001. Mapping porphyry-skarn alteration at Yerington, Nevada, using airborne hyperspectral VNIR–SWIR–TIR imaging data. In: IGARSS 2001. Scanning the Present and Resolving the Future. Proceedings. IEEE 2001 International Geoscience and Remote Sensing Symposium (Cat. No. 01CH37217), vol. 2. IEEE, pp. 631–633.

Farmer, V.C., 1968. Infrared spectroscopy in clay mineral studies. Clay Miner. 7 (4), 373–387.

Farooq, S., Govil, H., 2014. Mapping regolith and gossan for mineral exploration in the eastern kumaon himalaya, india using hyperion data and object oriented image classification. Adv. Space Res. 53 (12), 1676–1685.

Gaffey, S.J., 1987. Spectral reflectance of carbonate minerals in the visible and near infrared (0.35–2.55 μm): anhydrous carbonate minerals. J. Geophys. Res. Solid Earth 92 (B2), 1429–1440.

Green, A.A., Berman, M., Switzer, P., Craig, M.D., 1988. A transformation for ordering multispectral data in terms of image quality with implications for noise removal. IEEE Trans. Geosci. Remote Sens. 26 (1), 65–74.

Green, R.O., Eastwood, M.L., Sarture, C.M., Chrien, T.G., Aronsson, M., Chippendale, B.J., et al., 1998. Imaging spectroscopy and the airborne visible/infrared imaging spectrometer (AVIRIS). Remote Sens. Environ. 65 (3), 227–248.

Guha, A., Singh, V.K., Parveen, R., Kumar, K.V., Jeyaseelan, A.T., Rao, E.D., 2013. Analysis of ASTER data for mapping bauxite rich pockets within high altitude lateritic bauxite, Jharkhand, India. Int. J. Appl. Earth Obs. Geoinf. 21, 184–194.

Guha, A., Kumar, K.V., Dhananjaya Rao, E.N., Parveen, R., 2014. An image processing approach for converging ASTER-derived spectral maps for mapping Kolhan limestone, Jharkhand, India. Curr. Sci. (00113891) 106 (1).

Guha, A., Ghosh, B., Kumar, K.V., Chaudhury, S., 2015. Implementation of reflection spectroscopy based new ASTER indices and principal components to delineate chromitite and associated ultramafic–mafic complex in parts of Dharwar Craton, India. Adv. Space Res. 56 (7), 1453–1468.

Guha, A., Kumar, K.V., Porwal, A., Rani, K., Sahoo, K.C., Kumar, S.A., et al., 2019a. Reflectance spectroscopy and ASTER based mapping of rock-phosphate in parts of Paleoproterozoic sequences of Aravalli Group of rocks, Rajasthan, India. Ore Geol. Rev. 108, 73–87.

Guha, A., Yamaguchi, Y., Chatterjee, S., Rani, K., Vinod Kumar, K., 2019b. Emittance spectroscopy and broadband thermal remote sensing applied to phosphorite and its utility in geoexploration: a study in the parts of Rajasthan, India. Remote Sens. 11 (9), 1003.

Hedenquist, J.W., Lowenstern, J.B., 1994. The role of magmas in the formation of hydrothermal ore deposits. Nature 370 (6490), 519.

Hellman, M.J., Ramsey, M.S., 2004. Analysis of hot springs and associated deposits in Yellowstone National Park using ASTER and AVIRIS remote sensing. J. Volcanol. Geotherm. Res. 135 (1–2), 195–219.

Hörig, B., Kühn, F., Oschütz, F., Lehmann, F., 2001. HyMap hyperspectral remote sensing to detect hydrocarbons. Int. J. Remote Sens. 22 (8), 1413–1422.

Huang, S., Chen, S., Wang, D., Zhou, C., van der Meer, F., Zhang, Y., 2019. Hydrocarbon micro-seepage detection from airborne hyper-spectral images by plant stress spectra based on the PROSPECT model. Int. J. Appl. Earth Obs. Geoinf. 74, 180–190.

Hubbard, B.E., Crowley, J.K., 2005. Mineral mapping on the Chilean–Bolivian Altiplano using co-orbital ALI, ASTER and Hyperion imagery: data dimensionality issues and solutions. Remote Sens. Environ. 99 (1-2), 173–186.

Hunt, G.R., 1979. Near-infrared (1.3–2.4) μm spectra of alteration minerals—potential for use in remote sensing. Geophysics 44 (12), 1974–1986.

Johnson, J.R., Larson, S.M., Singer, R.B., 1991. Remote sensing of potential lunar resources: 1. Near-side compositional properties. J. Geophys. Res. Planets 96 (E3), 18861–18882.

Khan, S.D., Jacobson, S., 2008. Remote sensing and geochemistry for detecting hydrocarbon microseepages. Geol. Soc. Am. Bull. 120 (1–2), 96–105.

Kishida, A., Kerrich, R., 1987. Hydrothermal alteration zoning and gold concentration at the Kerr–Addison Archean lode gold deposit, Kirkland Lake, Ontario. Econ. Geol. 82 (3), 649–690.

Kokaly, R.F., Clark, R.N., Swayze, G.A., Livo, K.E., Hoefen, T.M., Pearson, N.C., et al., 2017. USGS Spectral Library Version 7: U.S. Geological Survey Data Series 1035, 61 p. Available from: https://doi.org/10.3133/ds1035.

Kruse, F.A., 2012. Mapping surface mineralogy using imaging spectrometry. Geomorphology 137 (1), 41–56.

Kruse, F.A., Perry, S.L., Caballero, A., 2002. Integrated multispectral and hyperspectral mineral mapping, Los Menucos, Rio Negro, Argentina, Part II. EO-1 Hyperion/AVIRIS comparisons and landsat TM/ASTER extensions. In: Proceedings of the 11th JPL Airborne Geoscience Workshop, pp. 4–8.

Kumar, C., Chatterjee, S., Oommen, T., Guha, A., 2020. Automated lithological mapping by integrating spectral enhancement techniques and machine learning algorithms using AVIRIS-NG hyperspectral data in Gold-bearing granite-greenstone rocks in Hutti, India. Int. J. Appl. Earth Obs. Geoinf. 86, 102006.

Kusuma, K.N., Ramakrishnan, D., Pandalai, H.S., 2012. Spectral pathways for effective delineation of high-grade bauxites: a case study from the Savitri River Basin, Maharashtra, India, using EO-1 Hyperion data. Int. J. Remote Sens. 33 (22), 7273–7290.

Laakso, K., Rivard, B., Peter, J.M., White, H.P., Maloley, M., Harris, J., et al., 2015. Application of airborne, laboratory, and field hyperspectral methods to mineral exploration in the Canadian Arctic: recognition and characterization of volcanogenic massive sulfide-associated hydrothermal alteration in the Izok Lake deposit area, Nunavut, Canada. Econ. Geol. 110 (4), 925–941.

Lucey, P.G., 2004. Mineral maps of the Moon. Geophys. Res. Lett. 31, 8.

Magendran, T., Sanjeevi, S., 2014. Hyperion image analysis and linear spectral unmixing to evaluate the grades of iron ores in parts of Noamundi, Eastern India. Inter. J. Appl. Earth Obser. Geoinfor 26, 413–426.

Mars, J.C., Rowan, L.C., 2006. Regional mapping of phyllic- and argillic-altered rocks in the Zagros magmatic arc, Iran, using Advanced Spaceborne Thermal Emission and Reflection Radiometer (ASTER) data and logical operator algorithms. Geosphere 2 (3), 161–186.

Molan, Y.E., Davood, R., Ali Hoseinmardi, Tarashti, 2014. Mineral mapping in the Maherabad area, eastern Iran, using the HyMap remote sensing data. Inter. J. Appl. Earth Obser. Geoinform 27, 117–127.

Mondal, S., Guha, A., Pal, S.K., Porwal, A., Chatterjee, S., Rani, K., et al., 2019. Conjugate utilization of Landsat-8 OLI, ground gravity and magnetic data for targeting mafic cumulates within anorthositic-layered complex of Sittampundi, India. Geocarto Int. 1–18.

Mustard, J.F., Sunshine, J.M., 1999. Spectral analysis for earth science: investigations using remote sensing data. Remote Sens. Earth Sci.: Man. Remote Sens. 3, 251–307.

Piranjo, Franco. 2008. Hydrothermal Processes and Mineral Systems. Springer Science & Business Media.

Pour, A.B., Hashim, M., 2011. Identification of hydrothermal alteration minerals for exploring of porphyry copper deposit using ASTER data, SE Iran. J. Asian Earth Sci. 42 (6), 1309–1323.

Pour, A.B., Hashim, M., 2012. The application of ASTER remote sensing data to porphyry copper and epithermal gold deposits. Ore Geol. Rev. 44, 1–9.

Rani, K., Guha, A., Pal, S.K., Kumar, K.V., 2018. Broadband reflectance, emittance spectroscopy and self-potential geophysical survey for targeting gold sulphide lode deposit in Bhukia, Rajasthan, India. Geocarto Int. 1–20.

Robb, L., 2013. Introduction to Ore-Forming Processes. John Wiley & Sons.

Schellmann, W., 1994. Geochemical differentiation in laterite and bauxite formation. Catena 21 (2–3), 131–143.

Tompkins, S., Pieters, C.M., 1999. Mineralogy of the lunar crust: results from Clementine. Meteorit. Planet Sci. 34 (1), 25–41.

Turner, D., Rivard, B., Groat, L., 2014. Rare earth element ore grade estimation of mineralized drill core from hyperspectral imaging spectroscopy. In: 2014 IEEE Geoscience and Remote Sensing Symposium. IEEE, pp. 4612–4615.

van der Meer, F., De Jong, S., Bakker, W., 2002. Imaging spectrometry: basic analytical techniques. Imaging Spectrometry. Springer, Dordrecht, pp. 17–61.

Van der Meer, F.D., Van der Werff, H.M., Van Ruitenbeek, F.J., Hecker, C.A., Bakker, W.H., Noomen, M.F., et al., 2012. Multi- and hyperspectral geologic remote sensing: a review. Int. J. Appl. Earth Obs. Geoinf. 14 (1), 112–128.

van Ruitenbeek, F.J., Cudahy, T.J., van der Meer, F.D., Hale, M., 2012. Characterization of the hydrothermal systems associated with Archean VMS-mineralization at Panorama, Western Australia, using hyperspectral, geochemical and geothermometric data. Ore Geol. Rev. 45, 33–46.

Veganzones, M.A., Grana, M., 2008. Endmember extraction methods: a short review. International Conference on Knowledge-Based and Intelligent Information and Engineering Systems. Springer, Berlin, Heidelberg, pp. 400–407.

Zadeh, M.H., Tangestani, M.H., Roldan, F.V., Yusta, I., 2014. Sub-pixel mineral mapping of a porphyry copper belt using EO-1 Hyperion data. Adv. Space Res. 53 (3), 440–451.

Zhang, X., Pazner, M., 2007. Comparison of lithologic mapping with ASTER, hyperion, and ETM data in the southeastern Chocolate Mountains, USA. Photogramm. Eng. Remote Sens. 73 (5), 555–561.

Further reading

Asadzadeh, S., de Souza Filho, C.R., 2016. A review on spectral processing methods for geological remote sensing. Int. J. Appl. Earth Obs. Geoinf. 47, 69–90.

Clouts, E.A., Gaffey, M.J., 1991. Pyroxene spectroscopy revisited: spectral-cornpositional correlations and relationship to geothermometry. J. Geophys. Res. 96, 22 809–22 826.

Cracknell, M.J., Reading, A.M., 2013. The upside of uncertainty: identification of lithology contact zones from airborne geophysics and satellite data using random forests and support vector machines. Geophysics 78 (3), WB113–WB126.

Gersman, R., Ben-Dor, E., Beyth, M., Avigad, D., Abraha, M., Kibreab, A., 2008. Mapping of hydrothermally altered rocks by the EO-1 Hyperion sensor, Northern Danakil Depression, Eritrea. Int. J. Remote Sens. 29 (13), 3911–3936.

Hymap. https://airbornescience.nasa.gov/instrument/HyMap (accessed 16.10.19.).

Kumar, C., Chatterjee, S., Oommen, T., & Guha, A. (2020). Automated lithological mapping by integrating spectral enhancement techniques and machine learning algorithms using AVIRIS-NG hyperspectral data in Gold-bearing granite-greenstone rocks in Hutti, India. International Journal of Applied Earth Observation and Geoinformation, 86, 102006.

Papp, E., Cudahy, T., 2002. Hyperspectral remote sensing. Geophys. Remote Sens. Methods Regolith Explor. 144, 13–21.

Rajendran, S., Nasir, S., 2014. Hydrothermal altered serpentinized zone and a study of Ni–magnesioferrite–magnetite–awaruite occurrences in Wadi Hibi, Northern Oman Mountain: discrimination through ASTER mapping. Ore Geol. Rev. 62, 211–226.

Rajendran, S., Nasir, S., 2017. Characterization of ASTER spectral bands for mapping of alteration zones of volcanogenic massive sulphide deposits. Ore Geol. Rev. 88, 317–335.

Sunshine, J.M., Pieters, C.M., 1998. Determining the composition of olivine from reflectance spectroscopy. J. Geophys. Res. Planets 103 (E6), 13675–13688.

Strashimirov, S., Petrunov, R., Kanazirski, M., 2002. Porphyry-copper mineralisation in the central Srednogorie zone, Bulgaria. Mineral. Depos. 37 (6–7), 587–598.

Yamaguchi, Y., Kahle, A.B., Tsu, H., Kawakami, T., Pniel, M., 1998. Overview of advanced spaceborne thermal emission and reflection radiometer (ASTER). IEEE Trans. Geosci. Remote Sens. 36 (4), 1062–1071.

16

Metrological hyperspectral image analysis through spectral differences

Hilda Deborah[1], Noël Richard[2], Jon Yngve Hardeberg[1], Jon Atli Benediktsson[3]

[1]DEPARTMENT OF COMPUTER SCIENCE, NORWEGIAN UNIVERSITY OF SCIENCE AND TECHNOLOGY, GJØVIK, NORWAY

[2]LABORATORY XLIM, UMR CNRS 7252, UNIVERSITY OF POITIERS, POITIERS, FRANCE

[3]DEPARTMENT OF ELECTRICAL AND COMPUTER ENGINEERING, UNIVERSITY OF ICELAND, REYKJAVÍK, ICELAND

16.1 Introduction

Decades after the first development of hyperspectral imaging for remote sensing and Earth observation, imaging technology has enjoyed increasing popularity, especially due to its high spectral and spatial resolutions. The high resolution images it provides have made it possible to carry out land cover classification and change detection, for example (Ghamisi, et al., 2018; Marpu et al., 2011). Additionally, nowadays, it has also been extensively used in various other fields for quality control and material inspection (Serranti et al., 2012; Mirschel et al., 2018; Deborah et al., 2019). The common thread between these applications that span from Earth observation to cultural heritage is the need for highly accurate results, a potential that hyperspectral imaging has to offer. For example, climate change research monitors atmospheric and surface temperature, ocean color, and leaf area index to detect small changes in the Earth system (Fox et al., 2017). In cultural heritage applications, precise identification of materials is needed in order to carry out, for example, a restoration project. In manufacturing and food industries, quality control is important in making sure that consumers are not at risk.

As mentioned previously, however, the gain of accuracy offered by hyperspectral imaging technology is a potential. Highly accurate images obtained in the acquisition step do not automatically entail highly accurate and relevant results at the end of the processing chain. The acquired accuracy must be maintained, managed, and controlled throughout the entire

Hyperspectral Remote Sensing. DOI: https://doi.org/10.1016/B978-0-08-102894-0.00020-6

image processing chain, from the acquisition to postprocessing steps. This shows an apparent need for *metrology*, that is, the science of measurement and its application (Joint Committee for Guides in Metrology; JGCM, 2012). In metrology, measurements are dependent upon the object being measured and not the approach to carry them out. As an illustration, the average length of an object will not change depending on whether a typical metal measuring tape or a laser-based one is used to conduct the measurement. What will vary is the accuracy of the measurement.

Taking a point of view derived from metrology, a hyperspectral image is considered as a measure[1] and the assessment of differences becomes central within the framework of its processing and analysis. The determination of, for example, uncertainty, trueness, measurement error, etc., relies on the assessment of the statistics of the measured values relative to an average value, a reference value, etc. And aside from metrology, the notion of differences or distance is also at the core of many image processing and analysis algorithms including machine learning. Therefore if metrology is to be enforced in hyperspectral image processing, it is crucial to metrologically assess the validity of a difference function.

In remote sensing, efforts in achieving a metrological level of processing of hyperspectral images can mostly be found in the acquisition step (Datla et al., 2010; Zibordi et al., 2012; Honkavaara et al., 2014; Vandermeulen et al., 2017; Talone and Zibordi, 2018). A limited number of efforts are available at the data processing step, i.e., in image modelling and unmixing (Engel et al., 2016; Bhatia et al., 2018). Subsequent processing of hyperspectral images would typically involve complex nonlinear algorithms in which it is difficult to enforce or even introduce metrological assessments into the system. Aiming to fill the gap, a metrological analysis framework for basic hyperspectral image analysis by means of spectral differences will be introduced in this chapter.

16.2 Is metrology justified for remote sensing?

Oil spill detection is a typical application of remote sensing, especially for the Deepwater Horizon (DW) oil spill in the Gulf of Mexico (Fingas and Brown, 2018). Even if there is no ground truth data easily available for the DW oil spill, some reference results have been presented in a technical report (Clark et al., 2010). In Fig. 16–1B, the results are expressed in terms of classification, relative to various measurements obtained in situ. Then, it can also be observed that the expected result was a linear measurement relative to the oil concentration in the water. Observing Fig. 16–1C, the spectral variations induced by the concentration of oil can be seen in terms of both shape and intensity variations. From the spectra, it seems that modifications in the spectra are monotonic.

Due to the lack of metrological solutions for hyperspectral image analysis in the past, many existing scientific propositions were and are based on classification (Salem and Kafatos, 2001),

[1] Or, in strict metrological terms, a *measured quantity value*, that is, quantity value representing a measurement result [Joint Committee for Guides in Metrology (JGCM), 2012]. However, in this article, the term *measure* will be used instead for simplicity.

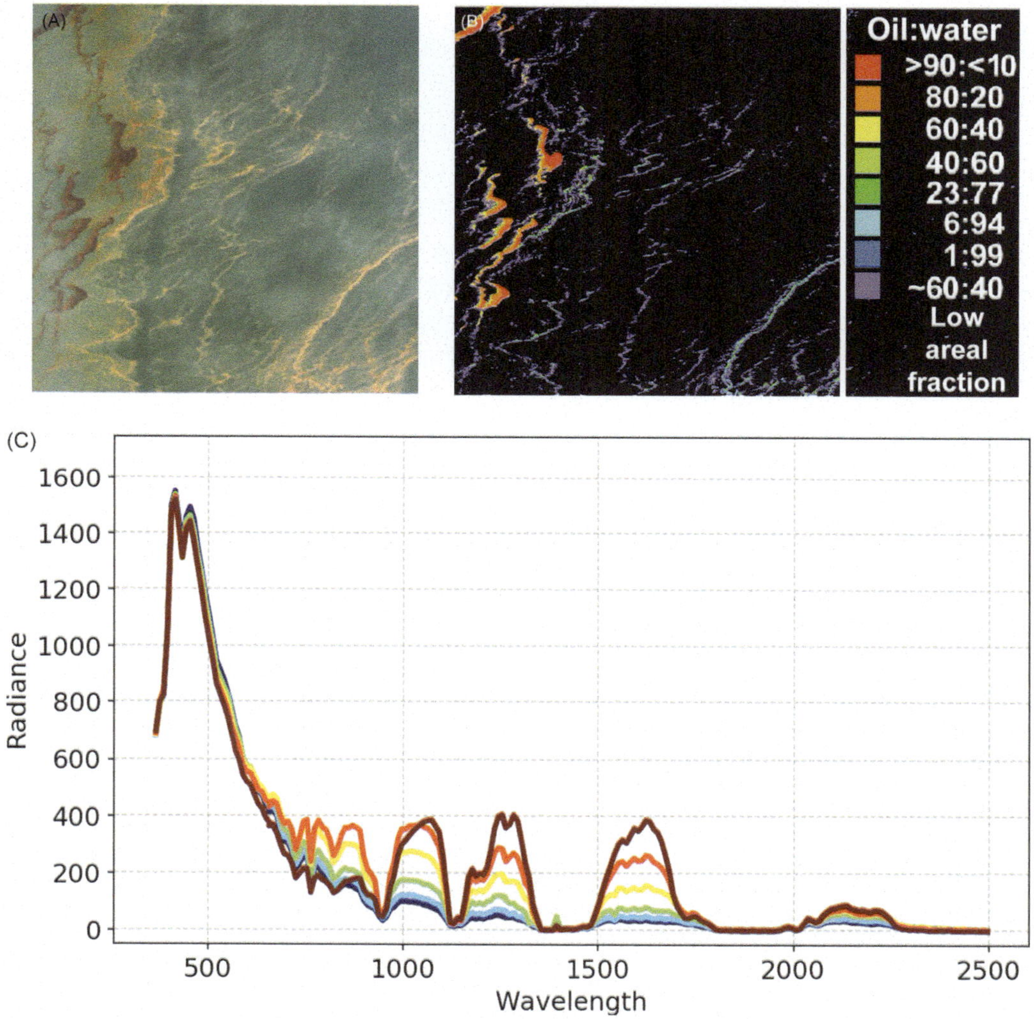

FIGURE 16–1 A classical question in remote sensing, oil spill detection. The classification result image is derived from Clark et al. (2010).

typically carried out after band selection or dimensionality reduction strategies (Lu et al., 2013; El-Rahman and Zolait, 2018). Observing the perceived monotonic variations in Fig. 16–1C, there is indeed the possibility of obtaining results through the use of only certain spectral bands. Unfortunately, such an approach removes the direct relationship that exists in the spectral measurement, linking the physical content of the observed surface and its digital image processing results. As a consequence, a new classification scheme must be developed every time the need for a new application emerges. On the contrary, a metrological tool is expected to be directly usable and applicable on any arbitrary object or surface, to measure the associated physical properties, for example, length or surface area. For remote sensing applications, metrology means that the provided measured values must be:

- Valid regardless of the spectral range.
- Independent from the spectral resolution, since it should only affect measurement accuracy.
- Independent from the context, meaning that spectral measurement between two spectra depends only on them and not, for example, on their neighborhood.

The first measurement required by metrology is the *difference*[2] between two spectra. Despite sounding like a basic question, multitudes of distance and similarity functions exist (Deza and Deza, 2013; Deborah, 2016). Thus the main question lies in identifying or defining a spectral difference function suitable for spectral metrology. Several levels of proof are required to enable a spectral difference function to be metrologically valid. First, the theoretical requirements to validating a function as a measure of distance or similarity must be satisfied. Then, the monotonicity of the function must be proven through the use of a suitable set of spectra, that is, with a sufficiently high number of different spectra that are both well-identified and reproducible.

Metrology also takes into account the fact that the world is not uniform as illustrated by the oil slick image in Fig. 16–1. The magnitude and characteristics of these nonuniformities are also representative of the observed materials, in their optical or physical properties in the visible and beyond visible spectral ranges, respectively. However, it is not possible to directly assess the diversity of a set of spectra or to directly define its standard deviation or other statistical parameters in the radiance or reflectance spaces. When the problem is translated into the domain of spectral differences, an alternative solution can be obtained. Furthermore, with the use of a metrologically validated spectral difference function, the statistical parameters obtained in the space will be representative of the actual uncertainty and diversity of the data.

16.3 Towards a metrological spectral difference function

Multitudes of distance and similarity functions exist, each typically constructed under a certain assumption and purpose. For example, some of the f-divergence functions are meant to measure the difference between two probability density functions. This means that each of their inputs is assumed to have an integral that is one. If they are used to compute the difference between, for example, two histograms, the result obtained might not be relevant or even meaningful. In the case of hyperspectral image analysis, one wants to measure the difference between two measurements coming from a sensor of light. When deciding which distance function to use, one has to also know what these measurements are and how to mathematically consider them, for example, as a vector, probability distribution, and so forth.

[2] The term *difference* follows the definition made by the CIE in the sense of color difference assessment by the human visual system and/or measurement tools. It includes the notions of distance and similarity measures.

Existing difference functions were classified into five categories based on their implicit assumption of a spectrum (Deborah et al., 2015; Richard et al., 2016). Spectral angle mapper (SAM) (Kruse et al., 1993), the Euclidean distance, and other Minkowski-based distance functions are examples of difference functions that consider a spectrum as a vector in the Euclidean space. When a spectrum is considered as n-dimensional data in manifold, then goodness-of-fit coefficient (GFC) (Hernández-Andrés et al., 1998) and Isomap (Tenenbaum et al., 2000) are examples of difference functions to use. Jeffrey divergence (Jeffreys, 1946), also known as spectral information divergence (SID) (Chang, 1999), is a commonly used difference function that assumes that a spectrum is a distribution, either as a probability density function, cumulative distribution function, or discrete probability distribution (histogram). The other two groups of difference functions are those that consider a spectrum as a sequence, for example, Hamming distance (Hamming, 1950), and as a series, for example, Kullback–Leibler pseudo-divergence (KLPD) (Richard et al., 2016). Taking the definition of a hyperspectral image as having contiguously sampled wavelengths over a certain range, not all of the mentioned difference function categories will be metrologically valid for hyperspectral data. A more complete hypothesis of their validity can be found in the work of Deborah (2016).

16.3.1 Expected theoretical properties

Mathematically speaking, there are several ways in which the differences between two objects can be defined and quantified. Among them are *distance* (or *dissimilarity*), *similarity*, and *divergence*. In general, a difference function can be defined on a set of spectra $\mathcal{S}$ as:

$$d : \mathcal{S} \times \mathcal{S} \rightarrow (0, \infty),$$

where $\forall\, S_1, S_2, S_3 \in \mathcal{S}$ the following holds:

- Reflexivity, $d(S_1, S_1) = 0$
- Nonnegativity, $d(S_1, S_2) \geq 0$
- Symmetry, $d(S_1, S_2) = d(S_2, S_1)$
- Identity of indiscernibles, $d(S_1, S_2) = 0 \Leftrightarrow S_1 = S_2$
- Triangular inequality or subadditivity, $d(S_1, S_2) \leq d(S_1, S_3) + d(S_3, S_2)$

Depending on whether a difference function satisfies all of the mentioned properties, it will either be a less or more robust difference function. In addition to those properties, the nonnegativity and triangular inequality properties induce the property of *monotonicity*. Given a spectrum S, a transformation function t, and two arbitrary parameters θ_1 and θ_2, the monotonicity property can be described as:

$$d(S, t(S, \theta_1)) \leq d(S, t(S, \theta_2)) \Leftrightarrow \theta_1 \leq \theta_2. \tag{16.1}$$

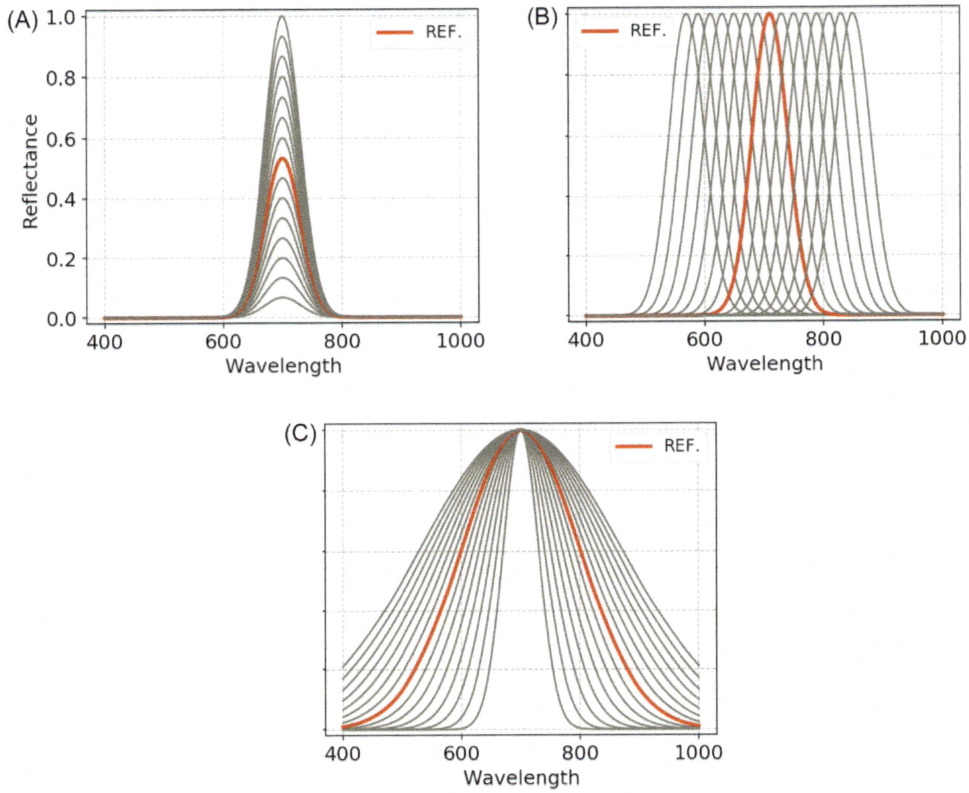

FIGURE 16–2 Illustration of three simulated spectral sets generated with a single variation in their magnitude, standard deviation, or peak wavelength. The *Y*-axes of all plots are identical.

16.3.2 Metrological validation of monotonicity property

The first five theoretical properties, from reflexivity to subadditivity, can be validated through the initial mathematical expression of any difference function. This is not the case for monotonicity, for which a dedicated test must be applied.

Considering the expression of monotonicity in Eq. (16.1), different spectral transformations affecting the shape and/or intensity of a spectrum S must be considered. For this, Gaussian-like functions[3] are used to simulate three types of sets of spectra = S_i (Deborah et al., 2015; Deborah, 2016). Each set, as shown in Fig. 16–2, is constructed to have one type of transformation t over the spectra, that is, magnitude, peak location, and standard deviation. Using the middle spectrum shown in red as a reference spectrum, the progression of

[3] The choice of Gaussian over other distribution models is to allow the most simple and basic tests for difference functions and not to model spectra from the real physical world.

difference response from this reference to the rest of the spectra should be monotonic. To be more specific, the difference functions are expected to present the behavior shown in Fig. 16–3, without specifying that it has to be symmetrical.

The response of five difference functions when given the three kinds of spectra sets as illustrated in Fig. 16–2 are calculated and can be observed in Fig. 16–4. Note that each difference function has its own dynamic range, and in these plots, they are shown in a normalized scale for observational purposes. Concerning the spectral set varying in its magnitude (Fig. 16–4A), only the Euclidean distance and KLPD provide strictly monotonic responses. The behavior of SAM, SID, and GFC is as expected since their mathematical expressions mainly account for shape information while suppressing the impact of magnitude or intensity; a behavior suitable for many remote sensing applications. Due to this, however, they do not satisfy the identity of indiscernible property of a difference function. This means that there is always a possibility of two distinct spectra having zero difference values, and as such, the validity of the obtained zero values can never be certain. Spectral set with peak location variation reveals the limit of SAM, Euclidean distance, and GFC, in which the three functions deem two distinct spectra to be equally different from the reference, see Fig. 16–4B. To be exact, the observed saturation of difference response starts when two spectra do not have overlapping area under curves (Deborah, 2016). With regards to the variation in standard deviation depicted in Fig. 16–4C, all difference functions are strictly monotonic as observed in the close-up depicted in Fig. 16–4D.

Finally, from this assessment, it can be seen that only KLPD provides a strictly monotonic distance response given the three variations of simulated spectra. This does not mean that other difference functions are invalid for use in hyperspectral image processing. They must,

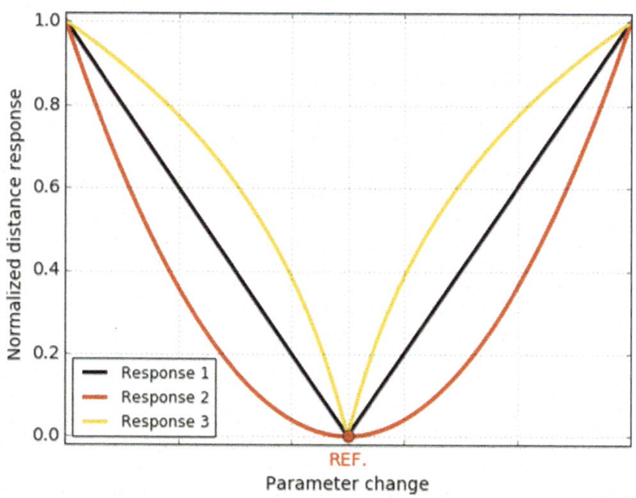

FIGURE 16–3 Illustration of expected responses from a difference function given any of the spectral sets in Fig. 16–2. The response does not have to be symmetrical as long as it is strictly monotonic.

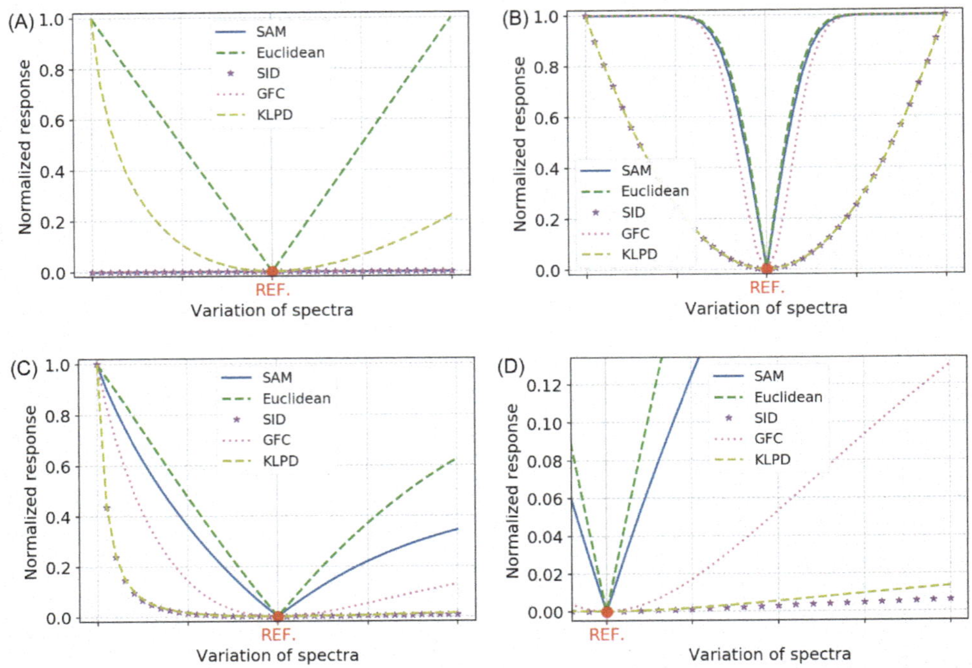

FIGURE 16–4 Responses of the five differences functions given the spectral sets in Fig. 16–2. A close-up of (C) is provided in (D) to show that SID and KLPD provide strictly monotonic responses. Note that the *Y*-axes are given in a normalized scale for observational purposes. *GFC*, Goodness-of-fit coefficient; *KLPD*, Kullback–Leibler pseudo-divergence; *SAM*, spectral angle mapper; *SID*, spectral information divergence.

however, be used in cases where they will not reach their limitations for the obtained values to be meaningful. Moreover, since this assessment was a theoretical one, it gives insight into the robustness of performance of these difference functions in a wide range of cases.

16.3.3 Metrological validation of discrimination performances

Albeit being a useful protocol, the previous assessment of difference functions cannot imply how a difference function would perform in real cases, where spectra can rarely be described as single Gaussian-like functions. Not only that, spectra acquired from the real world will have variations and noise.

Two hyperspectrally acquired pigment patches are shown in Fig. 16–5, with relatively stationary spectral content despite the existence of some variations. In both the color images and spectra, four relatively homogeneous groups can be observed. Having this knowledge, a test can be devised to evaluate the performance of difference functions in a real case. Using the spectra colored in red as a reference, a difference function must be able to provide a response resembling a step function as illustrated in Fig. 16–6. Here, each step corresponds to a color or spectral group. Since variations exist within the spectra, it follows that small

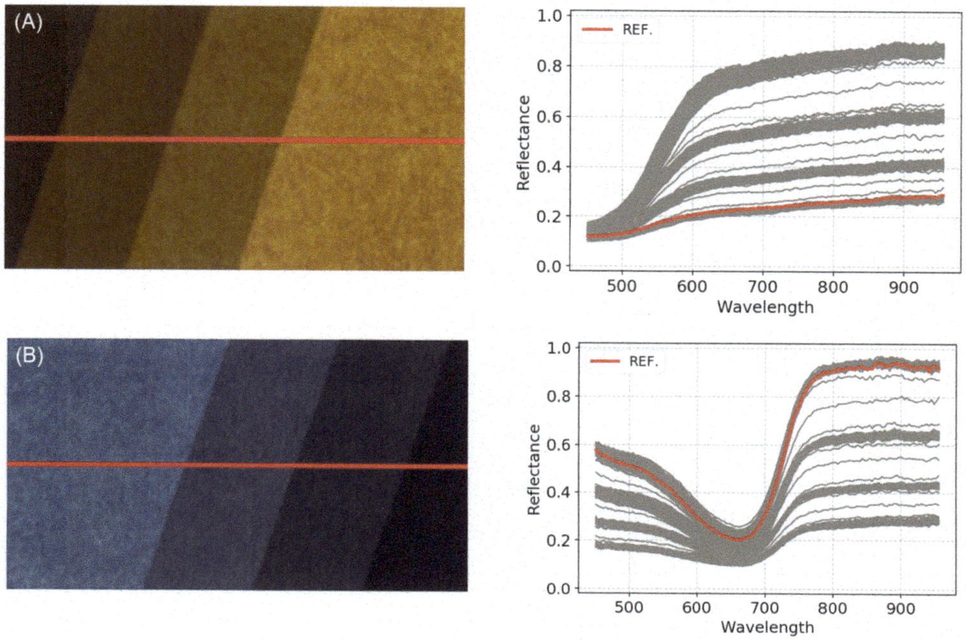

FIGURE 16–5 Two pigment patches with stationary content and their reflectance spectra obtained from pixels located on the red horizontal line of each patch. Spectra colored in red is associated to the leftmost pixel in the red horizontal line in the patch. Each image size varies by around 260 × 150 pixels, and has 140 spectral bands ranging from 450.97 to 956.15 nm.

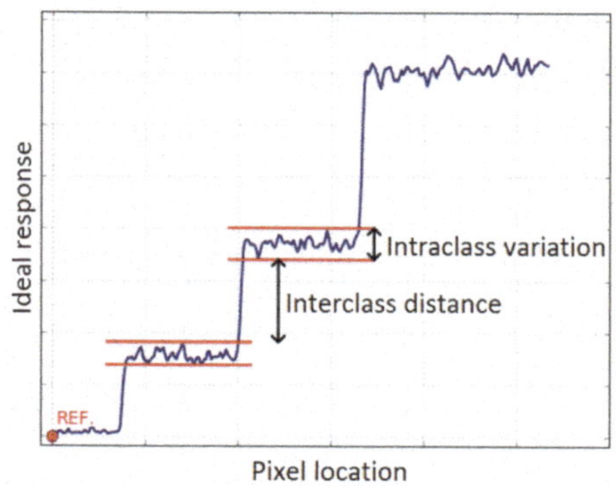

FIGURE 16–6 Ideal response of a difference function given the spectra from the pigment patches in Fig. 16–5, with interclass distance larger than intraclass variation.

fluctuations will also be perceived in the step function. An ideal response to these spectra sets will be one that maximizes the separation between each group, that is, interclass distance being larger than intraclass variation.

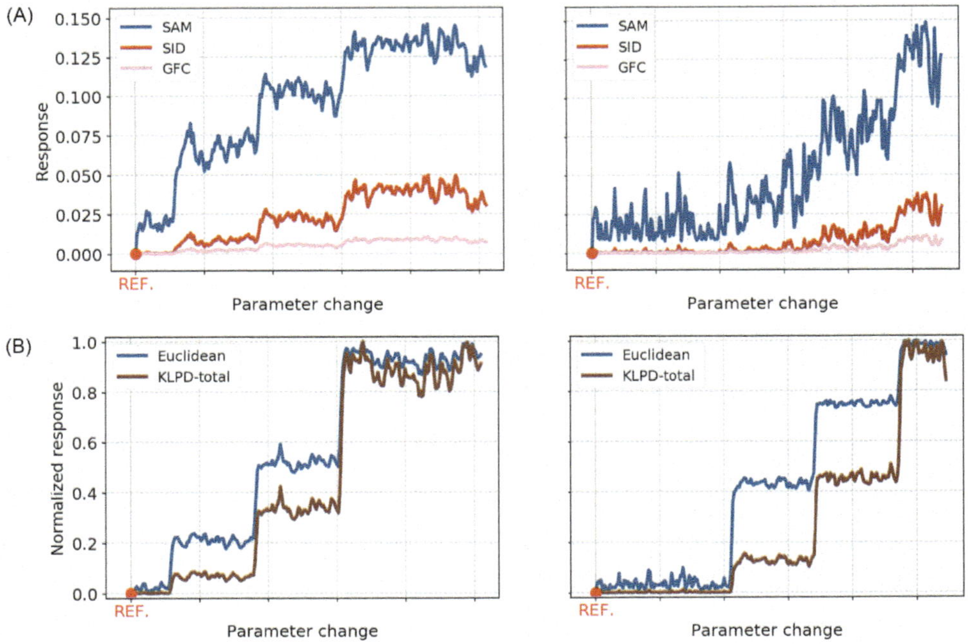

FIGURE 16–7 Responses obtained for five commonly used spectral difference functions when given the yellow and blue pigment patches in Fig. 16–5 as input. (B) The response of Euclidean distance and KLPD is given in a normalized scale for observation purposes. *GFC*, Goodness-of-fit coefficient; *KLPD*, Kullback–Leibler pseudo-divergence; *SAM*, spectral angle mapper; *SID*, spectral information divergence.

Fig. 16–7 presents the responses of the most commonly used difference functions for spectral data, that is, SAM (Kruse, et al., 1993), the Euclidean distance, SID (Chang, 1999), and GFC (Hernández-Andrés et al., 1998). Additionally, the response of KLPD (Richard et al., 2016) is also provided. It is the only function that satisfies metrological requirements. As the first observation, it can be remarked that SAM, SID, and GFC do not allow for the discrimination of the four spectral groups that exist in the two proposed images (Fig. 16–5), even if the four sets of spectra can be easily discriminated. Second, it can also be observed that the Euclidean distance and KLPD satisfy the expected constraint given in Fig. 16–3, despite having different dynamic ranges.

16.3.4 Demonstration of use in oil detection

Fig. 16–8 is provided to illustrate how the monotonicity test is applicable to a real case problem, showing a subset from Run 11 of the airborne visible/infrared imaging spectrometer (AVIRIS) Deepwater Horizon oil spills image. A spatially ordered sequence of pixels is taken from the image, that is, those located under the red vertical line. The corresponding classification result is also identified, showing that there is an order of oil to water concentration in these pixels, that is, from a low areal fraction to approximately 6:94, and back to a low areal

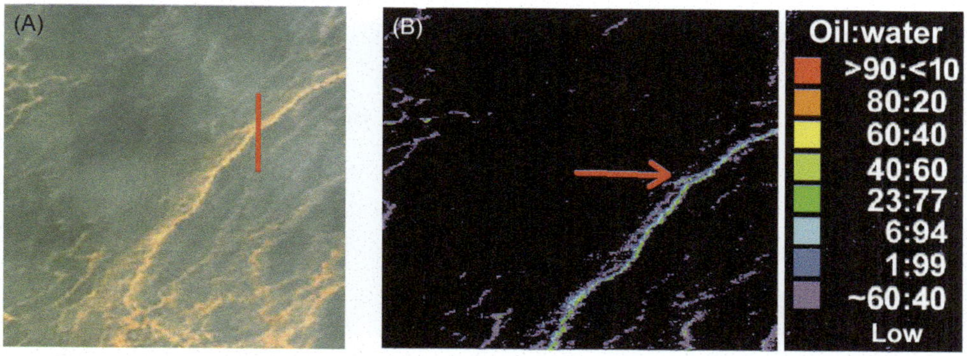

FIGURE 16–8 The area located under the red vertical line presents a spatially ordered sequence of pixels, corresponding to an ordered sequence of oil to water concentration ratio as shown by the classification result derived from the work of Clark et al. (2010).

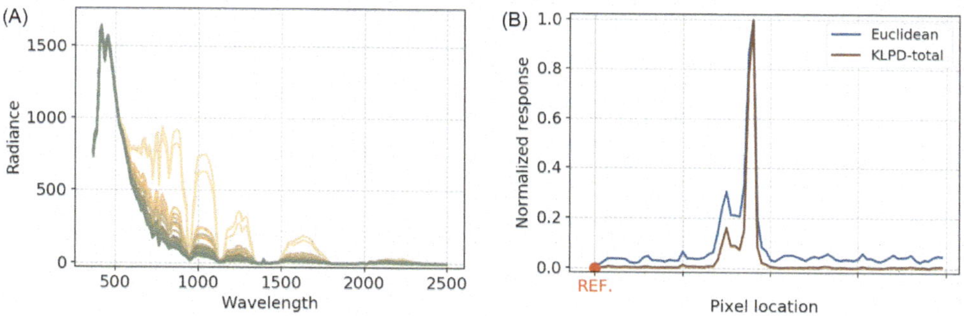

FIGURE 16–9 (A) Radiance spectra associated to the pixels shown in Fig. 16–8A and (B) the obtained responses of the Euclidean distance and Kullback–Leibler pseudo-divergence function when given these spectra. Note that the responses are given in a normalized scale for observation purposes.

fraction. Spectra associated to these pixels are also plotted in the colors of the pixels in Fig. 16–9A. The spectral plot shows that the higher oil to water ratio is reflected through the higher intensity around 500–1700 nm, where the yellow colored peaks are. If this ordered sequence of radiance spectra is given to a difference function, using pixels located above the red line as a reference, a peak located in the middle of the response curve is expected. The responses given by the Euclidean distance and KLPD function can be observed in Fig. 16–9B, with the Euclidean distance being more sensitive to the variations than the KLPD function.

Given a spectrum that corresponds to water, as seen in Fig. 16–10A, spectral difference maps can be produced and these can be compared in a qualitative manner with the classification result shown in Fig. 16–9B. Since the classification result provides a ratio of oil to water, a pixel with a higher oil concentration can be thought to also have a larger spectral difference to the spectrum of water. The difference maps obtained using the Euclidean

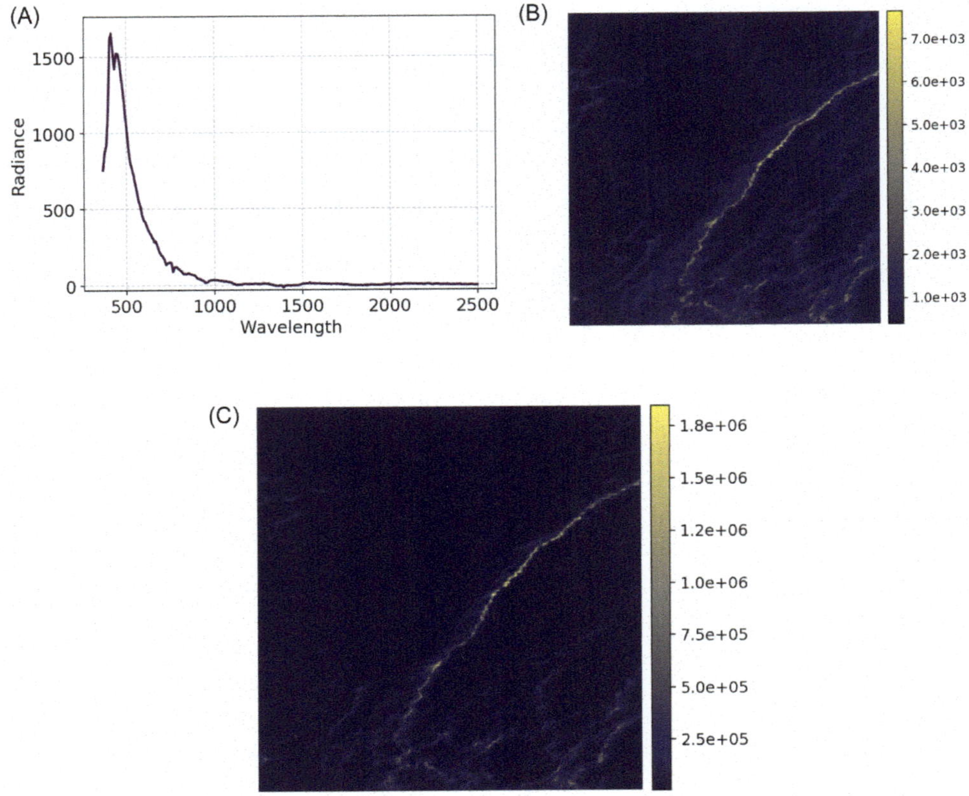

FIGURE 16–10 Radiance spectrum associated with water and spectral difference maps obtained relative to it, using the Euclidean distance and Kullback–Leibler pseudo-divergence function.

distance and KLPD function can be observed in Fig. 16–10B and C. In both cases, despite their different dynamic ranges, it can be observed that both maps agree with the classification result, where pixels with higher oil to water ratios also give higher difference values to water. Comparing the two maps, no significant difference can be perceived. However, recall that the Euclidean distance was shown to have a limit (Fig. 16–4B). The challenge in using this distance function then lies in the fact that the reliability and validity of the obtained values are limited, as opposed to those obtained with KLPD, which has been shown to be strictly monotonic.

16.4 Assessing the nonuniformity of the spectral world

A hyperspectral acquisition captures the interaction between light and a material at the atomic and molecular levels. It will then record the nonuniformities of the surface, which are induced by both the physical construction and optical behavior of this surface, in a set of spectra. However, the analysis of the set or distribution cannot be carried out directly in the radiance or reflectance spaces. Here, an alternative is developed based on the statistical analysis of spectral diversity in a spectral difference domain. Simply put, the analysis is based on histograms of spectral differences (HSDs) rather than of spectra.

16.4.1 Statistics on spectral differences

As previously mentioned, there is no direct way to assess the standard deviation, skewness, kurtosis, and other statistical moments from a set of spectra. Moreover, since all these statistical moments are normalized against the average, they are insensitive to changes in the average value itself. On the other hand, when the question of statistical moments is translated into the spectral difference domain as seen in Fig. 16–11, an alternative solution can be obtained.

In the case of spectral domain, a suitable reference point to characterize a spectral set $\mathcal{S}$ would be the spectrum $\mu_{\mathcal{S}}$ located at the center of the set. Following the definition of the median as the closest point to all other points in the distribution (Astola et al., 1990), it would then be the spectral median spectrum $\tilde{\mu}_{\mathcal{S}}$. When the distribution of the set is unimodal, the median will be located near the average. Thus in this case, $\tilde{\mu}_{\mathcal{S}}$ can be replaced by the marginal average $\hat{\mu}_{\mathcal{S}} = \left\{\frac{1}{n}\sum_{S_i \in \mathcal{S}} S_i(\lambda), \lambda \subset [\lambda_{\min}, \lambda_{\max}]\right\}$. In other cases, however, the use of $\hat{\mu}_{\mathcal{S}}$ will generate a false spectrum far from the ones originating from the initial set $\mathcal{S}$. Having obtained $\tilde{\mu}_{\mathcal{S}}$, or replacing it with $\hat{\mu}_{\mathcal{S}}$, obtaining the rest of the statistical moments will be straightforward.

16.4.2 Histogram of spectral differences

Fig. 16–12A shows an image of Sierra National Forest, where areas with trees as well as tree-barren areas can be observed. Pixels located within the blue square consist of spectra associated with trees, and the region seems to be relatively homogeneous. Reflectance spectra from this region can be observed in Fig. 16–12B. Pixels from the nonhomogeneous region within the red square are also extracted and their spectra can be observed in Fig. 16–12C. Using these two areas to illustrate the problem of deforestation, the detection of

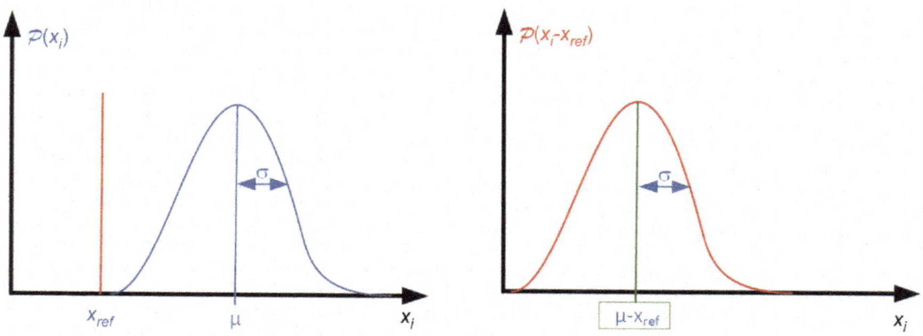

FIGURE 16–11 The translation of statistical processing from the initial distribution $P(x_i)$ to the distribution of difference values relative to a reference point $P(x_i - x_{\text{ref}})$, illustrated in a one-dimensional case.

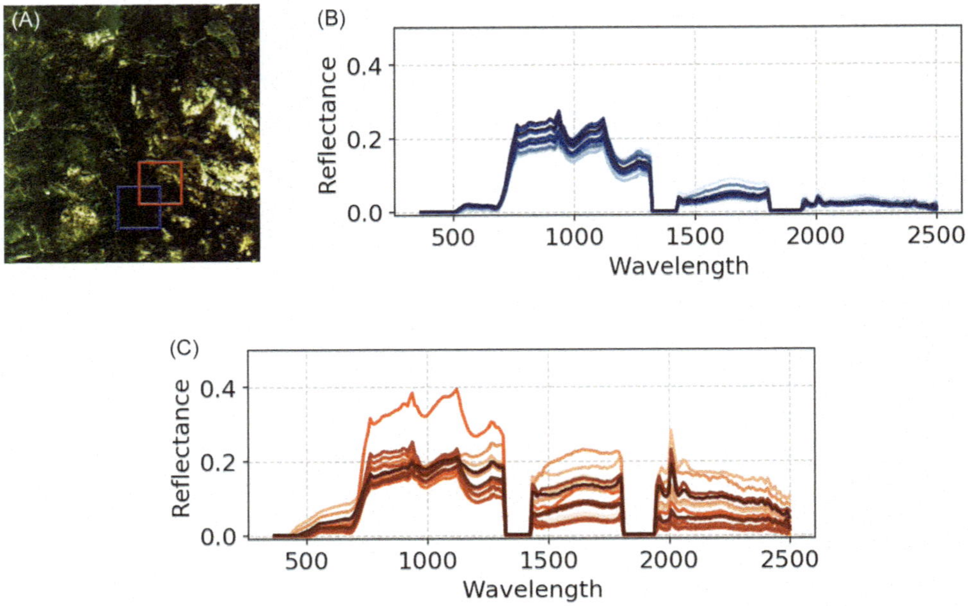

FIGURE 16–12 An airborne visible/infrared imaging spectrometer image showing areas in Sierra National Forest and 20 spectra randomly selected from the colored squares. The image is of 300 × 300 pixels and 224 spectral bands. Each square in the image is of 35 × 35 pixels.

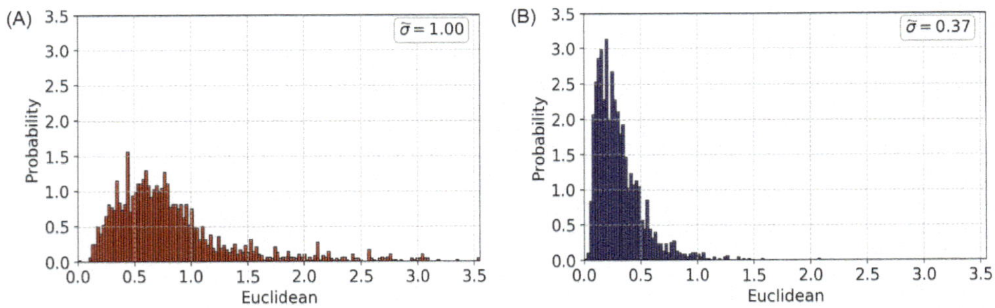

FIGURE 16–13 Histograms of spectral differences obtained for the areas in Fig. 16–12, obtained using the Euclidean distance and median of each corresponding set as reference.

deforestation can be described through the density of the spectral set. For the homogeneous region, representing thick forest area, a small standard deviation is expected. When an area suffers deforestation, its spectral distribution will be more dispersed since there will be pixels corresponding to the signature of, for example, bare soil. Thus a higher standard deviation value is expected.

Two HSDs corresponding to the squares in Fig. 16–12A are obtained using the Euclidean distance and these are shown in Fig. 16–13. The median of each square is used as the reference spectrum. As expected, the spectral difference distribution of the deforested area is more dispersed than that of the thick forest area. Pixels are more distributed across the dynamic range of difference values, which also induces a larger standard deviation value. Two HSDs are also obtained for the same forest region, albeit using KLPD function. Similar to the Euclidean distance, it also presents a reduced variation for the distribution of the thick forest region (Fig. 16–14A) compared to that obtained for the area suffering deforestation (Fig. 16–14B).

While there seems to be no significant difference between the performance of the Euclidean distance and KLPD as observed through their HSDs, KLPD is able to provide more information. KLPD was originally constructed such that the Kullback–Leibler divergence function could be used correctly for spectral data, where the integral of a spectrum is not necessarily a unity. The second objective driving the construction of KLPD was to naturally split, instead of empirically split, the measurement of differences into two parts, that is, shape and intensity differences. Since the two parts of KLPD are independent, a bidimensional histogram of spectral differences (BHSD) can be produced.

A certain level of tree density in the blue square in Fig. 16–12A induces a certain texture with reduced variations. If this is then compared to the deforested area located within the red square, a less uniform distribution can then be visually determined, due to the existence

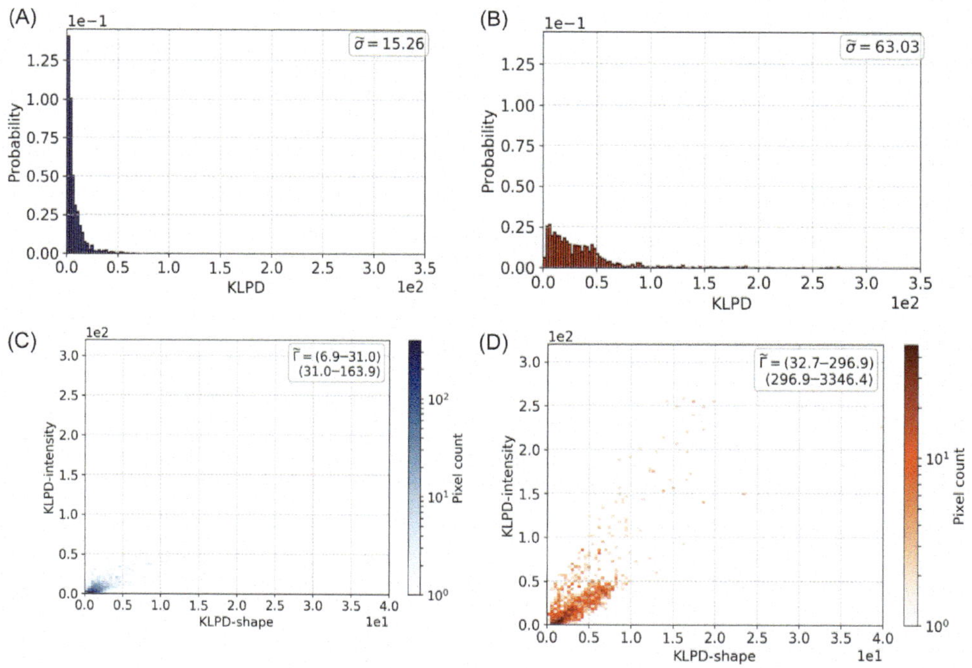

FIGURE 16–14 Histograms of spectral differences obtained using total Kullback–Leibler pseudo-divergence (KLPD) and bidimensional histograms of spectral differences obtained using the two components of KLPD for the areas shown in Fig. 16–12A.

of the yellowish soil pixels. Thus in the histogram associated with the deforested area, a less uniform distribution is also to be expected. Looking at the HSD of the Euclidean distance in Fig. 16–13B, a near-normal distribution is evident. KLPD, with its sensitivity to both shape and intensity variations, is able to provide the expected information. The shape and direction of the distribution of forest pixels can be obtained from Fig. 16–14C. Having that knowledge, it can then be observed in Fig. 16–14D that the main distribution of forest is still present, in addition to another distribution whose direction is leaning closer to the KLPD-intensity axis.

16.4.3 Variance–covariance in a spectral difference space

Taking the decomposition of KLPD into its two components in the BHSD space, the notion of variance-covariance in the spectral difference space can be defined as where $\tilde{\mu}_{\mathcal{S}}$ is the median spectrum of set $\mathcal{S}$.

$$\tilde{\Gamma}_{\mathcal{S}} = \begin{pmatrix} \alpha_{GG} & \alpha_{GW} \\ \alpha_{GW} & \alpha_{WW} \end{pmatrix},$$

where $\alpha_{GG} = \sum_{S_i \in \mathcal{S}} (\Delta G(S_i, \tilde{\mu}_{\mathcal{S}}))^2 f(S_i)$, $\alpha_{WW} = \sum_{S_i \in \mathcal{S}} (\Delta W(S_i, \tilde{\mu}_{\mathcal{S}}))^2 f(S_i)$,

$$\alpha_{GW} = \sum_{S_i \in \mathcal{S}} (\Delta G(S_i \tilde{\mu}_{\mathcal{S}}) \cdot \Delta W(S_i, \tilde{\mu}_{\mathcal{S}})) f(S_i)$$

Since variance–covariance is analogous to the one-dimensional standard deviation, similar responses to the thick forest and deforested areas from Fig. 16–12A can be expected. And, indeed, the variance–covariance measure of the thick forest area (Fig. 16–14C) is smaller than that of the deforested area (Fig. 16–14D).

Following the construction of spectral variance–covariance matrix $\tilde{\Gamma}_{\mathcal{S}}$, the Mahalanobis distance (MD) between any arbitrary spectrum S and a known spectral set $\mathcal{S}$ can be further obtained from:

$$d_M(S,\mathcal{S}) = \sqrt{d_{\mathrm{KLPD}}(S, \tilde{\mu}_{\mathcal{S}})^T \tilde{\Gamma}_{\mathcal{S}}^{-1} d_{\mathrm{KLPD}}(S, \tilde{\mu}_{\mathcal{S}})}$$

Note that what the MD does is essentially compute distance in a normalized BHSD space with normalization relative to the variance-covariance matrix $\tilde{\Gamma}$. Processing data in the normalized space further allows for the use of more simple statistical laws, for example, the empirical law for anomaly detection. Going back to the example of the deforested area in Fig. 16–12A, the variance–covariance measures of the two areas are given in Fig. 16–14C and D, where the variance–covariance values of the thick forest are smaller than those of the deforested area.

16.5 Application in forest mapping

Detailed and accurate forest maps represent crucial information needed to monitor forest disturbances, for example, fire and drought. Being able to identify forest type or even classify

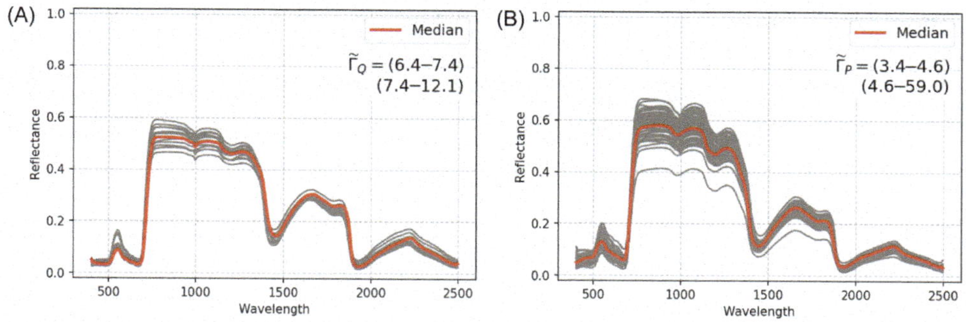

FIGURE 16–15 Reflectance spectra associated with *Quercus* and *Pinus*, with information regarding their median spectra and variance–covariance measures. Due to noise level around 350 nm, the spectra are only considered from 400 nm and higher.

tree species allows forest resources to be protected and managed (Ballanti et al., 2016). However, it is known that spectral variations in vegetation are more often than not due to the signature of chlorophyll and, thus, are not diagnostics of individual tree species. Here, using data available for natural plants from Sierra National Forest in California and an AVIRIS image covering the region, a mapping of the land cover in the forest is carried out.

16.5.1 Measuring variability within a known dataset

Measurements of reflectance spectra in the range of 350–2500 nm are available for natural plants from Sierra mountain forests in California (Serbin et al.). Among the plants, enough spectra are available for oak (*Quercus chrysolepis* and *Quercus kelloggii*), cedar (*Calocedrus decurrens*), pine (*Pinus lambertiana, Pinus jeffreyi,* and *Pinus ponderosa*), and fir (*Abies concolor*) trees. This allows for the variability within each set of spectra to be measured.

Considering that each set consists of spectra associated with trees within the same genus, the reflectance spectra of oak (genus *Quercus*) and pine (genus *Pinus*) trees are shown in Fig. 16–15, along with their variance–covariance measures. From Fig. 16-15, it can be observed that there are more intensity variations than shape variations for both *Quercus* and *Pinus*, with component $\alpha_{GG} > \alpha_{WW}$ in both $\tilde{\Gamma}_Q$ and $\tilde{\Gamma}_P$. Then, agreeing with the visual observation, the obtained $\tilde{\Gamma}$ are also able to reflect that with $\alpha_{GG} < \alpha_{WW}$. For *Abies* and *Calocedrus*, their variability is largely controlled by the intensity variations:

$$\tilde{\Gamma}_A = \begin{pmatrix} 5.9 & 13.9 \\ 13.9 & 353.1 \end{pmatrix}, \; \tilde{\Gamma}_C = \begin{pmatrix} 5.4 & 23.3 \\ 23.3 & 348.4 \end{pmatrix}$$

16.5.2 Mapping tree genus

Knowing that the laboratory spectral measurements of the four genera were taken for samples coming from Sierra National Forest, an AVIRIS image of the region from flight

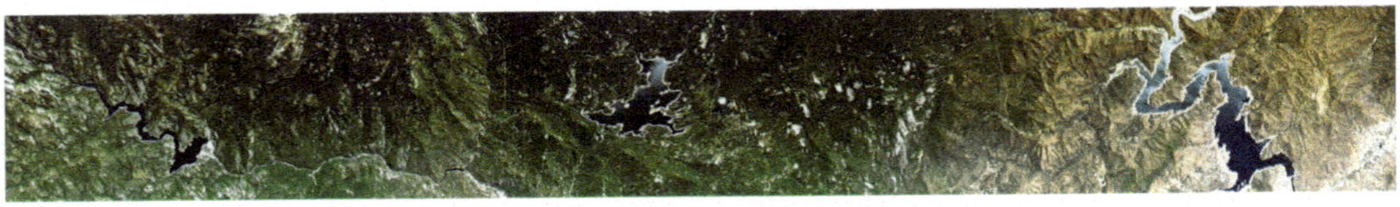

FIGURE 16–16 Subset of AVIRIS image from flight f130626t01p00r07, covering a part of Sierra National Forest.

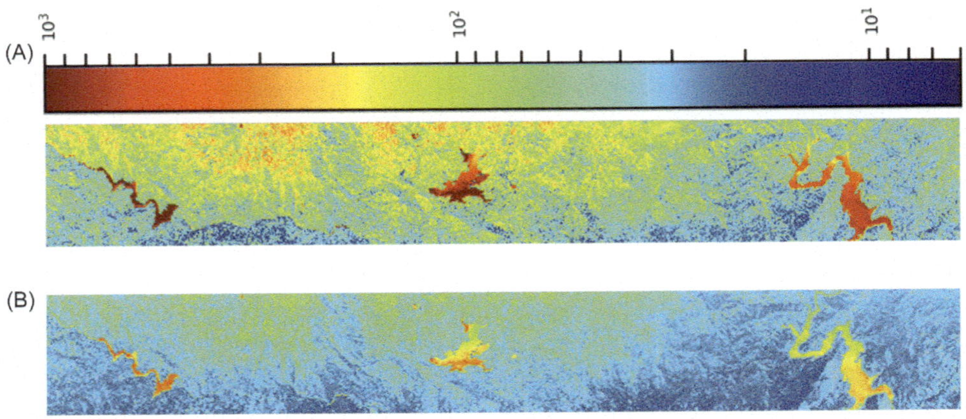

FIGURE 16–17 Mahalanobis distance of each pixel in the image relative to the distribution of *Quercus* and *Pinus*, presented in identical dynamic range.

f130626t01p00r07 is used in this experiment. A subset of the image that is used is shown in Fig. 16–16. It is of 660×5000 pixels and 224 spectral bands ranging from 365.93 to 2496.24 nm.

In order to map the tree genus in the target image, the MD is calculated between each pixel relative to the distribution of each tree genus. Recalling the variance–covariance associated to each genus, however, using the distribution model of *Abies* or *Calocedrus* is not going to provide meaningful results.This is because in both $\tilde{\Gamma}_Q$ and $\tilde{\Gamma}_P$, the components $\alpha_{WW} \gg \alpha_{GG}$ and $\alpha_{WW} \gg \alpha_{GW}$, which means that the models will render shape variations negligible. α_{WW} α_{GG} and α_{GW} Thus only the models associated to *Quercus* and *Pinus* are considered and the obtained MD maps can be observed in Fig. 16–17.

In the MD maps shown in Fig. 16–17, smaller MD values mean higher similarities to the corresponding tree genus. For *Quercus*, the pixels associated to it are in darker blue colors, located at the bottom and to the right of the image. Compared to the map associated to *Pinus*, it provides better separation as it is able to use the full dynamic range. This also allows a more confident identification of waterbody regions, which are colored from orange to red in the MD map. Taking into account the larger variance–covariance measure of *Pinus*, the remaining blue colored pixels in its corresponding map can be mapped to be associated to it. As a result, multiple thresholding can be applied, resulting in a map of

FIGURE 16–18 A map showing trees of *Quercus* (*yellow pixels*), *Pinus* (*brown*), and a waterbody region (*pink*). Regions marked in blue are undetermined.

Quercus, Pinus, and the waterbody region as shown in Fig. 16–18. Regions colored in dark blue are undetermined due to the limited reliable dataset of spectra available. However, if this map is compared with the color image in Fig. 16–16, the large undetermined area at the top of the image can possibly be due to nonuniform illumination or the effect of shading. Nevertheless, since no reference information is available on that, the pixels are set to be unrecognized.

16.6 Conclusion

A hyperspectral image must be processed with regards to its metrological aspects as measurements of physical and optical characteristics of a surface for its potential to be fully exploited. On the other hand, the sheer size of hyperspectral images often times poses the "curse of dimensionality" problem, which has led to the use of dimensionality reduction strategies prior to any processing or analysis of the data. In this chapter, it was shown that metrologically dealing with hyperspectral data is possible by means of spectral differences, where all information from all spectral bands is taken into account.

Throughout the development of the metrological analysis framework based on difference function, demonstration of its usability has been shown in oil detection and forest mapping. The application level shown in this chapter has been possible only through the use of simple statistics and thresholding, showing the potential of the framework for use in more advanced tasks, for example, classification and unmixing. Finally, the proposition of this framework is not to replace existing approaches, rather, it is to show what can be achieved through a generic metrological framework, where no new scheme will be required for every new application.

List of abbreviations

AVIRIS airborne visible/infrared imaging spectrometer
BHSD bidimensional histogram of spectral differences
GFC goodness-of-fit coefficient
HSD histogram of spectral differences
KLPD Kullback-Leibler pseudo-divergence
MD Mahalanobis distance
PDF probability density function

SAM spectral angle mapper
SID spectral information divergence

List of symbols

$\mathcal{S}$	an arbitrary set of spectra
S_i	any arbitrary spectrum
$d(S_i, S_j)$	difference (or, distance) function between two arbitrary spectra S_i and S_j
d_M	Mahalanobis distance
d_{KLPD}	Kullback Leibler pseudo-divergence (KLPD) function
$t(S, \theta)$	a transformation function modifying spectrum S with arbitrary parameter θ
$\mu_{\mathcal{S}}$	spectrum located at the center of set $\mathcal{S}$
$\tilde{\mu}_{\mathcal{S}}$	median spectrum of $\mathcal{S}$
$\hat{\mu}_{\mathcal{S}}$	marginal average spectrum of $\mathcal{S}$
n	number of spectra in set $\mathcal{S}$
λ	wavelength
$[\lambda_{\min}, \lambda_{\max}]$	wavelength range
$P(x)$	probability density function of x
x_{ref}	reference point
$\tilde{\Gamma}_{\mathcal{S}}$	variance-covariance matrix of $\mathcal{S}$ in spectral difference space
$\Delta G, \Delta W$	shape and intensity differences in KLPD function, respectively
α_{GG}	shape difference component of $\tilde{\Gamma}_{\mathcal{S}}$
α_{WW}	intensity difference component of $\tilde{\Gamma}_{\mathcal{S}}$
α_{GW}	shape-intensity difference component of $\tilde{\Gamma}_{\mathcal{S}}$

References

Astola, J., Haavisto, P., Neuvo, Y., 1990. Vector median filters. Proc. IEEE 78, 678–689.

Ballanti, L., Blesius, L., Hines, E., Kruse, B., 2016. Tree species classification using hyperspectral imagery: a comparison of two classifiers. Remote Sens. 8.

Bhatia, N., Iordache, M.-D., Stein, A., Reusen, I., Tolpekin, V.A., 2018. Propagation of uncertainty in atmospheric parameters to hyperspectral unmixing. Remote Sens. Environ. 204, 472–484.

Chang, C.-I., 1999. Spectral information divergence for hyperspectral image analysis. In: IEEE International Conference on Geoscience and Remote Sensing Symposium (IGARSS), vol. 1, pp. 509–511.

Clark, R.N., Swayze, G.A., Leifer, I., Livo, K.E., Kokaly, R., Hoefen, T., 2010. Airborne visible/infrared imaging spectrometer (AVIRIS) team. In: A Method for Quantitative Mapping of Thick Oil Spills Using Imaging Spectroscopy: US Geological Survey Open-File Report. Tech. Rep., United States Geological Survey (USGS).

Datla, R.V., Kessel, R., Smith, A.W., Kacker, R.N., Pollock, D.B., 2010. Review article: uncertainty analysis of remote sensing optical sensor data: guiding principles to achieve metrological consistency. Int. J. Remote Sens. 31, 867–880.

Deborah, H., 2016. Towards Spectral Mathematical Morphology (Ph.D. thesis). Norwegian University of Science and Technology, University of Poitiers.

Deborah, H., Richard, N., Hardeberg, J.Y., 2015. A comprehensive evaluation on spectral distance functions and metrics for hyperspectral image processing. IEEE J. Sel. Top. Appl. Earth Obs. Remote Sens. 8, 3224–3234.

Deborah, H., George, S., Hardeberg, J.Y., 2019. Spectral-divergence based pigment discrimination and mapping: a case study on The Scream (1893) by Edvard Munch. J. Am. Inst. Conserv. 90–107.

Deza, M.M., Deza, E., 2013. Encyclopedia of Distances. Springer.

El-Rahman, S.A., Zolait, A.H., 2018. Hyperspectral image analysis for oil spill detection: a comparative study. Int. J. Comput. Sci. Math. 9, 103.

Engel, D.W., Reichardt, T.A., Kulp, T.J., Graff, D.L., Thompson, S.E., 2016. Hierarchical multi-scale approach to validation and uncertainty quantification of hyper-spectral image modeling. Proc. SPIE 9840, 9.

Fingas, M., Brown, C.E., 2018. A review of oil spill remote sensing. Sensors 18.

Fox, N., Hueni, A., Honkavaara, E., Christen, W., Vaskuri, A., 2017. Publishable JRP Summary Report for ENV53 MetEOC2 Metrology for Earth Observation and Climate. Tech. Rep., EURAMET.

Ghamisi, P., Maggiori, E., Li, S., Souza, R., Tarablaka, Y., Moser, G., et al., 2018. New frontiers in spectral–spatial hyperspectral image classification: the latest advances based on mathematical morphology, markov random fields, segmentation, sparse representation, and deep learning. IEEE Geosci. Remote Sens. Mag. 6, 10–43.

J.C. Guides in Metrology (JCGM), 2012. International Vocabulary of Metrology—Basic and General Concepts and Associated Terms. Tech. Rep. International Bureau of Weights and Measures (BIPM).

Hamming, R.W., 1950. Error detecting and error correcting codes. Bell Syst. Tech. J. 29, 147–160.

Hernández-Andrés, J., Romero, J., García-Beltrán, A., Nieves, J.L., 1998. Testing linear models on spectral daylight measurements. Appl. Opt. 37, 971–977.

Honkavaara, E., Markelin, L., Hakala, T., Peltoniemi, J., 2014. The metrology of directional, spectral reflectance factor measurements based on area format imaging by UAVs. PFG Photogramm. Fernerkundung Geoinform. 2014, 175–188.

Jeffreys, H., 1946. An invariant form for the prior probability in estimation problems. Proc. Roy. Soc. Lond. A Math. Phys. Eng. Sci. 186, 453–461.

Kruse, F.A., Lefkoff, A.B., Boardman, J.W., Heidebrecht, K.B., Shapiro, A.T., Barloon, P.J., et al., 1993. The spectral image processing system (SIPS)—interactive visualization and analysis of imaging spectrometer data. Remote Sens. Environ. 44, 145–163.

Lu, Y., Tian, Q., Wang, X., Zheng, G., Li, X., 2013. Determining oil slick thickness using hyperspectral remote sensing in the Bohai Sea of China. Int. J. Digit. Earth 6, 76–93.

Marpu, P., Gamba, P., Benediktsson, J.A., 2011. Hyperspectral change detection using IR-MAD and feature reduction. In: IEEE International Geoscience and Remote Sensing Symposium, pp. 98–101.

Mirschel, G., Daikos, O., Scherzer, T., Steckert, C., 2018. Near-infrared chemical imaging used for in-line analysis of functional finishes on textiles. Talanta 188, 91–98.

Richard, N., Helbert, D., Olivier, C., Tamisier, M., 2016. Pseudo-divergence and bidimensional histogram of spectral differences for hyperspectral image processing. J. Imag. Sci. Technol. 60, 50402-1–50402-13.

Salem, F., Kafatos, M., 2001. Hyperspectral image analysis for oil spill mitigation. In: 22nd Asian Conference on Remote Sensing.

Serbin, S.P., Kruger, E.L., Townsend, P.A., n.d. NASA HyspIRI Airborne Campaign Leaf and Canopy Spectra and Leaf Traits. Available from: http://ecosis.org.

Serranti, S., Gargiulo, A., Bonifazi, G., 2012. Hyperspectral imaging for process and quality control in recycling plants of polyolefin flakes. J. Infrared Spectrosc. 20, 573–581.

Talone, M., Zibordi, G., 2018. Non-linear response of a class of hyper-spectral radiometers. Metrologia 55, 747–758.

Tenenbaum, J.B., Silva, V., Langford, J.C., 2000. A global geometric framework for nonlinear dimensionality reduction. Science 290, 2319–2323.

Vandermeulen, R.A., Mannino, A., Neeley, A., Werdell, J., Arnone, R., 2017. Determining the optimal spectral sampling frequency and uncertainty thresholds for hyperspectral remote sensing of ocean color. Opt. Express 25, A785–A797.

Zibordi, G., Ruddick, K., Ansko, I., Moore, G., Kratzer, S., Icely, J., et al., 2012. In situ determination of the remote sensing reflectance: an inter-comparison. Ocean. Sci. 8, 567–586.

SECTION V

Hyperspectral Remote Sensing: Multi-sensor, Fusion and Indices applications for Pollution Detection and Other Applications

17

Improving the detection of cocoa bean fermentation-related changes using image fusion

Ronald Criollo[1], Oswaldo Bayona[1], Daniel Ochoa[1], Juan Cevallos-Cevallos[1], Wenzhi Liao[2,3]

[1]ESCUELA SUPERIOR POLITÉCNICA DEL LITORAL, ESPOL, GUAYAQUIL, ECUADOR
[2]SUSTAINABLE MATERIALS MANAGEMENT, FLEMISH INSTITUTE FOR TECHNOLOGICAL RESEARCH (VITO), ANTWERP, BELGIUM
[3]IPI-TELIN, GHENT UNIVERSITY, GHENT, BELGIUM

17.1 Introduction

The main ingredient for chocolate is cocoa butter extracted from whole cocoa beans using the Broma process (Krysiak et al., 2013) which includes fermentation, drying, roasting, and separation steps. In plantations, beans are removed from the pods and stacked in wooden boxes to allow microorganisms to start the fermentation of the pulp (Schwan and Wheals, 2004). In the pulp, yeast converts sugars into ethanol that, in turn, is oxidized into acetic acid. The acetic acid and high temperatures break down cell walls in the embryo and the segregated substances mix. A series of complex chemical reactions determine the butter's flavor and smell (Lopez and Dimick, 2008). High-quality cocoa beans yield a more aromatic and less bitter butter than low-quality beans (Kadow et al., 2013).

Cocoa pods contain between 30 and 40 beans wrapped in a mucilaginous pulp. The seed has a coat and the embryo is protected by two cotyledons. The chemical composition of cocoa beans includes water, fat, proteins, polyphenols, starch, sucrose, and purine bases such as caffeine and theobromine. Among the polyphenols we found catechins, anthocyanins, and proanthocyanins. During fermentation the concentration of anthocyanins decreases, these pigments give a purple color to the cotyledon. The absence of anthocyanins makes cotyledons turn brown. The concentration of (−)-epicatechin, (+)-catechin, theobromine, caffeine, and moisture also decrease.

Varieties of cocoa include Criollo, Forastero Amazonico, and Trinitario trees. Cocoa quality is measured based on the mix of flavors it can provide. Low-quality, bulk cocoa, is mostly obtained from Forasteros trees. Fine cocoa is obtained from Criollos and Trinitarios trees (Caligiani et al., 2014). Bean mixing not only cause penalties in export prices it also makes it

Hyperspectral Remote Sensing. DOI: https://doi.org/10.1016/B978-0-08-102894-0.00013-9

difficult to reproduce the taste and smell of chocolate, even if the beans come from the same geographical area. Visual inspection for bean recognition is challenging because the pulp is white regardless of the bean type. During fermentation the embryo dies and the cotyledons gradually changes color from purple to brown, see Fig. 17–1A. Manual color tests are done to evaluate the fermentation process but not for bean quality assessment. hyperspectral (HS) imaging provides a quantitative description of the browning process and, more importantly, spectral information can be used for bean inspection.

Previous studies on bean classification include: Teye et al. (2016) in which near-infrared spectroscopy and chemometrics techniques are used to identify five kinds of cocoa beans—they achieved an optimal performance in their model; in Caligiani et al. (2014) proton nuclear magnetic resonance spectroscopy (1H-NMR) and chemometrics techniques were used to discriminate between different fermentation levels in various kinds of cocoa, including NT, detecting and quantifying some important substances such as amino acids, carbohydrates, epicatechin, and methylxanthines, and they obtained a classification rate of 85.2%. These works got good results, but their methodologies are destructive, time-consuming, and rather expensive for real environments. In Vargas et al. (2016) recognition, between Collection Castro Naranjal 51 (CCN51) and national (NT) beans is made in a nondestructive way using Raman spectroscopy and chemometrics approaches, and getting a 91.8% accurate detection rate.

The main problem in cocoa spectral characterization is to measure the bean surface reflectance reliably. Specular reflections in wet beans cause sensor saturation. Also, as fermentation progresses, cracks appear in the cotylendone surface and become larger and deeper. The low resolution and low signal to noice ratio (SNR) in close-range HS images objects' features difficult to detect. In this chapter we discuss image fusion techniques to enhance spatial and spectral information of the HS cube. By combining a high-resolution

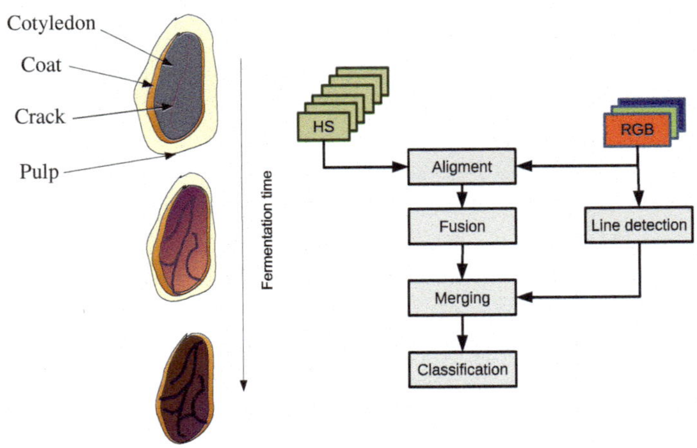

(A) Bean fermentation. (B) Proposed method.

FIGURE 17–1 Cocoa bean fermentation and method schematic.

red, green, and blue (RGB) image with the low-resolution HS cube we obtain a fused HS cube with the same spectral dimensions and increased spatial dimension. We exploit the enhanced spatial data to detect and remove specular regions in bean coats and linear features corresponding to cracks in bean-cuts.

The image fusion technique is explained in Section 17.2. Experimental results are presented in Section 17.3. Conclusions and future works are discussed in the final section.

17.2 Methodology

17.2.1 Image acquisition

The image acquisition system used in our experiments is described in detail in Ochoa et al. (2016). Fig. 17–2 shows the system that consists of a monochrome 12 bit CCD camera 1500M-GE Thorlabs with NIR sensitivity, a spectrograph SPECIM Imspector V10 with spectral sensitivity in the [364,1031] nm range and spectral resolution of 1.2 nm, and a lighting system with two 50 W halogen lamps placed on the sides of the camera. These elements are mounted on a motorized slider in a pushbroom configuration. For each scan, a hyperspectral cube of 520 images is obtained. Additionally, we used a high-dynamic range CCD camera to capture RGB images.

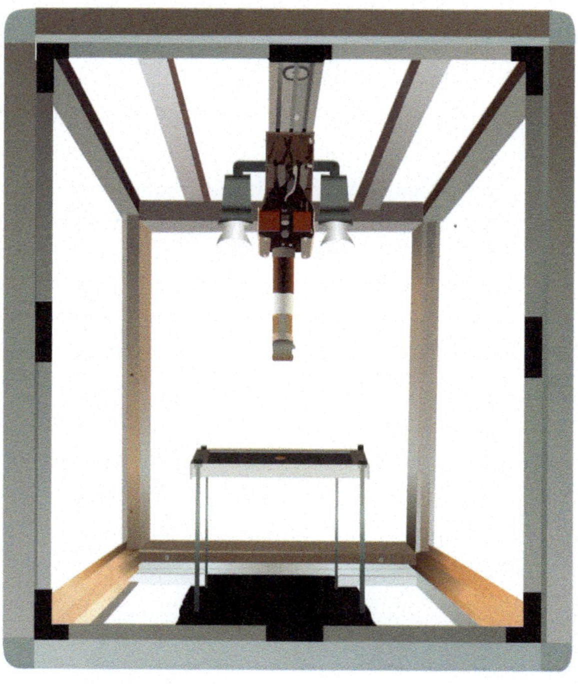

FIGURE 17–2 Hyperspectral imaging system.

17.2.2 Image preprocessing

To reduce the spectral distortion produced by changes in quantum efficiency of the CCD and the geometric configuration of the lightning system, the spectral data was initially calibrated using Eq. (17.1). Where R_λ is the reflectance in the λ wavelength, I_λ is the intensity of the sample, W_λ is the intensity of a white standard reference, and D_λ is the intensity measured when the sensor does not receive light.

$$R_\lambda = \frac{I_\lambda - D_\lambda}{W_\lambda - D_\lambda} \tag{17.1}$$

Spectral curves were smoothed using the Savitzky–Golay (SG) filter with a window size of 11. The SG filter is commonly used to reduce the random noise derived from read-out times of the camera, the data transfer, and analog to digital conversion (Sun, 2010). The 40 wavelengths at each end of the spectral curve were discarded due to sensor noisy in those regions. The effective spectral range was between 400 and 1000 nm. Since the bean coat is ellipsoidal, the distance form sample's points to the sensor and the relative angle of reflected light reduce the reflectance value. To compensate for these effects, geometric normalization based on standard normal variance transformation, linear optical models (Shahrimie et al., 2016), and photometric invariance (Stokman and Gevers, 1999) can be applied. We chose the photometric invariance approach of the Eq. (17.2) (Polder et al., 2004).

$$X_\lambda = \frac{R_\lambda}{\Sigma_\lambda R_\lambda} \tag{17.2}$$

Fig. 17–3A shows a RGB image of a cocoa bean, Fig. 17–3B is a pseudo-color image (generated by three images of the HS cube) of a cocoa bean marked with 5 regions. Fig. 17–3C shows the average spectra of the marked regions without geometric normalization, it can be observed that intensity of reflectance increase while a region is closer to the light. Finally in Fig. 17–3D can be observed the aligned spectral curves after normalization.

17.2.3 Image fusion

HS image fusion with complementary data is frequently used in remote sensing to enhance spatial information and, consequently, improve the performance of subsequent image processing steps. Fusion techniques are usually referred to as pan-sharpening because the input is a high-resolution panchromatic image. A comprehensible survey of these techniques can be found in (Thomas et al., 2008). In general, pan-sharpening approaches are based on component substitution, multiresolution analysis (Vivone et al., 2014), Bayesian methods (Wei et al., 2015), and hybrid methods. Hybrid methods have proven suitable for RGB-HS data fusion as spectral information is well preserved and due to its low-computational demands compared to other methods.

We exploit the guide filter principal component analysis (GFPCA) hybrid method (Liao et al., 2013). First, HS data is decorrelated using PCA and 95% of the image variance is found

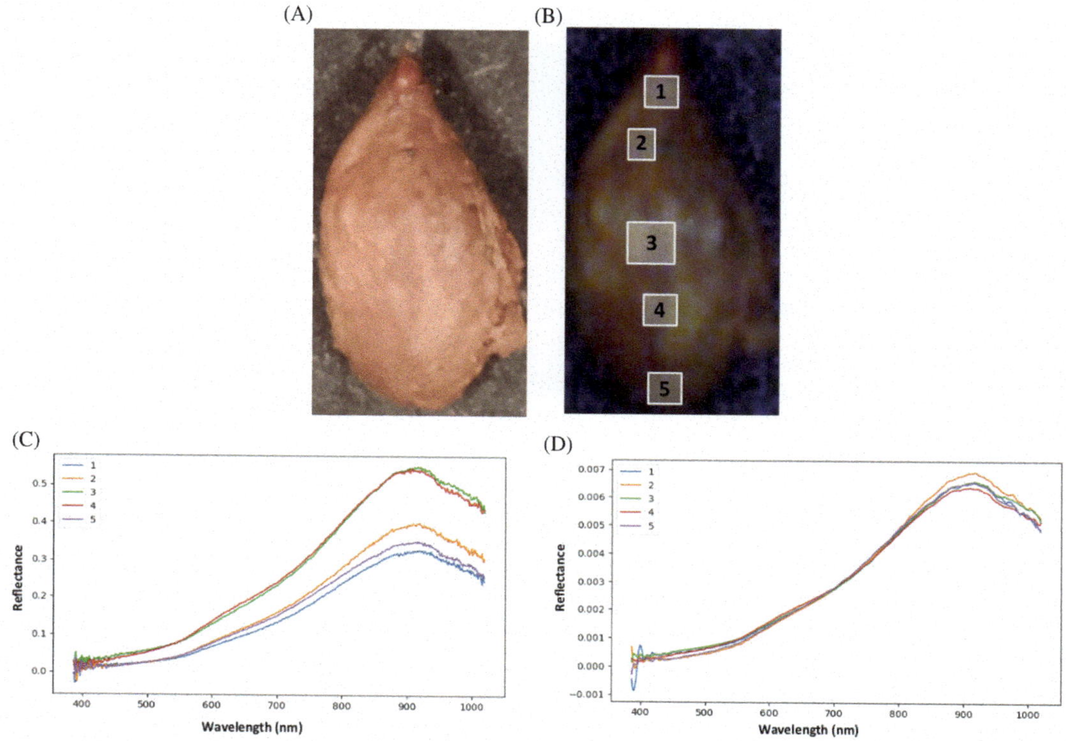

FIGURE 17–3 Normalization of a cocoa bean spectrum.

in the first $k = 6$ principal components (PC), the remaining components contain mostly noise. Then PC images are up-sampled using bi-cubic interpolation. Spatial structures are transferred to the first k PC images using a guided filter (He et al., 2013). This operator has the advantage of preserving the edges of the guide image, in our case the RGB image, in the output image. Let PC_i, be the up-sampled PC image at pixel, i, and the guided image, Y. The filter's output, PC_i^g, can be expressed as:

$$PC_i^g = a_j Y + b_j, \quad \forall\, i \in \omega_j \tag{17.3}$$

where ω is a local window of size $(2r + 1) \times (2r + 1)$. By taking the gradient on both sides of Eq. (17.3), we can verify that the filtered image has an edge only if the guided image has one. The following cost function is used to determine the coefficients a_j and b_j:

$$C(a_j, b_j) = \sum_{\omega} [(a_j Y + b_j - PC_i)^2 + \subset e\, a_j^2] \tag{17.4}$$

where e is the regularization parameter that determines the degree of blurring for the guided filter. The remaining principal components are not filtered because the guide filter will only amplify noise. A wavelet-based denoising routine is applied to those components. Finally, inverse PCA transform is used to obtain the enhanced high-resolution HS cube. In terms of cocoa image content, the GFPCA approach allows to keep small image structures as well as removing sensor noise.

17.2.4 Morphological feature data

To improve spectral data extraction, we want to remove dark linear features from the bean-cut image and focus the analysis on the cotyledons surface. A scale-space ridge detector (Steger, 1996) is used for this purpose. A ridge point is found when the maximum eigenvalue λ_+ of the Hessian matrix is larger than the minimum eigenvalue λ_- and the latter is close to zero. The scale-space derivatives are computed using Gaussian derivative kernels, G_σ. The scale parameter σ depends on the width of the line to be detected.

$$\lambda_{+-} = G_{xx,\sigma} + G_{yy,\sigma} \pm \sqrt{(G_{xx,\sigma} - G_{yy,\sigma})^2 - 4G^2_{xy,\sigma}} \tag{17.5}$$

The detector's output is then normalized based on the expected shape of the ridge. Assuming a parabolic ridge profile with contrast h, the normalized response at a ridge point is defined by:

$$D_\sigma = \frac{\sigma^2 \lambda_+}{h} \tag{17.6}$$

In low-resolution HS images the lack of ridge evidence and low SNR prevent the segmentation of cracks. Better results are obtained when ridge detection is performed on the RGB image at several scales. Each D_σ was added to the fused HS cube, and the resulting extended cube was used as input for classification.

17.3 Experiments

CCN51 and NT beans were used in our experiments. We prepared two samples to apply fusion with beans coats and the other with bean-cuts. In both cases, HS cubes were obtained and processed using the approach described in Section 17.2. The fusion algorithm requires the alignment of HS low-resolution and RGB high-resolution images. We aligned the RGB image to the HS cube using GeFolki nonrigid registration algorithm (Brigot et al., 2016). GeFolki returns the optical flow at every pixel position and, in turn, is used to guide the registration step. Guided filter parameters were set to $r = 2$ and $e = 1e - 4$. Experimentally we noticed that this combination of parameters prevented loss of details in flat areas. Fig. 17–4 shows the original and fused image at 745 nm. Structures visible in the RGB image are transferred to the HS image. The difference between the average spectra of the original and the fused cubes was small, as expected, in the order of $10e - 1$. An example is depicted in

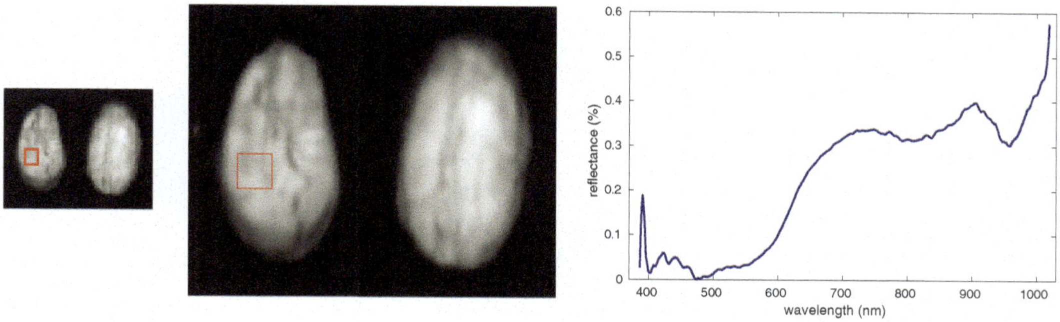

(A) HS image. (B) Fused HS image. (C) Difference.

FIGURE 17–4 Example of fusion output at 745 nm with the spectral difference computed inside the red square.

Fig. 17–4C. The average accuracy (AA) metric was used to evaluate classification results. AA is the ratio of correctly classified points to the number of test data points for each class.

The coat is the outer part of the cocoa bean and a cut is the results of cutting the bean in half. The browning of both is a sign of fermentation, yet in the bean-cuts the color change from purple to brown can be easily seen.

17.3.1 Bean coat

For bean coat analysis 30 beans of each variety at 0 h, 24 h, 48 h, and 72 h fermentation stages were randomly selected from bean batches in a fermentation center. The RGB images were segmented using a blob detector to filter out the specular patches produced by the moisture of the mucilage of the bean in the fused HS-cube. As input for bean coat classification the well-known spectral features used in other subjects were extracted. The first feature was the average spectrum of each bean in pixels. Other features were computed from the output of several band selection methods that keep the most significant bands according to a certain criterion. We extracted the first spectral derivative ($D_1(\lambda)$), the second spectral derivative ($D_2(\lambda)$), information entropy ($H(\lambda)$), and PCA ranking (PCAr) from a subset of the 10 most informative wavelengths.

Features associated to the reflectance properties of the chemical compounds were also studied. Some works (Kim and Keeney, 1984; Nazaruddin et al., 2006; Peláez et al., 2016) suggest that the concentration of anthocyanin, theobromine, caffeine, and epicatechine can be good indicators for bean variety identification. To measure the anthocyanin concentration, the vegetation index anthocyanin reflectance index 2 (ARI2) (Gitelson et al., 2001) was computed using Eq. (17.7). Where $\rho800$, $\rho550$, and $\rho700$ are ranges of 20 wavelengths of the spectrum that are centered in the wavelengths 800, 550, and 700 nm. This index was used in Reyes et al. (2015) as a quality indicator of the fermentation process in cut beans.

$$\mathrm{ARI2} = \rho_{800}\left[\frac{1}{\rho_{550}} - \frac{1}{\rho_{700}}\right] \tag{17.7}$$

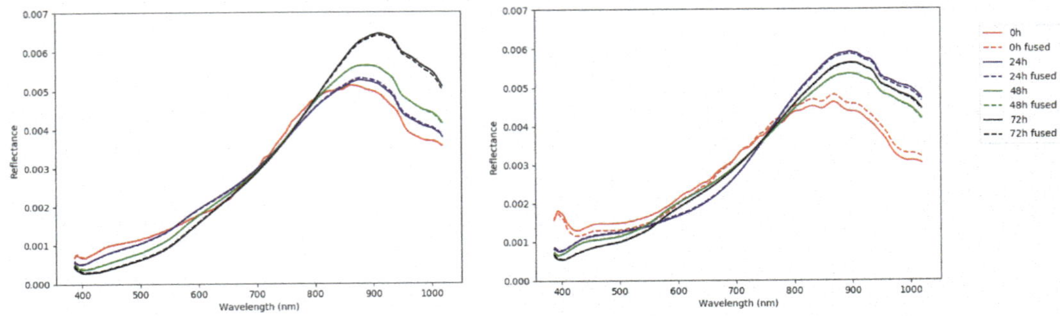

(A) NT (B) CCN51

FIGURE 17–5 Average reflectance spectral curves of low-resolution and fused-hyperspectral cubes.

Theobromine, caffeine, and epicatechine present peaks in the ultraviolet and infrared range of the absorbance spectrum [200–300] nm and [500–4000] nm^{-1}. Such wavelengths cannot be captured with the current acquisition system. The water index (WI) (Peñuelas et al., 1997) is a vegetative index that measures moisture, see Eq. (17.8). The average WI of each bean was calculated to measure the water content changes during fermentation.

$$WI = \frac{\rho_{900}}{\rho_{970}} \tag{17.8}$$

For supervised classification, K-nearest neighbors (KNN) algorithm with $k = 8$ was performed. Fig. 17–5 shows the average spectra of both varieties of cocoa for the 4 fermentation stages in low-resolution and fused-HS cubes. It can be observed that in the visible range the first stages reflect more light, while in the NIR range the lasts stages reflect more light.

Table 17–1 shows the AA of the classification performed on two classes, NT and CCN51, at each fermentation stage using the low-resolution and fused-HS cubes. The best results were obtained with the set of wavelengths extracted from the low-resolution cubes using $H(\lambda)$, because after 24 h and 72 h the accuracy values were the highest. Another outstanding result is that for the last fermentation 95% accuracy was obtained using $H(\lambda)$, ARI2, and ARI2 + WI. However, the ARI2 + WI feature is not useful for classification considering that the accuracy after the 48 h and 72 h stages are the same using just the ARI2 feature and only on the first stages are higher than ARI2 but lower than PCAr and $H(\lambda)$. Additionally, comparing the accuracy values between low-resolution and fused-HS cubes, the image fusion technique improves the accuracy values of most of the features at the 0 h, 24 h, and 48 h stages, but not in the same way at the 72 h stage.

The result of ARI2 seems to indicate that the concentration of anthocyanin in the coat after 72 h of fermentation is an informative feature. Fig. 17–6 shows the distribution of ARI2. It can be observed that as the fermentation progresses, the differences in anthocyanin concentrations of both varieties increase. The studies report that the anthocyanin concentration of the cotyledons is reduced during fermentation, but the anthocyanin behavior in the pulp has not been reported.

Table 17–1 Average accuracy of the classification of the two varieties of cocoa for each fermentation stage using different features.

	0 h		24 h		48 h		72 h	
Features	**LR**	**Fused**	**LR**	**Fused**	**LR**	**Fused**	**LR**	**Fused**
Avg. spectrum	0.60	0.73	0.87	0.88	0.78	0.78	0.90	0.90
$D_1(\lambda)$	0.63	0.70	0.82	0.87	0.82	0.87	0.87	0.85
$D_2(\lambda)$	0.70	0.80	0.78	0.78	0.55	0.58	0.82	0.82
$H(\lambda)$	0.67	0.72	0.88	0.75	0.80	0.63	0.95	0.85
PCAr	0.72	0.67	0.75	0.77	0.70	0.68	0.85	0.85
ARI2	0.42	0.65	0.50	0.43	0.78	0.80	0.95	0.95
WI	0.58	0.73	0.73	0.75	0.72	0.65	0.53	0.45
ARI2 + WI	0.45	0.68		0.62	0.52	0.78	0.80	0.95

17.3.2 Bean-cuts

To classify bean-halves 20 beans, 10 samples from each type were selected from the middle of the fermentation box for 72 h after the start of fermentation. After cutting them in half the excess water in each half surface was removed. Line detection was performed on the RGB image for $\sigma = [1.18, 1.76, 2]$, the local contrast, h, was measured within a linear region oriented in perpendicular direction to the ridge. A point was declared as ridge point if $D > 0.7$ and $\lambda_- < 0.1$. Maps containing D values were added to the fused-HS cube as additional channels forming an extended HS cube.

The KNN classification algorithm used $k = 4$ for NT cotyledon, CCN51 cotyledon, coat, and cracks. For training, 40 data points for each class were randomly selected from the extended cube. Classification results are shown in Table 17–2 where we show detection rates using only hyperspectral data (raw), multiscale line evidence (ridge), and the stacked fused data (fused). An example of the classification output is shown in Fig. 17–7; the RGB

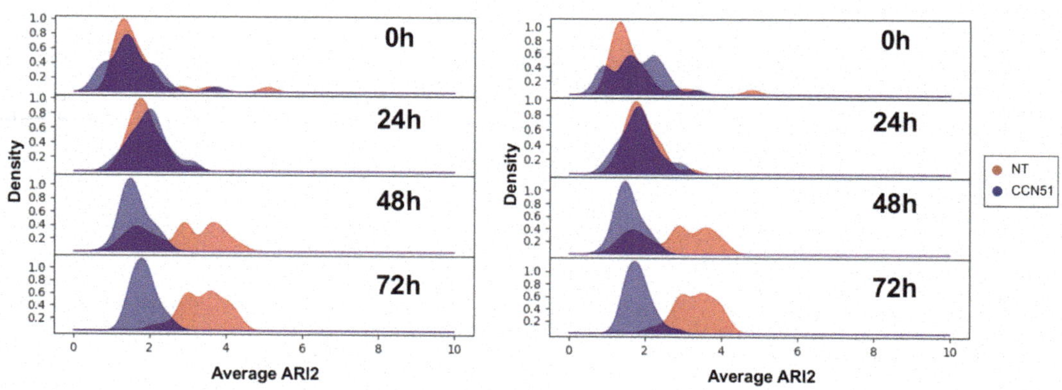

(A) Low Resolution (B) Fused

FIGURE 17–6 Scatter plot of anthocyanin reflectance index 2 for each fermentation stage.

Table 17–2 Detection rates.

	NT cotyledon	CCN51 cotyledon	Crack	Coat
Ridge	0.15	0.12	0.41	0.53
Raw	0.59	0.59	0.48	0.81
Fused	0.80	0.79	0.78	0.75

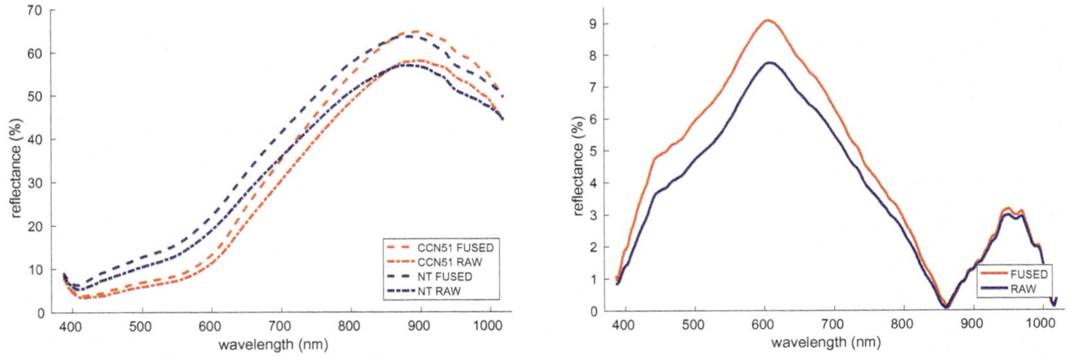

FIGURE 17–7 Example of classification results using *K*-nearest neighbors for four classes: national cotyledon (blue), CCN51 cotyledon (green), crack (orange), and coat (red).

image is included as a reference. We can detect more data points corresponding to the cotyledon of both types of beans using the fused data and the cracks are visible. The coat region is rather noisy as this effect is cause by the bean shadow on the sample holder.

Further evidence of improvements in spectral estimation of bean cotyledon is shown in Fig. 17–8(left) that depicts the average spectral curves of the beans' surface from the low--resolution HS cube and the curve built from data points classified as cotyledon in the fused cube. The second curve has consistently higher reflectance values in the visible and part of the infrared regions of the spectrum as we removed the points with lower reflectance. The difference of spectral curves between bean types increases, see Fig. 17–8(right). Although the maximum difference is relatively small it entails a better separability of spectra.

17.4 Discussion

Our experiments are the first steps toward separating low- and high-quality cocoa bean in a nondestructive way using HS imaging of the bean coat and bean-cuts. Regarding bean coats, the combination of high spatial resolution RGB image allowed to filter out specular regions. For beans after 24 and 72 h of fermentation it should be possible to build an optimized acquisition system to measure only those wavelengths in the visible spectrum and near infrared that report high entropy. In the last fermentation stage, the ARI2 can be used for classification because 95% of accuracy is obtained and this suggests that if it is used with other spectral

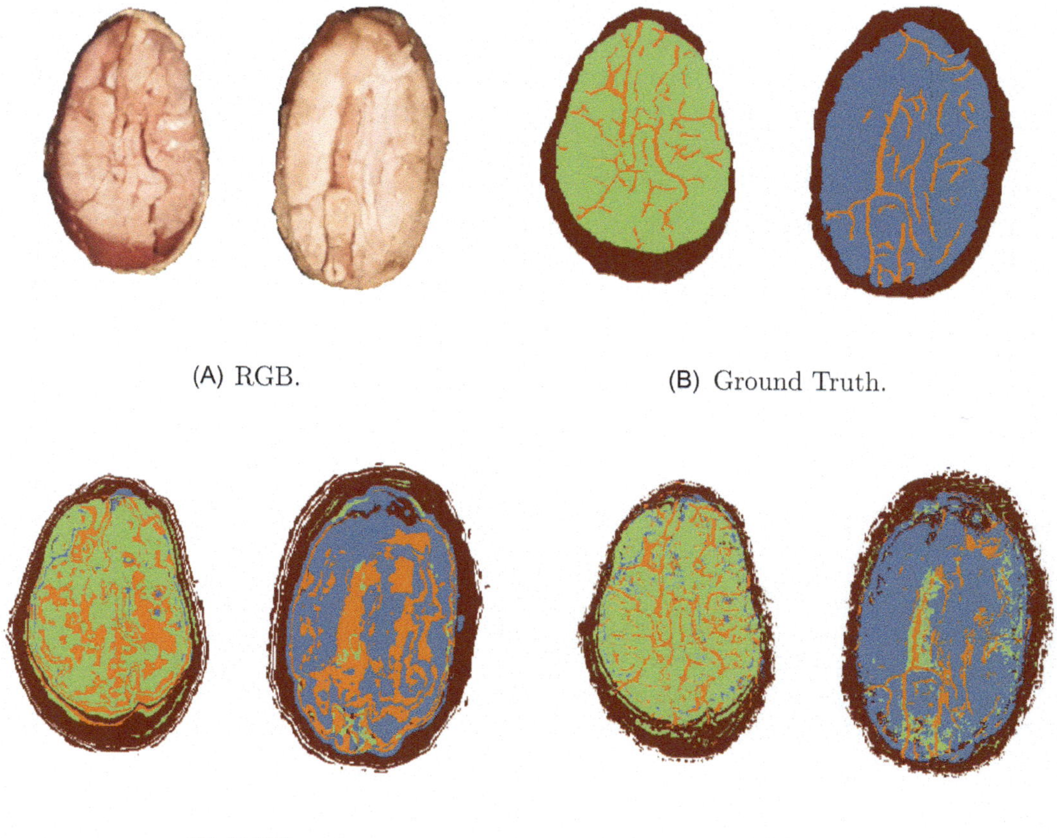

(A) RGB. (B) Ground Truth.

(C) RAW. (D) Fused.

FIGURE 17–8 Results: average spectral curves (left) and spectral difference between bean types (right).

indexes it could obtain good classification accuracies. The accuracy of the classification of cocoa beans without fermentation (0 h stage) is very low for both types of cubes. The reason is that the pulp spectral signatures for both types of beans are similar. To corroborate the measures performed with ARI2, we have to correlate them with anthocyanin concentration values in the bean coat and include features related to theobromine, caffeine, and epicatechine. The main effect of adding spatial information on classification can be observed at early fermentation stages; nevertheless classification accuracies are lower compared to later fermentation stages. An alternative would be to use spatial information to extract pulp-free regions.

Regarding bean-cuts, we found that classification accuracy was improved considerably when using the proposed method compared to the raw HS data. Image registration should be improved (i.e., using nonlinear methods as the guided filter proved very sensitive to border locations). Most detection errors appear in narrows zones between bean regions and may occur because the response of the line detector increases. The inherent color variability of cotyledons explains our results to a certain degree. Overall detection accuracy is still lower compared to data sets commonly used in remote sensing but we think that by adding additional spatial information, bean classification can be improved further. Knowing how browning develops is key to derive additional spatial features, for example, radial spectral measurements from the bean center or at increasing distances from the cracks. To the best of our knowledge, such approaches have not yet been explored. Therefore in our future work we will try to collect more evidence and include other types of spatial information either in the HS cube or its PCA components, which has not been tested in closed-range images. While complex feature sets and advanced classification algorithms can be used, our current results indicate that close-range images have the potential for bean analysis not only for cocoa but other important commercial crops as well.

Acknowledgments

This work was financially supported by the Flemish Interuniversity Council (VLIR), ESPOL research fund.

Abbreviations

HS hyperspectral
CCN51 Collection Castro Naranjal 51
NT national
RGB red, green, and blue
^{1}H-NMR proton nuclear magnetic resonance
SNR signal to noise ratio
CCD charged coupled device
M-GE monochrome gigabit ethernet
NIR near-infrared
SG Savitzky–Golay
SNV standard normal variance
GFPCA guided filter principal component analysis
PCA principal component analysis
PC principal component
AA average accuracy
WI water index
KNN *K*-nearest neighbors
ARI2 hyperspectral analysis-based anthocyanin index

Symbols

λ wavelength
R_λ reflectance in a specific wavelength
I_λ intensity of the sample in a specific wavelength
W_λ intensity of a white standard reference in a specific wavelength
D_λ intensity when the sensor does not receive light in a specific wavelength
nm nanometer
k number of principal components
PC_i principal component image at pixel i
Y guided image

PC_i^g output of GFPCA
ω local window
$a_j b_j$ coefficients of cost function
ϵ regularization parameter
λ_+ maximum eigenvalue of Hessian matrix
λ_- minimum eigenvalue of Hessian matrix
G_σ Gaussian derivative kernel
σ scale parameter
h contrast
D_σ normalized response at a ridge point
r window radius
$D_1(\lambda)$ first spectral derivative
$D_2(\lambda)$ second spectral derivative
$H(\lambda)$ information entropy
PCA_r PCA ranking
$\boldsymbol{k}$ number of nearest neighbors of KNN

References

Brigot, G., et al., 2016. Adaptation and evaluation of an optical flow method applied to co-registration of forest remote sensing images. IEEE J. Sel. Top. Appl. Earth Observ. Remote Sens. 6, July.

Caligiani, A., Palla, L., Acquotti, D., Marseglia, A., Palla, G., 2014. Application of 1H NMR for the characterisation of cocoa beans of different geographical origins and fermentation levels. Food Chem. 157 (Supplement C), 94–99.

Gitelson, A.A., Merzlyak, M.N., Chivkunova, O.B., 2001. Optical properties and nondestructive estimation of anthocyanin content in plant leaves. Photochem. Photobiol. 74 (1), 38–45.

He, K., Sun, J., Tang, X., 2013. Guided image filtering. IEEE Trans. Pattern Anal. Mach. Intell. 35, 1397–1409. June.

Kadow, D., Bohlmann, J., Phillips, W., Lieberei, R., 2013. Identification of main fine flavour components in two genotypes of the cocoa tree (*Theobroma cacao* L.). J. Appl. Bot. Food Qual. 86 (1).

Kim, H., Keeney, P., 1984. (−)-Epicatechin content in fermented and unfermented cocoa beans. J. Food Sci. 49.

Krysiak, W., Adamski, R., Żyżelewicz, D., 2013. Factors affecting the color of roasted cocoa bean. J. Food Qual. 36 (1), 21–31.

Liao, W., Goossens, B., Aelterman, J., Luong, H., Pizurica, A., Wouters, N., et al., 2013. Hyperspectral image deblurring with pca and total variation. In: Proceedings of IEEE GRSS Workshop Hyperspectral Image Signal Process: Evolution in Remote Sensing (WHISPERS), FL, pp. 1–4.

Lopez, A.S. and Dimick, P.S., 2008. Cocoa fermentation, Biotechnology. In: Rehm, H.-J., Reeds, G. (Eds)., 561–577.

Nazaruddin, R., Seng, L., Hassan, O., Said, M., 2006. Effect of pulp preconditioning on the content of polyphenols in cocoa beans (*Theobroma cacao*) during fermentation. Ind. Crop Prod. 24 (1), 87–94.

Ochoa, D., Cevallos, J., Vargas, G., Criollo, R., Romero, D., Castro, R., et al., 2016. Hyperspectral imaging system for disease scanning on banana plants. Proc. SPIE 9864, Sensing for Agriculture and Food Quality and Safety VIII, 98640M.

Peláez, P., Guerra, S., Contreras, D., 2016. Changes in physical and chemical characteristics of fermented cocoa (*Theobroma cacao*) beans with manual and semi-mechanized transfer, between fermentation boxes. Sci. Agropecuaria 7, 111–119.

Peñuelas, J., Piñol, J., Ogaya, R., Filella, I., 1997. Estimation of plant water concentration by the refectance water index WI (r900/r970). Int. J. Remote Sens. 18 (13), 2869–2875.

Polder, G., W.A.M. Van der Heijden, van der Voet, G.H., Young, I., 2004. Measuring surface distribution of carotenes and chlorophyll in ripening tomatoes using imaging spectrometry, Postharvest Bio. Tec. 34 (2), 117–129.

Reyes, J., Soto, J., William, I., 2015. Hyperspectral analysis based anthocyanin index (ARI2) during cocoa bean fermentation process. In: 2015 Asia-Pacific Conference on Computer Aided System Engineering, pp. 169–172.

Schwan, R.F., Wheals, A.E., 2004. The microbiology of cocoa fermentation and its role in chocolate quality. Crit. Rev. Food Sci. Nutr. 44 (4), 205–221.

Shahrimie, M., Mishra, P., Mertens, S., Dhondt, S., Wuyts, N., Scheunders, P., 2016. Modeling effects of illumination and plant geometry on leaf reflectance spectra in close-range hyperspectral imaging.

Steger, C., 1996. Extraction of curved lines from images. In: Proceedings of the 13th International Conference on Pattern Recognition, vol. 2. IEEE, pp. 251–255.

Stokman, H., Gevers, T., 1999. Deteccion and clasification of hyper-spectral edges. In: The Tenth Britisch Machine Vision Conference.

Sun, D.-W., 2010. Hyperspectral Imaging for Food Quality Analysis and Control. Elsevier.

Teye, E., Uhomoibhi, J., Hui, W., 2016. Nondestructive authentication of cocoa bean cultivars by FT-NIR spectroscopy and multivariate techniques. Focus Sci. 2 (3), 1–5.

Thomas, C., Ranchin, T., Wald, L., Chanussot, J., 2008. Synthesis of multispectral images to high spatial resolution: a critical review of fusion methods based on remote sensing physics. IEEE Trans. Geosci. Remote Sens. 46, 1301–1312. May.

Vargas, P., Ciobotă, V., Salinas, W., Kampe, B., Aponte, P., Rösch, P., et al., 2016. Distinction of Ecuadorian varieties of fermented cocoa beans using Raman spectroscopy. Food Chem. 211.

Vivone, G., Restaino, R., Licciardi, G., Dalla Mura, M., Chanussot, J., 2014. Multiresolution analysis and component substitution techniques for hyperspectral pansharpening. In: Geoscience and Remote Sensing Symposium (IGARSS), 2014 IEEE International. IEEE, pp. 2649–2652.

Wei, Q., Dobigeon, N., Tourneret, J.Y., 2015. Bayesian fusion of multi-band images. IEEE J. Sel. Top. Signal Process. 9, 1117–1127. September.

18

Noninvasive detection of plant parasitic nematodes using hyperspectral and other remote sensing systems

Uroš Žibrat, Saša Širca, Nik Susič, Matej Knapič, Barbara Gerič Stare, Gregor Urek

AGRICULTURAL INSTITUTE OF SLOVENIA, PLANT PROTECTION DEPARTMENT, LJUBLJANA, SLOVENIA

18.1 Introduction to noninvasive detection

> *We need better methods for diagnosis; none of the methods given are to be considered as 'standardized'. To think of them in such a way would put an end to efforts of improvement. They are useful only until better procedures can be developed.*
>
> Riker and Riker (1936)

Traditional visual detection of plant pests and diseases is based on characteristic plant symptoms or visible pathogen/pest reproduction structures such as lesions, changes in pigment ratios, galls, cysts, conidia, urediniospores, etc. Even though the introduction of common guidelines and standards has increased detection accuracy, this type of identification is performed by experts, skilled at diagnosing plant pests and diseases. Detection accuracy then depends on each individual's experience and is, thus, subject to human bias. Accuracy is further confounded by temporal variation of symptoms. This leads to significant interrater variability and low interrater repeatability (Bock et al., 2008, 2010; Steddom et al., 2005). According to various disease and pest detection guidelines and standards, each plant has to be checked individually, making this a time-consuming endeavor, especially at field level.

The occurrence of plant pests and diseases in a field is often heterogeneous; hence, traditional monitoring methods are performed on a per-plant basis. These methods are feasible in small fields or greenhouses where individual plants can be checked, but in agricultural fields, a systematic approach is impossible due to the large area and number of plants. Randomized spot checks of individual plants are utilized, but these can easily miss small infections or infestation foci, which can then function as a reservoir from which infections or infestations spread.

Hyperspectral Remote Sensing. DOI: https://doi.org/10.1016/B978-0-08-102894-0.00015-2

Furthermore, traditional detection often depends on visual symptoms, while presymptomatic detection usually depends on molecular methods. The detection of visible symptoms versus presymptomatic detection of individual plants leads directly to longer durations before identification, hence, higher yield loss or management costs. In addition, infection or infestation symptoms are often visually similar to symptoms of abiotic stress such as water or nitrogen deficiency. In the case of soilborne pathogens such as root-knot nematodes (RKNs) that parasitize the roots and don't cause any specific visual symptoms on the aboveground part of plants, visual detection of infestation is only possible by uprooting individual plants and detecting the formation of root-galls. Understanding the invasive nature of such visual detection, which is clearly impractical for large-scale use, it becomes evident that a presymptomatic and noninvasive approach for the detection of plant pests and diseases would represent a great leap forward. Remote sensing may give the tools to attain this goal.

Remote sensing can be defined as obtaining information about an object without having direct physical contact with it. Visual detection of disease, with or without cameras, can, thus, be considered as remote sensing (Nutter, 1990). But, remote sensing by definition is not necessarily noninvasive. Visual inspection of roots for nematode induced root-galls is an invasive remote sensing method since the sensor (the investigator's eye) and the plant don't touch, but the plant has to be uprooted. In noninvasive remote sensing for plant protection, one measures a plant's response to diverse stressors without affecting the plant in any way. Plants respond to stress by utilizing various mechanisms by adapting their physiology and/or morphology (e.g., changes in color, transpiration, leaf and canopy morphology, and the production of specific metabolites) (West et al., 2010). These changes occur at both the tissue and canopy levels, thus, affecting the absorption, reflectance, and transmittance of light, and, therefore, altering the spectral signatures of plants. By utilizing noninvasive remote sensing methods, the need to introduce a new stressor (e.g., uprooting the plant), which could confound the analytical results, can be avoided.

Detection and identification methods with high reliability, sensitivity, and specificity are needed to improve the detection of plant diseases and pests, so that they can replace traditional visual inspections. Spectral remote sensing methods assess changes in plant spectral signatures due to disease, abiotic stress, or ontogenetic development within and outside the visible part of the electromagnetic spectrum. For plant disease and pest detection, sensors measuring reflectance, temperature, and fluorescence are utilized (Mahlein et al., 2012; Sankaran et al., 2010). These sensors can be mounted on diverse platforms such as ground-based (e.g., tractors), unmanned aerial vehicles (UAVs), airplanes, and satellites. They enable automated data acquisition over comparatively large areas at relatively short time intervals with high spectral resolutions (depending on sensor type). Thus they generate large amounts of complex data. Appropriate data preprocessing steps need to be taken in order to obtain effectively useful data, which can then be analyzed using advanced methods (i.e., machine learning methods, neural networks, and deep learning algorithms).

Newly developed methods of remote sensing for the detection of plant pests and diseases have to be at least as accurate as traditional methods, and perform in a shorter amount of time with at most an equal cost. In order to accomplish this, remote sensing data have to fulfill several criteria (Mahlein, 2016), namely (1) enable early detection at different points in time (i.e., prior to the development of visible symptoms), (2) differentiation among diseases and pests, (3) differentiation between abiotic and biotic stresses, and (4) quantification of disease/infestation severity. Analyses using contemporary optical sensors fulfill these criteria, therefore, research focus shifts to data analysis and interpretation (Behmann et al., 2014). Most optical sensors do not measure plant physiological parameters, but instead measure the sum reflectance of various plant tissue and metabolic products (Jensen, 2007). Plant optical properties are characterized by three processes, namely (1) transmission through the leaves and stems, (2) absorption by chemicals inside the tissue (e.g., metabolites, water, pigments, proteins, cellulose, and lignin), and (3) reflectance from inside tissue structures and from leaf surfaces. Plant spectral signatures are, therefore, always a complex combination of these three processes. Plants have different defense mechanisms for preventing entrance and colonization by pests and pathogens such as the induction of hypersensitive reactions, the production of antimicrobial metabolites and proteins, and plant tissue structure (Voigt, 2014). These changes lead to highly specific changes in reflectance (Mahlein, 2016). At the beginning of the growing season, the root function of infested plants is not yet impaired and the symptoms of nematode infestation on plants are not visible to the naked eye, however, the reflectance of the electromagnetic spectrum changes (Hillnhütter et al., 2010). The visible symptoms on the aboveground parts of plants occur relatively late and are nonspecific; they can be similar to drought stress or lack of nutrients since plant parasitic nematodes cause decreases in the content of water, chlorophyll, and carotenoids. For example, a lower reflectance around 535 nm indicates an increase in zeaxanthin content, leading to the photoprotective state of the zeaxanthin cycle (Gamon et al., 1992), a decrease at 550 nm is related to adjustments of anthocyanins and other photoprotective pigments (Steele et al., 2009), while an increase of reflectance at 1500 nm is linked to a decrease of leaf water content (Eitel et al., 2006).

18.2 Introduction to plant parasitic nematodes

18.2.1 Plant parasitic nematodes

Nematodes are the most abundant group of multicellular animals distributed in marine, freshwater, and terrestrial habitats accounting for about 80% of all individual animals on Earth (Platt, 1994). Their lifestyles and trophic levels are various; from bacterial and fungal feeding to predators as well as parasites playing important roles in many ecosystems. From the human perspective, nematodes may play a beneficial role like entomopathogenic nematodes used as biocontrol agents for the management of insect pests, but they may also act as harmful organisms parasitizing agricultural crops, domestic animals, and humans. More than 4100 nematode species were recognized as plant parasites (Decraemer and Hunt, 2006), but only a small proportion of these plant parasitic nematode species is causing substantial economic damage in agricultural crop production, which is estimated to range between US$80 and US$157 billion annually (Sasser and Freckman, 1987; Nicol et al., 2011). Based on a global survey, Jones et al. (2013) composed a list of the ten most scientifically and economically important plant parasitic nematodes. Root-knot, cyst, and pine wilt nematodes are all on this list. In conditions affected by climate change, these species will benefit by developing more generations in a growing season, thus, their spread and high impact on agricultural production can be expected. These important pests, among others, have been studied using various noninvasive remote sensing methods. Furthermore, several patents have been published worldwide dealing with their detection by remote sensing. Therefore this chapter will focus on noninvasive remote sensing detection of root-knot, cyst, and pine wilt nematodes.

18.2.2 Cyst and root-knot nematodes

The most notorious nematodes are cyst nematodes belonging to the *Globodera* and *Heterodera* genera and RKNs of the genus *Meloidogyne*. Nematodes of these three genera are soilborne pests causing major damage to crops such as potato, soybean, sugar beet, cereals, and vegetables (Figs. 18–1 and 18–2). These nematodes are sedentary endoparasites (living within a host plant for most of their life cycle), which attack a host plant's roots and cause direct damage to the root tissue. After the infective second-stage juveniles (J2s) penetrate into the host root they induce complex feeding structures in the root's tissue. The feeding site supplies the nematode with a food source for the rest of its lifetime. Damage to the root system aggravates the nutrient and water uptake of the plant, resulting in a reduced crop performance and, consequently, in reduced crop yield and quality. Plants infested with sedentary nematodes do not develop specific visible symptoms on the aboveground crop canopy. Infested plants become chronically water-stressed and generally show wilting and stunting symptoms or even symptoms of decay in extensive infestations. The symptoms of nematode infestation on the aboveground parts of plants can be divided into two phases. In the first part of the

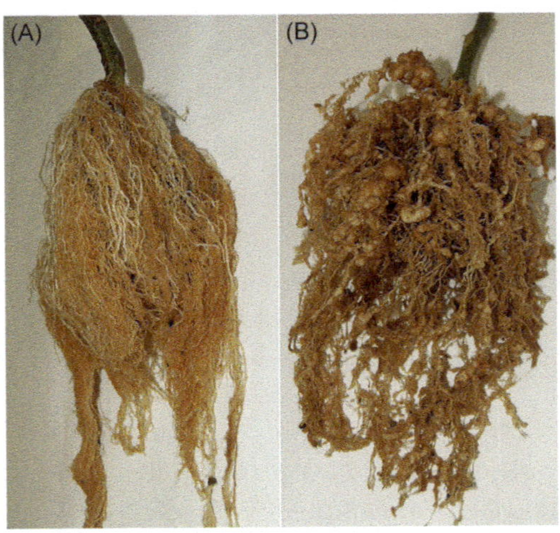

FIGURE 18–1 Formation of galls on roots infested with root-knot nematode *Meloidogyne luci* disrupts normal root function. (A) Roots of a healthy tomato plant and (B) infested tomato root system with typical root-galls.

FIGURE 18–2 Heavy infestation of potato tubers with root-knot nematode *Meloidogyne incognita* leads to the formation of blisters and unmarketable potato tubers.

growing season, affected plants show reduced photosynthetic activity due to reduced nutrient uptake and changes in the water regime of the plant (Melakeberhan et al., 1987; Kirkpatrick et al., 1991; Strajnar et al., 2012). The dry matter content of the root system increases, while the dry matter content of the canopy is reduced due to the decreased growth of the stem and leaf tissues. The leaves are smaller and thicker. At this stage, it is possible to detect general symptoms such as the loss of turgor pressure and the withering of the plant canopy. In the second part of the growing season, nematode infestation leads to premature defoliation, decreased leaf formation, decreased nutrient and water uptake due to root system damage, and, consequently, reduced crop yields.

18.2.3 Pinewood nematode

In addition to economic damage in agriculture, nematodes cause serious damage in forestry as well, namely by the pinewood nematode (PWN), *Bursaphelenchus xylophilus*, which causes pine wilt. The nematode is native to North America where it does not cause substantial damage due to the presence of adapted and tolerant conifer species. When PWN was introduced to new regions with susceptible conifer species, it caused huge damage. In Japan, PWN caused over 2 million cubic meters of timber loses between 1978 and 1979 in combination with abnormally warm weather conditions, and the economic damage in the region was estimated to be US$10 million each year (Nose and Shiraishi, 2007). In Portugal, PWN was detected for the first time in Europe in 1999 (Mota et al., 1999). From 1999 to 2013 the costs of attempting PWN eradication were estimated to be approximately €100 million (REPHRAME, 2015) in addition to direct losses from dead trees. PWN were, therefore, considered to be one of the most damaging emerging pest problems in forests around the world (Mota et al., 1999; Rodrigues, 2008; Jones et al., 2008; Sousa et al., 2013). The nematode is vectored by longhorn beetles of the genus *Monochamus* (Coleoptera: Cerambycidae), which transport nematodes into the crowns of healthy living trees during their feeding phase. Since PWN is highly pathogenic, the infested tree can die within the same season as infestation or in a consecutive year depending on the environmental conditions. Infestations of suitable host trees in summer may result in the rapid death of the trees within as little as 40–60 days due to the high summer temperatures, while spring infestation may take longer to develop. Later infestation in autumn or winter may result in no symptoms development, but the infested tree may die in the coming season (Kiyohara and Tokushige, 1971). The severity of the pine wilt symptoms is related to the host species and also to the temperature and weather conditions. Visible symptoms of pine needle discoloration and canopy wilting are caused by failure of water transport and consequent water stress (Suzuki, 1984). The spread of pine wilt disease is currently being controlled by taking preventative measures including frequent monitoring in order to exterminate PWN before their spread. The most important aspect of keeping pest-free areas or disease containment is early pest detection.

18.3 Examples of noninvasive detection of plant parasitic nematodes

18.3.1 Characteristics of nematode infestations

Early detection of nematode infestation is of utmost importance for timely and targeted management, in accordance with integrated pest management guidelines. Nematode infestations generally exhibit several typical characteristics, that is, (1) visible symptoms are indistinguishable from drought stress, (2) the juvenile active mobility of individuals in the soil is low, (3) clustered spatial distribution of infestations, (4) symptoms in the foliage are present throughout the growing season, and (5) introductions of new infestation foci are comparatively rare. Furthermore, integrated pest management guidelines envision site specific treatments, and machine learning methods enable generalizable detection models, thus, allowing for detection in different crops. These attributes favor the use of remote sensing methods to facilitate integrated pest management at field scale prior to the development of visible infestation symptoms. The temporary and spatially accurate application of appropriate management practices has the potential of significantly reduce yield losses and pesticide use, and has, therefore, the potential to reduce environmental impact and economic expense in crop production (Gebbers and Adamchuck, 2010).

Infestations of plant parasitic nematodes attacking plant roots reduce or completely destroy normal root functioning, leading to reduced water and nutrient uptake. This induces changes in pigmentation (changes in the amounts of various pigment types such as chlorophyll, anthocyanins, and carotenoids), evapotranspiration, tissue chemicals (e.g., metabolites and proteins) and physical structure (e.g., lesions), and leaf temperature. Different regions of the electromagnetic spectrum provide information on diverse plant physical and biochemical characteristics. The visible part of the spectrum (400–700 nm) is absorbed by leaf pigments, for example, chlorophyll (chl)-a and chl-b absorb light at 430 and 662 nm, and 453 and 642 nm, respectively. Accessory pigments (e.g., β-carotene and carotenoids) expand and shift the wavelengths of maximum absorption to approximately 450 and 650 nm (the exact wavelengths depend on the ratios between the different pigments). Because of this absorption, plants show a comparatively low reflectance in the visible part of the spectrum with a peak at the green part at approximately 550 nm. In the near-infrared region (NIR; 700–1000 nm), reflectance increases to around 50%. This difference between the visible and near-infrared regions of the spectrum is typical for plants and can be used as a simple identifier in multi- and hyperspectral images. The well-known and widely used normalized difference vegetation index (NDVI) uses this difference to assess plant "greenness," which can be used as a measure of photosynthetic activity. The NIR spectral region provides information on leaf tissue structure and can also be used for crop yield assessment. Water absorption is the main characteristic of the short-wave infrared (SWIR, 1000–2500 nm) part of the spectrum. In addition, leaf biochemical properties can also be ascertained in this region. Chemometric analyses utilize the SWIR region. The far infrared or thermal part of the spectrum starts at 5000 nm, and is important in thermometry, that is, assessing plant stress by measuring leaf temperature. Because plant responses to disease and pathogens are complex, the entire spectrum is often utilized, from the visible to SWIR and thermal regions.

18.3.2 Remote sensing of nematode infestations

Norman and Fritz (1965) were among the first researchers to use infrared sensors for presymptomatic detection of the plant parasitic nematode *Radopholus similis* in citrus trees. They found that burrowing nematode infestations can be detected in trees older than 4 years in areas where nematodes have been present for several years. On the other hand, in trees younger than 4 years, nematode infestations could not be detected using infrared photography. Several plant diseases (footrot, tristeza, psorosis, xyloporosis, and exocortis) were also tested, and a reduced infrared reflectance of up to 50% was found in infected or infested trees. Seven years later, Heald et al. (1972) used airborne infrared images of cotton fields to detect the suppression of reniform nematode *Rotylenchulus reniformis* by different management practices in cotton fields. Differences in plant growth between different treatments were observed at 61 and 111 days after planting; at a later time, early symptoms of root rot were detected in the infrared spectrum. This nematode species affects the growth and development of cotton plants causing a 40%–60% yield reduction (Lawrence et al., 1990). Gausman et al. (1975) also worked with cotton and *R. reniformis*, but used a spectroradiometer at leaf and canopy level, measuring wavelengths in the visible and near-infrared (VNIR) and SWIR regions (500–2500 nm). They also found a reduced reflectance in infested plants, as did Norman and Fritz (1965), and attributed this change to thinner leaves with a more compact mesophyll of infested plants.

Plant parasitic nematodes and many other soilborne diseases cause reduced water uptake or transport in infested plants. This leads to reduced evapotranspiration in leaves, and hence, increased leaf temperature. The first study of plant diseases using thermometry was performed in 1979 by Pinter et al. They showed that soilborne pathogens cause a presymptomatic increase in leaf temperature in infected plants of up to 5°C. The beet cyst nematode, *Heterodera schachtii*, was the first plant parasitic nematode to be studied using thermometry. Berg (1980) succeeded in identifying presymptomatic plants in sugar beet fields. The potato cyst nematode, *Globodera rostochiensis*, was successfully detected in potato plants by Gebhardt (1984). A remote sensing study of the root system of infested tomato plants using nuclear magnetic resonance imaging (NMRI) was performed by Johnson et al. (1986). This approach yields reliable results, but is time consuming, expensive, and can only be used at greenhouse level with potted plants.

Early remote sensing studies of plant diseases mostly focused on just one disease or pathogen, while their interactions, plant responses to complex stresses, and infestation severity were not considered. The first study of nematode infestation severity and its effect on leaf temperature was performed by Nicolas et al. (1991). They studied the effect of different levels of the cereal cyst nematode, *H. avenae*, infestation severity on winter wheat, and found that severe infestations led to significantly higher temperatures compared to mild infestations. The first study that considered the complex interactions between parasites and pathogens was performed by Cook et al. (1999). They used a three-band (red, yellow–green, and NIR) multispectral video imagery to discern between the southern RKN, *Meloidogyne incognita*, and root rot due to *Phytophthora omnivorum* in kenaf (*Hibiscus cannabinus*), separately and in combination. Vegetation indices were not calculated for this study, instead the detection of infested and infected plants was performed by visual inspection of black and white and RGB images of the three spectral bands. Heath et al. (2000) used spectroradiometer measurements of potatoes infested with *G. pallida* and *G. rostochiensis*, at various infestation severities. They found good correlations between the NDVI and the number of juveniles per gram of potato roots. The correlation between NDVI and potato cyst nematode population density was also observed by Stephens et al. (2000). This multispectral vegetation index approach was upgraded by Nutter et al. (2002) for detecting the soybean cyst nematode, *H. glycines*. The presence of *H. glycines* at low population densities on soybean fields cannot be easily observed as the pest needs several years to reach high population densities, after which areas with stunted plants become visible with the naked eye. They used three different platforms, ground-based, airborne, and satellite (Landsat 7), at different times during the growing season. They also showed that using traditional methods (i.e., collection and analysis of soil samples for soybean cyst nematode population density) explained less than a third of the variation in soybean yield among quadrats, while using aerial- and ground-based remote sensing measurements explained more than 80% of the variation in soybean yield. On the other hand, Wheeler and Kaufman (2003) used a three-band (green, red, and NIR) airborne multispectral sensor on several flight missions, but failed to achieve a good relation between *M. incognita* damage in cotton and NDVI and several band ratios. A single-band (at 850 nm) multispectral approach was used by Shah et al. (2004) to assess the effects of early blight *Alternaria solani* on potatoes grown with different nitrogen applications and infested by *G. rostochiensis*.

18.3.3 Hyperspectral remote sensing of nematode infestations

Hyperspectral imaging has become a staple in remote sensing of nematode infestations and plant health. Narrowband sensors provide a greater amount of data and their high spectral resolution (several tens to hundreds of bands with a bandwidth of approximately 4 nm) allows for chemometric analyses and direct identification of metabolites and proteins related to the immune system response of plants (Fig. 18–3). At larger spatial scales, for example at field level, airborne hyperspectral imagery can be combined with accurate geographic information systems (GIS) and global navigation satellite systems (GNSS) data, leading to targeted management practices. Lawrence et al. (2004) combined ground-based hyperspectral imaging with self-organizing maps (unsupervised neural networks) to detect *R. reniformis* in cotton, and obtained a classification accuracy between 83% and 97%. This approach of classifying infested plants was further developed by Doshi (2007). Rupe et al. (2003) used a different approach and identified four spectral bands out of 300 (at 550, 650, 750, and 850 nm) that were most responsive to soybean cyst nematode, *H. glycines*, infestations in soybean fields. A vegetation index approach was used by Bajwa et al. (2017) to detect soybean cyst nematodes and sudden death syndrome caused by a soilborne fungal pathogen *Fusarium solani* f. sp. *glycines* in soybean plants. They obtained mixed results; the detection of healthy plants was good with an accuracy of up to 93%, while the detection of infested and diseased plants reached a modest success with up to 68% correctly classified plants. Doshi et al. (2007a,b) developed a series of self-organizing map-based methods, which dealt with the prediction of nematode populations based on leaf reflectance, time series analysis,

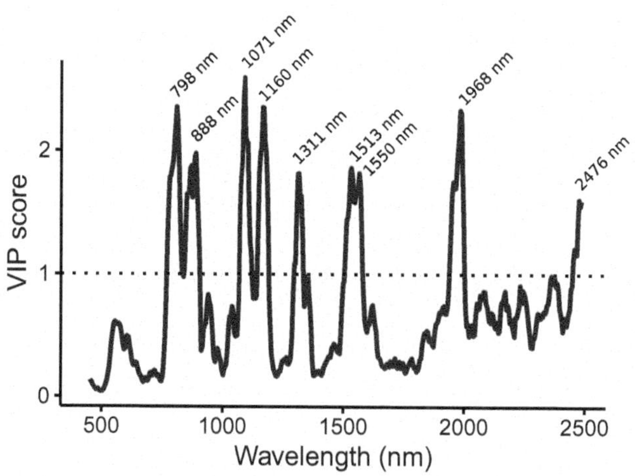

FIGURE 18–3 Variable in projection (VIP) (Chong and Jun, 2005) scores from a PLS-DA model of nematode infestation severity (solid line) based on hyperspectral imaging data. Wavelengths with VIP scores above 1 (dotted line) are considered to be important for the detection of nematode infestation severity. There are two peaks in the near-infrared spectral regions, and seven in the short-wave infrared region. These wavelengths are linked to the C–H and N–H stretching of proteins and carbohydrates. This indicates the importance of wavelengths linked to leaf chemistry and structure for the detection of nematode infestations. *PLS-DA*, partial least squares - discriminant analysis.

and wavelet feature extraction for the hyperspectral detection of nematodes. Doshi et al. (2010) combined principal component analysis and self-organizing maps, and found that the visible part of the spectrum yielded the worst classification results, while the red edge and NIR (651–1300 nm) and mid-infrared (IR) (1301–2500 nm) regions performed best. Hillnhütter et al. (2011b) used nuclear magnetic resonance imaging to assess root damage caused by *H. schachtii* and *Rhizoctonica solanii* on sugar beet. An inverted canopy reflectance model combined with support vector machines was proposed by Palacharla et al. (2011). The canopy reflectance model, PROSAIL, obtained sensor specific reflectance of cotton plants, and included seven biophysical parameters (leaf area index, mean leaf inclination angle, the hot spot parameter, a spectral soil parameter that controls the soil reflectance levels, a leaf structure parameter, chl-a and chl-b, and soil brightness) divided into seven classes. Even though this approach considers all possible combinations between the parameters, the obtained classification results were not satisfactory.

18.3.4 Remote sensing of cyst, root-knot, and pinewood nematodes

New studies show a high potential of combining different sensors, platforms, and parameters for plant phenotyping and detection of nematode infestations. Hillnhütter et al. (2011a) combined airborne hyperspectral remote sensing using two different sensors and a handheld spectroradiometer to detect plants infested with *H. schachtii* and *R. solanii*. Using a spectral angle mapper (SAM) classifier they obtained a classification accuracy of up to 60%. In a later study, Hillnhütter et al. (2012) tested three data preprocessing methods for data extraction to assess their suitability to generate the most sensitive spectral information for detecting *H. schachtii* and *R. solanii*. They found that spectral indices had a low specificity to characterize leaf symptoms, while a SAM classification achieved up to 80% accuracy for detecting leaf damage severity. Joalland et al. (2017) compared the ability of visible light imaging, thermometry, and spectrometry to detect beet cyst nematode, *H. schachtii*, infestations in two sugar beet varieties, one susceptible and one tolerant. They found that using visible imaging, canopy area can be determined, from which they could detect delays in leaf growth as well as benefits from nematicides as early as 15 days after sowing. Spectrometry was suitable when plants reached maturity and the canopy reached full coverage. Thermometry on the other hand, could detect infestations only in the susceptible cultivar. The visible imaging approach was further developed by Joalland et al. (2018a) who developed a plant phenotyping algorithm that evaluates sugar beet phyllotaxis (Milford et al., 1985). Joalland et al. (2018b) further expanded the work of Joalland et al. (2017) by including different sensors on diverse platforms and spectral vegetation indices under field conditions. They found that combinations of vegetation indices in multivariate analysis and decision trees were able to differentiate cultivars, and could be used as proxies for sugar yield prediction.

The PWN, *B. xylophilus*, is an invasive species causing pine wilt disease and may cause rapid tree decline. Ju et al. (2014) used ground-based hyperspectral remote sensing in the visible to NIR spectral range to determine the best wavelengths for identifying infested *Pinus massoniana* trees. They identified the first derivative spectrum at 759 nm as being the most effective at discriminating between healthy and infested trees. Properly processed remote sensing data from various airborne sensors with different spatial and spectral resolutions were also shown to detect individual pine trees showing canopy decline (Beck et al., 2015). Kim et al. (2018) used a vegetation index approach and introduced the green–red spectral area index, which proved to be optimal for pine wilt disease detection, and showed lower variability than all other indices used.

One of the first attempts at detecting RKNs of the genus *Meloidogyne* was performed using airborne infrared imagery to detect cotton plants infested by *M. incognita* (Orion et al., 1982). RKNs are one of the most significant agricultural pests among plant parasitic nematodes. Infestations with RKNs can be detected by visually examining the roots of plants for the presence of root knots or galls. However, at the early stage of infestation, the symptoms on the roots are hardly visible, while in the aboveground parts of plants, symptoms can be seen only later in the growing season, when the root system is damaged by higher nematode population density. The visual symptoms are indistinguishable from drought stress. A novel approach was used by Johnson et al. (1986) who used NMRI remote sensing to assess damage to the root system of infested plants. These early studies used visual comparisons between images to assess damage to plants. Using airborne three-band (green, red, and NIR) multispectral imaging, Wheeler and Kaufman (2003) attempted to detect infestations with *M. incognita* in sugar beet, but failed to find any vegetation indices or their combinations that would consistently detect infestations in fields. On the other hand, Strajnar et al. (2015) showed the potential of hyperspectral proximal remote sensing to detect presymptomatic infestations of *Meloidogyne* spp. in tomato plants. This work was further improved by Susič et al. (2018), who developed a method to not only detect infested plants, but also distinguish between them and plants in drought stress at different time intervals. By combining partial least squares-discriminant analysis (PLS-DA) and support vector machine classification in a time series, they successfully identified spectral regions (e.g., the C–H stretching of proteins at 1189–1222 and 1324–1346 nm) in the SWIR part of the spectrum (1000–2500 nm) that were best able to predict infested plants. They also found that the visible part of the spectrum wasn't important for the identification of infested plants since visible aboveground symptoms are identical to drought stress symptoms. Furthermore, they found that classification accuracy increased with time as nematodes developed and their effect on the plants became more pronounced. Ultimately, this work led to the setting of the guidelines for hyperspectral image preprocessing and analysis for the detection of *Meloidogyne* spp. infestation (Žibrat et al., 2019). These results are a good basis for further investigations and development of remote sensing applications for the detection of RKNs in open fields.

18.4 Patents in the field of remote sensing of nematode infestations

Both public and private research sectors rely on patents to protect their investments. Remote sensing of plant pests and diseases, especially nematode infestations, could be considered to be at the beginning of its development, with 32 currently published patents related to hyperspectral remote sensing of nematode infestations worldwide (data obtained from www.lens.org in April 2019). The majority of these patents are owned by private companies such as Monsanto, BASF, and Gen Electric.

Even though remote sensing of nematode infestations in plants began more than 50 years ago, the first patent was only submitted in 2005, and published in 2006 (Mississippi State University). This patent deals with three aspects of nematode infestation, namely (1) detection of plant parasitic nematodes using hyperspectral reflectance data, (2) a device for the determination of reniform nematodes in a target crop, and (3) a method for the management of nematode population with variable rate applications of nematicides. Inventors from the Nanjing Forestry University in China (2014) published a patent for the hyperspectral monitoring of pine wilt in black pine. Even though these patents deal with remote sensing, they are not necessarily designed for noninvasive detection. For example, Monsanto (2009) devised a system for hyperspectral imaging using artificial light for soybean cyst nematode detection. But their system requires destructive sample preparation in order to detect and count soybean cyst nematodes, and based on these counts, select plants for further breeding. In addition to plant parasitic nematodes, patents utilizing remote sensing for the detection of nematode fish parasites in fillets have also been published (Tecnologia Marina Ximo, 2015).

Generally, plant phenotyping systems for breeding purposes are more prevalent, and have some common components. These imaging systems contain hyper- or multispectral cameras, recording visible, NIR, and/or SWIR light. Data acquisition and analysis are also automated. Strictly plant phenotyping related patents were published by Monsanto (2011) and Pioneer (2013), while a patent by BASF (2017) also contains plant treatment subsystems. A system specializing in optimizing ethanol yield from grain was developed by Monsanto (2017), and BASF (2016) invented a high-throughput phenotyping system for measuring the inflorescence, seed, and seed yield of plants.

Precision agriculture patents such as the TAIR system (Targeted Agricultural Input Recommendation; Pioneer, 2014) use various sensors (e.g., VNIR, red edge, thermal, hyperspectral) to obtain information on localized usage context, and provide management recommendations based on these data. The University of Sydney (2015) developed a similar system that uses information from several sensors including hyperspectral for the detection of diseased plants and targeted application of pesticides. A system for the optimization of crop production using various machine learning methods in combination with agricultural, soil type and quality, and weather data, has also been patented by Indigo (2019). This system utilizes remote sensing data from various platforms (ground-based, airborne, UAV, satellite), but is limited to the VNIR and UV spectrums.

The success of research institutions and their scientists can be measured not only in acquired research grants and published research articles, but also in the impact and changes that research institutions bring to the world and society at large. In order to unleash the innovation potential of research institutions, basic research needs to be translated into commercially viable processes and technology. Patents and commercialization ensure direct revenue generation and further benefits for the public through economic development as well as further advantages for the inventors such as increased opportunities for research funding, sustaining high scholarship levels, student success, and increased prestige (Sanberg et al., 2014).

18.5 Conclusion

Over the years, a large amount of information on remote sensing for the determination of the spatial and temporal distribution of nematodes has been generated. The development of new sensors has led to the accumulation of large amounts of data, which led to the introduction of advanced data analysis methods (e.g., machine learning) in remote sensing of plant pests and pathogens. Nematodes are a good fit for remote sensing applications and precision agriculture since they have a clustered spatial distribution, low mobility, and characteristic aboveground symptoms (in NIR and SWIR). Depending on the crop, individual plants can be identified using UAV or airborne multi- and hyperspectral systems, making the even removal of infested plants a feasible management option depending on the infestation severity (number of infested plants in a given field) and field size. The implementation of remote sensing methodology in nematode pest management practices would have several benefits, namely (1) the targeted use of pesticides and other chemical or biological management practices, (2) early warning systems would enable timely application of management practices, (3) reduced use of chemical treatments, (4) distinction between abiotic and biotic stressors would lead to correct management, (5) economic benefits for the grower due to targeted management, and (6) lower environmental impact. The widespread implementation of remote sensing for the noninvasive detection of plant pests and pathogens in agricultural practice faces the challenges of relatively high costs of hyperspectral cameras and the required computational capacities as well as the need for highly specialized analysts to perform the advanced data analysis.

Any remote sensing application that aims to reliably identify nematode infestation in plants should also successfully discriminate between abiotic (e.g., drought and resource deficiencies) and biotic stresses (caused by nematodes or other pests/pathogens) since visible symptoms in the plant canopy are often similar, if not identical. Furthermore, presymptomatic detection of individual infested plants at field level would facilitate precision agriculture and integrated pest management approaches by enabling spatially accurate, timely, and appropriate management responses.

While individual remote sensing systems (multispectral, hyperspectral, and thermal) on various platforms (ground-based or handheld, airborne, or satellite) can provide comparatively accurate assessments of nematode infestations, the best results are obtained by combining different sensors and advanced data analysis methods. Furthermore, future research needs to combine complementary scientific fields such as plant pathology, pest/pathogen population modeling, machine learning and artificial intelligence, informatics, and sensor development (Mahlein, 2016). Such a highly interdisciplinary approach should also consider the input and experience of growers from agriculture production, which will enable the development of novel detection methods with higher accuracy.

Future remote sensing studies of plant pests and pathogens face several challenges. In order for noninvasive remote sensing detection of infestations to become truly implemented in precision agriculture and integrated pest management schemes, all of these challenges will have to be addressed to some degree. The areas of research that need further efforts in the future are:

- Advanced phenotyping is needed to obtain the characteristics of different plant varieties and to account for them in order for them not to confound remote sensing detection methods.
- Characterization of plant responses to pathogen groups (e.g., fungi, bacteria, viruses, soilborne and foliar pathogens) is needed.
- Mixed infestations/infections with different pests and/or pathogens as well as interactions with abiotic stress, should be researched.
- Plant molecular (genomic, transcriptomic, and metabolomic) responses to pests, pathogens, and abiotic stress and their influence on plant optical properties should be researched.
- Interdisciplinary research (e.g., by combining remote sensing and pests and pathogen population modeling) will provide additional information and, hence, better predictive models. This would also benefit early warning (presymptomatic) systems.
- The results of these studies have to be put into practice by developing appropriate decision support systems. These would help make the results of complex hyperspectral data analyses, machine learning, and their interpretation easier to understand and accessible to growers.
- Automating disease and pathogen detection in combination with robotics will significantly impact crop production and pest management systems.

Remote sensing for plant pest and pathogen detection has a vast potential in plant protection and the use of this technology is in its initial stages. As a growing interest in the use of remote sensing by governments and companies as well as growers is being witnessing, there is no doubt that this field will be thriving and exciting in the years to come.

Acknowledgments

The preparation of this chapter was financially supported by the Slovenian research agency (ARRS), grant P4-0072, and young researcher grant no. 38128 as well as by the Ministry of Agriculture, Forestry and Food of the Republic of Slovenia, grant no. C2337, and the European Food Safety Authority, grant GP/EFSA/ALPHA/2018/02. Some of the research was conducted using equipment financed by EU-FP7 project CropSustaIn, grant agreement FP7-REGPOT-CT2012-316205. The position and opinions presented in this chapter are those of the authors alone and do not necessarily represent the views of EFSA.

List of abbreviation

GIS geographic information systems
GNSS global navigation satellite systems
J2s second-stage juveniles
NDVI normalized difference vegetation index
NIR near-infrared

NMRI nuclear magnetic resonance imaging
PLS-DA partial least squares-discriminant analysis
PWN pinewood nematode
RGB red, green, and blue
RKNs root-knot nematodes
SAM spectral angle mapper
SWIR short-wave infrared
UAVs unmanned aerial vehicles
UV ultraviolet
VIP variable in projection
VNIR visible and near-infrared

References

Bajwa, S.G., Rupe, J.C., Mason, J., 2017. Soybean disease monitoring with leaf reflectance. Remote Sens. 9, 127. Available from: https://doi.org/10.3390/rs9020127.

Basf Plant Science Co. Gmbh., 2016. AU 2016/368598 A1, Australia.

Basf Plant Science Co. Gmbh., 2017. AU 2017/259723 A1, Australia.

Beck, P.S.A., Zarco-Tejada, P., Strobl, P., San Miguel, J., 2015. The feasibility of detecting trees affected by the pine wood nematode using remote sensing. Eur. Comm. JRC Tech. Rep. Available from: https://doi.org/10.2788/711975.

Behmann, J., Mahlein, A.-K., Rumpf, T., Römer, C., Plümer, L., 2014. A review of advanced machine learning methods for the detection of biotic stress in precision crop protection. Precis. Agric. 16, 239–260.

Berg, A., 1980. Remote sensing techniques applied to sugar beet diseases in Germany and Italy: introduction to the results of a European project. Int. Arch. Photogramm. 23, 457–462.

Bock, C.H., Parker, P.E., Cook, A.Z., Gottwald, T.R., 2008. Visual rating and the use of image analysis for assessing different symptoms of citrus canker on grapefruit leaves. Plant Dis. 92, 530–541.

Bock, C.H., Poole, G.H., Parker, P.E., Gottwald, T.R., 2010. Plant disease severity estimated visually, by digital photography and image analysis, and by hyperspectral imaging. Crit. Rev. Plant Sci. 29, 59–107.

Chong, I.G., Jun, C.H., 2005. Performance of some variable selection methods when multicollinearity is present. Chemom. Intell. Lab. Syst. 78, 103–112.

Cook, C.G., Escobar, D.E., Everitt, J.H., Cavazos, I., Robinson, A.F., Davis, M.R., 1999. Utilizing airborne video imagery in kenaf management and production. Ind. Crop. Prod. 9 (3), 205–210. Available from: https://doi.org/10.1016/S0926-6690(98)00033-8.

Decraemer, W., Hunt, D., 2006. Structure and classification. In: Perry, R., Moens, M. (Eds.), Plant Nematology. CAB International, Oxfordshire, 2006.

Doshi, R.A., 2007. Self-Organizing Maps for Classification and Prediction of Nemtatode Populations in Cotton (M.Sc. thesis). Mississippi State University, Mississippi State.

Doshi, R.A., King, R.L., Lawrence, G.W., 2007a. Self-organized maps-based spectral prediction of *Rotylenchulus reniformis* population numbers. In: Proceedings IEEE Geoscience and Remote Sensing Symposium (IGARSS), Barcelona, pp. 444–447.

Doshi, R.A., King, R.L., Lawrence, G.W., 2007b. Wavelet-SOM in feature extraction of hyperspectral data for classification of nematode species. In: Proceedings IEEE Geoscience and Remote Sensing Symposium (IGARSS), Barcelona, pp. 2818–2821. doi:10.1109/IGARSS.2007.4423429.

Doshi, R.A., King, R.L., Lawrence, G.W., 2010. Classification of *rotylenchulus reniformis* numbers in cotton using remotely sensed hyperspectral data on self-organizing maps. J. Nematol. 42 (3), 179–193.

Eitel, J.U.H., Gessler, P.E., Smith, A.M.S., Robberecht, R., 2006. Suitability of existing and novel spectral indices to remotely detect water stress in *Populus* spp. For. Ecol. Manag. 229, 170–182. Available from: https://doi.org/10.1016/j.foreco.2006.03.027.

Gamon, J., Penuelas, J., Field, C., 1992. A narrow-waveband spectral index that tracks diurnal changes in photosynthetic efficiency. Remote Sens. Environ. 41, 35–44. Available from: https://doi.org/10.1016/0034-4257(92)90059-S.

Gausman, H.W., Heald, C.M., Escobar, D.E., 1975. Effect of *Rotylenchulus reniformis* on reflectance of cotton plant leaves. J. Nematol. 7, 368–374.

Gebbers, R., Adamchuck, V.I., 2010. Precision agriculture and food security. Science 327, 828–831.

Gebhardt, A., 1984. Möglichkeiten zur Früherkennung phytopathologischer Erscheinungen mit Hilfe der Thermographie—Übersicht und erste Ergebnisse. Arch. Phytopathol. Plant Prot. 20, 437–445 (in German).

Heald, C.M., Thames, W.H., Wiegand, C.L., 1972. Detection of *Rotylenchulus reniformis* infestations by aerial infrared photography. J. Nematol. 4, 298–300.

Heath, W.L., Haydock, P.P.J., Wilcox, A., Evans, K., 2000. The potential use of spectral reflectance from the potato crop for remote sensing of infection by potato cyst nematodes. ASP Appl. Biol. 60, 185–188.

Hillnhütter, C., Schweizer, A., Kühnhold, V., Sikora, R.A., 2010. Remote sensing for the detection of soil-borne plant parasitic nematodes and fungal pathogens. In: Oerke, E.-C., Gerhards, R., Menz, G., Sikora, R.A. (Eds.), Precision Crop Protection—The Challenge and Use of Heterogeneity. Springer, The Netherlands. Available from: https://doi.org/10.1007/978-90-481-9277-9_10.

Hillnhütter, C., Mahlein, A.-K., Sikora, R.A., Oerke, E.-C., 2011a. Remote sensing to detect plant stress induced by *Heterodera schachtii* and *Rhizoctonia solani* in sugar beet fields. Field Crop Res. 122, 10–77.

Hillnhütter, C., Sikora, R.A., Oerke, E.-C., van Dusschoten, D., 2011b. Nuclear magnetic resonance: a tool for imaging belowground damage caused by *Heterodera schachtii* and *Rhizoctonia solani* on sugar beet. J. Exp. Bot. 63 (1), 319–327.

Hillnhütter, C., Mahlein, A.-K., Sikora, R.A., Oerke, E.-C., 2012. Use of imaging spectroscopy to discriminate symptoms caused by *Heterodera schachtii* and *Rhizoctonia solani* on sugar beet. Precis. Agric. 13, 17–32.

Indigo Ag. Inc., 2019. US 2019/0050948 A1, USA.

Jensen, J.R., 2007. Remote Sensing of the Environment: An Earth Resource Perspective, second ed. Pearson Prentice Hall, Upper Saddle River, NJ.

Joalland, S., Screpanti, C., Liebisch, F., Varella, H.V., Gaume, A., Walter, A., 2017. Comparison of visible imaging, thermography and spectrometry methods to evaluate the effect of *Heterodera schachtii* inoculation on sugar beets. Plant Methods 13, 73. Available from: https://doi.org/10.1186/s13007-017-0223-1.

Joalland, S., Pari, M., Bevilacqua, A., Screpanti, C., 2018a. Application of plant phenotyping algorithm to detect stress caused by nematodes. Rhizosphere 6, 86–88.

Joalland, S., Screpanti, C., Varella, H., Reuther, M., Schwind, M., Lang, C., et al., 2018b. Aerial and ground based sensing of tolerance to beet cyst nematode in sugar beet. Remote Sens. 10, 787. Available from: https://doi.org/10.3390/rs10050787.

Johnson, G.W., Bailey, J.E., Bruck, R.I., Matyac, C.A., 1986. Studies of diseased root tissue using nuclear magnetic resonance imaging. Phytopathology 76, 1067.

Jones, J.T., Li, H., Moens, M., Mota, M., Kikuchi, T., 2008. *Bursaphelenchus xylophilus*: opportunities in comparative genomics and molecular host–parasite interactions. Mol. Plant Pathol. 9, 357–368.

Jones, J.T., Haegeman, A., Danchin, E.G.J., Gaur, H.S., Helder, J., Jones, M.G.K., et al., 2013. Top 10 plant–parasitic nematodes in molecular plant pathology. Mol. Plant Pathol. 14, 946–961. Available from: https://doi.org/10.1111/mpp.12057.

Ju, Y., Pan, J., Wang, X., Zhang, H., 2014. Detection of *Bursaphelenchus xylophilus* infection in *Pinus massoniana* from hyperspectral data. Nematology 16 (10), 1197–1207. Available from: https://doi.org/10.1163/15685411-00002846.

Kim, S.-R., Lee, W.-K., Lim, C.-H., Kim, M., Kafatos, M.C., Lee, S.-H., et al., 2018. Hyperspectral analysis of pine wilt disease to determine an optimal detection index. Forests 9, 115. Available from: https://doi.org/10.3390/f9030115.

Kirkpatrick, T.L., Oosterhuis, D.M., Wullschleger, S.D., 1991. Interaction of *Meloidogyne incognita* and water stress in two cotton cultivars. J. Nematol. 23, 462–467.

Kiyohara, T., Tokushige, Y., 1971. Inoculation experiments of a nematode *Bursaphelenchus* sp. onto pine trees. J. Jpn. For. Soc. 53, 210–218.

Lawrence, G.W., McLean, K.S., Batson, W.E., Miller, D., Borbon, J.C., 1990. Response of *Rotylenchulus reniformis* to nematicide applications on cotton. Ann. Appl. Nematol. 22, 471–476.

Lawrence, G.W., Kelley, A.T., King, R.L., Vickery, J., Lee, H.K., McLean, K.S., 2004. Remote sensing and precision nematicide application of *Rotylenchulus reniformis* management in Mississippi cotton. Nematode Monogr. Perspect. 2, 13–21.

Mahlein, A.-K., 2016. Plant disease detection by imaging sensors—parallels and specific demands for precision agriculture and plant phenotyping. Plant Dis. 100 (2), 241–251.

Mahlein, A.-K., Oerke, E.-C., Steiner, U., Dehne, H.-W., 2012. Recent advances in sensing plant diseases for precision crop protection. Eur. J. Plant Pathol. 133, 197–209.

Melakeberhan, H., Webster, J.M., Brooke, R.C., D'Auria, J.M., Cackette, M., 1987. Effect of *Meloidogyne incognita* on plant nutrient concentration and its influence on the physiology of beans. J. Nematol. 19, 324–330.

Milford, G.F.J., Pocock, T.O., Riley, J., 1985. An analysis of leaf growth in sugar-beet. I. Leaf appearance and expansion in relation to temperature under controlled conditions. Ann. Appl. Biol. 106, 163–172. Available from: https://doi.org/10.1111/j.1744-7348.1985.

Mississippi State University, 2006. US 2006/0006335 A1, USA.

Monsanto Technology Llc., 2009. WO 2009/067622 A1, World Intellectual Property Organisation.

Monsanto Technology Llc., 2011. US 2011/0135161 A1, USA.

Monsanto Technology Llc., 2017. US 9842252 B2, USA.

Mota, M.M., Braasch, H., Bravo, M.A., Penas, A.C., Burgermeister, W., Metge, K., et al., 1999. First report of *Bursaphelenchus xylophilus* in Portugal and in Europe. Nematology 1, 727–734.

Nanjing Forestry University, 2014. CN 102507453 B, China.

Nicol, J., Turner, D., Coyne, L., den Nijs, L., Hockland, S., Maafi, Z., 2011. Current nematode threats to world agriculture. In: Jones, J., Gheysen, G., Fenoll, C. (Eds.), Genomics and Molecular Genetics of Plant–Nematode Interactions. Springer Science Business Media, Berlin, pp. 21–43.

Nicolas, H., Rivoal, R., Duchesne, J., Lili, Z., 1991. Detection of *Heterodera avenae* infestations on winter wheat by radiothermometry. Rev. de Nématologie 14, 285–290.

Norman, G.G., Fritz, N.L., 1965. Infrared photography as an indicator of disease and decline in citrus trees. Proc. Fla. State Horticult. Soc. 78, 59–63.

Nose, M., Shiraishi, S., 2007. Breeding for resistance to pine wilt disease. In: Futai, K., Sutherland, J. (Eds.), Pine Wilt Disease. Springer, Tokyo, pp. 334–350.

Nutter, F.W., 1990. Remote sensing and image analysis for crop loss assessment. Crop Loss Assessment in Rice. International Rice Research Institute, Manila, pp. 93–105, ISBN 971-22-0001-9.

Nutter, F.W., Tylka, G.L., Guan, J., Moreira, A.J.D., Marett, C.C., Rosburg, T.R., et al., 2002. Use of remote sensing to detect soybean cyst nematode-induced plant stress. J. Nematol. 34 (3), 222–231.

Orion, D., Nessim-Bistritsky, B., Hochberg, R., 1982. Using color infrared aerial photography to study cotton fields infested with *Meloidogyne incognita*. Plant Dis. 66 (2), 105–108.

Palacharla, P.K., Durbha, S.S., King, R.L., Gokaraju, B., Lawrence, G.W., 2011. A hyperspectral reflectance data based model inversion methodology to detect reniform nematodes in cotton. In: 6th International Workshop on the Analysis of Multi-temporal Remote Sensing Images (Multi-Temp), Trento, pp. 249–252. Available from: https://doi.org/10.1109/Multi-Temp.2011.6005095.

Pioneer Hi Bred Int., 2013. WO 2013/055657 A1, World Intellectual Property Organisation.

Pioneer Hi Bred Int., 2014. WO 2014/036281 A2, World Intellectual Property Organisation.

Pinter, P.J., Stanghellini, M.E., Reginato, R.J., Idso, S.B., Jenkins, A.D., Jackson, R.D., 1979. Remote detection of biological stresses in plants with infrared thermometry. Science 205 (4406), 585–587. Available from: https://doi.org/10.1126/science.205.4406.585.

Platt, H.M., 1994. Forward in phylogenetic systematics of free-living nematodes. In: Lorenzen, S. (Ed.), London, The Ray Society. p. 383.

REPHRAME, 2015. REPHRAME Project Final Report. <www.rephrame.eu> (accessed 14.02.19.).

Riker, A.J., Riker, R.S., 1936. Introduction to Research on Plant Stress—A Guide to the Principles and Practice for Studying Various Plant–Disease Problems. John C. Swift Co., Inc., Buffalo Grove, IL.

Rodrigues, M., 2008. National eradication programme for the pinewood nematode. In: Mota, M.M., Vieira, P. (Eds.), PineWilt Disease: A Worldwide Threat to Forest ecosystems. Springer, Kato Bunmeisha, Japan, pp. 5–14.

Rupe, J., Kirkpatrick, T., Bajwa, S., Cartwright, R., 2003. Application of Precision Agriculture Technology to Define and Manage Nematodes and Diseases of Soybean. Research Report. University of Arkansas and Arkansas Cooperative Extension Service, USA.

Sanberg, P.R., Gharib, M., Harker, P.T., Kaler, E.W., Marchase, R.B., Sands, T.D., et al., 2014. Changing the academic culture: valuing patents and commercialization toward tenure and career advancement. Proc. Natl Acad. Sci. U.S.A. 111 (18), 6542–6547. Available from: https://doi.org/10.1073/pnas.1404094111.

Sankaran, S., Mishra, A., Ehsani, R., Davis, C., 2010. A review of advanced techniques for detecting plant diseases. Comput. Electron. Agric. 72, 1–13.

Sasser, J., Freckman, D., 1987. A world perspective on nematology: the role of the society. In: Veech, J., Dickson, D. (Eds.), Vistas on Nematology. Society of Nematologists, Inc., Hyattsville, MD, pp. 7–20.

Shah, S.F.A., McKenzie, B.A., Gaunt, R.E., Marshall, J.W., Frampton, C.M., 2004. Effect of early blight (*Alternaria solani*) on healthy area duration and healthy area absorption of potatoes (*Solanum tuberosum*) grown in Canterbury New Zealand with different nitrogen application and stress from potato cyst nematode (*Globodera rostochiensis*). N. Zeal. J. Crop Horticult. Sci. 32 (1), 85–102. Available from: https://doi.org/10.1080/01140671.2004.9514283.

Sousa, D., Naves, P., Vieira, M., 2013. Prevention of pine wilt disease induced by *Bursaphelenchus xylophilus* and *Monochanmus galloprovincialis* by trunk injection of emamectin benzoate. Phytoparasitica 41, 143–148.

Steddom, K., Bredehoeft, M.W., Khan, M., Rush, C.M., 2005. Comparison of visual and mutispectral radiometric disease evaluations of *Cercospora* leaf spot of sugar beet. Plant Dis. 89, 153–158.

Steele, M.R., Gitelson, A.A., Rundquist, D.C., Merzlyak, M.N., 2009. Nondestructive estimation of anthocyanin content in grapevine leaves. Am. J. Enol. Viticult. 60, 87–92.

Stephens, M., Wood, G.A., Taylor, J.C., 2000. Opportunities for targeted control of potato cyst nematode using remote sensing. ASP Appl. Biol. 60, 213–218.

Strajnar, P., Širca, S., Urek, G., Šircelj, H., Železnik, P., Vodnik, D., 2012. Effect of *Meloidogyne ethiopica* parasitism on water management and physiological stress in tomato. Eur. J. Plant Pathol. 132, 49–57.

Strajnar, P., Žibrat, U., Knapič, M., Širca, S., Gerič Stare, B., Urek, G., 2015. Hyperspectral remote sensing as a new approach for early root-knot nematode detection. In: Širca, S., Gerič Stare, B., Razinger, J. (Eds.), Plant Health for Sustainable Agriculture: Book of Abstracts. Agricultural Institute of Slovenia, Ljubljana, p. 120.

Susič, N., Žibrat, U., Širca, S., Strajnar, P., Razinger, J., Knapič, M., et al., 2018. Discrimination between abiotic and biotic drought stress in tomatoes using hyperspectral imaging. Sens. Actuators B: Chem. 273, 842–852. Available from: https://doi.org/10.1016/j.snb.2018.06.121.

Suzuki, K., 1984. General Effect of Water Stress on the Development of Pine Wilting Disease Caused by *Bursaphelenchus xylophilus*. Bulletin, Forestry and Forest Products Research Institute, Japan, pp. 97–126.

Tecnologia Marina Ximo, 2015. ES 2552405 A1, Spain.

University of Sydney, 2015. AU 2015/362069 A1. Australia.

Voigt, C.A., 2014. Callose-mediated resistance to pathogenic intruders in plant defence-related papillae. Front. plant Sci. 5, 168. Available from: https://doi.org/10.3389/fpls.2014.00168.

West, J.S., Bravo, C., Obert, R., Moshou, D., Ramon, H., McCartney, H.A., 2010. Detection of fungal diseases optically and pathogen inoculum by air sampling. In: Oerke, E.-C., Gerhards, R., Menz, G., Sikora, R.A. (Eds.), Precision Crop Protection—The Challenge and Use of Heterogeneity. Springer, The Netherlands, pp. 135–149.

Wheeler, T.A., Kaufman, H.W., 2003. Relationship of aerial broad band reflectance to *Meloidogyne incognita* density in cotton. J. Nematol. 35 (1), 48–57.

Žibrat, U., Susič, N., Knapič, M., Širca, S., Strajnar, P., Razinger, J., et al., 2019. Pipeline for imaging, extraction, pre-processing, and processing of time-series hyperspectral data for discriminating drought stress origin in tomatoes. Methods X 6, 399–408. Available from: https://doi.org/10.1016/j.mex.2019.02.022.

19

Evaluating the performance of vegetation indices for detecting oil pollution effects on vegetation using hyperspectral (Hyperion EO-1) and multispectral (Sentinel-2A) data in the Niger Delta

Nkeiruka N. Onyia[1], Heiko Balzter[2,3], Juan Carlos Berrío[3]

[1] CENTRE FOR LANDSCAPE AND CLIMATE RESEARCH, SCHOOL OF GEOGRAPHY, GEOLOGY AND THE ENVIRONMENT, UNIVERSITY OF LEICESTER, LEICESTER, UNITED KINGDOM

[2] NATIONAL CENTRE FOR EARTH OBSERVATION, UNIVERSITY OF LEICESTER, UNIVERSITY ROAD, LEICESTER, UNITED KINGDOM

[3] SCHOOL OF GEOGRAPHY, GEOLOGY AND ENVIRONMENT, UNIVERSITY OF LEICESTER, LEICESTER, UNITED KINGDOM

19.1 Introduction

Oil exploration in the Niger Delta began in the 1950s (Ogri, 2001; Aroh et al., 2010) and encompassed drilling, production, transportation, processing, and storage. Oyinloye and Olamiju (2013) and other researchers noted that oil production accounts for up to 96% of the Nigerian national economy. This growth has been at an enormous cost to the environment of the Niger Delta. Oil spillages, gas flaring, inappropriate waste disposal (solids and liquids), discharge of toxic chemicals, land use changes including forest fragmentation and degradation, flooding and soil erosion, and so on adversely impact on the environment (Emoyan et al., 2008; Opukri and Ibaba, 2008). Evidence of oil spill impacts on the mangrove forest within the Niger Delta abound. Corcoran et al. (2007) reported that the mangrove within the Niger Delta has been reduced by up to 26% since the oil boom, whereas Osuji et al. (2004) and United Nations Environmental Programme (2011) revealed the disastrous impacts of oil pollution on mangrove and other vegetation, ranging from extreme stress to destruction.

Hyperspectral Remote Sensing. DOI: https://doi.org/10.1016/B978-0-08-102894-0.00018-8

In a well-endowed biodiversity hotspot like Rivers State in Nigeria that is constantly subjected to oil spillages, biodiversity monitoring is crucial for several reasons including the need to meet the strategic goals of the Aichi 2020 Biodiversity Targets (Turnhout et al., 2017; Convention on Biodiversity, 2010). However, the dearth of baseline data and difficulties associated with field data collection have constantly sabotaged efforts to meet these targets. The Sixth National Report for the Convention on Biodiversity (CBD, 2018a,b) identified the lack of trained and skilled workforce, appropriate technologies, and adequate funding as significant constraints of biodiversity conservation in Nigeria. In the Nigerian National Biodiversity Strategy and Action Plan (NBSAP) report, the Federal Ministry of Environment proposed the assessment of biodiversity status at appropriate spatial scales to enhance biodiversity conservation (CBD, 2017). The government relies on data from conventional survey methods as well as secondary data from research publications to achieve this. However, due to constraints of time, equipment, and accessibility, conventional field survey methods seldom provide sufficient baseline data for implementing conservation policies (Thomson and Willerslev, 2015; Mulatu et al., 2017).

The integration of remote sensing data with field data for biodiversity monitoring and conservation has gained momentum, particularly in advanced countries where high-quality remote sensing datasets are available. Several successful ecological and habitat models such as MAXENT, habitat quality index, environmental niche models rely on a combination of remotely sensed and field data (Phillips et al., 2006). For developing and poorer countries such as Nigeria to achieve their biodiversity targets, the availability of free or low-cost satellite data is imperative. Such valuable resources will ensure the allocation of scarce funds toward meeting the needs of other crucial aspects of the NBSAPs such as creating awareness among the population on the importance of biodiversity.

Presently, few optical satellite sensors with global orbitals provide their data for free. These include MS sensors such as ESA's Sentinel-2A and B (available and downloaded from the Copernicus Services Data Hub) and HS sensors such as Hyperion EO-1 [available and downloaded from the United States Geological Survey (USGS) data distribution repository]. The spatial and spectral resolutions of these sensors vary and this affects their application in vegetation studies.

This study compared the performance of free satellite datasets characterized by high spectral resolution (Hyperion) on the one hand and high spatial resolution (Sentinel-2A) on the other hand to determine the best value open source imagery for vegetation study in the data scarce, oil-polluted Niger Delta region. The study also evaluated the performance of combined HS and MS datasets (fused data; FS). Based on the attributes of these datasets, HS data are expected to show strong sensitivity to the biochemical parameters (leaf pigments) of vegetation, whereas the high spatial resolution MS data are expected to should show strong sensitivity to the abundance and composition of vegetation. The FS data are expected to outperform both the MS and HS datasets.

This study aims to identify the most useful free satellite data for studying the VR to oil pollution in Nigeria. To achieve this, several questions are asked including:

1. Can polluted and unpolluted (control) vegetation be differentiated using vegetation indices (VIs) derived from satellite data?

2. Is the performance of VIs dependent on sensor type (MS or HS)?
3. Is the performance of VIs improved by the fusion of MS and HS data?

19.2 Materials and methods

19.2.1 Study area

The study area is an oil contaminated site located on the Shell Petroleum Development Corporation, Nkpoku-Bomu Pipeline in Kporghor, Tai Local Government Area of Rivers State in Nigeria (Fig. 19–1). The scarcity of HS satellite data for the study area limited the investigation to this location.

Kporghor is in the Niger Delta region of Nigeria, which is one of the most extensive wetlands (Agbagwa and Ndukwu, 2014). Yearly rainfall of up to 3000 mm (Omo-Irabor et al., 2011), a temperature range of 20°C–30°C, and humidity above 75% (Elenwo and Akankali, 2014) define the environment. High cloud cover affects satellite image quality (Omo-Irabor et al., 2011) and limits data availability. Vegetation growth is constant throughout the year, although agricultural activities occur between March and October.

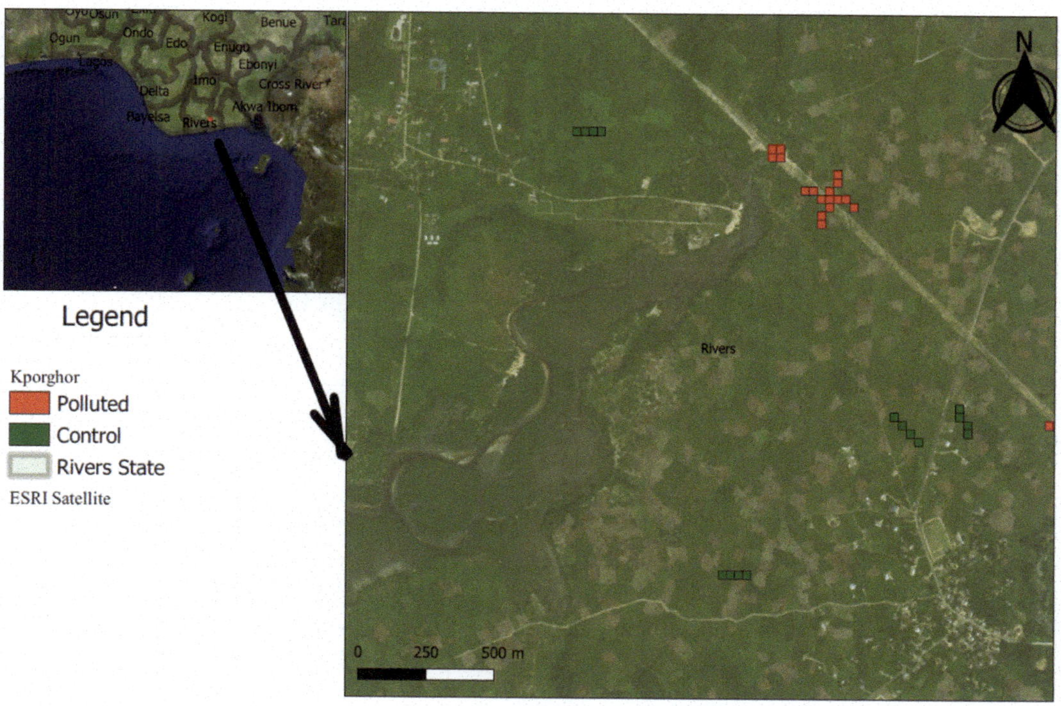

Oil spill location in Kporghor town, Rivers State of Nigeria

FIGURE 19–1 (A) Rivers State overlaid with the Hyperion image strip. (B) Study area in Tai Local Government Area. (C) Hyperion image subset of the study area in false color composite (red = band 20, green = band 36, blue = band 45) showing locations of polluted and control transects and other features.

The ecology of Kporghor is that of a coastal plain and freshwater rainforest populated by forest tree species, palms, shrubs, ferns, lianas, and so on (Ugochukwu and Ertel, 2008; World Bank, 1995). Dominant species in the study site include oil palm (*Elaeis guineensis*), raphia palm (*Raphia vinifera*), and timber species (*Symphonia globulifera*) (Ugochukwu and Ertel, 2008). Others are *Eichornia crassipies* (Ebigwa and Akomaye, 2014), *Calamus* and *Alchornea* species, *Alstonia boonei*, and *Berlinia* species (Adegbehin and Nwaigbo, 1990). Osuji and Ezebuiro (Osuji and Ezebuiro, 2006) reported the presence of *Antidesma* species, *Paulina pinnata*, *Ouretea* species, *Chassalia* species, *Cuvaria* species, *Dryopteris* species, *Memsecylon* species, *Blackoides* species, *Agelea oblique*, and *Psychotria manii*.

The topography of the area is generally flat with an elevation range of 14–25 m above mean sea level (Musa et al., 2014). The location of the spill epicenter (SS0) is 4.71842°N latitude and 7.22523°E longitude (WGS 84 Datum). The Nigerian Oil Spill Monitor website (NOSDRA, 2015) reported that a spill incident occurred on September 12, 2015 following illegal bunkering activities. About 864 barrels of crude oil (137,365 L) spilt onto a land area of 4190 m^2 out of which only about 557 barrels (88,556 L) were recovered through the containment measures implemented by the company.

19.2.2 Establishment of study transects

The study involved polluted and unpolluted (control) locations. The selection of these locations from the Nigerian Oil Spill Records website (NOSDRA, 2015) was based on satellite data availability, particularly HS data. Four transects were established on the polluted location (polluted transects), and four on the unpolluted locations (control transects) within Kporghor town. The polluted and control transects were within the same ecological zone having similar environmental variables (Ugochukwu and Ertel, 2008; World Bank, 1995). The only distinguishing factor is the concentration of petroleum hydrocarbon in the soil samples collected from the investigated transects.

Transects were 100 m long and transversed the polluted location in order to study the effect of crude oil on vegetation composition. Polluted transects labeled A, B, C, and D originated from the spill epicenter, SS0, and proceeded in the four cardinal directions, respectively. The location of the control transects was random in unpolluted areas. The randomization of the control transects was to minimize error from bias.

In addition to the spill epicenter, SS0, each transect was subdivided into five segments of 20 m length labeled SS1 to SS5, hence, there were 41 segments in total; 21 on polluted transects and 20 on control transects. For each segment, the species richness and diversity (α) values were recorded along with vegetation abundance, leaf chlorophyll content, and the total petroleum hydrocarbon (TPH) concentration in the soil. However, to correct spatial autocorrelation effects, only data from 25 segments (13 polluted and 12 control) were analyzed in this study.

19.2.3 Field survey of vascular plant species

A vegetation survey following the line-intercept method discussed by Cummings and Smith (2000) was conducted in February and March 2016. Due to the unique circumstances prevailing in the study area during this campaign, the line-intercept method offered the most potential at capturing adequate field data to determine species composition and abundance. Cummings and Smith (2000) and Buckland et al. (2007) agree that vegetation sampling based on the line intercept method, although less intensive than quadrant-based sampling methods, provides sufficient data to determine relative estimates of vegetation frequency, coverage, abundance, and other relevant vegetation characteristics. Furthermore, according to Cummings and Smith (2000) and Kercher et al. (2003), the line intercept method offers a handy tool for detecting community transition (such as ecotones) and ecological gradients in habitats. It is efficient for assessing the influence of environmental variables on species compositions with minimal error.

At each segment of polluted and control transects, the frequency and abundance of the occurring species were recorded. The identification of species was facilitated using species inventory sheets compiled from previous literature (Ubom, 2010; World Bank, 1995; Ebigwa and Akomaye, 2014; Ugochukwu and Ertel, 2008; Agbagwa and Ekeke, 2011). In order to cause as little damage as possible to the ecosystem, photographs of unidentified species were studied at herbariums in the local university and research center (College of Natural Resources and Environmental Management, Michael Okpara University of Agriculture and National Root Crops Research Centre, both in Umudike Umuahia, Nigeria) for identification. Although the location of both polluted and control transects was within the same ecological zone, a similarity analysis was performed to determine the degree of agreement between the species composition of both sites using the Sorenson similarity index. The Sorenson index is used by ecologists to compare ecological units (Sattout and Caligari, 2011; Petrovic et al., 2010). In this study, the similarity of species composition on segments of polluted and control transects are determined.

Furthermore, the species richness of the investigated transects was determined using Menhinick's and Chao1 richness indices. Species richness refers to the number of species present in a given area without reference to the abundance or distribution pattern of the species (Hurlbert, 1971). It is one of the cardinal concepts in ecology and the most commonly used index of species diversity in a given habitat (Peet, 1974). Species richness is reportedly high in tropical rainforests and positively correlated to the size of the area studied (Vasilevich, 2009). The Sorenson similarity index of transects, Menhinick's species richness index, and Chao1 richness index were computed in the Paleontological Statistics Software (Hammer et al., 2001) for selected segments.

19.2.4 Soil sampling and analysis

Composite soil samples collected from segments of the investigated transects were taken to the laboratory for analysis. At each segment, a composite soil mixture from three randomly collected samples was stored in sterilized and labeled airtight glass containers and transported in cooler boxes to the laboratory. The analysis of the soil samples was done according to national and international quality standards. The method adopted for soil TPH determination was method 8015 of the United States Environmental Protection Agency, which involves a gas chromatography (GC) technique using flame ionization detector (FID) electrodes (USEPA, 2003).

19.2.5 Data acquisition and preprocessing

19.2.5.1 Hyperspectral (Hyperion) image

The HS image was acquired over the study area on November 23, 2015, by the Hyperion sensor onboard the Earth Orbiter 1 (EO-1) satellite launched in November 2000 following a data acquisition request. On acquisition, the image was processed to level 1, which includes geometric and radiometric corrections (Barry, 2001; Pal and Porwal, 2015). The level 1T product downloaded from the USGS website using the Earth-explorer data tool had only 196 useful bands written as a 16-bit signed integer in scaled radiance units ($W/m^2/\mu m/sr$). The radiance values were converted into surface reflectance using the Fast line-of-sight Atmospheric Analysis of Hypercubes (FLAASH) module in Environment for Visualizing Images (ENVI) software version 4.4 to remove atmospheric interferences from aerosols, vapour and gases (Scheffler and Karrasch, 2013; Adler_Golden et al., 2013). Minimum noise fraction transformation was performed to address internal sensor defects such as smile and scan lines (stripes), consequently, increasing the signal to noise ratio (Goodenough et al., 2003; Datt et al., 2003; Dadon et al., 2010). Additionally, the smile effect was removed using the cross-track illumination correction procedure in ENVI 4.4. Further processing of the Hyperion image included spatial and spectral subsetting to reduce computational time. The resulting product comprised 68 columns, 66 lines, 4,488 pixels, and 164 spectral bands, although only wavelengths in the visible and near infrared (VNIR) channels were used in the subsequent analysis.

19.2.5.2 Multispectral (Sentinel-2A) image

The acquisition of the MS image over the study area was on December 22, 2015, approximately 1 month after the Hyperion overpass. The MS imager on board the Sentinel-2A satellite launched in 2015 by the European Space Agency (ESA) acquired images in 13 spectral bands with wavelengths ranging from 443 to 2190 nm. Spatially, Sentinel-2A images boast of 10 m (VNIR), 20 m (red edge), and 60 m (not used in this study) resolutions. These are some of the best spatial resolutions for freely available satellite data. Spectral resolution varies according to the band, but ranges between 15 nm (red edge) and 115 nm (near infrared; NIR). A level 1C (geometrically and radiometrically corrected) image was downloaded from the Copernicus Services Data Hub website (https://cophub.copernicus.eu/) and subjected to atmospheric correction using the Sen2Cor plugin tool in the ESA image processing platform for sentinel data (ESA SNAP).

Specifications of the Hyperion EO-1 and Sentinel-2A images used in this study are shown in Table 19−1.

19.2.6 Image to image registration

The two images were geometrically aligned to enhance comparison. This procedure was performed automatically in ENVI 5.3 using the Image Registration Workflow tool. Image registration ensures that corresponding pixels of two or more images represent the same object on the ground surface (Brown, 1992; Khanna et al., 2018). This automated procedure

Table 19–1 Specifications of HS (Hyperion) and MS (Sentinel-2A) images used in this study.

Attributes	Hyperspectral	Multispectral
Sensor	Hyperion EO-1	Sentinel-2A
Spectral channels	220	13
Spatial resolution	30	10 and 20
Radiometric resolution	16	12
Wavelength range (nm)	427.55–2395.53	443–2190
Date of acquisition	November 23, 2015	December 22, 2015

involves finding the optimal transformation, which generates tie points that correlate to corresponding points in the second image with minimal pixel error. For this study, first-order polynomial transformation was used to map the location of the tie points with a minimum matching score of 0.7 and a maximum allowable error of 1 pixel. One hundred tie points were initially selected automatically, but this was reduced to 25 by manual selection of points with the least root mean square error (RMSE). The overall RMSE after registration of the two images was 0.44. The output image was resampled to the spatial resolution of the base image (10 m) using bilinear interpolation.

19.2.7 Gram–Schmidt fusion of multispectral and hyperspectral images

The HS and MS images were fused together to create a synthetic image with improved spatial resolution and preserved spectral information. Data fusion is a technique that combines two or more images from different sensors using advanced image processing tools and is widely used by remote scientists to extract information from satellite images (Pohl and Genderen, 1998; Pande et al., 2009; Yakoya et al., 2016). This procedure is useful in several applications including in ecosystem monitoring and the identification of plant species at large spatial scales (Yakoya et al., 2016). There are several image fusion techniques discussed in the literature, Yakoya et al. (2016) compared several of these techniques in their article. However, in this study, the Gram–Schmidt procedure was applied to the coregistered images in ENVI 5.3. Each of the eight hyperspatial Sentinel-2A VNIR bands was used to pansharpen the corresponding range of the HS Hyperion bands. The appropriate HS wavelength range was determined by combining the MS central wavelength and bandwidth for each band, as shown in Eq. (19.1).

$$\text{HS wavelength range} = \text{MS central wavelength} \pm \text{bandwidth} \tag{19.1}$$

The Hyperion wavelength ranges selected for the pansharpening procedure using individual Sentinel-2A bands are listed in Table 19–2.

The quality of the fused image was assessed visually by comparing the color composites of the original and fused images and statistically by computing the bias of the mean and

Table 19–2 Selected range of Hyperion wavelengths pansharpened using individual Sentinel-2A bands determined using Eq. (19.1).

Spectral channel	MS central Wavelength (nm)	MS bandwidth (nm)	HS wavelength Range (nm)
Blue	496.6	98	426–594.9
Green	560	45	515–605
Red	664.5	38	626.5–702.5
NIR	835.1	145	690.1–980.1
RE1	703.9	19	684.9–722.9
RE2	740.2	18	722.2–758.2
RE3	782.5	28	754.5–810.5
RE4	864.8	33	831.8–897.8

HS, Hyperspectral; *MS*, multispectral; *NIR*, near infrared; *RE*, red edge .

correlation coefficient (r) of both images. The mean bias is the normalized difference in the means of the original and fused images (Abdikan et al., 2012), while r measures the correlation between the original and fused images. The ideal value for mean bias is zero, while higher r values (up to 1) indicate a good correlation between the spectral values of both images.

19.2.8 Vegetation indices

VIs to detect canopy chlorophyll content and stress pigments were derived from the three datasets, namely HS, MS, and FS. The indices calculated include:

1. Normalized difference vegetation index (NDVI), which is a broadband greenness index computed from the Sentinel-2A image.
2. Red-edge normalized difference vegetation index (RENDVI), which is a narrowband equivalent of the NDVI. This index was computed using the Hyperion and FS datasets.
3. Red-edge position index (REP) , which was manually computed as the maximum first derivative of reflectance at Hyperion wavelengths between 671 and 782 nm (Clark et al., 2010). It is a chlorophyll related index with reduced sensitivity to variations in leaf/ canopy chlorophyll content as well as environmental conditions (Gholizadeh et al., 2016).
4. Anthocyanin reflectance index 2 (ARI2), which was used to estimate the concentration of anthocyanins in the leaf canopy. This index is often used to detect plant stress.
5. Structural insensitive pigment index (SIPI), which maximizes sensitivity to carotenoids while minimizing sensitivity to variation in canopy structure (Peñuelas et al., 1993). The index is commonly used for plant physiological stress detection.

ARI2 and SIPI are used to detect stressed and unhealthy vegetation due to their ability to respond to changes in plant physiological status. Past studies have shown that the reflectance at the absorption maxima of these pigments decreases in stressed vegetation (Merzylak et al., 2008). These indices were automatically computed in ENVI 5.4 except the

Table 19–3 Vegetation indices analyzed in this study with the formulae and references.

Index	Formula	References
Normalized difference vegetation index	$\frac{n-r}{n+r}$	Pearson and Miller (1972)
Red-edge normalized difference vegetation index	$\frac{R750-R705}{R705+R750}$	Gitelson and Merzlyak (1996)
Red-edge position index	$(705+35)\left[\frac{((R783-R665)/2)-R705}{R740-R705}\right]$	Gholizadeh et al. (2016)
REP for hyperspectral and fused images	Eqs. (19.2) and (19.3)	Savitzky (1964)
Anthocyanin reflectance index 2	$R800\left(\frac{1}{R550}-\frac{1}{R700}\right)$	Gitelson et al. (2001)
Structure insensitive pigment index	$\frac{R800-R445}{R800-R680}$	Peñuelas et al. (1995)

REP, which was calculated in Excel for HS and FS, and in SNAP for MS. The VIs, formulae, and references used in this study are shown in Table 19–3.

The computation of REP from the wavelengths of the HS and FS images was manually performed in Excel using the formula proposed by Savitzky (1964) as shown here:

$$\text{REP} = \max\frac{dy}{dx} \tag{19.2}$$

$$\frac{dy}{dx} = \frac{R_i - R_j}{Y_i - Y_j} \tag{19.3}$$

where R_i is the reflectance at Y_i; R_j is the reflectance at Y_j; Y_i is the wavelength at the start of the slope segment; Y_j is the wavelength at the end of the slope segment.

19.2.9 Vegetation response to oil pollution

Plant responses to oil pollution can vary in severity depending on certain factors including life form (herbs, shrubs, trees, or graminoids), lifecycle (annuals, biennials, or perennials), and volume of the oil spill. Osam et al. (2011), Gighi et al. (2012), and Tanee and Albert (2015) reported the devastating effects of oil pollution on the ecosystem and its floristic and faunistic distribution. Kinako (1981) observed a species loss of at least 50% in oil-polluted habitats, while Baruah and Sarma (1996) reported that 80% of plant species were lost in crude oil spill sites. Similarly, Nkwocha and Duru (2010) and Tanee and Albert (2015) revealed the adverse effects of oil pollution on the abundance and density of vegetation in the Niger Delta region. Additionally, Lin and Mendelssohn (2012) reported significantly lower levels of life above ground biomass (a measure of abundance) at oil-impacted sites than at reference sites in their study of the impacts of an oil spill on vegetation structure in the coastal salt marshes of the northern part of the Gulf of Mexico.

In this study, plant response to oil pollution was determined by assessing the physical and biochemical parameters of vegetation. These parameters, which include vegetation abundance, species richness, and leaf chlorophyll (estimated using a hand-held chlorophyll meter), were measured on polluted and control transects for comparison.

$$\text{Vegetation abundance} = \frac{\text{sum}(S_i)}{\text{sum}(SS_i)} \tag{19.4}$$

$$\text{Menhinick's richness index} = \frac{S}{\sqrt{N}} \tag{19.5}$$

where S_i is the individuals of species i; SS_i is the segments in which species i occurred; N is the total number of individuals in a segment; S is the number of species observed.

19.2.10 Statistical analysis

Due to the small sample size of the data ($N = 29$), and the non-normal distribution of the data-set, the analysis was performed using nonparametric statistical procedures. First, a test for differences in the median was performed to compare the index values of the polluted and control transects. VIs extracted from Hyperion (HS), Sentinel-2A (MS), and FS were subjected to a Mann–Whitney U test to check for differences between the polluted and control transects. The research hypothesis that a random pixel from the control transects has significantly different ($\alpha = 0.05$) index values from a polluted pixel is tested against a null hypothesis of no difference (both groups have the same index values). The U test was performed in R using the formula:

$$U = \min(U1, U2) \tag{19.6}$$

$$U_1 = n1n2 + \frac{1}{2}n1(n1 + 1) - \text{sum}(R1) \tag{19.7}$$

$$U_2 = n1n2 + \frac{1}{2}n2(n2 + 1) - \text{sum}(R2) \tag{19.8}$$

where U is the Mann–Whitney test statistic; n is the size of sample groups; R is the ranks for sample groups.

Second, a nonparametric effect size measure was employed to compare the performance of the indices derived from MS, HS, and FS datasets in detecting VR to oil pollution. Researchers report aspects of their experimental results that are not obvious from P-values (Macbeth et al., 2010), for instance, the magnitude and direction of the difference using effect size measures (ESM). Khanna et al. (2018) used a parametric ESM (Cohen's d) to identify the set of sensor-based indices that best differentiated between healthy and stressed vegetation in an oil-polluted location. In this study, the Cliff delta metric was used to measure effect size because limitations of data normality, homogeneity of variance, and outliers do not constrain its use (Kromrey et al., 2005). Hogarty and Kromrey (2001) reported that the Cliff's delta metric provided the least bias estimate of effect size out of several indices that they evaluated. According to Cliff (1996), the statistic relies on the concept of dominance analysis, which refers to the degree of overlap between two groups. The greater the overlap, the lower the dominance and vice versa. The metric ranges between −1, when all the values of one group are lower than the values of the other group, and +1 when the reverse is true. A Cliff delta of 0 indicates complete overlap and no difference between the two groups (Macbeth et al., 2010). Since the metric is a standard unit, it is suitable for multiple comparisons of variables and sensors (Khanna et al., 2018). Cliff's delta was computed in *R* from the *U* test score using the formula:

$$\delta = 1 - \frac{2U}{n1n2} \tag{19.9}$$

where U is the Mann–Whitney test statistic; n is the size of sample groups.

Finally, the relationship between the VIs and VRs was investigated using the Spearman's rank correlation coefficient. The VIs with the largest effect size were selected for this procedure in order to identify the VR parameter most affected by oil pollution. A strong positive correlation was expected between chlorophyll VIs and field chlorophyll data, vegetation abundance and species richness, whereas the reverse was expected between these parameters and stress indices. A combination of the correlation coefficient and effect size determined the parameter most affected by oil pollution.

19.3 Results

19.3.1 Total petroleum hydrocarbon concentration in soil

The results of the soil sample analysis showed significant differences in the concentrations of TPH in the polluted and control transects. The highest concentration of TPH (16,063 mg/kg) was found in soil collected at the spill epicenter (SS0). This value was over 90% higher than the average TPH concentration from other segments of both polluted and control transects. The median (mg/kg), mean (mg/kg), and standard error (SE) of the TPH concentrations, respectively, were in Transect $A = 767$, 973.33, and 349.57; in Transect $B = 996$, 895.67, and 304.1; in Transect $C = 875$, 744, and 160.87; and in Transect $D = 1030$, 980, and 247.22. These values were well above the recommended Environmental Guidelines and Standards for the Petroleum Industry in Nigeria (EGASPIN) intervention values. The EGASPIN intervention values (mg/kg of soil) documented for different aromatic hydrocarbons that make up crude oil range from 1 for benzene to 130 for toluene (DPR, 2002).

Conversely, the TPH concentration levels were low at the control transect. The median (mg/kg), mean (mg/kg), and SE of mean concentration at Control 1 were 26.25, 26.73, and 0.944611913, respectively. Similarly, at Control 2, the median, mean, and SE values were 19.75, 20.2, and 2.66, respectively; whereas, on Control 3, the values were 44.7, 43.6, and 8.28, respectively. Finally, at Control 4, the median, mean, and SE values were 35.5, 36.25, and 3.75, respectively. While these values are below the EGASPIN intervention values, they are above the recommended target of 0.05 mg/kg in soils, suggesting that even the control transects were somewhat polluted.

19.3.2 Registration of images

The registration of the HS image (warp image) to the MS image (base image) was successful with a total RMSE of 0.44. The selection of 25 evenly distributed ground control points (GCPs) depended on the RMSE values, with GCPs with lower RMSE values being preferred. Details of the GCPs locations on both images, predicted locations, and error for each X and Y pair are shown in Table 19–4.

Table 19–4 Attributes of selected GCPs filtered from 100 autogenerated GCPs using the image registration workflow tool in ENVI 5.3. Number of GCPs selected = 25, selection manually performed based on tie-points with lowest root mean square error (RMSE). Overall RMSE = 0.44.

ENVI ground control points table								
ENVI image to image GCP table								
Total RMSE: 0.44								
Warp image		**Base image**		**Predict**		**Error**		**RMSE**
X	Y	X	Y	X	Y	X	Y	
128	19	997.34	3400.33	997.42	3400.11	0.08	− 0.23	0.24
130	37	997.67	3406.33	997.7	3406.74	0.03	0.41	0.41
113	38	992.01	3407.34	991.84	3406.91	− 0.17	− 0.42	0.46
145	141	1000.34	3445	1000.27	3445.03	− 0.07	0.03	0.07
167	122	1008.01	3438.67	1008.03	3438.22	0.02	− 0.45	0.45
55	69	972.05	3417.69	971.56	3417.75	− 0.49	0.06	0.49
24	36	961.26	3404.92	961.29	3405.11	0.03	0.19	0.19
127	21	996.77	3401.06	997.03	3400.83	0.25	− 0.23	0.34
140	36	1001.02	3405.99	1001.16	3406.5	0.14	0.51	0.53
144	67	1001.72	3417.64	1001.77	3417.91	0.05	0.27	0.27
38	102	965.43	3430.02	965.5	3429.88	0.07	− 0.14	0.16
58	100	972.06	3429.36	972.2	3429.3	0.14	− 0.07	0.15
59	164	972.04	3453.06	971.75	3453.08	− 0.29	0.03	0.29
144	102	1001.02	3431.39	1000.91	3430.73	− 0.12	− 0.65	0.66
43	85	967.08	3423.35	967.34	3423.59	0.25	0.24	0.35
128	19	997.25	3400.26	997.42	3400.11	0.16	− 0.15	0.22
130	37	997.46	3406.29	997.7	3406.74	0.24	0.45	0.51
111	22	992.24	3401.02	991.47	3400.99	− 0.77	− 0.03	0.77
145	68	1002.1	3418.01	1002.08	3418.28	− 0.02	0.27	0.27
113	38	992.09	3407.38	991.84	3406.91	− 0.25	− 0.47	0.53
59	100	972.41	3429.72	972.54	3429.31	0.13	− 0.42	0.44
78	100	978.08	3429.35	978.89	3429.46	0.81	0.12	0.82
94	162	983.41	3452.72	983.16	3452.5	− 0.24	− 0.22	0.33
113	134	990.1	3441.61	989.91	3442.27	− 0.2	0.66	0.68
148	166	1000.42	3453.96	1000.63	3454.2	0.2	0.24	0.32

19.3.3 Fusion of multispectral and hyperspectral images

The image fusion combined the single Sentinel-2A bands with a corresponding range of Hyperion bands to form a fused hyperspectral and hyperspatial (FS) image, which kept both spatial information from the high-resolution MS bands and the spectral characteristics of the lower resolution HS wavelengths. A visual comparison of the spectral profiles (Fig. 19–2) and color composite images (Fig. 19–3) of the fused and original images demonstrate that the fusion procedure was successful. Fig. 19–2 shows the spectral similarity of the polluted and control pixels from the HS and FS images. The mean bias of the FS image with respect to the HS image was 0.026, while the correlation coefficient for the different channels was also high ranging from 0.4 to 0.8. The highest correlation coefficient was from the NIR channel.

19.3.4 Effect of oil pollution on vegetation indices

Oil pollution affected the VI values as significant differences occurred between the polluted and control transects. This difference was especially true for NDVI computed from the MS image, but not for RENDVI from the HS and FS images. Conversely, the REPs from the HS and FS images appear significantly different between the polluted and control transects. Vegetation stress indicated by ARI2 and SIPI did not show a significant difference between the polluted and control transects, except for HS-based ARI2. Table 19–5 shows significantly different variables.

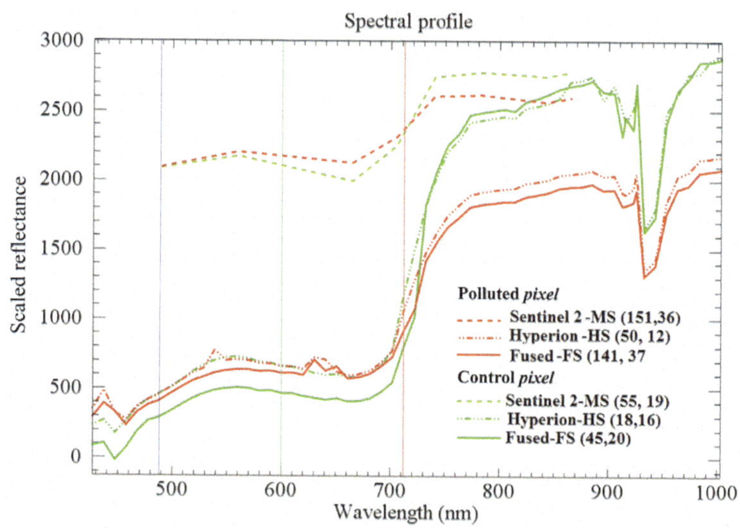

FIGURE 19–2 Spectral profile of polluted and control pixels from three sensors. The chart shows strong similarity in the spectral curves of polluted and control pixels from hyperspectral (HS) and fused HS and multispectral images.

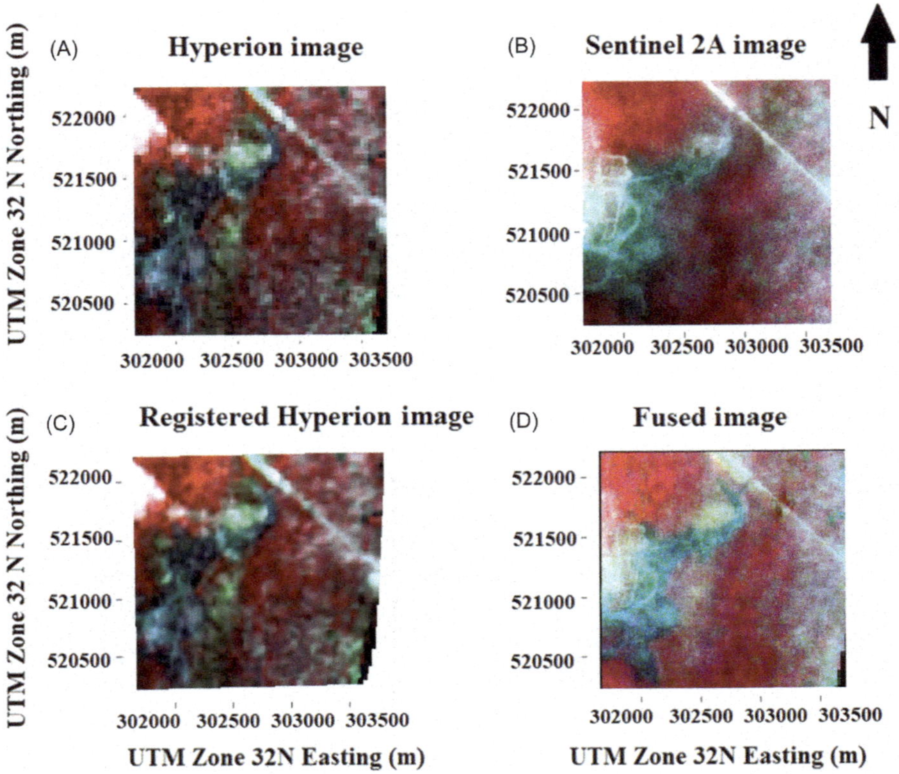

FIGURES 19–3 Composite plots of (A) hyperspectral (HS; Hyperion), (B) multispectral (MS; Sentinel-2A, hyperspatial), (C) registered Hyperion image, and (D) FS (fused HS and MS images). Both HS and FS images are plotted using bands 49, 28, and 10 for red, green, and blue (RGB) channels, respectively. The MS image is plotted using bands 8, 4, and 3 for RGB, respectively.

19.3.5 Response of vegetation parameters to oil pollution

VR from the polluted transects was more acute than that from the control transects. Vegetation abundance, species richness, and leaf chlorophyll significantly decreased in the polluted transects. The median values of these parameters are shown in Table 19–6.

A nonparametric regression analysis of the data revealed an inverse relationship between soil TPH concentrations and each of the vegetation parameters. The coefficient of determination (R-squared) values, which measure the degree to which variations in the parameters are explained by variations in soil TPH, were 0.99, 0.99, and 0.28, respectively, for leaf chlorophyll, species richness, and abundance. Interestingly, this pattern was observed along polluted transects as the soil TPH concentration reduced. The regression plots are shown in Fig. 19–4.

Table 19–5 Mann–Whitney *U* test results for differences in vegetation indices between polluted and control transects ($N = 29$; Median1 and Median2 = median values of control and polluted transects, respectively).

Index	Median1 (n=16)	Median 2 (n=13)	U	p-value
$NDVI_{ms}$	0.12	0.11	47	<0.05
$RENDVI_{hs}$	0.41	0.38	88	ns
$RENDVI_{fs}$	0.441	0.38	70	ns
REP_{ms}	51.22	50.13	104	ns
REP_{hs}	30.87	27.37	39	<0.05
REP_{fs}	45.68	28.49	37	<0.05
$ARI2_{ms}$	-1.29	-1.28	89	ns
$ARI2_{hs}$	0.23	0.31	54	<0.05
$ARI2_{fs}$	0.35	0.37	90	ns
$SIPI_{ms}$	-0.06	-0.004	69	ns
$SIPI_{hs}$	1.31	1.38	65	ns
$SIPI_{fs}$	1.31	1.31	94	ns

Table 19–6 Median values of parameters measuring vegetation response to oil pollution.

Response parameter	Polluted	Control
Menhinick's index	2.83	3.26
Abundance	1.4	1.8
Chlorophyll	36.06	50.33

19.3.6 Detecting vegetation response to oil pollution using vegetation indices

Chlorophyll related indices (NDVI, RENDVI, and REP) appeared to be more sensitive to changes in VR parameters following oil pollution, but indifferent to TPH concentrations in the soil. The results of the nonparametric regression analysis are summarized in Table 19–7.

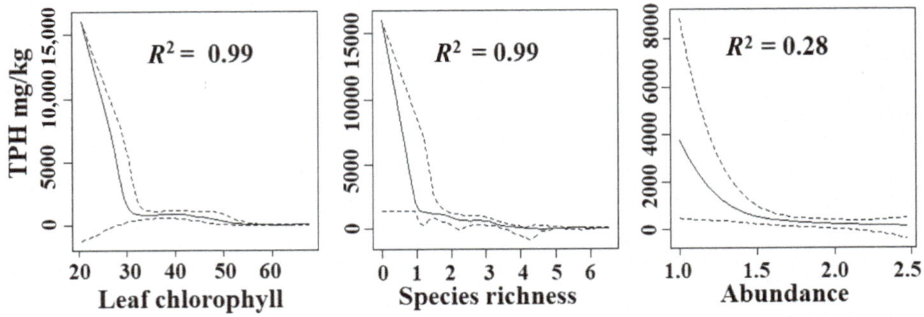

FIGURE 19–4 Regression plots showing inverse relationships between soil total petroleum hydrocarbon concentrations and vegetation response parameters (leaf chlorophyll, species richness, and vegetation abundance).

Table 19–7 Regression of vegetation responses on multisensor vegetation indices and effect size.

					delta	
$NDVI_{MS}$	0.16	0.14	0.02	0.05	**0.55**	large
$RENDVI_{HS}$	0.01	0.23	0.33	0.02	0.15	small
$RENDVI_{FS}$	0.02	0.17	0.41	0.2	0.33	small
REP_{MS}	0.06	0.26	0.45	0.14	<0.001	negligible
REP_{HS}	0.02	0.34	**0.66**	0.31	**0.63**	large
REP_{FS}	0.12	0.2	**0.53**	0.41	**0.64**	large
$ARI2_{MS}$	**0.98**	0.19	0.01	0.09	0.11	negligible
$ARI2_{HS}$	0.01	0.11	0.12	**0.35**	**-0.48**	large
$ARI2_{FS}$	0.22	0.03	0.21	**0.63**	-0.14	negligible
$SIPI_{MS}$	**0.99**	**0.41**	0.14	0.21	-0.34	medium
$SIPI_{HS}$	0.01	0.05	0.05	0.17	**-0.38**	medium
$SIPI_{FS}$	0.01	**0.4**	0.24	0.16	0.1	negligible

The most detectable response parameter from the VIs was from Menhinick's species richness index, which showed a strong positive relationship with $RENDVI_{HS}$ ($R^2 = 0.33$); $RENDVI_{FS}$ ($R^2 = 0.41$); REP_{MS} ($R^2 = 0.45$); REP_{HS} ($R^2 = 0.66$); and REP_{FS} ($R^2 = 0.53$). REP indices better detected leaf chlorophyll and vegetation abundance than NDVI/RENDVI indices.

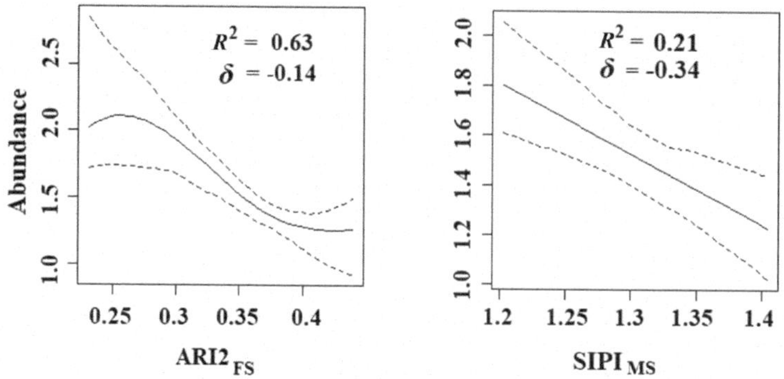

FIGURE 19–5 Regression plots showing the relationship between vegetation response parameter (abundance) and vegetation indices (stress-related anthocyanin reflectance index and structural insensitive pigment index) and the 95% confidence intervals, R^2 values, and Cliff's delta.

Alternatively, stress-related indices were able to detect changes in vegetation abundance due to oil pollution. The charts in Fig. 19–5 show the inverse relationship between two indices (ARI and SIPI) and abundance with R^2 values of up to 0.63. The pattern observed suggest that oil pollution effects are detectable using VIs related to plant health.

Sensor-wise, $NDVI_{MS}$ was the best at detecting pollution in transects with an R^2 of 0.16 and an effect size of 0.55 (large). Although both $ARI2_{MS}$ and $SIPI_{MS}$ regressed strongly with TPH ($R^2 = 0.98$ and 0.99, respectively), the effect sizes were negligible, indicating their inability to differentiate between polluted and control transects. The HS indices, however, outperformed the MS indices in detecting changes in VR due to oil pollution. REP_{HS}, which showed a strong correlation with species richness ($R^2 = 0.66$) had a large effect size of 0.63, demonstrating its ability to detect oil pollution effect on species richness. Similarly, $ARI2_{HS}$ regressed strongly with vegetation abundance ($R^2 = 0.34$), the relationship was inverse, and with an effect size of −0.48 (large) demonstrated its ability to detect oil pollution effect on vegetation abundance.

The fusion of MS and HS data to increase the spatial resolution of HS data while maintaining the spectral fidelity increased the performance of the HS indices. For instance, $RENDVI_{FS}$ correlated better with species richness and vegetation abundance than $RENDVI_{HS}$. The R^2 values for the FS index were 0.41 and 0.2, respectively, compared to the 0.33 and 0.02 obtained for the HS index. Although the effect sizes were small for both indices, the $RENDVI_{FS}$ effect size was larger with 0.33. Additionally, stress-related indices from the FS data ($ARI2_{FS}$) better correlated with vegetation abundance ($R^2 = 0.63$), but the effect size was negligible ($\delta = -0.14$) compared to that of $ARI2_{HS}$ ($\delta = -0.48$).

19.4 Discussion

The oil pollution effect on vegetation parameters (chlorophyll, species richness, and abundance) was evident throughout the study and was consistent with previous reports.

For instance, Ogbo et al. (2009) and Chima and Vure (2014) reported that oil pollution impeded plant growth in polluted sites. Furthermore, Kinako (1981) reported a species loss of at least 50% in oil-polluted habitats, whereas Baruah and Sarma (1996) found that 80% of plant species disappeared in a crude oil spill site in the Rudrasagar oil field. Nkwocha and Duru (2010) and Tanee and Albert (2015) affirmed that oil spills adversely influenced vegetation abundance in the Niger Delta region. Additionally, Lin and Mendelssohn (2012) also reported significantly lower levels of life above ground biomass (a measure of abundance) at oil-impacted sites than at reference sites in their study on the impacts of an oil spill on vegetation structure in coastal salt marshes in the northern part of the Gulf of Mexico.

Several studies attribute these effects to the unavailability of essential soil nutrients such as nitrogen and oxygen to plants (Njoku et al., 2008; Ogbo et al., 2009) or to oil constituents blocking stomata and intercellular spaces (Baker, 1970), thereby reducing transpiration rates and consequently plant uptake of nutrients (Novák and Vidovič, 2003). Regardless of the mechanism through which oil pollution affects vegetation, these effects have been shown to induce changes in vegetation reflectance measured remotely by satellite sensors. With limited resources and the increasing cost of remote sensing technology, the decision regarding the choice of a dataset to employ in environmental applications remains pertinent, and the results of this study to determine the best performing sensor for detecting oil pollution effects on vegetation provides valuable information to facilitate this process.

19.4.1 Changes in chlorophyll concentration of polluted vegetation

The NDVI, RENDVI, and REP indices derived from MS, HS, and FS datasets were compared for their ability to detect changes in chlorophyll concentration following an oil spill. The result shows that $NDVI_{MS}$ outperformed the other two, and increasing the spatial resolution of the original HS image using the Gram–Schmidt fusion technique improved its performance. This result contradicted the presumption that HS data offered a better prospect for monitoring vegetation due to the saturation limitations of MS data (Thenkabail et al., 2013) as well as its inability to detect subtle changes in the biophysical and biochemical parameters of vegetation (Gitelson, 2012). However, Teillet et al. (1997) found that the width of the bands in the red channel affects the performance of chlorophyll-related indices such as NDVI. They reported that at red bandwidths over 50 nm, these indices degraded in their capacity to monitor plant health. Since the red bandwidth of the Sentinel-2A data was 38 nm, the performance of the $NDVI_{MS}$ is consistent with expectations, considering the additional advantage of the higher spatial resolution of the Sentinel-2A dataset. Although, it is important to point out here that Thenkabail et al. (2013) found that finer spatial resolution did not improve the performance of MS indices in predicting plant biophysical parameters.

The superior performance of $NDVI_{MS}$ may be false due to the index computation method used. NDVI from MS data is computed from average spectral information over broad bandwidths, thus, may result in critical information loss that would otherwise be retained by specific narrow bands (Cho et al., 2007). However, Main et al. (2011) surmised that NDVI from MS data is useful for distinguishing large differences in vegetation condition such as greenness.

Other than this aberration, REP, ARI2, and SIPI indices derived from HS and FS datasets outperformed their MS-based counterparts in detecting changes in the vegetation chlorophyll concentration. Changes in plant chlorophyll concentration influence REP, defined as the point of the maximum slope between the red chlorophyll absorption trough and the NIR reflectance peak (Horler et al., 1983). Rock et al. (1988) asserted that environmental stress events could cause a decrease in leaf chlorophyll concentration. The results of the present study support this assertion and show that REP_{HS} and REP_{FS} are more sensitive to changes resulting from oil pollution. Since anthocyanins and carotenoids concentrations increase in stressed vegetation when photosynthetic activities are impeded, these pigments are indicators of vegetation health. In the present study, both ARI2 and SIPI computed from the HS and FS datasets correlated positively with soil TPH and negatively with leaf chlorophyll, whereas the MS-derived indices showed mixed patterns (Supplementary data 1).

19.4.2 Changes in species richness of polluted vegetation

The $NDVI_{MS}$ performed poorly in detecting changes in species richness in polluted transects. This failure is consistent with previous research such as that of Madonsela et al. (2017) who asserted that NDVI is insufficient in explaining tree species diversity. They opined that remote sensing applications in biodiversity research hinge more on spectral resolution than on spatial resolution. This phenomenon may validate the superior performance of $RENDVI_{HS}$ and $RENDVI_{FS}$ in detecting changes in species richness on the polluted transects. Additionally, the results also reveal that indices combining reflectance from the red edge and NIR regions are competent for estimating vascular plant species richness on polluted transects due to their ability to track changes. All the REP indices explained at least 50% of the variation in species richness of the investigated transects, although only REP_{HS} and REP_{FS} showed large effect sizes. The efficiency of the REP indices in tracking this event is attributable to their sensitivity to plant chlorophyll concentration, an essential plant biochemical parameter that determines vegetation productivity and health (Mishra et al., 2011; Curran et al., 1990). Several researchers found a strong positive correlation between chlorophyll concentration and species richness and diversity (Arellano et al., 2017; Behl et al., 2011; Carlson et al., 2007). Hence it is no surprise that oil pollution induced stress in vegetation affects chlorophyll concentration and species richness, as demonstrated in this study. Noomen et al. (2012) reported changes in vegetation patterns in polluted fields, while Robson et al. (2004) found lower diversity values for contaminated sites than for uncontaminated sites. The pattern of relationships observed between the VIs and species richness is in line with the productivity–diversity hypothesis, which asserts a positive linear relationship between species diversity and productivity that is dependent on an environmental gradient (Madonsela, 2017). Mandoloma et al. (2014) noted that the productivity–diversity relationship exhibits bidirectional causality. This pattern was observed in the present research as vegetation abundance, a measure of ecosystem productivity, correlated positively with measures of species diversity and richness.

19.4.3 Changes in vegetation abundance on polluted transects

Vegetation abundance decreased significantly on polluted transects compared to control transects. These results were consistent with those reported by Nkwocha and Duru (2010) and Tanee and Albert (2015) who affirmed that oil spills impacted negatively on the abundance of vegetation in the Niger Delta region. Additionally, Lin and Mendelssohn (2012) also reported significantly lower levels of life above ground biomass at oil-impacted sites than at reference sites in their study on the impacts of an oil spill on vegetation structure in coastal salt marshes of the northern part of the Gulf of Mexico. VIs from different datasets (MS, HS, and FS) performed variably in their ability to detect these changes. Stress-related indices (ARI2 and SIPI) from HS and FS datasets correlated better with abundance and outperformed the chlorophyll-related indices. Anthocyanins (AnC) and carotenoids (CaR) are accessory pigments that perform protective functions during photosynthesis in plants. Both AnC and CaR prevent photoinhibition and photodamage in plants by absorbing the excessive incident light that would otherwise damage chlorophyll pigments. In most literature, an increase in the concentration of these pigments in plants is an indication of vegetation stress. For instance, Merzlyak et al. (2008) reported the intensification of AnC occurrence due to environmental factors such as increased solar radiation, extreme temperatures, drought, nutrient deficiency, and other stress factors. Hence the observed increase in AnC and CaR absorption in polluted vegetation, which significantly differed from that of control vegetation, may explain the superior performance of stress-related indices in detecting changes in vegetation abundance.

19.5 Conclusion

The objective of this study was to compare the performance of VIs derived from HS and MS datasets for detecting the effect of oil pollution on vegetation in the Niger Delta. Furthermore, the performance of VIs derived from a fused dataset generated by combining HS and MS data were evaluated.

The results showed that both MS- and HS-based indices were adequate for monitoring changes in leaf chlorophyll in polluted vegetation, but a fusion of both datasets provided better detection of changes in species richness and vegetation abundance following oil pollution. Additionally, chlorophyll-related indices showed a strong correlation with leaf chlorophyll concentration and species richness, whereas stress-related indices better-detected changes in vegetation abundance, which was attributed to the increased anthocyanin concentration in polluted vegetation.

Acknowledgments

This study was financed partly by the Tertiary Education Trust Fund (TETFUND), Michael Okpara University of Agriculture, Umudike, Nigeria; the National Centre for Earth Observation (NCEO); and the Centre for Landscape and Climate Research (CLCR), University of Leicester, United Kingdom. The authors are grateful to the traditional ruler of the Kporghor community in Tai Local Government Area of Rivers State and the executive members of the Kporghor Youth Assembly as well as the field team in Port Harcourt, Nigeria for their support during the fieldwork campaign.

References

Abdikan, S., Balik Sanli, F., Sunar, F., Ehlers, M., 2014. A comparative data-fusion analysis of multi-sensor satellite images. Inter. J. Digital Earth 7 (8), 671–687.

Adler-Golden, S., Richtsmeier, S., Conforti, P., Bernstein, L., 2013. Spectral image destriping using a low-dimensional model. In: Shen, S.S., Lewis, P.E. (Eds.), SPIE Defense, Security, and Sensing, Baltimore, Maryland, United States. SPIE, Baltimore, MA, United States.

Arellano, P., Tansey, K., Balzter, H., Tellkamp, M., 2017. Plant family-specific impacts of petroleum pollution on biodiversity and leaf chlorophyll content in the amazon rainforest of Ecuador. PLOS ONE 12 (1), e0169867.

Baker, J.M., 1970. The effects of oils on plants. Environ. Pol 1 (1), 27–44. 1970.

Barry, P., 2001. EO1/Hyperion Science data User's Guide Level 1_B, NASA, California, US.

Baruah, D., Sarma, S.K., 1996. Impact of crude oil pollution on species number and live standing herbaceous crop biomass. Environmentalist 16 (4), 291–295.

Behl, S., Donval, A., Stibor, H., 2011. The relative importance of species diversity and functional group diversity on carbon uptake in phytoplankton communities. Limn. Ocean 56 (2), 683–694.

Brown, L., 1992. A survey of image registration techniques. ACM Comput. Surv. (CSUR) 24 (4), 325–376.

Carlson, K.M., Asner, G.P., Hughes, F.R., Ostertag, R., Martin, R.E., 2007. Hyperspectral remote sensing of canopy biodiversity in hawaiian lowland rainforests. Ecosystems 10 (4), 536–549.

Chima, U.D., Vure, G., 2014. Implications of crude oil pollution on natural regeneration of plant species in an oil-producing community in the Niger Delta Region of Nigeria. J. Forestry Res 25 (4), 915–921.

Cho, M.A., Skidmore, A., Corsi, F., van Wieren, S.E., Sobhan, I., 2007. Estimation of green grass/herb biomass from airborne hyperspectral imagery using spectral indices and partial least squares regression. Inter. J. Appl. Earth Obser. Geoinfor 9 (4), 414–424.

Cliff, N., 1996. Ordinal Methods for Behavioural Data Analysis. Psychology Press, New York. Available from: https://doi.org/10.4324/9781315806730.

Corcoran, E., Ravilious, C., Skuja, M., 2007 Mangroves of Western and Central Africa, UNEP-WCMC/UNEP, Cambridge, United Kingdom.

Curran, P.J., Dungan, J.L., Gholz, H.L., 1990. Exploring the relationship between reflectance red edge and chlorophyll content in slash pine. Tree Physiology 7 (1_2_3_4), 33–48.

Datt, B., McVicar, T.R., Van Niel, T.G., Jupp, D.L.B., Pearlman, J.S., 2003. Preprocessing EO-1 Hyperion hyperspectral data to support the application of agricultural indexes. IEEE Trans. Geosci. Remote Sens. 41 (6), 1246–1259.

Dadon, A., Ben-Dor, E., Karnieli, A., 2010. Use of derivative calculations and minimum noise fraction transform for detecting and correcting the spectral curvature effect (smile) in hyperion images. IEEE Trans. Geosci. Remote Sens. 48 (6), 2603–2612.

Department of Petroleum Resources, (DPR), 2002. Environmental Guidelines and Standards for the Petroleum Industry in Nigeria, Department of Petroleum Resources (DPR), Lagos.

Gholizadeh, A., Misurec, Jan, Kopackova, Veronika, Mielke, Christian, Rogass, Christian, 2016. Assessment of red-edge position extraction techniques: a case study for norway spruce forests using hymap and simulated sentinel-2 data. Forests 7 (10), 226.

Gitelson, A.A., Merzlyak, M.N., 1996. Signature analysis of leaf reflectance spectra: Algorithm development for remote sensing of chlorophyll. J. Plant Phy 148 (3-4), 494–500.

Gitelson, A.A., Merzlyak, M.N., Zur, Y., Stark, R., Gritz, U., 2001. Non-destructive and remote sensing techniques for estimation of vegetation status, papers. Nat. Res 273, 205.

Gitelson, A.A., Peng, Y., Masek, J.G., Rundquist, D.C., Verma, S., Suyker, A., et al., 2012. Remote estimation of crop gross primary production with landsat data. Remote Sens. Environ. 121 (0), 404–414.

Goodenough, D.G., Dyk, A., Niemann, K.O., Pearlman, J.S., Chen, Hao, Han, Tian, Murdoch, M., West, C., 2003. Processing hyperion and ali for forest classification, geoscience and remote sensing. IEEE Trans. Geosci. Remote 41 (6), 1321–1331.

Kromrey, J.D., Hogarty, K.Y., Ferron, J.M., Hines, V.C., Hess, M.R. 2005 Robustness in Meta-Analysis: An Empirical Comparison of Point and Interval Estimates of Standardized Mean Differences and Cliff's Delta. Joint Statistical Meetings, August 7 - 11 2005 Minneapolis:.

Hogarty, K., Kromrey, J., 2001. We have been reporting some effect sizes: Can you guess what they mean. American Educational Research Association, April 10-14 Seattle: American Educational Research Association.

Horler, D.N.H., Dockray, M., Barber, J., Barringer, A.R., 1983. Red edge measurements for remotely sensing plant chlorophyll content. Advan. Space Res 3 (2), 273–277.

Hurlbert, S.H., 1971. The Nonconcept of species diversity: a critique and alternative parameters. Ecology 52 (4), 577–586.

Khanna, S., Santos, M.J., Ustin, S.L., Shapiro, K., Haverkamp, P.J., Lay, M., 2018. Comparing the potential of multispectral and hyperspectral data for monitoring oil spill impact. Sensors (Basel, Switzerland) 18 (2), 558.

Kinako, P.D.S., 1981. Short-term effects of oil pollution on species numbers and productivity of a simple terrestrial ecosystem. Environ. Poll. Series A, Ecolo. Biol. 26 (2), 87–91.

Lin, Q., Mendelssohn, I.A., 2012. Impacts and recovery of the deepwater horizon oil spill on vegetation structure and function of coastal salt marshes in the Northern Gulf of Mexico. Environ. Sci. Tech 46 (7), 3737–3743.

Macbeth, G., Razumiejczyk, E., Ledesma, R.D., 2011. Cliffs Delta Calculator: A non-parametric effect size program for two groups of observations. Universitas Psychologica 10 (2), 545–555.

Madonsela, S., Cho, M.A., Ramoelo, A., Mutanga, O., Naidoo, L., 2018. Estimating tree species diversity in the savannah using NDVI and woody canopy cover. Inter. J. Appl. Earth Obser. Geoinfor 66, 106–115.

Mandoloma, L., Kazembe, J., Moyo, B., 2014. Testing the Ecosystem Productivity- Diversity Hypothesis in a Grassland. I. J. Sci.: Basic Appl. Res 14 (2), 305–322.

Main, R., Cho, M.A., Mathieu, R., O'Kennedy, M.M., Ramoelo, A., Koch, S., 2011. An investigation into robust spectral indices for leaf chlorophyll estimation. ISPRS J. Photogrammetry Remote Sens. 66 (6), 751–761.

Merzlyak, M.,N., Chivkunova, O., Solovchenko, B., Naqvi, A.E., Razi, K., 2008. Light absorption by anthocyanins in juvenile, stressed, and senescing leaves. J. Exp. Bot 59 (14), 3903–3911.

Mishra, D.R., Mishra, S., Cho, H.J., Ghosh, S., Fox, A., Downs, C., et al., 2012. Post-spill state of the marsh: Remote estimation of the ecological impact of the Gulf of Mexico oil spill on Louisiana Salt Marshes. Remote Sens. Environ. 118, 176–185.

Njoku, K., Akinola, M., Oboh, B., 2008. Growth and performance of glycine max l. (merrill) grown in crude oil contaminated soil augmented with cow dung. Life Sci. J 5 (3), 89–93.

Nkwocha, E.E., Duru, P.O., 2010. Micro-analytic study on the effect of oil pollution on local plant species and food crops. Advances Bioresearch 1 (1), 189–198.

Noomen, M.F., van der Werff, H.M.A., van der Meer, F.D., 2012. Spectral and spatial indicators of botanical changes caused by long-term hydrocarbon seepage. Eco. Informatics 8 (1), 55–64.

Novák, V., Vidovič, J., 2003. Transpiration and nutrient uptake dynamics in maize (Zea mays L.). Ecological Model 166 (1–2), 99–107.

Ogbo, M.E., Zibigha, M., Odogu, G., 2009. The effect of crude oil on growth of the Weed (Paspalum scrobiculatum L.) -phytoremediation potential of the plant. African. J. Environ. Sci. Tech. 3 (9), 229–233.

Osam, M.U., Wegwu, M.O., Uwakwe, A.A., 2011. The Omoku old pipeline oil spill: Total hydrocarbon content of affected soils and the impact on the nutritive value of food crops. Archives Appl. Sci. Res 3 (3), 514–521.

Pal, M.K., Porwal, A., 2015. Destriping of Hyperion images using low-pass-filter and local-brightness-normalization. IEEE 3509.

Pande, H., Tiwari, P., Dobhal, S., 2009. Analyzing hyper-spectral and multi-spectral data fusion in spectral domain. J. Indian Soc. Remote Sens. 37 (3), 395–408.

Pearson, R.L., Miller, L.D., 1972. Remote mapping of standing crop biomass for estimation of the productivity of the short-grass Prairie, Pawnee National Grasslands, Colorado. 8th Int. Symp. on Remote Sensing of Environment Ann Arbor, Michigan USA, October 2-6, 1972 University of Michigan, USA: Environmental Research Institute of Michigan, 1357.

Peñuelas, J., Filella, I., Biel, C., Serrano, L., Savé, R., 1993. The reflectance at the 950–970 nm region as an indicator of plant water status. Inter. J. Remote Sens 14 (10), 1887–1905.

Peñuelas, J., Filella, I., Gamon, J.A., 1995. Assessment of photosynthetic radiation-use efficiency with spectral reflectance. New Phytologist 131 (3), 291–296.

Pohl, C., Van Genderen, J.L., 1998. Review article Multisensor image fusion in remote sensing: Concepts, methods and applications. Inter. J. Remote Sens 19 (5), 823–854.

Robson, D.B., Knight, J.D., Farrell, R.E., Germida, J.J., 2004. Natural revegetation of hydrocarbon-contaminated soil in semi-arid grasslands. Canadian J.Botany 82 (1), 22–30.

Rock, B.N., Hoshizaki, T., Miller, J.R., 1988. Comparison of in situ and airborne spectral measurements of the blue shift associated with forest decline. Remote Sen. Environ 24 (1), 109–127.

Savitzky, A., Golay, M.J.E., 1964. Smoothing and differentiation of data by simplified least squares procedures. Anal. Chem. 36 (8), 1627–1639.

Scheffler, D., Karrasch, P., 2013. Preprocessing of hyperspectral images: a comparative study of destriping algorithms for EO1-hyperion. In: Lorenzo, Bruzzone (Eds.), SPIE Remote Sensing Dresden, Germany, 23-26 September. Dresden, Germany: SPIE.

Tanee, F.B.G., Albert, E., 2015. Reconnaissance assessment of long-term effects of crude oil spill on soil chemical properties and plant composition at kwawa, ogoni, Nigeria. J. Environ. Sci. Tech. 8 (6), 320–329.

Teillet, P.M., Staenz, K., William, D.J., 1997. Effects of spectral, spatial, and radiometric characteristics on remote sensing vegetation indices of forested regions. Remote Sens. Environ. 61 (1), 139–149.

Thenkabail, P.S., Mariotto, I., Gumma, M.K., Middleton, E.M., Landis, D.R., Huemmrich, K.F., 2013. Selection of hyperspectral narrowbands (hnbs) and composition of hyperspectral two-band vegetation indices (hvis) for biophysical characterisation and discrimination of crop types using field reflectance and hyperion/eo-1 data. IEEE J. Selected Topics Appl. Earth Obser. Remote Sens. 6 (2), 427–439.

Yokoya, N., Grohnfeldt, C., Chanussot, J., 2017. Hyperspectral and Multispectral Data Fusion: A comparative review of the recent literature. IEEE Geosci. Remote Sensing Magazine 5 (2), 29–56.

20 Hyperspectral vegetation indices to detect hydrocarbon pollution

Paul Arellano[1,2], Dimitris Stratoulias[3,4]

[1]SCHOOL OF EARTH SCIENCES, ENERGY AND ENVIRONMENT—CENTER OF EARTH OBSERVATION, YACHAY TECH UNIVERSITY, URCUQUI, ECUADOR
[2]DEPARTMENT OF POSTGRADUATE STUDIES, FACULTY OF BIOLOGICAL SCIENCES, CENTRAL UNIVERSITY OF ECUADOR, QUITO, ECUADOR
[3]DEPARTMENT FOR MANAGEMENT OF SCIENCE AND TECHNOLOGY DEVELOPMENT, TON DUC THANG UNIVERSITY, HO CHI MINH CITY, VIETNAM
[4]FACULTY OF APPLIED SCIENCES, TON DUC THANG UNIVERSITY, HO CHI MINH CITY, VIETNAM

20.1 Introduction

Over the past decades, the Amazon forest has been threatened by deforestation, selective logging, hunting, fire, and global and regional climate variations (Malhi et al., 2008; Davidson et al., 2012; Asner et al., 2004). Furthermore, the Amazon region consists of sedimentary basins in which vast petroleum reservoirs have been found (Kvenvolden and Rogers, 2005), some of which have been exploited during the past decades (Finer et al., 2008). Little or no attention has been given to the impacts of petroleum and hydrocarbons on the Amazonian forests, and their effects on tropical vegetation still remain poorly understood.

The Amazon rainforest plays an essential role in the ecological, climate, and biogeochemical processes of the Earth. In this highly diverse environment, complex interactions between plant communities and their surrounding environment need to be better understood. Remote sensing can provide a better understanding of the interactions between hydrocarbons and vegetation in the tropical forest due its ability to evaluate leaf and canopy chemical properties, which are the principal determinants of plant physiology and biochemical processes in terrestrial ecosystems. Due to its high spectral resolution, hyperspectral remote sensing has proven to be a promising tool to assess forest degradation through the investigation of biophysical and biochemical traits such as chlorophyll and other pigment concentrations at leaf and canopy levels. (Yang, 1999; van der Meer et al., 2006; Noomen, 2007; Smith et al., 2004a,b; Smith et al., 2005; Blackburn, 2007).

The global demand for fossil energy and the high prices of crude oil are triggering oil exploration and production projects all around the Amazon basin; even remote areas have already been leased out or are being negotiated for petroleum and gas exploration and production (Finer

Hyperspectral Remote Sensing. DOI: https://doi.org/10.1016/B978-0-08-102894-0.00019-X

et al., 2008). In Ecuador, petroleum productive areas located in the Amazon region have been exploited for oil and gas since 1967 (Bernal, 2011). During the first few decades of production, a considerable number of oil spills from pipeline networks and leakages from unlined open pits were frequent (Bernal, 2011; Hurtig and San-Sebastián, 2005). Some polluted areas have been identified and cleaned up, however, others are still undetected and covered by tropical vegetation. Currently, these polluted areas from the early stages of the petroleum industry are the focus of international legal actions by local residents against international oil companies.

20.1.1 Hydrocarbons and vegetation interactions

Different plant species respond differently to particular stressors. Furthermore, the nature, intensity, and length of exposure are factors that define the stress level on vegetation. Baker (1970) summarized several pieces of research related to the effects of crude oil on plants and showed that the toxicity of petroleum oil depends on the concentration of unsaturated, aromatic, and acid compounds; the higher their concentration, the more toxic the oil is for plants. Molecules of crude oil can penetrate into a plant through its leaf tissue, stomata, and roots. The rate of penetration depends on the oil type, the contact part (leaves, roots), the time of exposure, the thickness of the cuticle, and the density of the stomata. After penetrating into the plant, the oil may travel into the intercellular space and possibly also into the vascular system. Cell membranes are damaged by the penetration of hydrocarbon molecules leading to the leakage of cell contents, and the possible entrance of oil into the cells.

Plant transpiration, respiration, and photosynthetic rates are affected by hydrocarbon pollution (Baker, 1970). The effects of hydrocarbons on plants reduce plant transpiration rates. On the other hand, plant respiration may either decrease or increase depending on the plant species and the oil type. Hydrocarbons reduce the rate of photosynthesis, and the amount of reduction varies with the type and amount of oil and with the species of plant. Cell injury may be the principal cause of photosynthesis inhibition because hydrocarbons tend to accumulate in the chloroplasts, which explains the reduced levels of chlorophyll content in vegetation affected by hydrocarbons. For a thorough review of the effects of hydrocarbons on vegetation, see the work of Arellano Mora (2014).

20.1.2 Vegetation indices for vegetation stress caused by hydrocarbons

Some vegetation indices (VIs) derived from remote sensing data have been specifically developed and tested to identify chlorophyll content across multiple vegetation species using specific spectral bands sensitive to chlorophyll content changes. Other indices have been applied to remote sensing images to detect the effects of hydrocarbon pollution on vegetation. Some of these indices are explained here.

Three primary types of VIs (simple ratio, normalized difference, and red edge) have been developed for the estimation of chlorophyll content, however, in most cases these indices have been tested on a few species only and, therefore, it is not clear whether they can be applied across a range of species with varying leaf structure and phenology. Sims and Gamon (2002) developed specific indices to estimate chlorophyll across species and found that indices based on reflectance at 680 nm such as the pigment specific simple ratio for chlorophyll-a index (also called SR680) were largely insensitive to variations in chlorophyll content. On the other hand, normalized difference indices such as $NDVI_{705}$ using the reflectance at 705 nm are sensitive to the variation of chlorophyll content across several species ($R^2 = 0.65$), moreover the VIs mSR445 and $mNDVI_{445}$ using the reflectance at 445 nm produced substantially improved correlation with total chlorophyll content ($R^2 = 0.74$ and $R^2 = 0.83$, respectively). This improved performance was attributed to the suppression of the surface reflectance at 445 nm, which appears to be the most important factor in the variation of chlorophyll content derived from remote sensing across species.

Narrowband VIs have been applied to detect the effects of hydrocarbon pollution on vegetation (Yang, 1999; Smith et al., 2004a,b; Noomen et al., 2008; Zhu et al., 2013). It is assumed that hydrocarbon pollution affects vegetation adversely by the displacement of the soil oxygen needed for respiration. This causes changes in plant and canopy reflectance. Noomen et al. (2008) applied 54 VIs based on canopy characteristics and chlorophyll content in a simulated natural hydrocarbon gas leak in order to detect changes in maize (*Zea mays*) and wheat (*Triticum aestivum*). Six VIs performed best, namely $NDVI_{705}$, $mNDVI_{705}$, carter index 2, lichtenthaler index 1, lichtenthaler index 3, leaf area index, and optimized soil-adjusted vegetation index.

In another study, 20 VIs were applied to detect the detrimental effects of oil pollution on reed (*Phragmites australis*) communities around oil wells that had been producing petroleum for approximately 10 years in the Yellow River Delta, eastern China (Zhu et al., 2013). Two VIs best approximated soil total petroleum hydrocarbon concentrations; the modified chlorophyll absorption ratio index had the highest coefficient of determination ($R^2 = 0.73$) followed by the NDVI ($R^2 = 0.70$).

The red-edge position (REP) is a unique feature of green plants related to leaf chlorophyll content and LAI. REP is defined as the inflection point (or sharp change) of the low red reflectance caused by chlorophyll absorption near 680 nm and high infrared reflectance governed by the internal structure of leaves near 750 nm (Clevers and Jongschaap, 2006). Smith et al. (2004a,b) performed a derivative analysis to locate the position and height of the inflection point of the red-edge that may indicate stress in plants (decrease in growth and chlorophyll content reduction) caused by hydrocarbon gases in the soil.

REP has been used as an indicator of vegetation stress in areas affected by hydrocarbon gases (hydrocarbon seepages and leaks from pipelines). Noomen and Skidmore (2009) found that REP moved to shorter wavelengths in vegetation growing in polluted soil. On the other hand, Yang et al. (1999, 2000) reported a shift of the red-edge toward longer wavelengths in an agricultural field affected by hydrocarbon seepage. In this study, a linear interpolation method was applied to estimate the REP. The method first located the midpoint between the near infrared (NIR) at 780 nm and the reflectance minimum of the chlorophyll absorption feature at about 670 nm. Second, a linear interpolation was applied between the measurements at 700 and 740 nm for estimating the wavelength at the inflection point.

This study aims to identify the most suitable band combination and consequently to propose new VIs to detect the effects of hydrocarbon pollution in the Amazon forest using field spectroscopy and hyperspectral satellite images.

20.2 Materials and methods

20.2.1 Study area

Fieldwork was undertaken from April to July, 2012, at three sites in the Amazon tropical rainforest of Ecuador (Fig. 20–1). The first and second study sites are located in a lowland evergreen secondary forest in the Sucumbios province of the Tarapoa region (0°11'S; 76°20'W). The mean annual rainfall in the region is 3800 mm and the average annual temperature is 23°C with relative humidity of approximately 90% (Fitton, 2000). The area is located 232–238 m above the mean sea level. The third study site is a highly diverse lowland evergreen primary forest located at the Orellana province, in the northern section of Yasuní National Park (0°41'S; 76°24'W). The area lies 216–248 m above the mean sea level and receives an average of 3081 mm annual rainfall

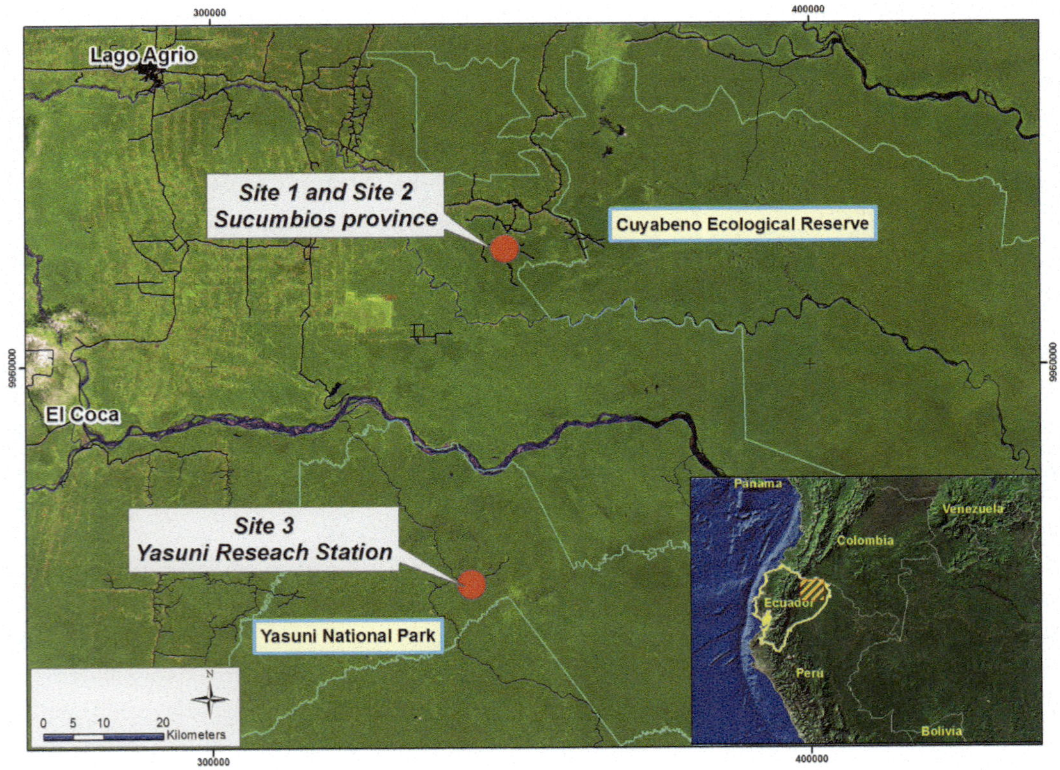

FIGURE 20–1 Map of the study area in the northeast Amazon region of Ecuador. Sites 1 and 2 are located in Sucumbios province and site 3 is located at Orellana province. Background is a Landsat image. Source of zoom-in map: Color Shaded Relief Image, Used with permission from WorldSat International, Inc.

Table 20–1 Description of the sites in the study area.

	Site 1	Site 2		Site 3
		Site 2a	Site 2b	
Forest classification	A seasonal lowland evergreen rainforest			
Forest type	Secondary/regrowth following abandoned farming			Pristine forest
Pollution	Petroleum-hydrocarbon pollution	Not affected by pollution		Not affected by pollution
Coordinate system	UTM-WGS84-Z18N			
Latitude	9979706	9980060	9978875	9924894
Longitude	348523	348694	347525	344220
Altitude (meters above sea level)	232	238	229	235
Mean annual temperature (°C)	27			28
Mean annual rainfall (mm)	3300			3081
Soil type	Histosols/fibrits/tropofibrits			Inceptisols/tropots/distrops

with peaks in October and November. Mean monthly temperatures vary from 22°C to 34°C (Valencia, 2004). In this site, the Pontifical Catholic University of Ecuador established and manages permanent forest dynamics plots of 50 ha where over 150,000 mapped trees ≥ 1 cm in diameter at breast height from over 1100 species have been identified (Valencia et al., 2004). A detailed description of the sites is provided in Table 20–1.

In spite of the study site being located in the lowland Amazon forest, there are substantial differences in plant species and forest structure. Sites 1 and 2 are disturbed forests that were exposed to selective logging, agricultural activities, petroleum industry impacts, and secondary forest regrowth over the past 20 years following diminishing human influence. Site 3 is a pristine primary tropical rainforest with legal protection status where a research project on plant and animal species diversity is currently conducted. Studies consider that the plant species richness in this area is among the highest in the world (Tedersoo et al., 2010).

The fieldwork campaign of this project was granted with the appropriate permits and approvals from the Environmental Ministry of Ecuador and the Non-Renewable Resource Ministry of Ecuador.

20.2.2 Sampling process

Well-developed branches were carefully selected and collected with a telescopic pruner, tree-climbing techniques, and canopy towers at different levels of the vertical profile of the forest. The collected branches were sealed in large polyethylene bags to maintain their moisture content and stored in ice coolers. The foliar material was transported to a local site, and fully expanded mature leaves with no damage by herbivorous or pathogens were selected for analysis. A total of 1134 samples was collected in the three fieldwork sites. The sampling process accounted for three levels of the vertical profile of the forest and included a wide range of vegetation heterogeneity related to species distribution, phenological stage, and leaf structure.

20.2.3 Chlorophyll meter readings

Depending on the size and shape of the leaf, different cork borers of variable sizes between 2.5 and 8.5 cm in diameter were used to clip leaf disks from the central and widest portions of the leaf blade, avoiding the major veins (Fig. 20–2).

FIGURE 20–2 Leaf samples and clipped disks from plants of different species and health status. The last photo shows the SPAD-502 meter.

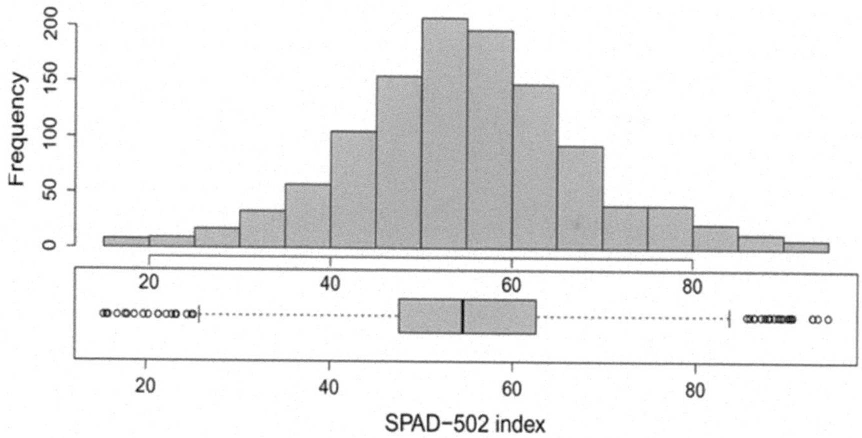

FIGURE 20–3 Histogram and boxplot of the SPAD-502 chlorophyll meter. Total samples collected in both sites of the Amazon forest in Ecuador.

Table 20–2 Number of foliar samples and descriptive statistics of the SPAD-502 chlorophyll meter readings from two sites of the study area.

Number of samples	Maximum	Minimum	Mean	SD
1134	94.47	15.37	54.88	12.83

All leaf disks were clipped from the midpoint of the leaves since it has been documented that this is the best position from which to take chlorophyll readings (Hoel, 1998). Three readings were taken from each disk using a portable SPAD-502 chlorophyll meter at different positions of each leaf disk, and a mean index value was used in further analysis.

Different readings per leaf took into account any possible variations in leaf thickness. A histogram of the dataset shows a normal distribution and a boxplot identified values outside the 95% confidence interval considered as outliers (Fig. 20–3). Table 20–2 shows descriptive statistics of the dataset.

20.2.4 Foliar biophysical and biochemical measurements

20.2.4.1 Foliar water content

The water content (Cw) of leaves is regulated by water loss during transpiration and water uptake from soil. During photosynthesis, the stomata of the leaves are open to allow carbon dioxide (CO_2) uptake, but at the same time, water vapor is lost from the leaf during transpiration. Stomata tend to close when the leaf water potential drops below a certain point (Bonan, 2008); Cw of the leaf affects stomata opening and, hence, respiration. Cw is closely correlated to leaf thickness, intercellular space, and infrared reflectance and transmittance (Castro-Esau et al., 2006).

Using the same leaf disks of known surface (S), fresh weight (Fw) was measured with a precision weighing scale, and then the samples were dried at approximately 60°C for a period of at least 48 hours. They were then reweighted to get the dry weight (Dw).

Cw in g/cm^2 (Gerber et al., 2011) or equivalent water thickness (Féret et al., 2011; Datt, 1999; Hunt and Rock, 1989) was estimated using the equations:

$$\mathrm{Cw} = \frac{\mathrm{Fw} - \mathrm{Dw}}{S} \tag{20.1}$$

$$\mathrm{Ewt} = \frac{\mathrm{Fw} - \mathrm{Dw}}{\mathrm{dw} \times S} \tag{20.2}$$

where dw is the density of water (=1 g/mL).

20.2.4.2 Dry matter content

Dry matter content (Cm) is associated with leaf density and leaf thickness and can also be estimated by the product of leaf density and thickness. It includes two compounds; the first being a non-nitrogenous compound (cell walls, membranes, and storage carbohydrates and lipids) and the second being nitrogenous compounds (proteins and chlorophyll). Cm is closely correlated to photosynthetic rates and atmospheric CO_2 (White and Montes-R, 2005).

Cm in g/cm^2 (Féret et al., 2011; Datt, 1999; Gerber et al., 2011), also known as specific leaf mass (Ws) (Marenco et al., 2009), leaf specific mass (White and Montes-R, 2005), or leaf mass per area (Wright et al., 2004) was computed using:

$$\mathrm{Cm} = \frac{\mathrm{Dw}}{\mathrm{S}} \tag{20.3}$$

20.2.5 Spectroradiometer measurements

Reflectance and "transflectance" (a term used in this study to describe the measurement of "double" transmittance) were measured for each leaf disk using an ASD FieldSpec HandHeld-2 spectroradiometer (Analytical Spectral Devices Inc., Boulder, Colorado). This instrument provides a wavelength range of 325–1075 nm with a resolution of full width at a half maximum of 3.5 nm and sampling interval of 1 nm. The spectrometer is attached with a plant probe to an internal 4.05 W halogen light source and a leaf clip that includes a rotating head with both white and black reference panels (Fig. 20–4).

This mechanism holds the leaves, excludes ambient direct and scattered light, and ensures a constant field of view of 10 mm geometry for the target sample. Reflectance was estimated by recording the radiance reflected from the leaf with the black reference panel, and transflectance was estimated by recording the radiance reflected from the leaf with the white panel. Spectrum averages of 10 scans were taken for each leaf disk using the white reference radiance panel. The appropriate number of scans was established based on a compromise between the time required for each spectrum collection and noise reduction. The

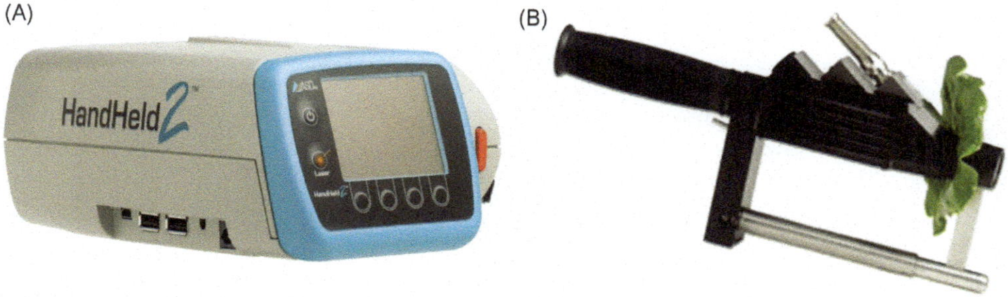

FIGURE 20–4 (A) ASD HandHeld-2 spectroradiometer and (B) plant probe and leaf clip.

measured differences were tested in the specific wavelength range used by reflectance indices for the estimation of chlorophyll content. The difference between 10 and 20 scans in the range of 700–750 nm was 2.0% and in the range 751–800 nm it was 0.9%. The differences between 10 and 30 scans in the same ranges were 3.3% and 1.8%, respectively. These low differences were considered acceptable for the reflectance index calculation. This can be considered as an advantage of using the plant probe and leaf clip mechanism, which provides a constant light source in a closed environment to allow the number of scans to be reduced and a comparable signal-to-noise ratio to be maintained.

20.2.6 PROSPECT model

The inversion of the PROSPECT model using leaf reflectance and transmittance was applied in order to estimate chlorophyll concentration. First, a Savitzky–Golay filter was computed to the reflectance and transflectance signals in order to smooth the signals and increase the signal-to-noise ratio. Second, the measurements of transflectance obtained from the leaf-clip attached to the analytical spectral devices (ASD) FieldSpec HandHeld-2 spectrometer were converted to transmittance based on a procedure developed for this research. This was founded on the Kubeika–Munk theory of light scattering and light absorption based on communications with Professor Stephane Jacquemoud and Dr. Sebastien Marcq from the Institute of Physique du Globe of Paris and Universidad Paris Diderot (ISO, 2012; Jacquemoud and Marcq, 2012).

$$L_1 = R_W \times L_{inc} \tag{20.4}$$

$$L_{inc} = \frac{L_1}{R_w} \tag{20.5}$$

$$L_2 = R_b \times L_{inc} \tag{20.6}$$

$$L_3 = R_w^{each} \times L_{inc} \tag{20.7}$$

$$L_4 = R_b^{each} \times L_{inc} \tag{20.8}$$

$$R(\lambda)_b = \frac{L_2 R_w}{L_1} \tag{20.9}$$

$$R(\lambda)_w^{each} = \frac{L_3 R_w}{L_1} \tag{20.10}$$

$$R(\lambda)_b^{each} = \frac{L_4 R_w}{L_1} \tag{20.11}$$

$$R_{est} = \frac{R(\lambda)_b R(\lambda)_w^{each} - R(\lambda)_b^{each} R_w}{R(\lambda)_b - R_w + R_w R(\lambda)_b \left(R(\lambda)_w^{each} - R(\lambda)_b^{each}\right)} \tag{20.12}$$

$$T = \sqrt{\frac{(R(\lambda)w^{each} - R_{measured})(1 - R_{measured} R_w)}{R_w}} \tag{20.13}$$

where L_1 is the luminance measured with the white background disk only; L_2, is the luminance measured with the black background disk only; L_3 is the luminance of the leaf mounted on the white background disk; L_4 is the luminance of the leaf mounted on the black background disk; $R_{measured}$ is the leaf reflectance measured; R_{est} is the leaf reflectance estimated; T is leaf transmittance; R_b is the reflectance of the blank background disk; R_w is the reflectance of the white background disk; R_w^{each} is the reflectance of each leaf on the white background disk; R_b^{each} is the reflectance of each leaf on the black background disk; L_{inc} is the lamp luminance.

Reflectance was estimated by applying Eq. (20.11) in order to test the procedure applied to estimate transmittance. Fig. 20–5 illustrates the close agreement between the measured and estimated reflectance spectra.

Foliar chlorophyll content (C_{ab}) was computed by the inversion process of the PROSPECT 5 model, which was modified by its author (Jacquemoud and Marcq, 2012) to work in the range of 400–1075 nm using reflectance and transmittance in the sampling interval of 1 nm for the 1134 leaf samples. Brown pigments (C_{bp}) and C_w were neutralized since foliar samples were of green vegetation and the spectra do not demonstrate water absorption features. Fig. 20–6 illustrates the process to estimate chlorophyll content of leaves.

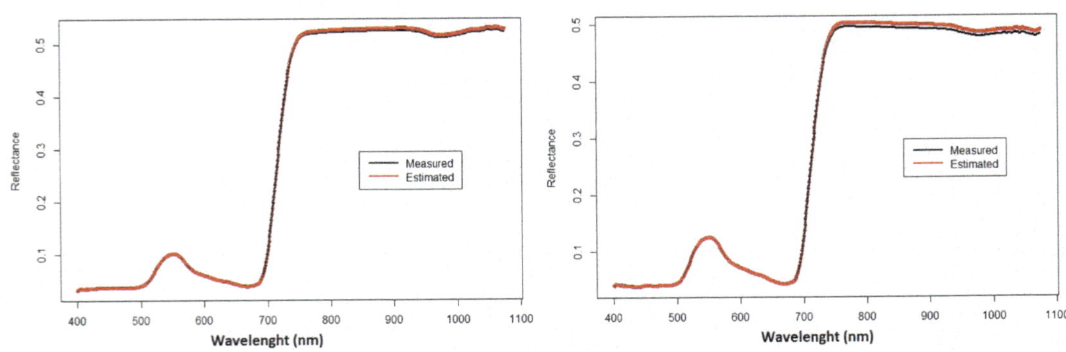

FIGURE 20–5 Measured and estimated reflectance of two representative samples of the foliar dataset.

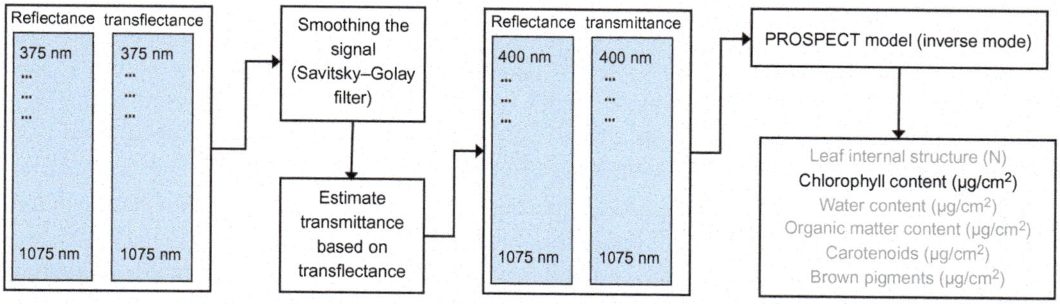

FIGURE 20–6 Process to estimate chlorophyll content (μg/cm^2) by applying the inversion process of the PROSPECT leaf reflectance model.

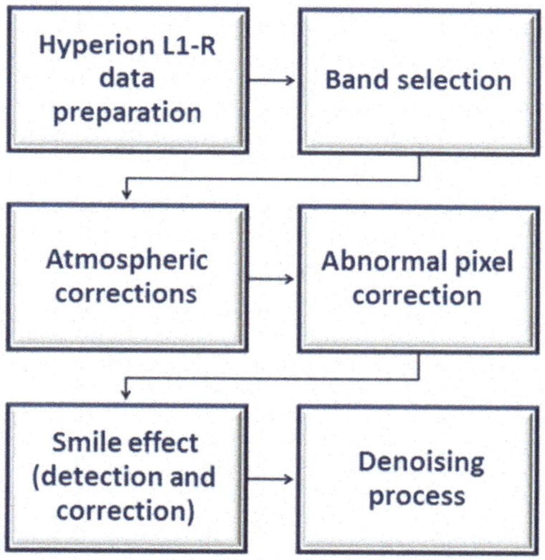

FIGURE 20–7 Process for preprocessing Hyperion images developed in this study.

20.2.7 Hyperion EO-1 hyperspectral satellite image: data acquisition and preprocessing

A Hyperion EO-1 image was acquired on February 15, 2005, and this was the focus of this research. Fig. 20–7 describes the procedure used for preprocessing Hyperion EO-1 images, which was followed in this study. The preprocessing procedure is described in the work of Arellano et al. (2015).

20.2.8 MTCI vegetation index

The medium resolution imaging spectrometer (MERIS) terrestrial chlorophyll index (MTCI) is a standard product of the European Space Agency (ESA), which provides estimations of the chlorophyll contents of vegetation (amount of chlorophyll per unit area of ground) at a global level. MTCI is simple to calculate, sensitive to high values of chlorophyll content (Dash and Curran, 2004; España-Boquera et al., 2006), and estimations are independent of soil and atmospheric conditions, spatial resolution and illumination, and observation geometry (Curran and Dash, 2005). Validation of the MTCI and ground chlorophyll content across a range of crop types and environmental conditions resulted in a strong relationship of $R^2 = 0.8$ and root mean square error (RMSE) = 192 g per MERIS pixel (Dash et al., 2010). Moreover, the strong relationship between MTCI and canopy chlorophyll content has been used to estimate gross primary production (GPP) across a range of ecosystems. Boyd et al. (2012) applied the MTCI together with radiation information (specifically photosynthetically active radiation and the fraction of photosynthetically active radiation), in models that extended the accuracy of the GPP estimated.

MTCI was computed using the ratio of the difference in reflectance between band 10 and band 9 and the difference in reflectance between band 9 and band 8 of the MERIS standard band setting:

$$\text{MTCI} = \frac{R_{\text{Band10}} - R_{\text{Band9}}}{R_{\text{Band9}} - R_{\text{Band8}}} = \frac{R_{753.75} - R_{708.75}}{R_{708.75} - R_{681.25}}$$

where $R_{753.75}$, $R_{708.75}$ and $R_{681.25}$ are the MERIS reflectance at wavelengths 753.75 nm, 708.75 nm, and 681.25 nm, respectively.

20.2.9 Vegetation indices suitable to detect hydrocarbon pollution

Several VIs were applied to the atmospherically corrected Hyperion EO-1 image in order to determine their ability to discriminate biophysical and biochemical changes in vegetation affected by hydrocarbon pollution. The results showed that almost all indices could clearly differentiate between secondary forest and pristine forest. The interest of this research was to explore VIs suitable to discriminate between the secondary tropical forest affected by hydrocarbon pollution and the secondary tropical forest not affected by pollution. From 29 indices applied to the study sites, the pair-wise comparison of means using the adjustment method of Holm established 11 indices that proved strong significant differences (99.9%) between the two secondary forests (polluted and nonpolluted); 3 indices are broadband and 8 are narrowband indices.

Discriminant function analysis applied to areas detected as suffering from vegetation stress by the 11 VIs showed that 3 indices (sum green, normalized difference vegetation index, and $NDVI_{705}$) explained 83% of the separability between the 3 study sites. Indeed, the detection of vegetation stress by applying those 3 indices illustrated similar results to those obtained with the 11 indices approach (Arellano Mora, 2014; Arellano et al., 2015).

20.2.10 Exploring new vegetation indices for detecting hydrocarbon pollution in the tropical rainforest

The methodology used in this section is based on the studies presented by Inoue et al. (2012), Stratoulias (2015), and Stratoulias et al. (2015). Spectral indices (SIs) using the exhaustive combinations of the complete spectral band set were calculated using contour maps of relevant statistical indicators such as the R^2. The degree of correlation between a trait and a SI indicate the predictive ability of the specific SI, and hence the two hyperspectral bands that construct the latter.

Two types of SI approaches were applied in this study to explore suitable band combinations of hyperspectral data to detect changes in the biophysical and biochemical parameters of vegetation affected by hydrocarbon pollution in the tropical forest. The first approach considers a RSI:

$$\mathrm{RSI} = \frac{R_i}{R_j}$$

where R_i and R_j are the reflectance response for bands i and j, respectively. The second approach explores a normalized difference spectral index (NDSI) that is effective in reducing the influence of errors or uncertainty due to sensor specifications and atmospheric and background differences. For instance, NDVI is derived from such a mathematical formula and is a widely used normalized VI applied for studies of plant growth, vegetation cover, and biomass and vegetation vigor. NDSI is defined by:

$$\mathrm{NDSI} = \frac{R_i - R_j}{R_i + R_j}$$

Maps of NDSI and RSI are useful for the evaluation of the relative significance of various combinations of wavelengths and for selecting a significant index with optimized wavelength and effective bandwidth. RSI and NDSI were estimated from reflectance samples collected at leaf level and correlated with the biophysical and biochemical parameters of the same leaves. The parameters considered were (1) chlorophyll content estimated from the SPAD-502 readings, (2) chlorophyll content derived from the inversion of the PROSPECT model, (3) foliar Cw, and (4) foliar organic matter content. The approach computes the R^2 for all possible combinations of available bands and the corresponding plant parameters. The computational process used an R programming language (R Core Team) generated by Stratoulias et al. (2015). In order to reduce the computer processing time, a subset of 127 foliar samples was selected, which represents a multispecies tropical rainforest of the Amazon region. For each sample, the reflectance signal at leaf level was averaged every three channels. The results are presented as raster maps of R^2 between the SIs and the specific biophysical or biochemical parameters of vegetation.

20.2.11 Results

20.2.11.1 *Exploring new vegetation indices for vegetation affected by petroleum pollution*

Maps of the R^2 for foliar parameters and the SIs (RSI and NDSI) are presented in Figs. 20–8 and 20–9, respectively. The maps use a complete combination of two wavebands at *i* and *j* nm for the foliar dataset from the Amazon forest. The results highlight the most significant regions of the spectrum to assess foliar parameters.

The results of R^2 for RSI and NDSI show similar patterns. Higher correlations are observed for chlorophyll content from SPAD-502 ($R^2 = 0.74$ and $R^2 = 0.73$, respectively)

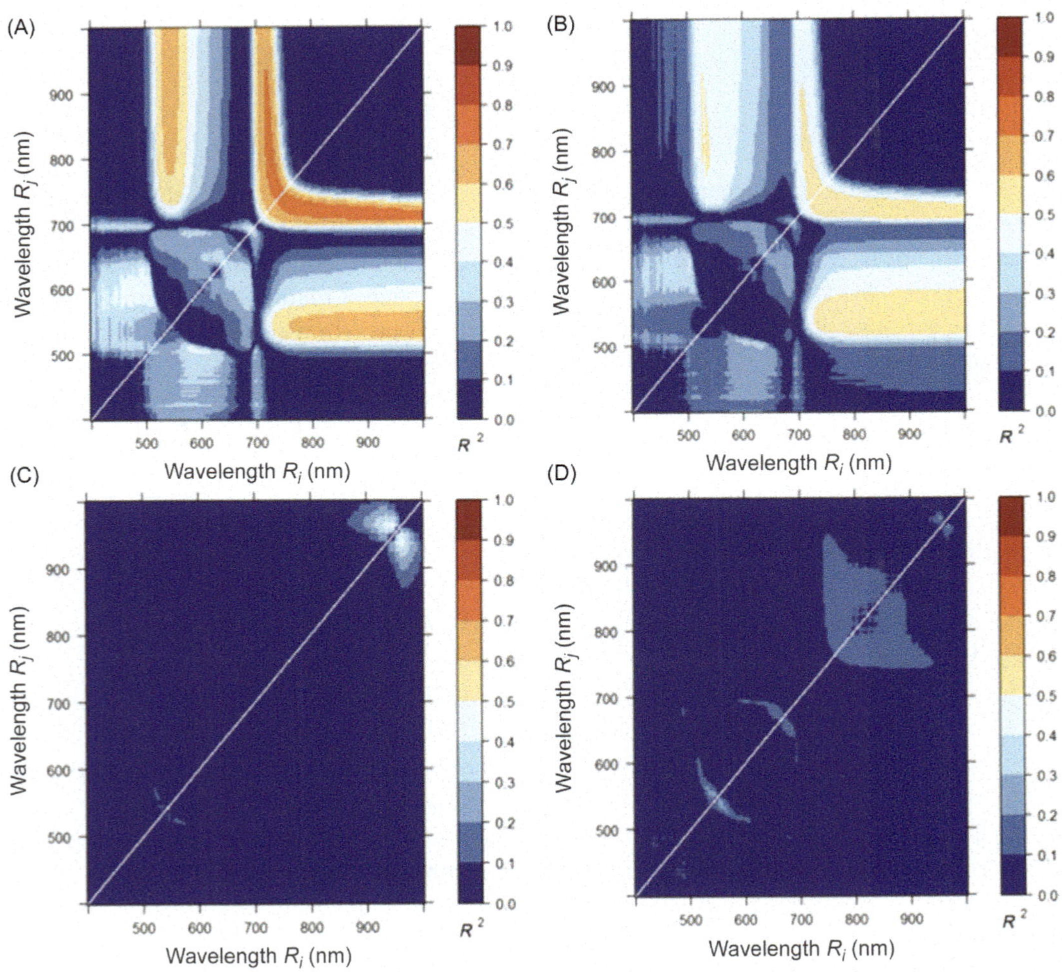

FIGURE 20–8 Coefficient of determination for each wavelength combination between RSI and foliar parameters. (A) Chlorophyll content from SPAD-502 readings, (B) chlorophyll content from the PROSPECT model, (C) water content, and (D) organic matter content.

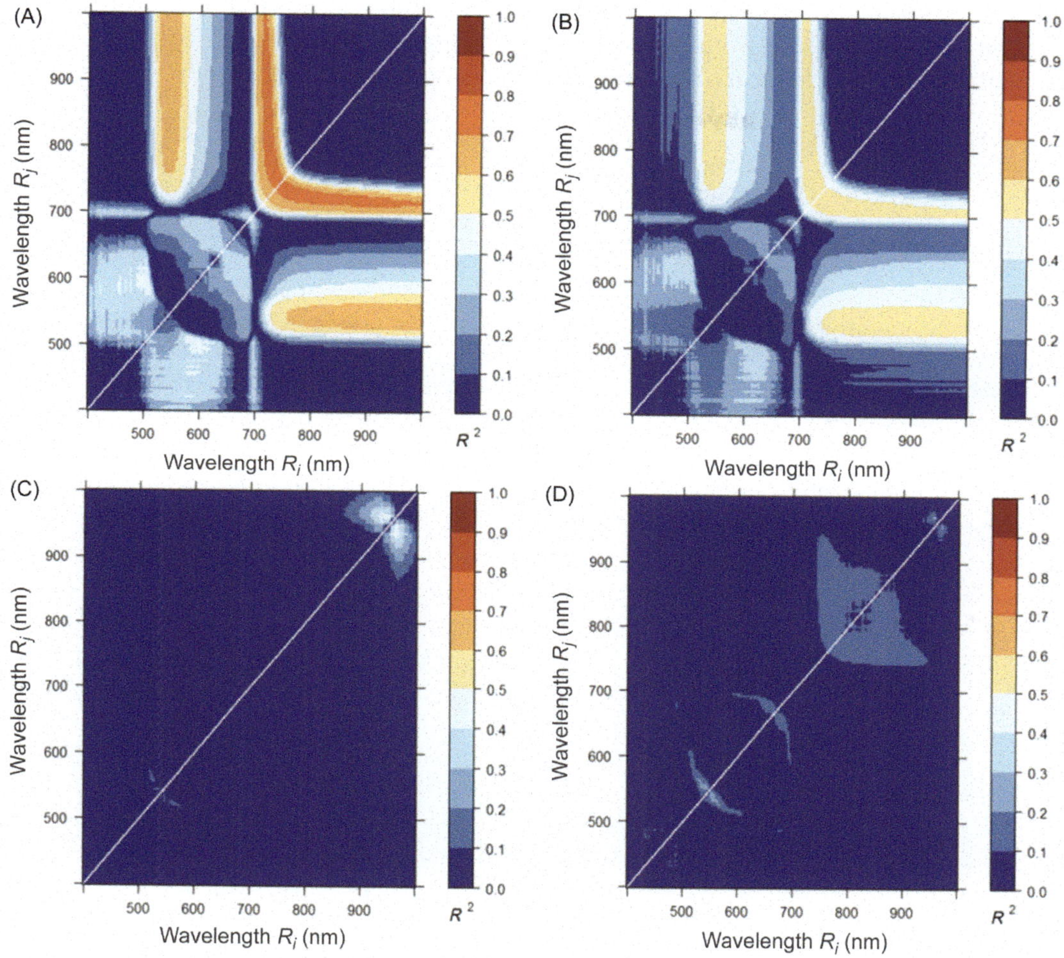

FIGURE 20–9 Coefficient of determination for each wavelength combination between normalized difference spectral index and foliar parameters. (A) Chlorophyll content from SPAD-502 readings, (B) chlorophyll content from the PROSPECT model, (C) water content, and (D) organic matter content.

followed by chlorophyll content from the PROSPECT model ($R^2 = 0.59$ and $R^2 = 0.53$, respectively). Lower R^2 are illustrated for Cw ($R^2 = 0.47$ and $R^2 = 0.47$, respectively) and organic matter content ($R^2 = 0.25$ and $R^2 = 0.25$, respectively). Table 20–3 presents the maximum R^2 and the corresponding R_i and R_j wavelength bands.

The higher R^2 (RSI = 0.74 and NDSI = 0.73) shown for chlorophyll content from SPAD-502 were selected to assess the chlorophyll content at canopy level in vegetation of the tropical rainforest in the study area using the reflectance spectral data from Hyperion EO-1 images.

Table 20–3 Maximum coefficient of determination and spectral bands for ratio spectral index (RSI) and normalized difference spectral index (NDSI) for foliar biophysical and biochemical parameters of tropical forest.

	RSI = x/y		NDSI = $(x-y)/(x+y)$	
	Coefficient of determination (R^2)	Spectral range (nm)	Coefficient of determination (R^2)	Spectral range (nm)
Chlorophyll content (μg/cm^2) from SPAD-502	0.74	$x=900$; $y=716$	0.73	$x=726$; $y=797$
Chlorophyll content (μg/cm^2) from PROSPECT model	0.59	$x=927$; $y=704$	0.53	$x=711$; $y=803$
Foliar water content	0.47	$x=948$; $y=962$	0.47	$x=948$; $y=962$
Foliar organic matter content	0.25	$x=555$; $y=542$	0.25	$x=543$; $y=554$

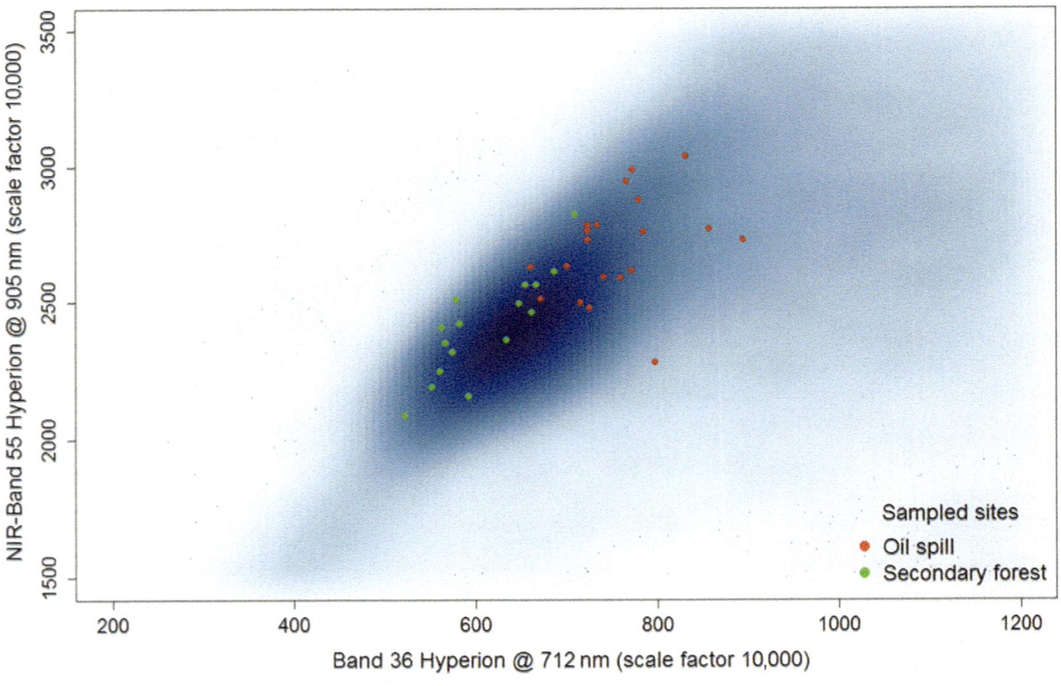

FIGURE 20–10 Ratio spectral index band combination. Smooth blue region represents a scatter plot of band 36 (712 nm) and band 55 (905 nm) of Hyperion EO-1 image (red dots = polluted site and green dots = nonpolluted site).

Fig. 20–10 illustrates the band combination of RSI adapted to the Hyperion EO-1 bands (band 36 and band 55) and Fig. 20–11 shows the NDSI band combination (band 37 and band 45 of Hyperion EO-1) in the areas of interest. The figures show a clear differentiation between forested areas affected and those not affected by hydrocarbon pollution.

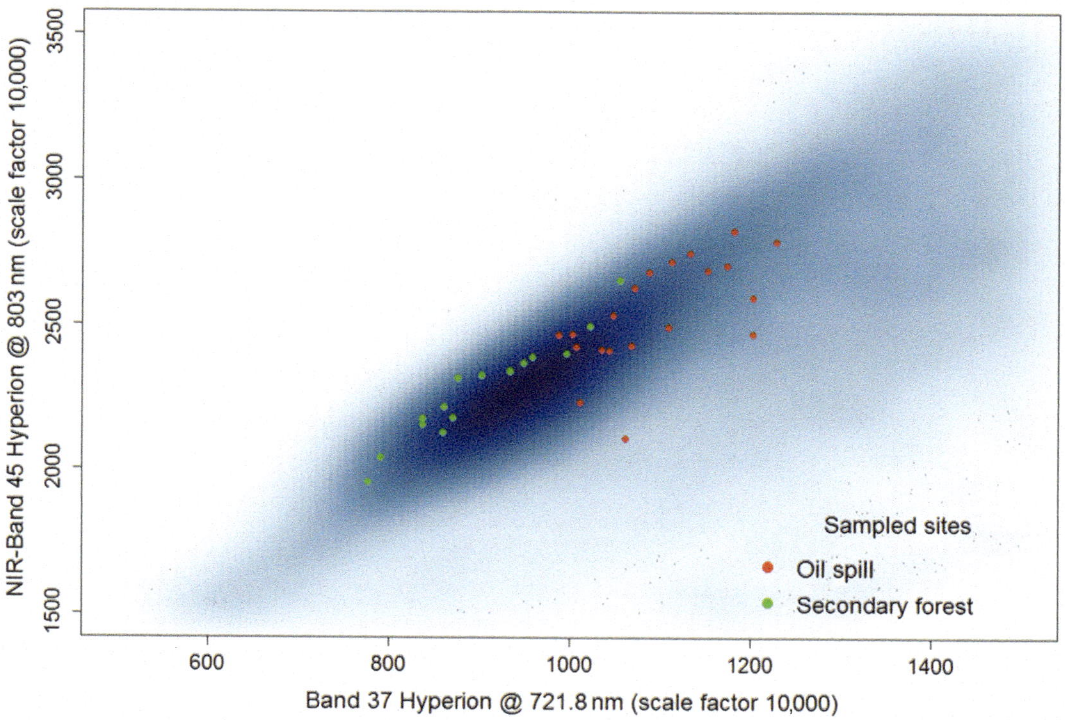

FIGURE 20–11 Normalized difference spectral index band combination. Smooth blue region represents a scatter plot of band 37 (721 nm) and band 45 (803 nm) of Hyperion EO-1 image (*red dots* = polluted site and *green dots* = nonpolluted site).

Plotting the relationship of spectral response of RSI and NDSI band combinations for the three study sites highlights an important difference between them (see Fig. 20–12). The pristine forest has a strong relationship between RSI and NDSI ($R^2 = 0.92$ and $R^2 = 0.96$, respectively), followed by the secondary nonpolluted site ($R^2 = 0.76$ and $R^2 = 0.93$, respectively), and finally the polluted site has a low relationship ($R^2 = 0.10$ and $R^2 = 0.40$, respectively). This suggests that in the tropical forest areas with pristine environments, the increase of RED reflectance is associated with an increase in NIR reflectance, meanwhile in polluted environments this relationship is not clear.

Tropical rainforest health was assessed by a combination of the derived VIs proposed in this study (SIs) and the MTCI, which is a standard product of the ESA and provides estimations of vegetation chlorophyll contents (amount of chlorophyll per unit area of ground) at a global level. A hyperspectral Hyperion EO-1 image was used to identify the status of three areas of tropical rainforest, namely (1) an area under vegetation stress caused by hydrocarbon pollution (detected by the three indices approach), (2) a secondary forest area nonpolluted by hydrocarbons, and (3) an area of pristine tropical rainforest in the Yasuni National Park.

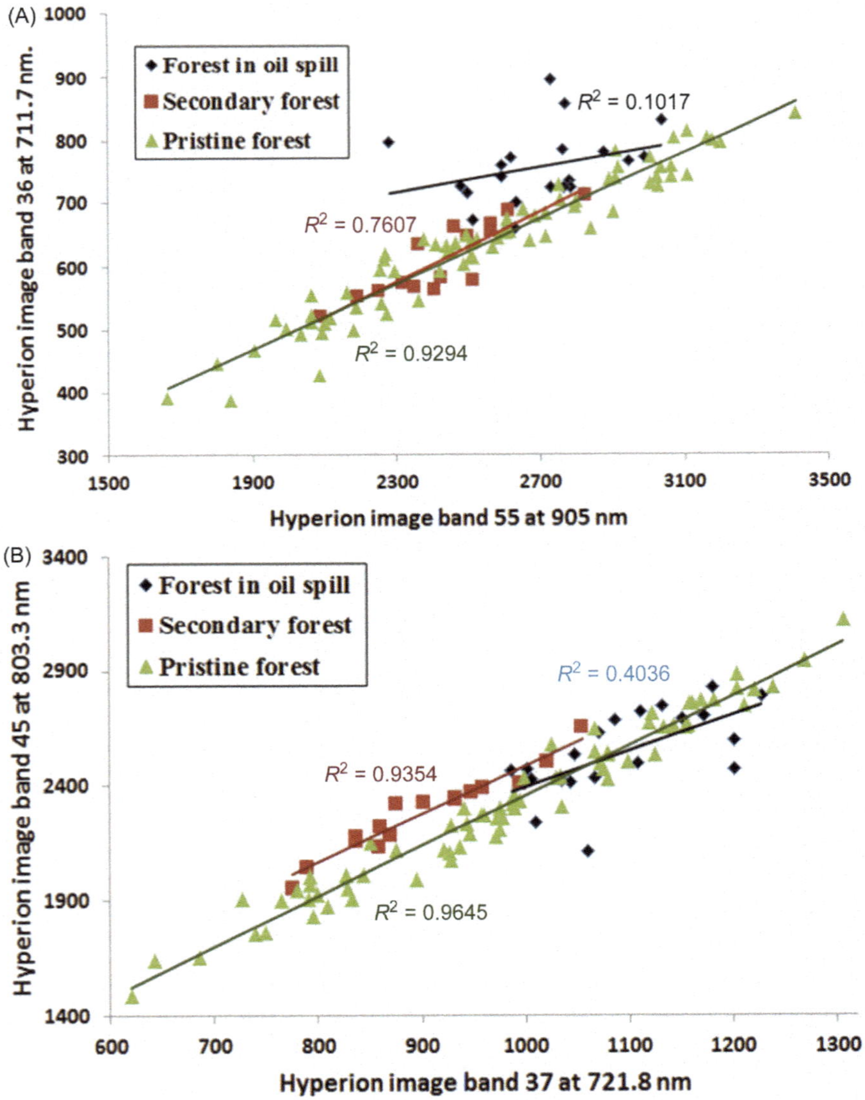

FIGURE 20–12 Correlation between band combinations applied to Hyperion EO-1 image bands in the study area for (A) ratio spectral index band combination and (B) normalized difference spectral index band combination.

Fig. 20–13 displays the sensitivity of RSI and MTCI indices to detect the status levels of the three tropical forest areas. The MTCI clearly differentiates higher chlorophyll levels in the pristine forest area, but showed less sensitivity to the polluted and nonpolluted secondary forest areas. Meanwhile RSI showed better sensitivity in detecting pristine forest than NDSI (see the same range of NDSI for pristine and secondary forest areas). On the other hand, NDSI provides a better identification of polluted and nonpolluted secondary forest areas.

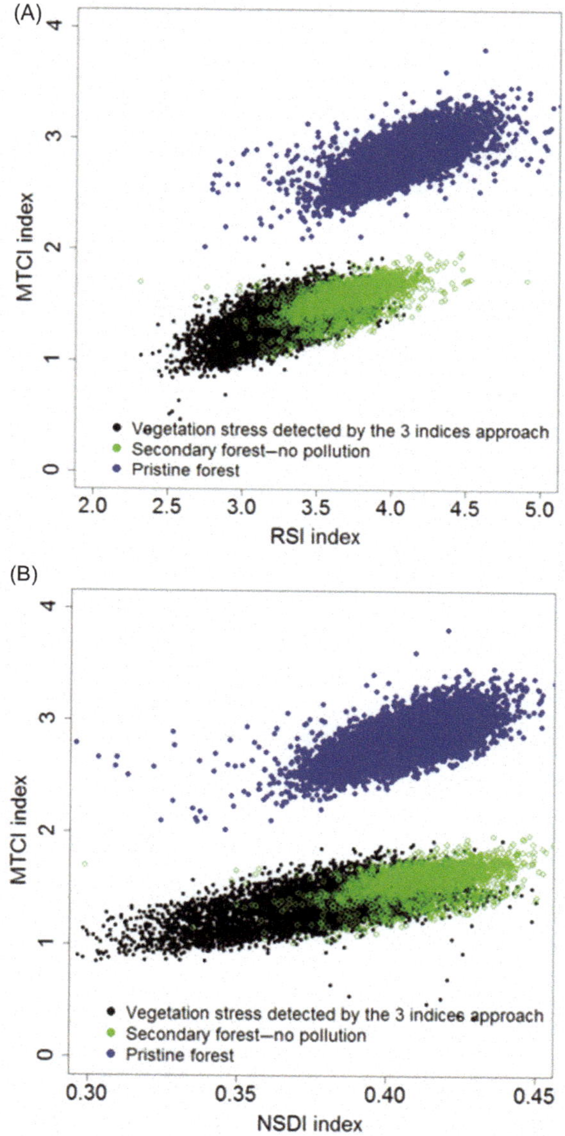

FIGURE 20–13 Identification of tropical rainforest health based on the vegetation indices proposed in this study and medium resolution imaging spectrometer terrestrial chlorophyll index (MTCI). (A) Ratio spectral index versus MTCI and (B) normalized difference spectral index versus MTCI.

20.3 Discussion

20.3.1 Exploring new vegetation indices for vegetation affected by petroleum pollution

From all possible combinations of two wavelengths, RSI (R_{900}/R_{716}) showed the most significant relationship with chlorophyll content ($R^2 = 0.74$) followed by the NDSI combination

($R_{726}-R_{797}/R_{726}+R_{797}$) with chlorophyll content ($R^2=0.73$). These are the new chlorophyll indices proposed in this study. It is worth noting that both indices were constructed with a combination of one wavelength in the red edge spectral region and one in the NIR spectral region. Moreover, Figs. 20–8 and 20–9 clearly depict that the broadband spectral region at the beginning of the NIR plateau (approximately 700–730 nm) strongly correlates with chlorophyll content. As shown in Table 20–3, the least significant relationship was found for chlorophyll content derived from the PROSPECT model ($R^2=0.59$ and $R^2=0.53$ for RSI and NDSI, respectively), which may be explained by the fact that the transmittance signal used for the inverse process of the model could not be collected from the field and was inferred from "transflectance".

Significant relationships were observed between the SIs and foliar Cw, which were $R^2=0.47$ for both cases and is related to the reflectance response driven by Cw around 834 nm in the NIR region. The low relationship of $R^2=0.25$ of organic matter content can be explained by the fact that the spectral range used for mapping coefficient correlations (400 and 1000 nm) is poorly correlated to vegetation parameters.

Figs. 20–10 to 20–12 illustrate higher reflectance values in the visible range of the polluted area, and are related to decreased levels of foliar chlorophyll content. On the other hand, higher reflectance values in the NIR region in the polluted site suggests lower foliar Cw. Indeed, the internal structure of leaves defines the air–water spaces and it has been proven that higher levels of intercellular water in leaves causes increased transmittance and reduced reflectance (Kumar et al., 2006). However, increased NIR reflectance signals have been found in plants growing in soils with high levels of pollution caused by heavy metals such as copper, lead, and zinc (Howard et al., 1971; Press, 1974).

A comparison of sensitivity in detecting the statuses of different tropical forest areas based on chlorophyll content levels showed that MTCI better identified the pristine forest area than the secondary forest areas. Meanwhile, NDSI and RSI illustrated a major sensitivity to differentiate between the polluted and nonpolluted secondary forest areas (Fig. 20–13).

Fig. 20–14 shows scatter plots of the best band combination of Hyperion EO-1 images (band 37 and band 45) resulting from the NDSI, which best differentiated between the polluted and nonpolluted secondary forest areas. Red dots represent the polluted area and green dots represent nonpolluted pixels. This figure confirms that vegetation growing in areas affected by hydrocarbons experience changes in biophysical and biochemical parameters, which can be detected using hyperspectral satellite images. Higher reflectance in the visible range stands for reduced levels of chlorophyll content and higher reflectance in the NIR region, which is related to changes in the internal structure of the leaves in areas affected by pollution. Increased reflectance in the NIR region has been reported to be correlated to decreased Cw, but the results of this study indicate that higher levels of Cw reported in the polluted area correlates with higher reflectance in the NIR. The increased reflectance in the polluted site agrees with other studies of plants growing in soils with high levels of pollution caused by heavy metals such as copper, lead, and zinc (Howard et al., 1971; Press, 1974).

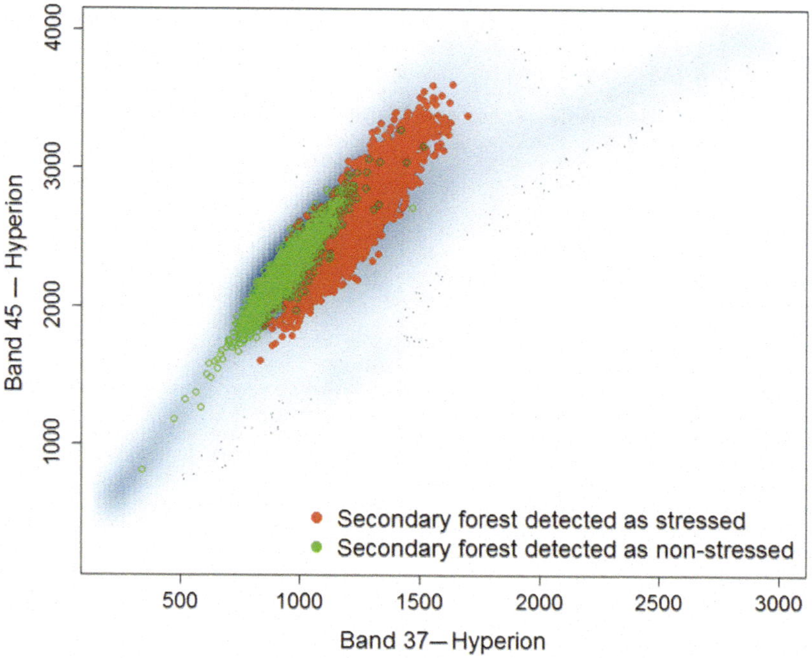

FIGURE 20–14 Scatter plot of Hyperion EO-1 normalized difference spectral index band combination (*blue smoothed*). Secondary forest detected as polluted by the three vegetation indices (SG, NDVI, and $NDVI_{705}$) in red and vegetation detected as nonpolluted in green.

20.4 Conclusion

New VIs were explored for the detection of vegetation stress caused by hydrocarbon pollution. Maps of coefficient of determination (R^2) between foliar parameters and two SIs (RSI and NDSI) were computed. The maps use a complete combination of two wavebands at i and j nm for a foliar dataset from the Amazon forest. The results highlight the most significant regions of the spectrum to assess foliar parameters.

Higher correlations were observed between RSI and NDSI, and chlorophyll content derived from SPAD-502 ($R^2 = 0.74$ and $R^2 = 0.73$, respectively), followed by chlorophyll content from the PROSPECT model ($R^2 = 0.59$ and $R^2 = 0.53$, respectively). Lower coefficients of determination are illustrated for Cw ($R^2 = 0.47$ for both SIs) and organic matter content ($R^2 = 0.25$ for both SIs). Based on that, two new indices sensitive to chlorophyll content in the tropical rainforest are proposed in this study, namely RSI (R_{900}/R_{716}) and NDSI ($R_{726}-R_{797}/R_{726} + R_{797}$).

The new indices were applied to a Hyperion EO-1 satellite image in order to assess the chlorophyll content at canopy level in vegetation of the tropical rainforest. In the pristine forest sampled during fieldwork, the results revealed a strong correlation of 0.92 and 0.96 by RSI and NDSI, respectively, followed by the secondary nonpolluted site ($R^2 = 0.76$ and

$R^2 = 0.93$, respectively). In the secondary forest polluted site, the correlation of band combination is weak ($R^2 = 0.10$ and $R^2 = 0.40$, respectively). This suggests that the effects of hydrocarbon pollution in the tropical forest can be detected by Hyperion EO-1 satellite images by observing an increase of RED reflectance, associated to an increase in NIR reflectance.

List of abbreviations

Cm dry matter content
CO_2 carbon dioxide
Cw water content
Dw dry weight
dw water density
ESA European Space Agency
Ewt equivalent water thickness
Fw fresh weight
GPP gross primary production
LAI leaf area index
MERIS medium resolution imaging spectrometer
MTCI terrestrial chlorophyll index
NDSI normalized difference spectral index
NIR near infrared
REP red-edge position
UTM Universal Transverse Mercator
VI vegetation index
WGS World Geodetic System
Ws specific leaf mass

List of symbols

°C degree Celsius
C_{ab} foliar chlorophyll content
C_{bp} brown pigment
cm centimeter
g gram
L luminance
L_1 luminance measured with the white background disk only
L_2 luminance measured with the black background disk only
L_3 luminance of the leaf mounted on the white background disk
L_4 luminance of the leaf mounted on the black background disk
m meter
nm nanometer
R reflectance
R^2 coefficient of determination
$R_{measured}$ leaf reflectance measured
R_{est} leaf reflectance estimated
T leaf transmittance
R_b reflectance of blank background disk

R_w	reflectance of the white background disk
R_w^{each}	reflectance of each leaf + white background disk
R_b^{each}	reflectance of each leaf + black background disk.
L_{inc}	lamp luminance
R_i	reflectance at wavelength *i*

References

Arellano, P., Tansey, K., Balzter, H., Boyd, S.D., 2015. Detecting the effects of hydrocarbon pollution in the Amazon forest using hyperspectral satellite images. Environ. Pollut. 205, 225–239.

Arellano Mora, P.N., 2014. Hyperspectral Remote Sensing for Detecting Vegetation Affected by Hydrocarbons in the Amazon Forest (Doctoral dissertation, Department of Geography).

Asner, G.P., Nepstad, D., Cardinot, G., Ray, D., 2004. Drought stress and carbon uptake in an Amazon forest measured with spaceborne imaging spectroscopy. Proc. Natl. Acad. Sci. U.S.A. 101 (16), 6039–6044.

Baker, J.M., 1970. The effects of oil on plants. Environ. Pollut. 1, 27–44.

Bernal, A.M., 2011. Power, powerlessness and petroleum: indigenous environmental claims and the limits of transnational law. N. Polit. Sci. 33 (2), 143–167.

Blackburn, G.A., 2007. Hyperspectral remote sensing of plant pigments. J. Exp. Bot. 58 (4), 855–867.

Bonan, G., 2008. Ecological Climatology, second ed. Cambridge University Press, United Kingdom.

Boyd, D.S., Almond, S., Dash, J., Curran, P.J., Hill, R.A., Foody, G.M., 2012. Evaluation of envisat MERIS terrestrial chlorophyll index-based models for the estimation of terrestrial gross primary productivity. IEEE Geosci. Remote Sens. Lett. 9 (3), 457–461.

Castro-Esau, K.L., Sánchez-Azofeifa, G.A., Rivard, B., Wright, S.J., Quesada, M., 2006. Variability in leaf optical properties of mesoamerican trees and the potential for species classification. Am. J. Bot. 93 (4), 517–530.

Clevers, J.P.G.W., Jongschaap, R., 2006. Imaging spectrometry for agricultural applications. In: van der Meer, F., de Jong, S.M. (Eds.), Imaging Spectrometry. Springer, The Netherlands, pp. 157–199.

Curran, P.J., Dash, J., 2005. Algorithm Theoretical Basis Document ATBD 2.2: Chlorophyll Index. University of Southampton, Southampton.

Dash, J., Curran, P.J., 2004. The MERIS terrestrial chlorophyll index. Int. J. Remote Sens. 25 (23), 5403–5413.

Dash, J., Curran, P.J., Tallis, M.J., Llewellyn, G.M., Taylor, G., Snoeij, P., 2010. Validating the MERIS terrestrial chlorophyll index (MTCI) with ground chlorophyll content data at MERIS spatial resolution. Int. J. Remote Sens. 31 (20), 5513–5532.

Datt, B., 1999. Remote sensing of water content in Eucalyptus leaves. Aust. J. Bot. 47 (6), 909–923.

Davidson, E.A., De Araüjo, A.C., Artaxo, P., Balch, J.K., Brown, I.F., Bustamante, C., et al., 2012. The Amazon basin in transition. Nature 481 (7381), 321–328.

España-Boquera, M.L., Cárdenas-Navarro, R., López-Pérez, L., Castellanos-Morales, V., Lobit, P., 2006. Estimating the nitrogen concentration of strawberry plants from its spectral response. Commun. Soil. Sci. Plant Anal. 37 (15–20), 2447–2459.

Féret, J., François, C., Gitelson, A., Asner, G.P., Barry, K.M., Panigada, C., et al., 2011. Optimizing spectral indices and chemometric analysis of leaf chemical properties using radiative transfer modeling. Remote Sens. Environ. 115 (10), 2742–2750.

Finer, M., Jenkins, C.N., Pimm, S.L., Keane, B., Ross, C., 2008. Oil and gas projects in the Western Amazon: threats to wilderness, biodiversity, and indigenous peoples. PLoS One 3 (8).

Fitton, L.J., 2000. Helminthiasis and culture change among the cofán of ecuador. Am. J. Hum. Biol. 12 (4), 465–477.

Gerber, F., Marion, R., Olioso, A., Jacquemoud, S., Ribeiro da Luz, B., Fabre, S., 2011. Modeling directional–hemispherical reflectance and transmittance of fresh and dry leaves from 0.4 μm to 5.7 μm with the PROSPECT–VISIR model. Remote Sens. Environ. 115 (2), 404–414.

Hoel, B.O., 1998. Use of a hand-held chlorophyll meter in winter wheat: evaluation of different measuring positions on the leaves. Acta Agric. Scand.— B Soil Plant Sci. 48 (4), 222–228.

Howard, J.A., Watson, R.D., Hessin, T.D., 1971. Spectral reflectance properties of Pinus ponderosa in relation to copper content of the soil—Malachite Mine, Jefferson Country, Colorado. In: Proceedings of the 7th International Symposium on Remote Sensing of Environment, May 1971, pp. 285.

Hunt Jr, E.R., Rock, B.N., 1989. Detection of changes in leaf water content using near- and middle-infrared reflectances. Remote Sens. Environ. 30 (1), 43–54.

Hurtig, A.K., San-Sebastián, M., 2005. Epidemiology vs epidemiology: the case of oil exploitation in the Amazon basin of Ecuador [5]. Int. J. Epidemiol. 34 (5), 1170–1172.

Inoue, Y., Sakaiya, E., Zhu, Y., Takahashi, W., 2012. Diagnostic mapping of canopy nitrogen content in rice based on hyperspectral measurements. Remote Sens. Environ. 126, 210–221.

ISO, 2012. Determination of transmittance by diffuse reflectance measurement. ISO 22891:2012(E), International Standard Organization.

Jacquemoud, S., Marcq, S., 2012. Personal Comunication. Institut de Physique du Globe de Paris & Université Paris Diderot.

Kumar, L., Schmidt, K., Dury, S., 2006. Imaging spectrometry and vegetation science. In: van der Meer, F., de Jong, S. (Eds.), Imaging Spectrometry. Basic Principles and Prospective Applications, fourth ed. Springer, Dordrecht, The Netherlands, pp. 111–155.

Kvenvolden, K.A., Rogers, B.W., 2005. Gaia's breath—global methane exhalations. Mar. Pet. Geol. 22 (4), 579–590.

Malhi, Y., Roberts, J.T., Betts, R.A., Killeen, T.J., Li, W., Nobre, C.A., 2008. Climate change, deforestation, and the fate of the Amazon. Science 319 (5860), 169–172.

Marenco, R.A., Antezana-Vera, S.A., Nascimento, H.C.S., 2009. Relationship between specific leaf area, leaf thickness, leaf water content and SPAD-502 readings in six Amazonian tree species. Photosynthetica 47 (2), 184–190.

Noomen, M.F., 2007. Hyperspectral Reflectance of Vegetation Affected by Underground Hydrocarbon Gas Seepage, PhD ed. International Institute for Geo-Information Science & Earth Observation-ITC, Enschede, The Netherlands.

Noomen, M.F., Skidmore, A.K., 2009. The effects of high soil CO_2 concentrations on leaf reflectance of maize plants. Int. J. Remote Sens. 30 (2), 481–497.

Noomen, M.F., Smith, K.L., Colls, J.J., Steven, M.D., Skidmore, A.K., Van Der Meer, F.D., 2008. Hyperspectral indices for detecting changes in canopy reflectance as a result of underground natural gas leakage. Int. J. Remote Sens. 29 (20), 5987–6008.

Press, N.P., 1974. Remote sensing for detect the toxic effects of metals on vegetation of mineral exploration. In: Proceesings of the 9th International Symposium on Remote Sensing of Environment, Michigan, 15–19 April, pp. 2027.

Sims, D.A., Gamon, J.A., 2002. Relationships between leaf pigment content and spectral reflectance across a wide range of species, leaf structures and developmental stages. Remote Sens. Environ. 81 (2–3), 337–354.

Smith, K.L., Steven, M.D., Colls, J.J., 2004a. Remote sensing techniques for monitoring plant stress responses to gas leaks. In: International Gas Research Conference Proceedings.

Smith, K.L., Steven, M.D., Colls, J.J., 2004b. Use of hyperspectral derivative ratios in the red-edge region to identify plant stress responses to gas leaks. Remote Sens. Environ. 92 (2), 207–217.

Smith, K.L., Steven, M.D., Colls, J.J., 2005. Plant spectral responses to gas leaks and other stresses. Int. J. Remote Sens. 26 (18), 4067.

Stratoulias, D., 2015. Assessment of Physiological Status and Spatial Distribution of Emergent Macrophytes Based on Imaging Spectroscopy (Doctoral Dissertation, Department of Geography).

Stratoulias, D., Balzter, H., Zlinszky, A., Tóth, V.R., 2015. Assessment of ecophysiology of lake shore reed vegetation based on chorophyll fluorescence, field spectroscopy and hyperspectral airborne imagery. Remote Sens. Environ. 157, 72–84.

Tedersoo, L., Sadam, A., Zambrano, M., Valencia, R., Bahram, M., 2010. Low diversity and high host preference of ectomycorrhizal fungi in Western Amazonia, a neotropical biodiversity hotspot. ISME J. 4 (4), 465–471.

Valencia, R., 2004. Yasuni forest dinamic plot, Ecuador. In: Losos, E., Leigh, J., Giles, E. (Eds.), Tropical Forest Diversity and Dynamism: Findings From a Large-Scale Plot Network. University of Chicago Press, Chicago, IL, pp. 609–628.

Valencia, R., Foster, R.B., Richard-Condit, G.V., Svenning, J.C., Hernández, C., Romoleroux, K., et al., 2004. Tree species distributions and local habitat variation in the Amazon: large forest plot in eastern Ecuador. J. Ecol. 92, 214–229.

van der Meer, F., de Jong, S., Bakker, W., 2006. Imagining spectrometry: basic analytical techniques. In: van der Meer, F., de Jong, S. (Eds.), Imaging Spectrometry. Basic Principles and Prospective Applications, third ed. Springer, p. 35.

White, J.W., Montes-R, C., 2005. Variation in parameters related to leaf thickness in common bean (*Phaseolus vulgaris* L.). Field Crop. Res. 91 (1), 7–21.

Wright, I.J., Reich, P.B., Westoby, M., Ackerly, D.D., Baruch, Z., Bongers, F., et al., 2004. The worldwide leaf economics spectrum. Nature 428 (6985), 821–827.

Yang, H., 1999. Imagining Spectrometry for Hydrocarbon Microseepage. Delf University of Technology.

Yang, H., Zhang, J., Van Der Meer, F., Kroonenberg, S.B., 1999. Spectral characteristics of wheat associated with hydrocarbon microseepages. Int. J. Remote Sens. 20 (2–4), 807–813.

Yang, H., Zhang, J., Van Der Meer, F., Kroonenberg, S.B., 2000. Imaging spectrometry data correlated to hydrocarbon microseepage. Int. J. Remote Sens. 21 (1), 197–202.

Zhu, L., Zhao, X., Lai, L., Wang, J., Jiang, L., Ding, J., et al., 2013. Soil TPH concentration estimation using vegetation indices in an oil polluted area of Eastern China. PLoS One 8 (1).

SECTION

VI

Hyperspectral Remote Sensing: Challenges, Future Pathway for Research & Emerging Applications

21

Future perspectives and challenges in hyperspectral remote sensing

Prem Chandra Pandey[1], Heiko Balzter[2,3], Prashant K. Srivastava[4,5], George P. Petropoulos[6], Bimal Bhattacharya[7]

[1]CENTER FOR ENVIRONMENTAL SCIENCES & ENGINEERING, SCHOOL OF NATURAL SCIENCES, SHIV NADAR UNIVERSITY, GAUTAM BUDDHA NAGAR, INDIA

[2]CENTRE FOR LANDSCAPE AND CLIMATE RESEARCH, SCHOOL OF GEOGRAPHY, GEOLOGY AND THE ENVIRONMENT, UNIVERSITY OF LEICESTER, LEICESTER, UNITED KINGDOM

[3]NATIONAL CENTRE FOR EARTH OBSERVATION, UNIVERSITY OF LEICESTER, LEICESTER, UNITED KINGDOM

[4]REMOTE SENSING LABORATORY, INSTITUTE OF ENVIRONMENT AND SUSTAINABLE DEVELOPMENT, BANARAS HINDU UNIVERSITY, VARANASI, INDIA

[5]DST-MAHAMANA CENTRE FOR EXCELLENCE IN CLIMATE CHANGE RESEARCH, BANARAS HINDU UNIVERSITY, VARANASI, INDIA

[6]DEPARTMENT OF GEOGRAPHY, HAROKOPIO UNIVERSITY OF ATHENS, ATHENS, GREECE

[7]SPACE APPLICATIONS CENTRE, INDIAN SPACE RESEARCH ORGANIZATION, AHMEDABAD, INDIA

21.1 Introduction

Remote sensing (RS) technologies have advanced significantly and led to the employment of new technologies enabling novel data processing methods and better quality data with increased spatial, spectral, radiometric, and temporal resolutions for a wide variety of research needs (Pandey et al., 2019b). There has been parallel growth in instrument development, methods for improving data processing, information retrieval, integration/fusion of different datasets (signal, pixel, feature, and decision-level), and interpretation of products for dissemination in different research fields. RS techniques offer an economical and practical means to assess features required by users for applications such as mineral and soil mapping, plant species or crop type identification, snow and ice mapping, and others (Lillesand et al., 2014; Vaiphasa et al., 2007). RS enables large-area coverage over inaccessible regions, making field sampling more focused, efficient, and reliable (Lillesand et al., 2014; Pandey et al., 2019a). RS provides users with detailed information about target features at different scales (regional to global) with varying spectral responses at different temporal

Hyperspectral Remote Sensing. DOI: https://doi.org/10.1016/B978-0-08-102894-0.00021-8

scales (Wulder, 1998). However, often RS techniques fail to capture the finer details about different objects within similar features (such as different crop types or tree species).

To address this shortcoming, hyperspectral imaging (HSI) systems have been developed. Sometimes hyperspectral images have a low spatial resolution, which makes their interpretation difficult. To deal with this problem, different methods, for example, multisensor concepts combining multispectral and hyperspectral sensors through multisource analysis, can assist in providing the level of detail required for planning, management, and policy implementation (Pandey et al., 2019b). The prospect of upcoming satellite missions with higher spatial and spectral resolution sensors in orbit in the near future (Pandey et al., 2019b) will increase the utility of hyperspectral data in several research domains and will likely increase the amount of users of HSI for soil, forest, agriculture, urban, and snow research.

Hyperspectral RS has led to a giant leap in the field of RS. The increasing demand for hyperspectral images across the globe and their adoption in several research applications has increased their visibility. The availability of hyperspectral images has enriched RS research through better data quality and a much stronger ability to discriminate different features easily and identify them from their spectral profile, which is unique to them (like human fingerprints are unique to each individual). However, approaches for identifying hyperspectral image characteristics are sometimes not as successful during processing. Numerous spectral bands and voluminous data not only cause difficulties in data storage and transmission, but also pose new challenges in image processing (Nhaila et al., 2014).

HSI technologies are undergoing a technological revolution in the sense of moving toward high spatial resolution, high spectral resolution as well as frequent revisit intervals (Pandey et al., 2019b; Colkesen and Kavzoglu, 2019). To achieve all these technical improvements, RS is undergoing rapid development and improvement in sensor designs to launch new missions, incorporating satellites, aircrafts, and unmanned aerial vehicles (UAVs). The outcomes of these changes are facilitating the increased availability of very-high spatial resolution imagery at meter and submeter pixel resolutions (Benediktsson et al., 2013) and also the extraction of extremely pure pixels for analysis (Krishna et al., 2018). Spatial resolution and spectral resolution play important roles in different hyperspectral research applications and have attracted researchers aiming to develop techniques to provide super-resolution hyperspectral images in both the spatial and spectral domains (Mianji et al., 2008). Similarly, target detection also remains one of the major challenges of HSI, as the inherent variability in target and background spectra is a severe obstacle in the development of effective target detection algorithms for hyperspectral images. It may be due to the problem of unknown backgrounds or a lack of sufficient target data, making it more challenging and an issue that needs to be resolved by more sophisticated techniques (Mianji et al., 2008). Most researchers find major challenges in the complex and high-load computational algorithms of RS, which pose a hurdle for achieving the desired higher outputs or accuracies from hyperspectral processing.

The evolution from aerial photographs (Avery and Berlin, 1992) to multispectral and hyperspectral RS illustrates the emerging trend of the RS technology from low to high spatial resolution and much better spectral resolution from spaceborne to airborne acquisition of hyperspectral images from 30 m (~Hyperion data) down to 2–4 m resolution [Airborne

Visible/Infrared Imaging Spectrometer Next Generation (AVIRIS NG)]. Employing UAVs, the spatial resolution can be increased to much less than a meter, but data volume and storage can become problems (Hardin et al., 2019). In order to supplement RS data, in situ HSI spectroradiometers have been used and have generated accurate results for the distinction between different features (phytoplankton, plant species, species distribution, carbon mapping, heavy metal detection, soil, and minerals) (Manjunath et al., 2011; Schmidt and Skidmore, 2003; Lamine et al., 2019; Anand et al., 2020; Malhi et al., 2020). Further, the combination of spectral and spatial information extraction techniques using image integration shows promise for increased and improved accuracy of results. Therefore the integration of RS techniques with field measurements helps to achieve more accurate results over large areas.

In order to identify different features within a similar class such as different plant/crop types or species, the spectral resolution must be sufficiently high (Pandey et al., 2019a; Adar et al., 2014; Fauvel et al., 2008; He et al., 2011). Hyperspectral RS plays an important role in several research activities and identifies target features based on the spectral signatures and their analysis, for example, heavy metal contamination (Lamine et al., 2020), grass discrimination (Naicker et al., 2020), soil (Yu et al., 2020), feature classification (Pal, 2020), and parasites (Žibrat et al., 2020).

21.2 Challenges of hyperspectral imaging systems

Hyperspectral techniques are used extensively in a variety of research areas and environmental applications due to their capability to detect and discriminate between the specific features of objects with several contiguous spectral channels. HSI systems (hundreds of thousands of narrow bands) provide more detailed information than multispectral images (3–10 wider bands). Hyperspectral images provide better capability to see the previously unseen through their higher spectral detail. For example, if HSI can differentiate three minerals in a region, it is only because of its high spectral resolution, whereas multispectral imaging could not do the same due to its broad spectral width. Thus the high spectral resolution details in HSI offers the capability to see the unseen. But the addition of several narrow contagious spectral bands (approximately more than 200) in the data will increase the level of complexity, hence, the challenges associated with data redundancy between spectral channels (refer to Hughes phenomenon discussed earlier).

Another advantage of HSI is that because an entire spectrum is acquired at each point, the data analyst needs no prior knowledge of the sample, and postprocessing allows all available information from the dataset to be extracted. HSI can also take advantage of the structural relationships among the different spectra in a neighborhood, allowing more elaborate spectral–structural models for a more accurate analysis and classification of an image (Nguyen, 2017).

Imaging a large number of spectral bands and the advantages that this brings over multispectral imaging comes with several limitations and drawbacks. The prime disadvantages of hyperspectral data are cost and data complexity. To simplify hyperspectral data analysis, fast computers and large data storage capacities are needed and can be rented on-demand via cloud computing services, for example. Another disadvantage is finding ways to program hyperspectral instrumentation with onboard processing that transmits only quality-assured images, as both the transmission and storage of that much data could prove difficult and costly (Nguyen, 2017).

Increased demands for spectral resolution in forestry, soil science, agriculture, and many more applications have gained momentum due to their robustness in change detection. Spectral signatures employing continuum removal techniques, first, or second derivatives for change detection are generally robust. Spectral libraries have been created for groups of features by The United States Geological Survey and international agencies for crops (Rao et al., 2007), medicinal plants, soil types, and different minerals (Rao et al., 2007) to help ease signature-matching and analysis (Rajadell et al., 2009). To address the data dimensionality, segmentation techniques are employed to extract more meaningful information from hyperspectral data for analysis in different research domains (Lanthier et al., 2008; Pandey et al., 2014, 2018).

Classification algorithms involving hyperspectral images employ several easy as well as complex techniques, which are utilized in multidisciplinary applications. But these classification algorithms are complex in nature and associated with image acquisition and processing with high dimensionality, with the necessity of data mining techniques (Nhaila et al., 2014). Thus, it can be concluded, how to eliminate the redundant spectral bands (attributes or features) and how to choose the really informative relevant spectral bands are two main challenges associated with the hyperspectral images.

Finding the most appropriate hyperspectral image classification algorithm is a challenging task (Kakhani and Mokhtarzade, 2019) because traditional statistical classification algorithms (maximum likelihood) are not efficient in obtaining accurate results with hyperspectral images due to their high data dimensionality (Pandey et al., 2014; Richards and Richards, 1999; Pandey et al., 2018). These traditional classification algorithms require a large number of training samples proportional to the number of input data dimensions, which is generally referred as the Hughes phenomenon (Hughes, 1968). The Hughes phenomenon occurs when training sites or reference sample numbers are inadequate compared with the number of input target features (and, therefore, of classifier parameters). This causes a decline in the accuracy of results as a component of the information dimensionality increase with an increased number of spectral channels in digital images. Thus the lack of a sufficient number of training samples during classification causes a singularity in most statistical processing methods. Pandey et al. (2014) demonstrated methods to reduce the Hughes phenomenon utilizing segmented principal component analysis (S-PCA) techniques that segregate spectral ranges into several informative discrete spectral groups. This SPCA techniques is modified version of principal component analysis (PCA) in terms of spectral division.The highly informative principal component bands of each group (first three principal components only) were considered for further processing and the incorporation of information as much as possible, giving rise to the enhanced visualization of the resultant images. The resultant images were used for classification, which generated improved results compared to individual images. This technique reduces the Hughes phenomenon because it separates full spectral ranges into several groups.

To deal with this, nonstatistical algorithms such as the spectral angle mapper (SAM) and multiple endmember spectral mixture methods (which mainly focus on the spectral patterns) are employed and are widely applied to hyperspectral images (Petropoulos et al., 2015). Again, these algorithms are extremely dependent upon the endmember extraction or exact determination of pure pixels from hyperspectral images. Therefore another challenging issue with hyperspectral images seems to be the endmember extraction due to the low spatial resolution resulting in mixed pixels. Endmember extraction is challenging due to the presence of many mixed pixels in hyperspectral images.

Spectral unmixing is a different technique for classification that separates a pixel into extents of several spectral segments known as endmembers, and further information of their extent to the general spectral response of the pixel is mandatory (Zhang et al., 2006; Okin et al., 2001; Hostert et al., 2003). In general, the number of endmembers chosen must be less than the spectral channels of the image being employed for the classification methods. Further, all chosen endmembers available in the image ought to be utilized with the specific end goal of getting consistent and reliable end products. The spectral unmixing outcome is one for each particular endmember, with pixel values ranging from zero and one, demonstrating the fraction of the first image attributed to the specific endmembers (Pandey et al., 2018). Subpixel classification techniques are by and large separated into linear and nonlinear unmixing methods, depending upon whether each pixel's reflectance is a linear or a nonlinear combination of target features within the pixel. Hyperspectral information is proper for linear methods as it has substantially more spectral channels. Mixing methods are additionally useful for small target feature identification and the discrimination of alien or intrusive species. Miao et al. (2006) demonstrated an overview of four distinct linear mixing models and their limitations for classification methods.

A drawback affecting Hyperion images is linked to the spectral and spatial artifacts related to array malfunctions in some sensors such as flag pixels >4095, which represent pixels affected by echo and smear, dark current and background removal of the residual charge in the detector, sensor bias effects, and bad pixels (uncalibrated pixels). [Echo removal from short wave infrared (SWIR) data occurs when the signal from one image echoes into the neighbor pixels and smear correction to SWIR data is the leakage of a signal from one pixel into the next readout pixel in the spectra direction.]

Another problem that exists with hyperspectral data is spectral smile. Spectral smile, also known as the "smile" or "frown" curve, is a spectral distortion referring the spectral distortion to an across-track wavelength shift from the central wavelength, which is due to a change of the dispersion angle with the field position. It is typically found in pushbroom sensors. It was named as such because the curve looks like a mixture of a "frown" and a "smile" resulting in a "wave" shape. Therefore it is a shift in wavelength in the spectral domain, which is a function of an across-track pixel (column) in the swath. For VIS–NIR–SWIR wavelength regions, it shifts slightly to some extent. In visible-near infrared (VNIR) bands the shift range is between 2.6 and 3.5 nm, with the maximum shift occurring at column 256 in band 10. In SWIR bands, the spectral shift is less than 1 nm (Goodenough et al., 2003). This problem is mostly talked about in the context of the Hyperion sensor, and little is heard or talked about in relation to the other sensors. The Hyperion instrument is susceptible to the smile effect because of the arrangement of its two-dimensional detector arrays, which is suspected due to the mechanical vibration occurring at the time of launching and the decaying of sensors over time as probable reasons for the mismatch in prelaunch smile calibration. Over time, the calibrated sensor may become uncalibrated; decay is also one of the reasons for smile effects. Dadon et al. (2010) first attempted a smile correction to the Hyperion sensor based on prelaunch parameters of the detector arrays, but the procedure did not properly correct the effect and, in fact, significant errors were increased, specially in the first VNIR bands (Dadon et al., 2010). The smile effect may affect Hyperion images to different degrees in the spectral range and may vary from scene to scene. A spectral smile is not obvious to the naked eye, so an indicator is necessary to make it visible for researchers.

There are two methods to detect images affected by the smile effect; the first is through the effects of the absorption features of oxygen (O_2) in specific spectra regions, while the second is the use of minimum noise fraction (MNF) transformation. Vertical stripes are less severe effects of the pushbroom sensor calibration problem, thus, one is able to quantify the smile effect per column of hyperspectral data, and this effect can be removed from the images. The smile effect alters the pixel spectra and reduces classification accuracies, which can be removed using techniques per column in hyperspectral images. Detailed information about correction techniques can be found in the literature (Aktaruzzaman, 2008; Arellano Mora, 2014).

To sum up, the main challenges with hyperspectral images to date are the high cost of acquisition, data storage and analysis, the complexity of the data, and high dimensionality and mixed pixels when extracting endmembers.

21.3 Conclusion

The major challenges and limitations with hyperspectral data are their availability for the region of interest and the operational use of hyperspectral data (unavailability of data for most regions or limited availability). Even for different real-time mineral exploration researches, there is limited availability of hyperspectral data with desired specifications (swath, spatial resolution, spectral resolution, and signal to noise ratio). Further, the main problem encountered with hyperspectral image classification is the high dimensionality of hyperspectral data; for example, AVIRIS hyperspectral sensors have 224 bands.

Some of the challenges are illustrated by the authors in this book. The main challenges in anomaly detection are the definition of the typical/normal data for complex data sources and the limited number or lack of reference (training) data for exact anomaly definition (Głomb and Romaszewski, 2020). Regarding the problem of identification and the challenges in the application of hyperspectral RS for the extraction of soil properties, including mineral, nutrient, organic carbon, moisture, salinity, and soil texture identification, are demonstrated by Yu et al. (2020). Reducing the impact of target components and establishing a relationship between spectral reflectance and monitoring objectives for a variety of soils remain challenging issues.

While there are noteworthy challenges in soil research, these also provide a base for challenges in heavy metal detection as discussed by Lamine et al. (2020). The authors described the complex spectral features of soil in the spectral range of 350–2500 nm in detail. Identifying exact spectral wavebands corresponding to Pb, Cu, Zn, and Cd concentrations unaffected by the chemical composition and surviving in the physical conditions at the surface of the soil is a noteworthy challenge.

Moreover, Naicker et al. (2020) discussed the issues and challenges that hyperspectral images often pose with the problem of multicollinearity and multidimensionality (Ling et al., 2014; Sibanda et al., 2015). The high correlation between the contiguous bands within a hyperspectral dataset causes matrix inversion and hampers the attainment of accurate parameter estimates (Huang et al., 2004; Shao and Li, 2012; Wolter et al., 2009; Boulesteix and Strimmer, 2006). The issues with classification and discrimination problems are examples of challenges with HSI, and are well discussed by Naicker et al. (2020). Partial least squares discriminant analysis is designed for regression analyses, but it performs well for classification and discrimination problems (Huang et al., 2004; Dorigo et al., 2007).

Pal (2020) discussed the challenges associated with HSI for classification analysis due to the availability of a large number of features in hyperspectral data due to data redundancy and the use of too many features requiring a larger training set. Further, some authors demonstrated that the availability of a large number of features/bands in hyperspectral data represents a challenge in terms of storing and handling the large datasets, data redundancy because of the fact that the information contained in many adjacent bands are similar, and classification analysis. For example, the use of too many features may require a large number of training datapoints to avoid the Hughes phenomenon and the estimation of parameters during the classification process, especially with statistical and neural classifiers. The size of the training dataset required for accurate parameter estimation may exceed that available during classification.

Detailed methods and techniques are discussed by Koloniatis et al., (2020), who proposed a new technique called the trend line smile correction method instead of the usual radiative transfer model. Koloniatis et al., (2020) examined the effects of sensor illumination conditions and land cover radiance variations on the spectral smile indicators. The limitations and challenges faced by mineral exploration researchers were discussed by Guha (2020). The authors discussed the limited availability of HSI data with the desired specifications (swath, spatial resolution, spectral resolution, and signal to noise ratio) for mineral discrimination and exploration research in most of the region, where there is availability of hyperspectral Hyperion data. But with the launch of HySIS, (India observation satellite) this limitation in the context of India's regions will soon be solved. This upcoming hyperspectral mission will provide spatial and spectral resolution sensors launched from several countries (Pandey et al., 2018). In future, Ultra-spectral sensors will be in limelight and may be able to get its position widely for all research themes due to a large number of spectral bands (more than 2700). Therefore advanced knowledge about data preprocessing and the implementation of appropriate techniques and methods are required to fully exploit all the potentials of hyperspectral data.

In summary, the authors hope that this book has contributed to a deeper understanding of the huge potential arising from advances in hyperspectral sensors, processing algorithms, and computing power. It sets the scene for new and exciting applications of hyperspectral data, which can measure features that were previously not detectable. Hyperspectral RS has opened the door to mapping the unmappable.

Acknowledgments

The authors also wish to express their gratitude to the anonymous reviewers for their useful and constructive comments, which resulted in an overall improvement of the originally submitted manuscript. Dr. Pandey is thankful to Shiv Nadar University for help and support.

List of abbreviations

HSI hyperspectral imaging
MNF minimum noise fraction
PCA principal component analysis
RS remote sensing
SAM spectral angle mapper
SPCA segmented principal component analysis
SNR signal to noise ratio
SWIR short wave infrared
UAVs unmanned aerial vehicles
VNIR visible-near infrared

References

Adar, S., Shkolnisky, Y., Ben Dor, E., 2014. A new approach for thresholding spectral change detection using multispectral and hyperspectral image data, a case study over Sokolov, Czech republic. Int. J. Remote Sens. 35, 1563–1584.

Aktaruzzaman, M., 2008. Simulation and Correction of Spectral Smile Effect and its Influence on Hyperspectral Mapping. ITC.

Anand, A., Pandey, P.C., Petropoulos, G.P., Palvides, A., Srivastava, P.K., Sharma, J.K., Malhi, R.K.M., et al., 2020. Use of Hyperion for Mangrove Forest Carbon Stock Assessment in Bhitarkanika Forest Reserve: A Contribution towards Blue Carbon Initiative. Remote Sensing 12, 597. https://doi.org/10.3390/rs12040597.

Arellano Mora, P.N., 2014. Hyperspectral Remote Sensing for Detecting Vegetation Affected by Hydrocarbons in the Amazon Forest. Department of Geography.

Avery, T.E., Berlin, G.L., 1992. Fundamentals of Remote Sensing and Airphoto Interpretation. Macmillan, New York.

Benediktsson, J.A., Chanussot, J., Moon, W.M., 2013. Advances in very-high-resolution remote sensing. Proc. IEEE 101, 566–569.

Boulesteix, A.-L., Strimmer, K., 2006. Partial least squares: a versatile tool for the analysis of high-dimensional genomic data. Brief. Bioinform. 8, 32–44.

Colkesen, I., Kavzoglu, T., 2019. Comparative evaluation of decision-forest algorithms in object-based land use and land cover mapping. Spatial Modeling in GIS and R for Earth and Environmental Sciences. Elsevier.

Dadon, A., Ben-Dor, E., Karnieli, A., 2010. Use of derivative calculations and minimum noise fraction transform for detecting and correcting the spectral curvature effect (smile) in Hyperion images. IEEE Trans. Geosci. Remote Sens. 48, 2603–2612.

Dorigo, W.A., Zurita-Milla, R., de Wit, A.J.W., Brazile, J., Singh, R., Schaepman, M.E., 2007. A review on reflective remote sensing and data assimilation techniques for enhanced agroecosystem modeling. Int. J. Appl. Earth Observ. Geoinf. 9, 165–193.

Fauvel, M., Benediktsson, J.A., Chanussot, J., Sveinsson, J.R., 2008. Spectral and spatial classification of hyperspectral data using SVMs and morphological profiles. IEEE Trans. Geosci. Remote Sens. 46, 3804–3814.

Głomb, P., Romaszewski, M., 2020. Anomaly detection in hyperspectral remote sensing images. In: Pandey, P.C., Srivastava, P.K., Balzter, H., Bhattacharye, B., Petropoulos, G. (Eds.), Hyperspectral Remote Sensing: Theory and Applications. Elsevier.

Goodenough, D.G., Dyk, A., Niemann, K.O., Pearlman, J.S., Chen, H., Han, T., et al., 2003. Processing Hyperion and ALI for forest classification. IEEE Trans. Geosci. Remote Sensing 41, 1321–1331.

Guha, A., 2020. Mineral exploration using hyperspectral data. In: Pandey, P.C., Srivastava, P.K., Balzter, H., Bhattacharye, B., Petropoulos, G. (Eds.), Hyperspectral Remote Sensing: Theory and Applications. Elsevier.

Hardin, P.J., Lulla, V., Jensen, R.R., Jensen, J.R., 2019. Small unmanned aerial systems (sUAS) for environmental remote sensing: challenges and opportunities revisited. GIScience Remote Sens. 56, 309–322.

He, K.S., Rocchini, D., Neteler, M., Nagendra, H., 2011. Benefits of hyperspectral remote sensing for tracking plant invasions. Divers. Distrib. 17, 381–392.

Hostert, P., Roder, A., Hill, J., 2003. Coupling spectral unmixing and trend analysis for monitoring of long-term vegetation dynamics in Mediterranean rangelands. Remote Sens. Environ. 87, 183–197.

Huang, W., Wang, J., Wang, Z., Zhaochun, J., Liu, L., Wang, J., 2004. Inversion of foliar biochemical parameters at various physiological stages and grain quality indicators of winter wheat with canopy reflectance. Int. J. Remote. Sensing 25, 2409–2419.

Hughes, G., 1968. On the mean accuracy of statistical pattern recognizers. IEEE Trans. Inf. Theory 14, 55–63.

Kakhani, N., Mokhtarzade, M., 2019. A new neuro-fuzzy-based classification approach for hyperspectral remote sensing images. J. Earth Syst. Sci. 128, 30.

Koloniatis, K., Andronis, V., Karathanassi, V., 2020. Spectral smile correction for airborne imaging spectrometers. In: Pandey, P.C., Srivastava, P.K., Balzter, H., Bhattacharye, B., Petropoulos, G. (Eds.), Hyperspectral Remote Sensing: Theory and Applications. Elsevier.

Krishna, G., Sahoo, R.N., Pradhan, S., Ahmad, T., Sahoo, P.M., 2018. Hyperspectral satellite data analysis for pure pixels extraction and evaluation of advanced classifier algorithms for LULC classification. Earth Sci. Inform. 11, 159–170.

Lamine, S., Petropoulos, G.P., Brewer, P.A., Bachari, N.-E.-I., Srivastava, P.K., Manevski, K., et al., 2019. Heavy metal soil contamination detection using combined geochemistry and field spectroradiometry in the United Kingdom. Sensors 19, 762.

Lamine, S., Pandey, M.K., Petropoulos, G., Brewer, P.A., Srivastava, P.K., Manevski, K., et al., 2020. Spectroradiometry as a tool for monitoring soil contamination by heavy metals in a floodplain site. In: Pandey, P.C., Srivastava, P.K., Balzter, H., Bhattacharye, B., Petropoulos, G. (Eds.), Hyperspectral Remote Sensing: Theory and Applications. Elsevier.

Lanthier, Y., Bannari, A., Haboudane, D., Miller, J.R., Tremblay, N., 2008. Hyperspectral data segmentation and classification in precision agriculture: a multi-scale analysis. In: IGARSS 2008–2008 IEEE International Geoscience and Remote Sensing Symposium, IEEE, II-585-II-588.

Lillesand, T., Kiefer, R.W., Chipman, J., 2014. Remote Sensing and Image Interpretation. John Wiley & Sons.

Ling, B., Goodin, D.G., Mohler, R.L., Laws, A.N., Joern, A., 2014. Estimating canopy nitrogen content in a heterogeneous grassland with varying fire and grazing treatments: Konza Prairie, Kansas, USA. Remote Sens. 6, 4430–4453.

Manjunath, K., Ray, S.S., Panigrahy, S., 2011. Discrimination of spectrally-close crops using ground-based hyperspectral data. J. Indian Soc. Remote Sens. 39, 599–602.

Malhi, R.K.M., Anand, A., Mudaliar, A.N., Pandey, P.C., Srivastava, P.K., Sandhya, K.G., 2020. Synergetic use of *in situ* and hyperspectral data for mapping species diversity in Shoolpaneshwar Wildlife Sanctuary, Gujarat. Tropical Ecology, Special issue (Ecosystem Function and Remote Sensing. https://doi.org/10.1007/s42965-020-00068-8.

Mianji, F.A., Zhang, Y., Sulehria, H.K., Babakhani, A., Kardan, M.R., 2008. Super-resolution challenges in hyperspectral imagery. Inf. Technol. J. 7, 1030–1036.

Miao, X., Gong, P., Swope, S., Pu, R., Carruthers, R., Anderson, G.L., et al., 2006. Estimation of yellow starthistle abundance through CASI-2 hyperspectral imagery using linear spectral mixture models. Remote Sens. Environ. 101, 329–341.

Naicker, R., Mutanga, O., Sibanda, M., Peerbhay, K., 2020. Discriminating tropical grasses grown under different nitrogen fertilizer regimes in KwaZulu-Natal, South Africa. In: Pandey, P.C., Srivastava, P.K., Balzter, H., Bhattacharye, B., Petropoulos, G. (Eds.), Hyperspectral Remote Sensing: Theory and Applications. Elsevier.

Nguyen, R., 2017. What are the advantages and disadvantages of using hyperspec (online). Prezi. Available from: <https://prezi.com/gaygsbqa789k/what-are-the-advantages-and-disadvantages-of-using-hyperspec/> (accessed 23.05.19.).

Nhaila, H., Sarhrouni, E., Hammouch, A., 2014. A survey on fundamental concepts and practical challenges of Hyperspectral images. In: 2014 Second World Conference on Complex Systems (WCCS), IEEE. pp. 659–664.

Okin, G.S., Roberts, D.A., Murray, B., Okin, W.J., 2001. Practical limits on hyperspectral vegetation discrimination in arid and semiarid environments. Remote Sens. Environ. 77, 212–225.

Pal, M., 2020. Hyperspectral image classifications and feature selection. In: Pandey, P.C., Srivastava, P.K., Balzter, H., Bhattacharye, B., Petropoulos, G. (Eds.), Hyperspectral Remote Sensing: Theory and Applications. Elsevier.

Pandey, P., Manevski, K., Srivastava, P.K., Petropoulos, G.P., 2018. The use of hyperspectral earth observation data for land use/cover classification: present status, challenges and future outlook. In: Thenkabail, P. (Ed.), Hyperspectral Remote Sensing of Vegetation, first ed. pp. 147–173.

Pandey, P.C., Anand, A., Srivastava, P.K., 2019a. Spatial distribution of mangrove forest species and biomass assessment using field inventory and earth observation hyperspectral data, Biodiversity and Conservation, 28. *Springer*, pp. 2143–2162.

Pandey, P.C., Koutsia, N., Petropoulos, G.P., Srivastava, P.K., Ben Dor, E., 2019b. Land use/land cover in view of earth observation: data sources, input dimensions and classifiers—a review of the state of the art. Geocarto Inter. 1–32. https://doi.org/10.1080/10106049.2019.1629647.

Pandey, P.C., Tate, N.J., Balzter, H., 2014. Mapping tree species in coastal portugal using statistically segmented principal component analysis and other methods. IEEE Sens. J. 14, 4434–4441.

Petropoulos, G.P., Kalivas, D.P., Georgopoulou, I.A., Srivastava, P.K., 2015. Urban vegetation cover extraction from hyperspectral imagery and geographic information system spatial analysis techniques: case of Athens, Greece. J. Appl. Remote Sens. 9, 096088.

Rajadell, O., García-Sevilla, P., Pla, F., 2009. Textural features for hyperspectral pixel classification. In: Pattern Recognition and Image Analysis, pp. 208–216.

Rao, N.R., Garg, P., Ghosh, S., 2007. Development of an agricultural crops spectral library and classification of crops at cultivar level using hyperspectral data. Precis. Agricul. 8, 173–185.

Richards, J.A., Richards, J., 1999. Remote Sensing Digital Image Analysis. Springer.

Schmidt, K., Skidmore, A., 2003. Spectral discrimination of vegetation types in a coastal wetland. Remote Sens. Environ. 85, 92–108.

Shao, X., Li, Y., 2012. Classification and prediction by LF NMR. Food Bioprocess. Technol. 5, 1817–1823.

Sibanda, M., Mutanga, O., Rouget, M., Odindi, J., 2015. Exploring the potential of in situ hyperspectral data and multivariate techniques in discriminating different fertilizer treatments in grasslands. J. Appl. Remote Sens. 9, 096033.

Vaiphasa, C., Skidmore, A.K., De Boer, W.F., Vaiphasa, T., 2007. A hyperspectral band selector for plant species discrimination. ISPRS J. Photogram. Remote Sens. 62, 225–235.

Wolter, P.T., Townsend, P.A., Sturtevant, B.R., 2009. Estimation of forest structural parameters using 5 and 10 meter SPOT-5 satellite data. Remote Sens. Environ. 113, 2019–2036.

Wulder, M., 1998. Optical remote-sensing techniques for the assessment of forest inventory and biophysical parameters. Prog. Phys. Geogr. 22, 449–476.

Yu, H., Kong, B., Wang, Q., Liu, X., Liu, X., 2020. Hyperspectral remote sensing applications in soil: a review. In: Pandey, P.C., Srivastava, P.K., Balzter, H., Bhattacharye, B., Petropoulos, G. (Eds.), Hyperspectral Remote Sensing: Theory and Applications. Elsevier.

Zhang, J.K., Rivard, B., Sanchez-Azofeifa, A., Castro-Esau, K., 2006. Intra and inter-class spectral variability of tropical tree species at La Selva, Costa Rica: implications for species identification using HYDICE imagery. Remote Sens. Environ. 105, 129–141.

Žibrat, U., Širca, S., Strajnar, P., Susič, N., Knapič, M., Gerič Stare, B., et al., 2020. Non-invasive detection of plant parasitic nematodes using hyperspectral and other remote sensing systems. In: Pandey, P.C., Srivastava, P.K., Balzter, H., Bhattacharye, B., Petropoulos, G. (Eds.), Hyperspectral Remote Sensing: Theory and Applications. Elsevier.

Author Index

Note: Page numbers followed by "*f*" and "*t*" refer to figures and tables, respectively.

M

Subject Index

Note: Page numbers followed by "*f*" and "*t*" refer to figures and tables, respectively.

D

I

S

T